ISBN 978-1-330-37780-2
PIBN 10044708

1 MONTH OF
FREE
READING

at

www.ForgottenBooks.com

By purchasing this book you are eligible for one month membership to ForgottenBooks.com, giving you unlimited access to our entire collection of over 700,000 titles via our web site and mobile apps.

To claim your free month visit:
www.forgottenbooks.com/free44708

English
Français
Deutsche
Italiano
Español
Português

www.forgottenbooks.com

Mythology Photography **Fiction**
Fishing Christianity **Art** Cooking
Essays Buddhism Freemasonry
Medicine **Biology** Music **Ancient**
Egypt Evolution Carpentry Physics
Dance Geology **Mathematics** Fitness
Shakespeare **Folklore** Yoga Marketing
Confidence Immortality Biographies
Poetry **Psychology** Witchcraft
Electronics Chemistry History **Law**
Accounting **Philosophy** Anthropology
Alchemy Drama Quantum Mechanics
Atheism Sexual Health **Ancient History**
Entrepreneurship Languages Sport
Paleontology Needlework Islam
Metaphysics Investment Archaeology
Parenting Statistics Criminology
Motivational

CAMBRIDGE PHYSICAL SERIES.

General Editors:—F. H. Neville, M.A., F.R.S.
and W. C. D. Whetham, M.A., F.R.S.

THE STUDY

OF

CHEMICAL COMPOSITION

London: C. J. CLAY AND SONS,
CAMBRIDGE UNIVERSITY PRESS WAREHOUSE,
AVE MARIA LANE,

AND

H. K. LEWIS,
136, GOWER STREET, W.C.

Glasgow: 50, WELLINGTON STREET.
Leipzig: F. A. BROCKHAUS.
New York: THE MACMILLAN COMPANY.
Bombay and Calcutta: MACMILLAN AND CO., Ltd.

THE STUDY

OF

CHEMICAL COMPOSITION

AN ACCOUNT OF ITS METHOD AND HISTORICAL DEVELOPMENT

WITH ILLUSTRATIVE QUOTATIONS

BY

IDA FREUND

STAFF LECTURER AND ASSOCIATE OF NEWNHAM COLLEGE, CAMBRIDGE

90407
21/8/08

CAMBRIDGE
AT THE UNIVERSITY PRESS
1904

𝕮𝖆𝖒𝖇𝖗𝖎𝖉𝖌𝖊:

PRINTED BY J. AND C. F. CLAY,

AT THE UNIVERSITY PRESS.

PREFACE.

THE study of composition, constituting as it does so fundamentally important a part of the science of chemistry, may be considered a subject well suited for separate treatment. In my presentation of the doctrine of chemical composition I have tried to show how the empirical knowledge comprised in it has been obtained, what the initial discoveries were and how these have been established. It has been my aim to demonstrate in the earlier part of the book that the notation by which we represent chemical composition can be developed from a purely empirical basis, independent of any hypothesis concerning the ultimate constitution of matter. In the subsequent treatment of the subject of composition on the basis of the atomic and molecular theory, I have endeavoured to keep sharp and clear the boundary between facts and hypotheses, to direct attention to the existence and position of this separating line, and to emphasise those special features of the argument which bring out the nature and function of hypotheses, their place and importance in the science of chemical composition. Although anxious to trace separately the *historical development* in the discovery and in the establishment of certain laws and classes of phenomena, I have made no attempt to produce anything sufficiently complete or even sufficiently proportioned to deserve the name of *history*. I have preferred to deal in greater detail with a few researches, especially such as I could repeatedly utilise from various points of view, than to treat a greater number more cursorily, believing in the truth and wisdom of what Lavoisier said more than a century ago, that "in such matters as these the choice of proofs is more important than their number."

In my desire to render the account I had to give real and vivid, I have made special use of two means. *First:* In dealing with quantitative researches—and from the nature of the case these constitute the greatest and most important part of my subject—I have not merely stated the final results, but have reproduced the values obtained in the actual measurements made, so as to indicate in each case the scope of the work involved and the degree of accuracy attained. *Second:* I have quoted extensively from classical memoirs and from great writers on the science, and, restricting within the narrowest possible limits my own share in the exposition, I have given the actual words of those who have announced discoveries, who have described their own experimental investigations or who have propounded new theoretical views. The series of reprints of important papers on chemistry which are being published in different countries are a proof of the growing interest in the history of the science and of the increased importance assigned to first-hand study. I venture to hope that the form of treatment adopted in this book may be found to be in conformity with these tendencies. My presentation cannot take the place of extensive reading of the actual papers; but there are students who have not access to libraries, whilst others whose time is limited may derive some compensating advantage from being taken by means of important and striking extracts over a far greater amount of ground than they could cover by direct recourse to the originals. In quotations from foreign authors of which English translations recognised as standard exist, I have often availed myself of these. Of the translations specially made for the purposes of this book, those of the quotations from Aristotle's works (other than the Metaphysics, Book I) were kindly provided by Miss E. M. Sharpley, Classical Lecturer of Newnham College, to whom I wish to express my thanks.

I have made considerable use of explanatory interpolations and footnotes, in order that those with little or no previous knowledge of chemistry, but with mental training otherwise acquired, might be able to follow the argument; and I have also had in mind possible readers who without intending to make a study of chemistry are interested in the fundamental problems of physical science. I can only hope that these incidental explanations may not prove a source of annoyance to students who are not in need of such help.

A few words remain to be said concerning the scope of the subjects included under the title "The Study of Chemical Composition." In my omission of a chapter dealing with the determination of molecular weight on the basis of the laws of dilute solutions, I was guided by the following considerations: the kind of historical treatment adhered to throughout the book seemed in this case unsuitable; space was an object; and as a matter of fact this subject is adequately dealt with in a number of recent text-books on physical chemistry, besides which a reprint renders the important original papers easily accessible. On the other hand, the part of Chapter XV which deals with crystallography has been introduced because the average student of chemistry is as a rule ignorant of this subject, and does not find in the current text-books even the minimum amount of information required to appreciate the results obtained in the study of the relation between crystalline form and chemical composition.

Finally I wish to express my best thanks to Mr F. H. Neville, one of the editors of the series, for the corrections and improvements suggested by him in reading the proofs, and to Mr A. Hutchinson, Demonstrator in Mineralogy, for valuable help given me with the chapter on Isomorphism.

IDA FREUND.

NEWNHAM COLLEGE LABORATORY,
    CAMBRIDGE.
      November, 1904.

ABBREVIATIONS OF REFERENCES TO JOURNALS

(THE FORM BEING USUALLY THAT ADOPTED IN THE INTER-
NATIONAL CATALOGUE OF SCIENTIFIC LITERATURE).

Ann. chim. phys., Paris. Annales de chimie et de physique, Paris.

Ann. Phil. (Thomson). Annals of Philosophy (Thomson).

Ann. Physik. Annalen der Physik, Leipzig.

Baltimore, Md., Amer. Chem. J. American Chemical Journal, Baltimore, Md.

Bruxelles, Bull. Acad. Belgique. Bulletin de l'Académie royale de Belgique, Bruxelles.

Bull. Min. Bulletin de la société française de minéralogie, Paris.

Chem. News, London. Chemical News and Journal of Science, London.

Edinb. J. Sci. Edinburgh Journal of Science, Edinburgh.

Geneva, Bibl. Univ. Archives. Archives des sciences physiques et naturelles, Supp^t à la Bibliothèque Universelle, Genève.

J. phys., Paris. Journal de physique théorique et appliquée, Paris.

J. prakt. Chem., Leipzig. Journal für praktische Chemie, Leipzig.

Liebig's Ann. Chem., Leipzig. J. v. Liebig's Annalen der Chemie, Leipzig.

London, J. Chem. Soc. Journal of the Chemical Society, London.

London, Phil. Trans. R. Soc. Philosophical Transactions of the London Royal Society.

London, Proc. R. Soc. Proceedings of the London Royal Society.

Paris, C.-R. Acad. Sci. Comptes-rendus hebdomadaires des séances de l'académie des sciences, Paris.

Phil. Mag., London. London, Edinburgh, and Dublin Philosophical Magazine, and Journal of Science.

Poggend. Ann., Leipzig. Annalen der Physik und Chemie, v. J. C. Poggendorff, Leipzig.

Zs. Chem. Zeitschrift für Chemie, Leipzig.

Zs. physik. Chem., Leipzig. Zeitschrift für physikalische Chemie, Leipzig.

ERRATA.

p. 145, line 10 from end; *for* " Chap. XVI " *read* " Chap. XIX."

p. 178, line 13; *for* 1827 *read* 1825.

p. 190, footnote; *for* 1827 *read* 1825.

TABLE OF CONTENTS.

Chapter XVII. Kekulé and the Doctrine of Valency . . . 506

Chapter XVIII. Berzelius and Isomerism 545

INTRODUCTION.

THE METHOD OF THE INDUCTIVE SCIENCES. OBSERVATION, GENERALISATION AND LAW. HYPOTHESIS AND THEORY.

"The footsteps of Nature are to be trac'd, not only in her ordinary course, but when she seems to be put to her shifts, to make many doublings and turnings, and to use some kind of art in endeavouring to avoid our discovery."

HOOKE, *Micrographia*, 1665.

THE object of all the Natural Sciences is the acquisition of knowledge concerning the natural objects surrounding us, as we apprehend them by our senses; of the changes

The scope of the Natural Sciences. occurring in these objects, together with the laws governing these changes; and of the more proximate or more ultimate causes, to the operation of which are due the individual phenomena and the general laws comprising these. The method now commonly employed for this object is that of proceeding from the observation and the study of the individual phenomena to the detection of uniformities in these, that is, to the law; from that which refers to one to that which refers to many; from the special to the general, by the process termed *Induction*. And the knowledge thus acquired is next utilised in the inverse process, in which from the general laws obtained by induction, inferences are drawn for the purpose of explaining the observed phenomena and of foretelling the occurrence of others. This process, termed *Deduction*, proceeds on the principle that what is asserted to be true of all similar phenomena of a special kind will also be true of any one individually; it argues from the many to the one; from the general to the special. But the inferences thus drawn according to the laws of thought are again checked and verified by appeal to the actual facts, by the study of the phenomena the course of which deduction foretells; and exact coincidence between what actually happens and what had been foretold theoretically, is made the test for the correctness of the inductive and deductive processes by which these inferences had been arrived at.

Knowledge of the objects surrounding us has been stated to be the common object of all natural sciences. But as knowledge grew, the need for classification and specialisation became evident, and thus there arose a division between the sciences, in which Chemistry has taken for its province one side of the study of the materials of which these objects are composed. Leaving Physics, the science most closely related to it, to investigate the properties common to all kinds of matter and differing only in degree, such as density, power of conducting an electric current, etc., Chemistry deals with the properties which belong to certain kinds of matter and not to others, which characterise one kind of matter and differentiate it from all other matter.

The province of Chemistry.

For instance, it is a common property of all kinds of matter to undergo change in volume on the application of heat, but each substance has its own characteristic coefficient of expansion; on the other hand it is a specific property of the solid called red precipitate, that above a certain temperature the further addition of heat transforms it into liquid mercury and gaseous oxygen.

A not uncommon description of Chemistry as " the science dealing with the study of all the different homogeneous kinds of matter met with in nature, and with the permanent changes these can undergo when transformed into other kinds of matter" gives the basis for the usual subdivision of this science into two parts, (I.) descriptive and classificatory, (II.) theoretical. Of these the first has to do with the investigation of the properties peculiar to each of the different kinds of matter, and the classification of all matter according to these properties; that is, the putting together of those kinds of matter which agree in having in common a greater or lesser number of properties, and the separation of these from all the other kinds of matter which do not possess those particular properties. The second is mainly concerned with the facts and laws observed in the study of chemical change, and of the composition of the substances undergoing or resulting from the change.

Thus theoretical chemistry has to deal with two kinds of problems: (i) with those which relate to the changes that matter can undergo, and to the laws regulating these changes; and since change can only be realised by a comparison of the initial and the final condition, (ii) with those referring to its composition. Function and composition are therefore the two kinds of phenomena studied in theoretical chemistry. Mechanical analogies are dangerous, but the comparison of these two aspects of theoretical chemistry to dynamics and statics respectively may be ventured on because it has at any rate become justified by long-established use. To keep these two sets of problems completely separate would be very difficult and unsatisfactory, whatever the form chosen for the presentation of the subject, but when that of the historical development is adopted it becomes practically impossible. And hence, though the subject of this book is professedly the theoretical chemistry of composition, the discussion of dynamical problems cannot and will not be altogether excluded.

Theoretical Chemistry deals with problems referring to function and composition.

The general remarks made at the outset concerning the method now followed in the building up of the Natural Sciences apply of course in every detail to the special case of theoretical chemistry. But before showing how this method has operated in the development of this particular branch of science, it may be advisable to discuss separately the processes involved, their sequence and interdependence.

The Method of the Natural Sciences.

The beginning is made by the recognition of individual phenomena, leading to what is called the knowledge of facts. This knowledge may be gained either by direct observation of the phenomena occurring in nature, or of those which have been caused by some act undertaken by ourselves for that purpose. "Observation" and "Experiment" are the names given to these two modes of collecting knowledge of facts.

I. The study of facts by observation and experiment.

"When we merely note and record the phenomena which occur around us in the ordinary course of nature we are said to *observe*. When we change the course of nature by the intervention of our will and muscular powers, and thus produce unusual combinations and conditions of phenomena, we are said to *experiment*. Sir John Herschel has justly remarked that we might properly call these two modes of experience 'passive' and 'active' observation —an experiment differs from a mere observation in the fact that we more or less influence the character of the events which we observe." (Jevons, *Principles of Science.*)

1—2

The different behaviour of iron, which rusts in air, and of gold, which remains unchanged, had no doubt been observed as a natural occurrence long before experiments were performed in which these metals were exposed to the influence of heat, of water, of acids, etc., and the comparative effect produced on them by these various agents noted. Important and valuable as is the observation of naturally occurring phenomena, yet for the advance of science, experiment is paramount.

"When Galileo let balls of a particular weight, which he had determined himself, roll down an inclined plane ; or when Torricelli made the air carry a weight, which he had previously determined to be equal to that of a certain column of water ; when at a still later stage Stahl changed metal into calx, and calx back again into metal, by first withdrawing something and then restoring it ; then a new light was flashed on all students of nature....Reason, holding in one hand its principles according to which concordant phenomena alone can be admitted as laws of nature, and in the other hand the experiment which it has devised according to those principles, must approach nature for instruction ; but not as a pupil, to be taught just what the master pleases, but as a judge, who forces the witnesses to answer the questions he puts to them....Thus after many centuries of groping, the study of nature was first made to walk along the sure path of a science." (Kant's *Critique of Pure Reason*, Second Preface.)

Before they can become material for the building up of a science, it is essential that the occurrences themselves should be correctly apprehended, and that the relation between an effect observed and that which caused it should be ascertained.

"In order that the facts obtained by observation and experiment may be capable of being used in furtherance of our exact and solid knowledge, they must be apprehended and analysed according to some Conceptions which, applied for this purpose, give distinct and definite results, such as can be steadily taken hold of and reasoned from." (Whewell, *Philosophy of the Inductive Sciences*.)

To illustrate the two distinct points involved in the above :

We all have heard about the sea-serpent, but should not find
The facts ob- anything about such an animal in a treatise on
served must be zoology, and that because the tales concerning it
correct. cannot be looked upon as trustworthy evidence. No doubt there is a great difference in the number of occurrences reported to the Society for Psychical Research and that used by it as the basis of its work. And to give a chemical example : Regnault had by the action of caustic potash on ethylene chloride

(a substance consisting of carbon, hydrogen, and chlorine, and prepared by the direct union of olefiant gas and chlorine) obtained a new substance differing from the parent one in that the elements of hydrochloric acid had been withdrawn, but still containing carbon, hydrogen, and chlorine. This substance was termed vinyl-chloride[1], and at one time it was of the utmost interest to chemists to establish beyond doubt whether another substance having a percentage composition identical with that of Regnault's compound but different properties, did or did not exist. It was maintained by certain chemists that by an altogether different process they had obtained a substance having the percentage composition of vinyl-chloride but entirely different properties[2]. The experiments relating to the production of this substance were repeated by Kekulé and Zincke, who found "that the most remarkable property of this remarkable compound was its non-existence." (Schorlemmer, *Rise and Development of Organic Chemistry.*)

And if caution is required as to what should and what should not be accepted as "facts," it is none the less so as regards the relation between causes and effects. A certain effect

The facts observed must be traced to their causes. undoubtedly does occur, but what has been its real cause? The correct correlation may be a matter of considerable difficulty, because the conditions under which a certain effect is observed to occur are always very complex; a large number of these may be effective at the same time, and it does not follow that those which are most easily apprehended are also those which are really determinant. The correct solution of such a problem, though relating to one fact only, the referring of an effect observed to the real cause producing it, involves the same mental and experimental processes, and the same sequence of these, as does the treatment of a whole collection of facts; and hence a detailed consideration of some such typical cases becomes important. A short account will therefore be given of certain investigations, undertaken with the object of assigning to a phenomenon observed its true cause, from which will be deduced the general method followed in all such cases.

[1] $C_2H_4Cl_2 + KOH = C_2H_3Cl + KCl + H_2O$
ethylene chloride   vinyl
                            chloride

[2] $C_2H_4O + COCl_2 = C_2H_3Cl + CO_2 + HCl$
aldehyde + phosgene
                    supposed different compound, proved to be a mixture of
                    aldehyde and phosgene.

· In 1770, very early in his career, Lavoisier presented to the Académie des Sciences a paper entitled "On the Nature of Water

and on the Experiments adduced in Proof of the Possibility of its Change into Earth[1]." This paper exhibits as well as any of his later ones the peculiar characteristics of Lavoisier's method and style, that is, it is marked by the display of extraordinary genius. Lavoisier thus enunciates the object of the investigation :

"I find myself confronted with the task of settling by decisive experiments a question of interest in physics, namely, whether water can be changed into earth, as was thought by the old philosophers, and still is thought by some chemists of the day."

He begins by investigating whether the fact stated is correct, whether earth (solid matter) is really produced in an operation in which, at any rate apparently, water plays the determining part. This fact he finds vouched for historically. Plants had been made to grow, deriving their increase in weight seemingly only from the water supplied to them. Van Helmont had planted a willow weighing 5 lbs. in 200 lbs. of earth thoroughly dried before weighing, then moistened with distilled water, and always fed with such water only. A suitable hood kept out dust, and after five years the willow was found to weigh 169 lbs. and 3 ozs., whilst the earth after again being dried and then weighed, had lost 2 ozs. only. Hence 164 lbs. of willow were assumed to have been produced from water only. Similar experiments seemed very popular and they all led to the same inference. But still more to the point were the observations of Boyle, Becher, Stahl and others, all of whom had found as the result of experiment that water, no matter how often it had been distilled previously,—that is, made to undergo an operation in which the gasifiable water could be separated from any non-volatile solid held by it in solution—yet left on evaporation an earthy residue. But Lavoisier is not content to simply accept the fact; he repeats the experiment and finds that in distilling rain water, a very pure form of water, from a glass vessel, he obtains an earthy residue; and he at once goes further, ascertaining a fact well calculated to throw some light on the cause of the phenomenon investigated. He compares the

[1]. *Œuvres*, ii. (p. 1).

density of the distillate with that of the original rain water and finds it practically identical:

"I thought that I might infer from this experiment one of two things, either that the earth which I had separated by the distillation was of such a nature that it could be held in solution in the water without increasing its density, or at least without increasing it as much as other substances would do; or else that this earth was not yet in the water when I had determined its density, that it had been formed during distillation, in short that it was a product of the operation. To decide with certainty which of these views I should adopt, no means has seemed to me more suitable than to repeat precisely the same experiment in hermetically sealed vessels, keeping exact count of the weight of the vessel and of that of the water used in the experiment.

For if it should be a case of the fire matter passing through the glass and combining with the water, there must needs occur after many distillations an increase in the total weight of the matter, that is to say, in the combined weight of the water, the earth and the vessel. The same thing should not occur if the earth had been formed at the expense of the water or of the vessel; but if so, there must needs also be found a diminution in the weight of one or the other of these two substances, and this diminution must be exactly equal to the quantity of earth separated."

Here then we find Lavoisier enumerating the various possible causes for the formation of the earth, and in each case drawing the inference as to what would be the influence of the operation of this particular cause on the weights of the whole system and of its component parts. The paragraph just quoted, when cast into tabular form, would present itself thus:

Earth is formed by the repeated distillation of water in a hermetically sealed glass vessel

cause: the earth has its origin in something external to the vessel and its contents		the earth has its origin in the vessel and its contents themselves
inference: as the earth forms, the weight of the vessel and its contents should increase.		as the earth forms, the weight of the vessel and its contents should remain the same.

cause: the earth comes from the vessel	the earth comes from the water	the earth comes from the vessel and the water
inference: the vessel loses weight, and this loss is exactly equal to the weight of earth formed.	the weight of the vessel remains the same.	the vessel loses weight, but not to the extent of the weight of earth formed.

It is evident then what quantities must be determined experimentally, in order to settle to which of these theoretical inferences the actually occurring phenomenon corresponds, and hence what is the cause sought for.

A special glass vessel termed a pelican, the use of which for repeated distillation goes back to alchemical times, was employed.

"A pelican is a flask devised for the circulation, the rising and falling back, of liquids, and therefore adapted for distillation, for which purpose it is provided with handle-like tubes reaching almost to the top, and curving back into the sides, like a pelican plucking at its own breast. The lower bulb is the larger of the two, and communicates with the neck, which terminates in a small top with an opening. But of this vessel also there are very many different varieties[1]."

The pelican was weighed empty in a balance specially constructed for the purpose, and surpassing in sensitiveness the instruments of that time. A certain amount of water purified by repeated distillation was introduced into the pelican, the whole heated gently on a sand-bath, and the stopper closing the vessel lifted from time to time to allow the air to escape. As soon as it

John French in *The Art of Distillation*, 1650, gives this illustration:

Form of a Pelican.

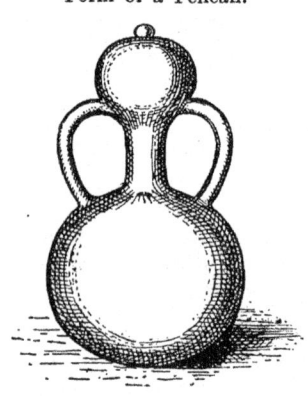

Fig. 1.

" The matter must be put in at the top, which afterwards must be closed up."

could be assumed that all the air had been expelled, the stopper was fixed in securely and the pelican with the water contained in it weighed. The whole apparatus was then surrounded by sand, and heating was begun on October 24th, 1768; for 25 days no change was noticed, on December 20th solid particles were observed floating about, the quantity of these was seen to increase until on February 1st, 1769, the experiment was stopped, lest by some accident the results of this long operation should be lost. The whole apparatus was then weighed again. The values obtained in the different weighings were:

[1] "Pelecanus est ampulla circulatoria, ascensui descensuiq.; atque ita vario discursui spirituum apta, cuius gratia ansata est canalibus, prope caput productis, et in latus reflexis, instar pelicani pectus suum fodientis. Venter inferius grandior est; inde quasi in collum coit, cui caput paruum cum foramine impositum est quamquam etiam in hac vase mira sit varietas." (Libavius, *Alchymia*, 1595.)

	Livres	Onces	Gros	Grains[1]
Weight before heating on Oct. 24, 1768, of the empty pelican	1	10	7	21·50
Weight before heating on Oct. 24, 1768, of the pelican and water	5	9	4	41·50
∴ Weight of the water	3	14	5	20·00
Weight after heating from Oct. 24 to Feb. 1st, of the pelican, water and earth	5	9	4	41·75
∴ Change in weight	0	0	0	0·25

"The weight at the end differs only by one quarter of a grain[2] from that determined before the operation; but so trifling a difference can be neglected because the accuracy of the balance is not great enough to allow me to answer for so small a quantity.......From the fact that no increase had been found in the total weight of the matter, it was natural to conclude that it was not fire matter, nor any other extraneous body, that had penetrated the substance of the glass and combined with the water to form the earth. It remained to determine whether the earth owed its origin to a destruction of a portion of the water, or of the glass; and nothing was easier. With the precautions I had taken it was only a case of determining whether it was the weight of the vessel or that of the water contained in it that had suffered diminution."

The pelican was next unstoppered, a process attended with difficulty, and thereby affording conclusive proof that the vessel had been securely closed, no air having been able to leak in. It was emptied, and the water, together with the solid suspended in it, carefully preserved in a glass vessel. The empty pelican was weighed with the following results:

	Livres	Onces	Gros	Grains
Weight of the vessel in which water had been distilled 100 days	1	10	7	4·12
Original weight of the vessel	1	10	7	21·50
∴ Loss of weight sustained by the vessel ...	0	0	0	17·38

"Therefore it was clearly shown that it was the substance of the glass itself which had supplied the earth separated from the water during the digestion, that what had happened was merely a solution of the glass; but in order to completely accomplish my object, it still remained for me to compare the weight of the earth which had separated from the water during the digestion, with the diminution in weight sustained by the pelican. These two quantities should of course be equal, and if there had been found a considerable excess in the earth, it would have become necessary to conclude from it that it had not been furnished by the glass alone."

[1] Old French measures: 1 livre=16 onces à 8 gros à 72 grains;
1 livre=489·5058 grams; 1 once=30·59 grams; 1 grain=0·053 gram.

[2] 1 quarter grain=·013 gram.

The weight of this earth was ascertained by adding together the weight of the solid actually suspended in the water, and that obtained from the water by evaporating it in another glass vessel.

Weight of earth   ....   .20·40 grains
Loss of weight of the pelican   17·38   „
Difference   3   „

"But the diminution in weight of the pelican was only $17\frac{4}{10}$ grains, and hence there is an excess of three grains in the weight of the earth which cannot be attributed to the solution of the particles of the pelican. A little reflection on the conditions of the operation will however make it easy to see what is the origin of this excess, and how indeed it was a necessity of the case. It will have been noted that on removal from the pelican, the water had been poured into a glass vessel, and that it had afterwards been transferred for evaporation to a glass retort. But these different operations could not have been accomplished without solution of a small portion of the substance of these two vessels."

He concludes, "It follows from the experiments described in this memoir that the greater part, possibly the whole, of the earth separated from rain water by evaporation, is due to the solution of the vessels in which it has been collected and evaporated."

In the Bakerian Lecture given by Sir H. Davy before the Royal Society in 1806, the subject of which was, "On Some Chemical Agencies of Electricity[1]," is found an investigation concerning the products of the electrolysis of water. Besides hydrogen and oxygen there are also formed acid and alkali. Davy states this as a fact:

(*b*) Davy on the acid and alkali produced in the electrolysis of water.

"The appearance of acid and alkaline matter in water acted on by a current of electricity at the opposite electrified metallic surfaces, was observed in the first chemical experiments made with the column of Volta."

The fact itself was therefore well established; it had been observed by Davy himself, as well as by other investigators. The problem requiring solution was, to ascertain whether the acid and alkali were derived from the water, and if they were not, whence they came. In pre-Lavoisierean times the action of the electric current itself might have been looked upon as a possible generating cause, but the day for such interpretations had gone. Davy, in his attempt to settle this question, had to make other plausible hypotheses for explaining the fact observed. He had to pass in review all the possible guesses as to the cause of the

[1] London, *Phil. Trans. R. Soc.* 1807 (p. 1).

production of the acid and the alkali. These substances might have been contained as constituents in the water itself, or they might have been derived from the vessels in which the electrolysis occurred, or from the surrounding air. These were practically the only possible, or at any rate they were the most obvious assumptions. Davy first set himself the task of accounting for the formation of the alkali, which always appeared at the negatively charged pole. He had observed before that when electrolysis in a glass vessel had proceeded for a considerable time, the vessel in which the alkali was formed seemed corroded, and probably it was this that led him to his first assumption, namely, that the alkali came from the vessel. If correct, this assumption would lead to the necessary inference that varying the material of the vessel should have an influence on the amount of alkali produced; and if such an influence could be actually proved, it would become very likely that the assumption made was correct. Following this out experimentally, Davy performed the electrolysis in agate cups (fig. 2) connected by a strand of amianthus (fine asbestos), the cups and the connecting material having been carefully cleaned by boiling in distilled water. There appeared a great deal of acid and very little alkali, the amount of alkali yielded under otherwise the same conditions in glass vessels being about twenty times as great; and moreover it was noted that whilst the amount of acid produced increased continuously, and depended mainly on the time the current had been passing, the amount of alkali produced in a glass vessel increased at first rapidly, then more and more slowly, and that when the same glass vessel was used for a second similar experiment very much less alkali was produced. Hence it would appear that a glass vessel favoured the production of alkali, but that a definite amount of glass could yield a limited amount of alkali only, whilst the production of the acid was not influenced by the substitution of agate for glass, and seemed to depend for its formation on some store of matter which, at any rate in the course of the experiments made, did not sensibly diminish. The electrolysis was next carried out in gold cones (fig. 3), with the result that the amount of the alkali produced was minute, that of the acid as great as ever. It should here be mentioned that in the course of the experiments Davy proved the acid to be nitrous acid, the alkali soda. The results obtained so far seemed to indicate that the production of the alkali was mainly, but not entirely, due to the material of the glass vessels. But whence came the minute trace of alkali obtained on electrolysis in the gold cones? The water used for electrolysis had been distilled in

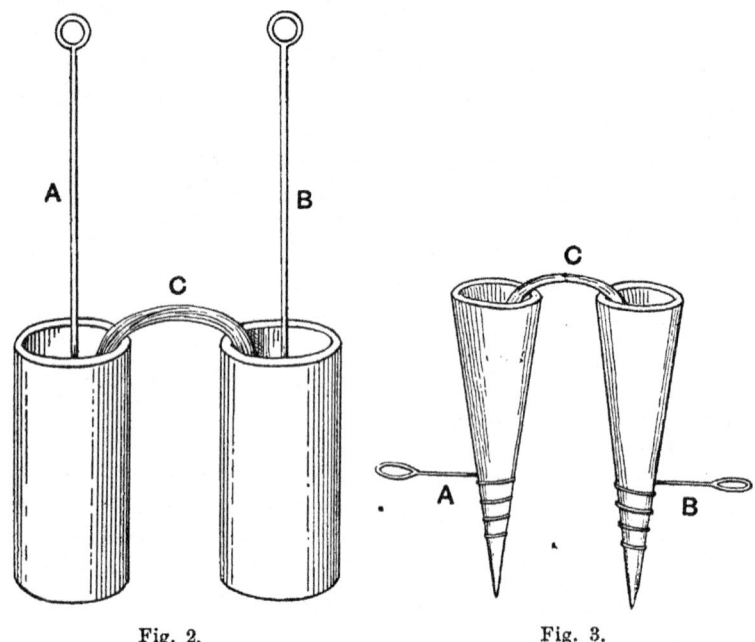

Fig. 2.   Fig. 3.

glass vessels, and this seemed to afford a possible explanation. Davy says:

"It was now impossible to doubt that the water contained some substance in very minute quantities, capable of causing the appearance of fixed alkali, but which was soon exhausted, and the question that immediately presented itself was: Is this substance saline matter carried over in distillation?"

Here then a second guess is made, namely, that a minute quantity of saline matter, yielding alkali on electrolysis, is contained in the water as an impurity, produced by the solvent action of the water on the glass of the still, and carried over mechanically with the steam. This supposition was put to an experimental test by evaporating in a silver vessel some of the distilled water used for electrolysis. A small quantity of solid matter was left. This solid might or might not have been the origin of the alkali produced. Further experiment must decide. Some of the solid was thrown into the gold cone in which electrolysis had produced its maximum effect of alkalinity; a great increase of alkalinity could be observed at once. Hence it was proved that when water originally distilled from glass vessels is electrolysed in glass vessels, the alkali produced is due mainly to the electrolysis of the saline matter dissolved from the glass by the water, and also in a small degree to the saline matter dissolved in water distilled from and preserved in glass vessels. A final test of the truth of these two assumptions was made by electrolysing in the agate and gold vessels water which had been distilled in silver vessels. No alkali whatever was formed, but as much acid as before, and Davy sums up the results by saying:

"To detail any more operations of this kind will be unnecessary; all the facts prove that the fixed alkali is not generated from the water but evolved either from the solid materials employed, or from saline matter in the water."

But all the same he performs a further experiment to test his explanation deductively.

"I was now able to determine distinctly that the soda produced in glass tubes came principally from the glass, as I had always supposed."

Into the gold cup in which electrolysis is being carried out with water producing no alkali, he drops a piece of glass, and the result is the immediate formation of alkali.

The source of the acid formed at the positive pole is next investigated.

"I had never made any experiments in which acid matter having the properties of nitrous acid was not produced, and the longer was the operation the greater was the quantity that appeared."

The experiments already made have shown that the material of the vessels and saline matter dissolved in the water do not account for this acid. Hence, of the most obvious guesses enumerated before as to its possible cause, there is left practically only the influence of the air, or generation from the water itself. The fact that the acid produced is nitrous acid makes it in itself probable that the oxygen liberated at the positive pole, together with the nitrogen of the air, should prove the true cause. Davy tests this supposition, from which follows the inference that removal of the air should prevent the formation of the acid, and he finds that when carrying out the electrolysis under the receiver of an air-pump, the yield of acid is diminished.

"I repeated the experiment under more conclusive circumstances. I arranged the apparatus as before [gold cones and water distilled in silver vessels] ; I exhausted the receiver and filled it with hydrogen gas from a convenient air-holder ; I made a second exhaustion and again introduced hydrogen that had been carefully prepared. The process was conducted for twenty-four hours, and at the end of this time neither of the portions of the water altered in the slightest degree the tint of litmus. It seems evident then that water chemically pure is decomposed by electricity into gaseous matter alone, into oxygen and hydrogen."

One more investigation shall now be described, one of comparatively recent date, in further illustration of the method used to establish the correct connection between an effect observed and the cause to which this effect is due.

In his wonderfully exact determinations of the densities of certain elementary gases, Lord Rayleigh found that nitrogen derived from the atmosphere had a density about $\frac{1}{2}$ per cent. greater than that of nitrogen obtained by the decomposition of chemical compounds. How he established that this small difference did actually exist, and was not simply due to experimental error,

(c) Rayleigh and Ramsay on the discrepancy between the densities of chemical and atmospheric nitrogen.

will be dealt with later in the chapter on the interpretation of the results of quantitative experiments (p. 91). This then was the question to which an answer had to be found and was found: What is the cause of the difference between the densities of the nitrogen derived from the two sources[1] ? Guesses had to be made ; the inferences drawn from all the possible answers had each to be passed in review, and put to the test of experiment. The causes producing the observed difference in density could be of two kinds: (i) the lighter gas might contain an admixture of some gas, known or unknown, of density less than ordinary nitrogen, or (ii) the heavier gas might contain an admixture of some gas, known or unknown, heavier than ordinary nitrogen. The lighter gas possibly contained in the chemical nitrogen might, for instance, have been hydrogen derived from the decomposition of the substances from which the nitrogen had been prepared, and which all contained hydrogen ; or it might have been a special form of nitrogen differing from ordinary nitrogen in being made up of less complex ultimate particles[2].

[1] Rayleigh and Ramsay, "Argon, a New Constituent of the Atmosphere." London, *Proc. R. Soc.* 57, 1895 (p. 265). *Nature*, London, 51, 1895 (p. 347).
[2] Dissociated nitrogen.

"When the discrepancy of weights was first encountered, attempts were naturally made to explain it by contamination with known impurities. Of these the most likely appeared to be hydrogen, present in the lighter gas in spite of the passage over red-hot cupric oxide[1]. But inasmuch as the intentional introduction of hydrogen into the heavier gas, afterwards treated in the same way with cupric oxide, had no effect upon its weight, this explanation had to be abandoned. ...At this stage it seemed not improbable that the lightness of the gas extracted from chemical compounds was to be explained by partial dissociation of nitrogen molecules into detached atoms. In order to test this suggestion both kinds of gas were submitted to the action of the silent electric discharge, with the result that both retained their weights unaltered. This was discouraging, and a further experiment pointed still more markedly in the negative direction....On standing, the dissociated atoms might be expected to disappear, in partial analogy with the known behaviour of ozone. With this idea in view, a sample of chemically prepared nitrogen was stored for eight months. But at the end of this time the density shewed no sign of increase, remaining exactly as at first. ...Regarding it as established that one or other of the gases must be a mixture, containing, as the case might be, an ingredient much heavier or much lighter than ordinary nitrogen, we had to consider the relative probabilities of the various possible interpretations. Except upon the already discredited hypothesis of dissociation, it was difficult to see how the gas of chemical origin could be a mixture....The simplest explanation in many respects was to admit the existence of a second ingredient in air from which oxygen, moisture and carbonic anhydride had already been removed. The proportional amount required was not great. If the density of the supposed gas were double that of nitrogen, $\frac{1}{2}$ per cent. only by value would be needed ; or if the density were but half as much again as that of nitrogen, then 1 per cent. would still suffice. But in accepting this explanation, even provisionally, we had to face the improbability that a gas surrounding us on all sides, and present in enormous quantities, could have remained so long unsuspected."

The next stage was to put the supposition that atmospheric nitrogen contains an admixture of a heavier gas to the test of experiment. The methods employed for the purpose of trying to separate out from atmospheric nitrogen something denser than chemical nitrogen were the physical one of diffusion, whereby it could not be expected that more than a very partial separation could be effected, and by means of which a gas denser than atmospheric nitrogen was actually obtained ; and the chemical one of the absorption of the nitrogen. Amongst substances which under suitable conditions combine directly with nitrogen—the number of these is very small—are oxygen and magnesium. Nitrogen

---

[1] The preparation and purification of the chemical nitrogen used for the density determination always involved passage over red-hot cupric oxide.

mixed with excess of oxygen and sparked in contact with an alkali which will absorb the nitrous acid as soon as it is formed, can be completely converted into the latter, the residual oxygen being of course easily removed in the ordinary way; nitrogen is also fairly readily absorbed when passed over heated magnesium, a nitride of magnesium being formed. The application of the two chemical methods yielded in each case a residual gas incapable of combining with oxygen or magnesium, and the amount of which was proportional to the amount of atmospheric nitrogen originally taken, thus doing away with the possibility of its being derived from the substances employed in the process of absorption. This gas was denser than atmospheric nitrogen, had a characteristic spectrum, and a definite boiling point and freezing point, different from those of any other hitherto known substance. Thus the suspected new constituent of air was isolated; but in order to further strengthen the proof that the greater density of atmospheric nitrogen was due to the presence in it of this new substance, which its discoverers named Argon, theoretical deductions concerning the relative properties of chemical nitrogen, atmospheric nitrogen and argon were tested by observation of the actual properties.

(i) From the data:

D = density of chemical nitrogen = $1\cdot2505$.

D_1 = „   „ atmospheric „   = $1\cdot2572$.

a = proportional volume of argon in atmospheric nitrogen
 = $\cdot0104$ [this value was determined incidentally in the course of the isolation of the argon].

d, the density of argon, can be calculated by the formula

$$ad + [1 - a]\, D = D_1$$

$$\therefore d = D + \frac{D_1 - D}{a} = 1\cdot8945,$$

which makes the specific gravity of argon (referred to $N = 14$ or $O = 16$) = $20\cdot6$.

The specific gravity of argon, as found experimentally for a sample obtained by means of magnesium, was $19\cdot9$, a very good agreement considering the not very great accuracy with which a was known, and the difficulty of getting the argon used in the actual density determinations quite free from nitrogen.

(ii) Chemically prepared nitrogen was submitted to the very same processes which in the case of atmospheric nitrogen had yielded about 1 per cent. by volume of the unabsorbable residue termed Argon, that is, chemical nitrogen was sparked over an alkali with excess of oxygen, or passed over heated magnesium, and was in every case practically completely absorbed.

Hence an element present in enormous quantities, but till then not suspected, was conclusively proved to exist, and to be the cause of the difference in density between chemical and atmospheric nitrogen, a difference revealed by most accurate measurements of a physical quantity, aptly described by Lord Rayleigh as "the triumph of the last place of decimals."

What is it that the three investigations just described have in common? Their object was to ascertain the causes producing

Steps in the process of referring an effect to its true cause.
certain effects observed: the production of earth from water, the production of acid and alkali on the electrolysis of water, the difference in density between two gases till then supposed to be the same.

In each case the actual occurrence of the effect had first to be proved beyond doubt.

Next taking into account all the pertinent phenomena observed incidentally (the identical density of the water before and after the

(a) Hypotheses are made.
distillations in which the earth is produced, the corrosion of the glass vessels in which the alkali is formed electrolytically, the specific difference in the substances from which the two kinds of nitrogen had been produced), and all the conditions of the experiment as far as they could be realised, hypotheses were framed, that is guesses were made concerning the possible causes of the effects observed. Each of these hypotheses had to be shown to afford an explanation of the particular effect it referred to, which is merely a thinking process; thus Lavoisier's possible explanations were that the earth formed came from the water (least likely after the result of the density determinations), or from the fire through the pores of the glass vessel, or from the vessel itself; Davy had the choice between the hypotheses of making the water, or the air, or the material of the vessels the origin of the acid and alkali formed; Lord Rayleigh had to pass in review the assumptions that a lighter or a heavier gas was mixed with one or other of his differing specimens of nitrogen.

F.

Then followed the testing of the adequacy of the hypotheses to the purpose for which they had been devised. The deductive inferences drawn from each of them were put to an experimental test, and, according as the result yielded was positive or negative, hypotheses were retained or rejected, rejection of course leading to a reduction in the number. Since the total weight of a closed vessel containing water in which earth had been formed, had not altered, the origin of the earth could not be in the fire matter; change of material of the containing vessel, and the substitution of purer water for some containing a solid in solution, was found to have no influence on the amount of the nitric acid formed by electrolysis of the water, and hence the acid was not likely to be derived from either the water itself or the containing vessel; hydrogen purposely added to nitrogen was completely removed by the process ordinarily followed in the preparation of the chemical nitrogen, and hence the cause sought could not be the presence of hydrogen as an impurity in the lighter gas. It may happen that all hypotheses but one are eliminated when tested by a first set of deductions. It may be that they are all eliminated; if so fresh ones must be sought, and if these should share the same fate, the phenomenon remains for the time being unexplained, though the work done will not be lost, future investigators being spared the framing and testing of hypotheses already proved unsuccessful. Or it may be that two or more hypotheses will all stand the test of a first set of inferences; when this happens the two hypotheses must be tested further and further, until an inference is drawn, which, according to the one, would give a result different and easily distinguished from that yielded by the other. An experiment, in this case called a *crucial* experiment (from *crux*, the fingerpost at a bifurcation of a road), is called upon to decide.

(b) Adequacy of the hypotheses tested by their deductive inferences, and consequent rejection of some of them.

"Instances of the Fingerpost [or of the Cross-roads], borrowing the term from the fingerposts which are set up where roads part, to indicate the several directions...I also call Decisive and Judicial.... I explain them thus. When in the investigation of any nature, the understanding is so balanced as to be uncertain as to which of two or more natures the cause of the nature in question should be assigned, on account of the frequent and ordinary concurrence of many natures, Instances of the Fingerpost show the union of the natures with the nature in question to be sure and indissoluble, of the other to be varied and separable; and thus the question is decided, and the former nature is admitted as the cause, while the latter is dismissed and rejected. ...These Instances of the Fingerpost...for the most part...are expressly and designedly sought for and applied, and discovered only by earnest and active diligence. ...Let the nature in question be Weight or Heaviness. Here the road will branch into two, thus. It must needs be that heavy or weighty bodies either tend of their own nature to the centre of the earth, by reason of their proper configuration; or else that they are attracted by the mass and body of earth itself as by the congregation of kindred substances, and move to it by sympathy. If the latter of these be the cause, it follows that the nearer heavy bodies approach to the earth, the more rapid and violent is their motion to it; and that the further they are from the earth, the feebler and more tardy is their motion.... With regard to this then, the following would be an instance of the Fingerpost. Take a clock moved by leaden weights, and another moved by the compression of an iron spring; let them be exactly adjusted...; then place the clock moving by weights on the top of a very high steeple.... Repeat the experiment in the bottom of a mine.... If the virtue of the weights is found diminished on the steeple and increased in the mine, we may take the attraction of the mass of the earth as the cause of weight." (Bacon, *Novum Organum*, Book II, xxxvi.)

The function of such crucial experiments may be illustrated from the investigations just described. In Lavoisier's experiment on the change of water into earth, the hypothesis of the extraneous origin of the earth was eliminated when the constancy of the weight of the whole system had been proved; but this left it open whether the earth was derived from the vessel, from the water, or from both. The weighing of the earth itself and of the vessel supplied the crucial experiments required to give the answer[1]. The inferences were: (i) if derived from the vessel, the weight of earth formed must be exactly equal to the loss of weight of the vessel; (ii) if derived from the water, the weight of the vessel must not have changed; (iii) if derived from both the vessel and the water, the weight of earth formed must exceed the loss in weight of the vessel. The result of the actual experiments gave the answer that the earth came from the material of the vessel only.

(c) **Crucial Experiments and consequent retention of one hypothesis as giving the required explanation.**

One hypothesis and one only having been found to give a satisfactory explanation of the phenomenon investigated, the hypothesis becomes an actually ascertained fact, which may or may not be further verified by the testing of more deductive inferences. Lavoisier could state that water is not changed into earth, and that

---

[1] Here, as always, the decision between three possible answers necessitates two experiments.

when earth appears, this is due to the solvent action of water on glass vessels; Davy was able to state that the alkali and acid formed when water is electrolysed were derived from the material of the vessels and from the air, and that under suitable conditions pure water yields oxygen and hydrogen only; Rayleigh and Ramsay could actually isolate a heavier gas from atmospheric nitrogen.

These investigations, described in illustration of general principles, were here utilised with reference to what is the first step in the acquisition of scientific knowledge, namely the correct recognition of individual facts, and the connection between an effect produced and the cause producing it. But isolated facts, however clearly recognised and however great in number, do not yet constitute a science. Classification steps in and makes these facts the basis of generalisations. Classification consists in selecting some one property, putting together all the objects or all the phenomena which possess this property, and separating them from all the others which do not; it consists in the putting together of what is like in some way and the separating of it from what is unlike in that respect. The chemist, in dealing with the permanent changes produced in matter, inquires into the various causes which produce them; for instance, he may select for separate consideration those changes which all have the common property of being the result of the action of heat. In the study of a number of facts brought together because of their being in some way of the same kind, we may recognise that the same cause produces in a greater or lesser number of cases the same effect, and such a recognition leads to a generalisation. The following may serve as definitions:

II. Classification and Generalisation.

"*Classification* is the arrangement together of any series of objects [or of occurrences] which are like and the separation of those which are unlike, in order to facilitate the operations of the mind in clearly conceiving and retaining in memory the character of the objects in question" (Huxley).

"*Generalisation* is the recognition of a certain common nature between a greater or lesser number of facts and the extension of what has been observed in a limited number of cases to a multitude of yet unexamined cases."

But in considering how classification and generalisation are actually used in physical science, it is found that practically the one always involves the other. When we put together the objects or phenomena which have something in common, *i.e.*, when we classify, we have in this very act recognised a common nature, *i.e.*, generalised. Again, when we have recognised that a certain cause produces with many or with all substances the same effect, that is, when we generalise, we had prior to this to supply ourselves with the necessary material for so doing; we had to bring together for separate consideration all the phenomena which possessed the class-characteristic of the operation of the same cause, and the generalisation itself is after all nothing but a class-characteristic, appertaining to a more or less extensive class according as to whether the generalisation is a more or less wide one. In classifying all metals into noble (not changed when heated in air) and base (changed when heated in air) the alchemists really founded a classification on the generalisation that some metals when heated in air are changed and others not ; and again the generalisation that all gases when heated expand equally, is bound up with the recognition of the class-characteristics of gases.

Classification and Generalisation go together.

Generalisation leads to laws. Law is the statement of a definite relation between cause and effect, which by observation and experiment we have ascertained to hold for a number of cases belonging to a certain class of facts, and which we therefore assume will hold for any other case belonging to the same class of facts. The law may be of a qualitative nature only. Thus the name of acid has been given to a number of substances which all agree in possessing certain definite properties, *e.g.* a certain effect on the sense of taste, the power of changing certain vegetable colouring matters to red, the destruction of the alkaline nature of substances such as potash and soda, effervescence with carbonates, solution of metals such as zinc and magnesium accompanied by the evolution of hydrogen. The investigation of the composition of a number of acids, and of the nature of the change termed neutralisation in which, by the inter-action with other substances, such as soda, carbonates, zinc or magnesium, these characteristic properties are destroyed, has led to a generalisation embodied in the law " all acids contain hydrogen replaceable by metal," with

The result of generalisation is termed "law."

the result, that now when a new substance is discovered which has the functional class-characteristics of an acid, *e.g.* sour taste, turning certain vegetable colouring matter red, etc., etc., we should assume that it also possessed the class-characteristic of composition embodied in the above law. We should expect it to contain hydrogen, and to be capable of yielding a compound which contained metal in place of some of this hydrogen at any rate.

But the generalisation to be embodied in a law may be of a quantitative nature. It is one thing to make the generalisation that all gases expand equally under the influence of heat, another to find the value for this common coefficient of expansion. Berthollet had in 1809 demonstrated that hydrogen diffuses much more rapidly than any other gas; but it remained for Graham first to show, in 1828, that the diffusion of all gases is inversely as some function of their density, apparently the square root; and then to definitely establish, in 1838, that "the diffusion, or spontaneous intermixture, of two gases in contact,...is, in the case of each gas, inversely proportional to the square root of the density of that gas."

"Experiments may be of two kinds: experiments of simple fact, and experiments of quantity. ...[In the latter] the conditions will vary, not in quality, but quantity, and the effect will also vary in quantity, so that the result of quantitative induction is also to arrive at some mathematical expression involving the quantity of each condition, and expressing the quantity of the result. In other words, we wish to know what function the effect is of its conditions. We shall find that it is one thing to obtain the numerical results, and quite another thing to detect the law obeyed by those results, the latter being an operation of an inverse and tentative character." (Jevons, *Principles of Science.*)

The discovery of such laws likewise necessitates the framing and the testing of hypotheses. The history of the Inductive Sciences affords many examples of how such laws have emerged, only after much labour, as the survival of the fittest of many hypotheses. The law of refraction lends itself admirably to showing the successive steps in such discoveries, and the place amongst these of hypotheses.

It was known to the ancients that a ray of light, in passing from one medium into another, was refracted, *i.e.*, that rectilinear propagation ceased and that the ray was bent. The amount of this bending was actually measured for certain cases by Ptolemy (2nd century A.D.), and the general fact recognised that when light passes from air to glass—from a less dense to a more dense medium—the angle of refraction is less than the angle of incidence, *i.e.* the ray is bent towards the perpendicular; and the corresponding angles were given. In the middle ages observations were made, and more or less correct tables constructed, in which angles of incidence and the corresponding angles of refraction were recorded. Kepler (1604) attempted to reduce to rule the measured quantities of refraction. He is known to have made as many as 17 suppositions as to the law connecting the value of the angle of incidence with that of the angle of refraction,—17 guesses as to what function the one was of the other. These suppositions, or guesses, or hypotheses, he had to test deductively by comparing the actually measured angles of refraction corresponding to certain angles of incidence with those calculated according to his hypothetical law. All his hypotheses proved erroneous, including one according to which the angle of refraction should be partly proportional to the angle of incidence and partly proportional to the secant of that angle, and which gave values agreeing to within $\frac{1}{2}$ degree with the experimental ones. In 1621 Snell discovered the real relation. If he knew of the work done on the subject before, he must of course have been much helped thereby, being saved from spending time on the testing of hypotheses already proved inadequate. His supposition that the sine of the angle of refraction is directly proportional to the sine of the angle of incidence rose from a hypothesis to a law, once it had been proved deductively that it always gave theoretical values identical with those actually measured, and that no other supposition did likewise[1].

Such a development of a hypothesis into a law is characterised by Mill in the following manner:

"It appears, then, to be a condition of a genuinely scientific hypothesis, that it be not destined always to remain an hypothesis, but be certain to be either proved or disproved by...comparison with observed facts. ...In hypotheses of this character, if they relate to causation at all, the effect must be already known to depend on the very cause supposed, and the hypothesis must relate only to the precise mode of dependence; the law of the variation of the effect according to the variations in the quantity or in the relations of the cause. With these may be classed the hypotheses which do not make any supposition with regard to causation, but only with regard to the law of correspondence between facts which accompany each other in their variations, though there

---

[1] These facts concerning the discovery of the law of refraction are taken from Whewell, *History of the Inductive Sciences.*

may be no relation of cause and effect between them. Such were the different false hypotheses which Kepler made respecting the law of the refraction of light. It was known that the direction of the line of refraction varied with every variation in the direction of the line of incidence, but it was not known how; that is, what changes of the one corresponded to the different changes in the other. In this case any law, different from the true one, must have led to false results. ...In all these cases, verification is proof; if the supposition accords with the phenomena there needs no other evidence of it." (Mill, *System of Logic.*)

A whole group of laws may furnish the material for further generalisation and may be embraced in a more general law. For instance the statement, "the elements combine in ratios which are those of their combining weights, or of simple whole multiples of these" is a further generalisation from the three laws of chemical combination, and includes them all (*post*, chap. VIII.).

Generalisation and the formulation of laws is the response to

III. Hypotheses to account for the facts and laws observed. the desire of the human mind for simplicity, for the power to comprise a number of apparently isolated facts under one aspect. But this desire goes further, and leads to attempts to find some ultimate reason for the phenomena observed. Some fundamental properties are assigned to that which is the vehicle of the phenomena studied; in chemistry, which is the science dealing with the composition and the transformation of matter, we postulate certain properties of that matter; in optics—a science in which the phenomena are not inseparably connected with matter, *i.e.*, that which exhibits mass or the property of being attracted by the earth—the hypothetical properties are those of an all-pervading medium called the luminiferous ether, the existence of which is assumed for the purposes of the case. It is by virtue of these hypothetical properties that the cause of the individual phenomena observed and of the more or less general laws deduced from these, must be such as it is and no other. Such an assumption devised

Nature of hypotheses. for such a purpose constitutes a scientific hypothesis. The nature of hypotheses, and their function in the discovery of individual facts and laws, has already been discussed, but they must now be considered in their bearings on great divisions or the whole of a science. It has been said several times in what has preceded, that a hypothesis is simply a guess which may be right or may be wrong, and that this process of guessing must be continued until a right one has been found.

But how shall we know which is right ? or put somewhat differently, what are the requirements of a good hypothesis ?

(1)   It must explain *all* the phenomena and laws which classification has brought together in a particular branch of science, that is, it must be inductively true.   The examples given

A good hypothesis must be :
(1) Inductively true at time of promulgation.

before show how the efficiency of a hypothesis, for explaining an individual fact (p. 18) or law (p. 23), can be tested, and within a larger scope, the method is the same.   The laws must seem natural consequences of the cause assumed in the hypothesis.

"To discover a Conception of the mind which will justly represent a train of observed facts is, in some measure, a process of conjecture,...and the business of conjecture is commonly conducted by calling up before our minds several suppositions, selecting that one which most agrees with what we know of the observed facts.   Hence he who has to discover the laws of nature may have to invent many suppositions before he hits upon the right one ; and among the endowments which lead to his success, we must reckon that fertility of invention which ministers to him such imaginary schemes, till at last he finds the one which conforms to the true order of nature.   A facility in devising hypotheses, therefore, is so far from being a fault in the intellectual character of a discoverer, that it is, in truth, a faculty indispensable to his task.   ...But if it be an advantage for the discoverer of truth that he be ingenious and fertile in inventing hypotheses which may connect the phenomena of nature, it is indispensably requisite that he be diligent and careful in comparing his hypotheses with the facts, and ready to abandon his invention as soon as it appears that it does not agree with the course of actual occurrences.   This constant comparison of his own conceptions and supposition with observed facts under all aspects forms the leading employment of the discoverer ; this candid and simple love of truth, which makes him willing to suppress the most favourite production of his own ingenuity as soon as it appears to be at variance with realities, constitutes the first characteristic of his temper.   He must have neither the blindness which cannot, nor the obstinacy which will not, perceive the discrepancy of his fancies and his facts.   He must allow no indolence, or partial views, or self-complacency, or delight in seeming demonstration, to make him tenacious of the schemes which he devises, any further than they are confirmed by their accordance with nature.   The framing of hypotheses is, for the enquirer after truth, not the end, but the beginning of his work.   Each of his systems is invented, not that he may admire it and follow it into all its consistent consequences, but that he may make it the occasion of a course of active experiment and observation.   And if the results of this process contradict his fundamental assumptions, however ingenious, however symmetrical, however elegant his system may be, he rejects it without hesitation.   He allows no natural yearning for the offspring of his own mind to draw him aside from the higher duty of loyalty to his sovereign, Truth, to her he not only gives his affections and his wishes, but strenuous labour and scrupulous minuteness of attention." (Whewell, *Philosophy of the Inductive Sciences.*)

(2) But in science there is no finality; facts accumulate, knowledge grows; and hence the second requirement of a good hypothesis is, that it should, without any or with only slight modifications in its original form, explain the phenomena and laws discovered after its promulgation in the particular branch of science to which it refers. Lavoisier explained the properties of acids by the hypothesis of the acidifying principle being oxygen, the name of which (ὀξύς, γεννάω = acid, I generate) still bears witness to this view. This hypothesis held sway until the time of the proof of the elementary nature of chlorine, which involved that of the absence of oxygen from hydrochloric acid, a compound of hydrogen and chlorine, and hence led to the complete abandonment of what in the history of science is known as Lavoisier's Oxygen Theory of Acids.

(2) Applicable to discoveries made after its promulgation.

Sometimes the original hypothesis is modified to adapt it to the new demands made upon it, but the less strain it need bear in this way, the greater is its inherent probability.

"When the hypothesis, of itself and without adjustment for the purpose, gives us the rule and reason of a class of facts not contemplated in its construction, we have a criterion of its reality, which has never yet been produced in favour of falsehood. ...[In *true* hypotheses] all the additional suppositions tend to simplicity and harmony ; the new suppositions...require only some easy modification of the hypothesis first assumed, the system becomes more coherent as it is further extended. The elements which we require for explaining a new class of facts are already contained in our system. ...In *false* theories, the contrary is the case. The new suppositions are something altogether additional ;—not suggested by the original scheme, perhaps difficult to reconcile with it. Every such addition adds to the complexity of the hypothetical system, which at last becomes unmanageable, and is compelled to surrender its place to some simpler explanation." (Whewell, *Philosophy of the Inductive Sciences.*)

(3) A good hypothesis must indicate the lines of future research in that its deductions referring to phenomena not known before are verified when put to the test of experiment ; that is, it must be deductively true and suggestive.

(3) Deductively suggestive.

"The hypotheses which we accept ought to explain phenomena which we have observed. But they ought to do more than this ; our hypotheses ought to foretell phenomena which have not yet been observed;...because if the rule prevails, it includes all cases; and will determine them all, if we can only calculate its real consequences. Hence it will predict the results of new combinations, as well as explain the appearances which have occurred in old ones. And that it does this with certainty and correctness, is one mode in which the hypothesis is to be verified as right and useful." (Whewell, *Philosophy of the Inductive Sciences.*)

"By deductive reasoning and calculation, we must endeavour to anticipate such new phenomena, especially those of a singular and exceptional nature, as would necessarily happen if the hypothesis be true." (Jevons, *Principles of Science.*)

The discovery of the three elements, gallium, scandium and germanium (*post*, chap. XVI.), the properties of which were foretold with the closest approximation to truth from the application of the system according to which the properties of all elements are assumed to be a periodic function of their atomic weights, is a striking example of the deductive application of hypotheses. Amongst other such brilliant results stands out prominently the discovery of radium[1]. M. Becquerel found in 1896 that compounds of uranium spontaneously and continuously emit some radiation which, among other properties, has that of making air a conductor of electricity. This effect, the quantity of which can be determined with great accuracy, was used by Mme. Curie to measure the amount of radiation produced by various compounds of uranium and of thorium, which latter had meanwhile been found to emit the same kind of radiation.

"The radio-activity of compounds of thorium and uranium, measured under different conditions, shows that it is not influenced by any change of physical state or chemical composition. ...The chemical combinations and mixtures containing uranium and thorium are active in proportion to the amounts of the metal contained."

The subsequent testing of a large number of rocks and minerals showed that certain minerals which contained uranium and thorium, *e.g.* pitchblende (oxide of uranium), chalcolite (double phosphate of copper and uranium), possess radio-activity much greater than that "theoretically" due to the amount of uranium present.

"These facts did not accord with previous conclusions, according to which no mineral should be as active as the element thorium or uranium."

[1] Madame Curie, "Radio-active Substances." *Chem. News,* London, LXXXVIII, 1903 (p. 85 *et seq.*).

Hence the inference:

"It appeared probable that if pitchblende, chalcolite, etc. possess so great a degree of activity, these substances contain a small quantity of a strongly radio-active body, differing from uranium and thorium and the simple bodies actually known. I thought that if this were indeed the case, I might hope to extract this substance from the ore by the ordinary methods of chemical analysis."

The search was made, and resulted (1898) in the proof of the existence of several hitherto unknown substances, characterised by their great radio-activity, and in fact discovered and isolated by this property. To the substance so far obtained in greatest amount was given the name radium.

Kolbe's prognosis from theory of secondary and tertiary alcohols will be dealt with in detail in a subsequent chapter (XVIII.).

· It should be specially noted that it is not a necessary requirement of a good hypothesis that the assumptions made should ever be capable of sensual realisation, that is of demonstrative proof. "The suppositions made must not in themselves be absurd, that is contradictory to the laws of nature or of mind held true" (Jevons), that is all. We need not even stipulate that the suppositions made should be true, it is not by its truth that we judge a hypothesis, but by its utility: by the simplicity of its postulates; by the extensiveness of the phenomena to which it applies; and most important of all, by its adaptability to deductive application. And nothing is more dangerous to the proper appreciation of scientific methods than confusion concerning the nature and relative importance of experimental data and of scientific hypotheses. The first is the real, the unalterable, the ruler; the second is the assumed, the changeable, the servant, the tool which has to be thrown away when it is no longer able to cope with the work demanded from it, or when a better, because a simpler and more adaptable one, is devised.

The hypothesis need not be capable of sensual realisation, need not even be true.

But when a hypothesis fulfils all these demands, when in a simple manner it correlates knowledge, when it is so elastic as to let the new at once fall into its proper place by the old, and when under its directions the quest for further knowledge becomes a direct advance along clearly indicated paths, then the hypothesis takes rank as a theory.

The development of hypotheses into theories.

Thus we have the Undulatory Hypothesis of Light, which assumes the existence of an all-pervading medium of definite properties in which light is propagated by transversal waves. This hypothesis explained satisfactorily the phenomena and laws of rectilinear propagation, of reflection and refraction, of the colours of thin plates and of diffraction fringes; and when confronted with the additional phenomena of polarisation and of double refraction, it could deal with these also without encumbering itself by new and recondite assumptions, but simply by settling a point which in its original conception had been left open, namely the direction of the vibration in the wave-front; and lastly it was able to foretell the phenomena of internal and external conical refraction, phenomena so strange and so unique that it is safe to say they might never have been discovered had they not been looked for, in order to test the result of deduction. Hence we have the Undulatory Theory of Light, which comprises very nearly the whole science of optics, all the phenomena and laws arrived at inductively and explained by the wave hypothesis, and all such as have been discovered as the result of deductive inference from it.

So also we have a Kinetic Hypothesis of Gases, which assumes that the constituent particles are perfectly elastic, that they are at such a distance apart as not to influence each other, that they occupy a space which is negligible when compared with that occupied by the gas as a whole, and that they are in a continuous state of motion, subject to the ordinary laws of dynamics. And there is the Kinetic Theory which comprises: (i) All the empirical laws of gaseous pressure, temperature, and diffusion, with their explanation in terms of the above hypothesis. (ii) Deductions from the hypothesis, such as equality of the numbers of constituent particles in equal volumes of different gases, or the difference of the ratio between the two specific heats of a gas according to the number of constituent parts contained in each molecule, inferences which are in perfect accordance with experimental results and the interpretation of these.

And we should differentiate between what is implied in the Atomic Hypothesis and the Atomic Theory, the exposition of which theory is the main object of this book.

To sum up the subject matter of this chapter. It consists in an attempt to characterise the method of the inductive sciences, to show that the sequence of the processes employed is: (1) the collection of facts, which corresponds with finding an Summary. answer to the question—*what* happens? (2) the classification of these facts, and the generalisation from these classified facts, which yields the laws and which answers the question —*how* do these things happen? (3) the explanation of all that has been found to occur in terms of a hypothesis devised for this purpose, which supplies an answer to the question, *why*[1] do these things happen? and finally the welding together of all these processes in the theory of the science.

[1] Objection has been raised against the separation of the processes here given under (2) and (3), and it has been urged that even in the devising and applying of hypotheses we are only following out the "how"; that the "why" is beyond the range of what science can deal with.

CHAPTER I.

THEORIES OF COMBUSTION.

" I know not what fatal calamity has invaded the sciences, for when an error is born with them and with the lapse of time becomes as it were fixed, those who profess the science will not suffer its withdrawal."

JEAN REY, 1630.

" Die Gewohnheit einer Meinung erzeugt oft völlige Ueberzeugung von ihrer Richtigkeit, sie verbirgt die schwächeren Theile davon, und macht uns unfähig, die Beweise dagegen anzunehmen."

BERZELIUS, 1827.

ONE of the most interesting chapters in the history of chemical science is that dealing with the study of the phenomena of combustion and their interpretation. It lends itself

Phlogistic theory of combustion affords good example of the use, development, and deposition of a theory.

specially well to the purpose of showing within the scope of not too complicated phenomena, how a theory arises, how it is applied, how the conservatism inherent in the human mind is reluctant to give up an accustomed interpretation of nature, even when it no longer answers to the first requirements of a theory, that is when it no longer explains the facts and laws observed in the class of phenomena to which it refers; but how after all facts are and always must be strongest, and hence how a theory is finally given up when no longer able to deal with the facts, and how its place is then taken by another better fitted to do so.

The effect of heat on matter had from early times been a subject for observation and experiment, which soon led to classification and generalisation[1]. It was observed that whilst some sub-

---

[1] Kopp's *Geschichte der Chemie*, vol. III. p. 102 *et seq.* has been closely followed in the short summary of early views on combustion about to be given.

stances are not permanently changed when heated (sand, noble metals, etc.) others are (wood, sulphur, base metals,

Observation and experiment lead to the classing of metals with combustible substances like wood.

etc.). The burning of substances, that is, the occurrence of a permanent change marked by the appearance of flame, *i.e.* great evolution of light and heat, and the remaining behind of ash, naturally arrested attention, and the view that substances were combustible in virtue of the common presence in them of "fire matter" goes back to the time of the Greek Philosophers. That the substances left behind when wood is burnt or

Burnt metals termed 'cineres' and 'calxes.'

when metals such as copper and lead are heated, were alike called "cineres" (ashes), bears witness to the fact that even then these two phenomena, outwardly not very similar, the burning of wood and the change produced by heating metals, were already classed together. The name of "calxes" (Latin calx = lime), for burnt metals, which up to about 1600 was used along with "cineres" and after that exclusively, is due to the Arabian Alchemists, and suggests an analogy with the burning of chalk, the burnt metal being produced from the metal by the same process as quick-lime from chalk, namely, by heating. All through the Middle·Ages, the idea was retained that what occurs when substances burn with flame, and when metals are changed to calxes, is of essentially the same nature and must therefore be explained by the same cause. For many centuries sulphur was looked upon as the principle of combustibility and metals which could be "burnt," *i.e.*, calcined, owed this to the common presence in them of sulphur. "Ubi ignis et calor, ibi sulphur" summed up this view.

J. J. Becher (1635—1682) a German physician, who led a very roving life, who was greater as a theorist than as an experimenter, further correlated the phenomena of combustion.

Becher's views on combustion.

His views on the subject were laid down in a book called *Physica Subterranea* (1669), and shortly stated, these are: Combustion is a destruction, a dissolution of the combustible substance into its components. Hence a substance incapable of being resolved into others, what we would now term an element, cannot burn. Every combustible substance must in itself contain the cause of combustibility. This cause Becher finds in the principle of combustibility, by him termed "terra pinguis," "fatty earth," and of which he distinctly says that it is not sulphur, but a constituent of sulphur just as of every other combustible substance. The calcination of metals consists in the driving out of this terra pinguis by fire.

"Fire dissolves and breaks up all things made up of different parts,—in metals the more volatile part is expelled."

Becher's views on the nature of combustion are kept in somewhat vague outline, and he often in his writings expresses the wish for a successor to complete his work. He found him in Stahl, with whose name is always associated the theory of combustion which reigned supreme for quite a century.

G. E. Stahl (1660—1734) was an investigator and teacher. As professor in the then newly founded University of Halle he transmitted his ideas to his many pupils. He made himself **Stahl extends Becher's views.** the exponent of Becher's views, whose *Physica Subterranea* he republished. He extended the scope of the phenomena to which Becher's explanation of combustion applied, by including in it the regeneration of metals from their calxes, that is the process of metallic reduction. It was of course well known, that metal calxes on being heated with carbon or with sulphur, that is with very combustible substances, regenerated the metals. Stahl in his " experimentum novum " definitely established the validity of the classing together of the burning of sulphur with the calcination of a metal, when he regenerated the sulphur from sulphuric acid (burnt sulphur) by a process analogous to the recovery of metals from their calxes on heating with carbon[1]. This analogy may be thus represented :

Metal heated = metal calx. Sulphur heated = sulphuric acid.
Metal calx + carbon = metal. Sulphuric acid + carbon = sulphur.

These diverse phenomena comprising combustion, calcination, and reduction, all found their common explanation **The nature of the Phlogistic hypothesis.** in a hypothesis which in principle does not differ from that of Becher. Combustible substances were

---

[1] The details of the process were:

(1) Combination of the acid with an alkali: a sulphate is formed.

(2) Heating of the substance so obtained with carbon : the sulphate, which owing to its comparative non-volatility lends itself to the process better than the sulphuric acid which boils at 338°, is reduced to sulphide.

(3) Treatment of the substance so obtained with dilute acid, when sulphur is liberated: alkaline sulphides readily pass to polysulphides, which with acid give sulphuretted hydrogen and sulphur.

assumed to consist of the product of combustion united with an inflammable principle; metals, to contain the metal calx and that principle:

"The base metals contain an inflammable substance, which by the action of fire goes into the air, leaving behind a metal calx."

This inflammable principle, the fire matter of the ancients, the sulphur of the mediaeval chemists, the terra pinguis of Becher is to Stahl not fire itself, but only the condition necessary for the production of fire, and he names it " phlogiston.' The metal and the principle of fire combine together.

"I called it Phlogiston [=burnt, from $\phi\lambda o\gamma i\zeta \epsilon \iota \nu$=to set on fire] as it is certainly in the first place a combustible and inflammable principle, and one especially capable of directly taking up heat."

Beyond this Stahl does not commit himself concerning the nature and the properties of phlogiston, and it is left an open question whether this phlogiston is capable of real existence, and whether it could be isolated. The more easily a substance burns, the richer it is in phlogiston, and Stahl looks upon soot as nearly pure phlogiston.

"The metals thus burnt cannot return to their metallic form by any experiment or addition whatever, except by what can again communicate and supply to them inflammable material."

Carbon, a substance peculiarly rich in phlogiston, is able to effect such a transfer, and hence the reducing action of carbon on metal calxes.

The phenomena of the combustion and of the regeneration of the combustible substance thus find an extremely simple and perfectly consistent explanation, which is:

1. Combustible substance —   Phlogiston   = Burnt substance
       e.g. metals, sulphur,   (composition)   ⎧ metal calxes, sulphuric
         phosphorus, &c. ⎱   — phlogiston   = ⎨ and phosphoric acids,
                          ⎰   ⎩ &c.

2. Burnt substance   +   phlogiston   = Combustible substance
       e.g. metal calxes, sul-⎱ ⎧ phlogiston   ⎱   ⎧ metals, sulphur, phos-
        phuric and phos- ⎬+⎨   ⎬ = ⎨ phorus, &c.
        phoric acids, &c. ⎰ ⎩supplied by carbon⎰   ⎩

The immense importance and value of this explanation cannot be overrated. A whole number of facts, some of them at first sight

very different, like the burning of carbon and the calcination of a
metal, such as lead, are by experimental study shown
to belong to the same class, and are all comprised in
the explanation given by means of a very simple
and plausible hypothesis. And many other facts
discovered or investigated later and shown to belong to the same class of phenomena fit in with it
equally well; for instance the solution of metals in
acids when the products are the gas hydrogen (then
called inflammable air), and a substance identical with that
obtained when the metal calx dissolves in acid without evolution
of gas. Hence the solution of the metal could be looked upon
as comprising two distinct processes, first that of calcination and
next that of solution of the calx in the acid. The calcination and
attendant elimination of phlogiston must therefore account for the
formation of the inflammable air, which might be phlogiston itself,
but which anyhow must be a substance rich in phlogiston. And
the experimentally established fact that a metal calx when heated
with inflammable air regenerates the metal, is in perfect agreement
with this view, and can rank as a valuable deductive verification
of the hypothesis. Hence the phenomena of the solution of metals
in acids became included in the same theory as those of combustion, calcination and reduction, to which was added before
long, respiration.

But, side by side with these positive achievements of the
hypothesis, which warrant its being referred to as the " phlogistic
theory," there soon grew up formidable difficulties,
some cases which could not be explained in terms of
the original simple hypothesis, others which were
in direct contradiction to its deductive inferences.
Subsidiary hypotheses had to be framed in quick succession, until
what at first had been simple and consistent, had become complicated and contradictory, and hence unmanageable. The chief of
these difficulties were that: (1) Metal calxes, and in fact all burnt
substances, are heavier than the original combustible substances.
(2) Burning requires the presence of air. (3) Substances burn
much better in the gas now named oxygen than in ordinary
air. (4) The identification by many chemists of phlogiston
with inflammable air (hydrogen) led to the substitution for a
vague principle of a substance whose properties were definite and

The phlogistic hypothesis correlates combustion, calcination and reduction, and also explains the solution of metals in acids.

Difficulties encountered by the phlogistic theory.

3—2

fixed, and capable of being investigated experimentally. (5) The substance named red precipitate, which is mercury calx, can be changed into the metal without addition of phlogiston.

Each of these points will now be dealt with separately, with the object of discussing the particular difficulty or difficulties it presented to the phlogistic theory, how the theory tried to overcome these, and the success or failure it met with in these attempts.

1. The fact that the calcination of metals was attended by increase in weight was not new and had been perfectly well established at the time when the phlogistic hypo-

(1) Increase of weight on calcination.

thesis was first promulgated. It had been noticed in the eighth century by the Arabian Alchemist Geber, and during the sixteenth and seventeenth centuries the observations concerning this occurrence were many and definite. Thus Lemery in his *Cours de Chymie* (1675) says:

"In the calcination of lead and of several other substances there occurs an effect, which well deserves that some attention should be paid to it; it is that although by the action of the fire the sulphurous or volatile parts of the lead are dissipated, which should make it decrease in weight, nevertheless after a long calcination it is found that instead of weighing less than it did, it weighs more."

Many were the explanations of this fact, that a substance giving up its inflammable principle, be this "fire matter," or "sulphur," or "terra pinguis," or "phlogiston," thereby became heavier. Some of these, like the observation of the fact itself, go back to prephlogistic times.

An attitude—the word explanation cannot be used—with which it is impossible not to sympathise, since it is so common a way of settling a difficulty, consisted in saying that this increase in weight was so unimportant a fact that

(a) Considered unimportant.

it need not be taken into account. This was the course followed by Stahl himself, who well knew of the increase in weight on calcination. Others realised the occurrence, and stated it as a fact, admitting their incapacity to deal with it in the then state of the science. The reporter on a memoir presented by Tillet to the Académie des Sciences in 1763 says on the subject of the increase in weight of calcined lead:

"The increase in weight falls therefore solely to the litharge, and this is a true chemical paradox, which the results of experiment place beyond doubt.

But though it is easy to prove the occurrence, it is not as easy to give a satisfactory reason for it; it lies outside all our physical conceptions, and we must leave to time the solution of this difficulty."

(*b*) Explanation left to future times.

This is a perfectly legitimate attitude, and it generally happens that at any stage in the development of a theory, there are a certain number of residual facts, which at the time do not find their explanation in terms of the hypothesis underlying the theory. Such facts may be explained later on as singular or disguised results of the very hypothesis with which they seemed to conflict, or they may be due to other extraneous causes not taken into account, and generally it is the further study of such residual facts that initiates progress along new lines. It must depend on their number and importance relatively to the positive achievements of the theory, whether their explanation and incorporation may be left to future work, or whether the theory itself must be modified or even abandoned.

A very common attempt at explaining the observed increase of weight was that of endowing heat with weight, and saying that fire matter had combined with the burning substance. Boyle's name is specially associated with this view, which however could not stand the test of experiment.

(*c*) Ponderable fire matter is supposed to be absorbed.

"It will not be irrational to conjecture that multitudes of these fiery corpuscles, getting in at the pores of the glass, may associate themselves with the parts of the mixt body whereon they work, and with them constitute new kinds of compound bodies, according as the shape, size and other affections of the parts of the dissipated body happen to dispose them....I have been induced to think that the particles of an open fire working upon some bodies may really associate themselves therewith ; and add to the quantity." (Boyle, *The Sceptical Chymist*, 1661.)

Others again, much admired by certain of their contemporaries for their supposed ingenuity; thought that the observed increase in weight was apparent only, due to condensation, decrease in volume, whereby the volume of air displaced got less, and hence the loss of weight on weighing in air less, and hence the apparent weight greater. But it had escaped the attention of this school that the increase in weight observed was in all cases very much greater than could be thus accounted for, and further, that since most

(*d*) Accounted for by an increase in density.

metal calxes have a lesser density than the metals, their formation must be attended by expansion and not contraction.

And finally there were those who held the view, promulgated even before the phlogistic theory had been seriously attacked, that phlogiston was the principle of levity, that it was endowed with the property of negative gravity, a view no doubt due to the observed upward tendency of flame.

(e) Phlogiston is endowed with negative weight.

2. It was equally well known that substances could not burn out of contact with air. Geber in the second half of the eighth century directs that the calcination of mercury should be carried out in open vessels, and from his time dates the explanation that the air is required to take up escaping moisture. Boyle (1672) shows that sulphur does not burn in a vacuum; Stahl knows that calcination is not possible in closed vessels or in vacua, and that even soot, according to him the purest phlogiston, will not burn out of contact with air.

(2) Air is necessary for combustion.

"Metallic antimony, copper, lead, not even tin can be burned in quite closed or full vessels."

Becher's explanation was that air is required to take up the "escaping particles" and Stahl follows him, holding that phlogiston cannot escape unless there is something to take it up, to absorb it; but he regards this power of absorption as limited, thus accounting for the fact that in a limited volume of air combustion is also limited. The name of "phlogisticated air" for that which remains behind (nitrogen) when ordinary air is deprived of the power of supporting combustion, that is when it is saturated with phlogiston, expressed this view. This explanation takes no account of the fact that the volume of the residual gas is less than that of the original air.

Air supposed to be necessary for absorbing the escaping phlogiston.

C. W. Scheele (1742—1786), an eminent Swedish chemist, the discoverer and investigator of chlorine, manganese and oxygen, and of a large number of most important organic substances, a staunch adherent of the phlogistic theory, clearly realised this difficulty. This is what he says on the subject:

The lesser volume of the residual air not accounted for.

"It is not, as may be seen, a trifling circumstance that phlogiston, whether it separates itself from substances and enters into union with air, with or without a fiery motion, still in every case diminishes the air so considerably in its external bulk. I have already stated that I was not able to find again the lost air. One might indeed object, that the lost air still remains in the residual air which can no more unite with phlogiston; for, since I have found that it is lighter than ordinary air, it might be believed that the phlogiston united with this air makes it lighter, as appears to be known already from other experiments. But since phlogiston is a substance, which always presupposes some weight, I much doubt whether such hypothesis has any foundation." (*Chemical Treatise on Air and Fire*, 1777. Alembic Club Reprint, No. 8.)

There were various other explanations for the necessity of air in combustion, which somewhat approximated to the correct one, in that they assumed a fixing by the burning substance of part of the air. Of these the most interesting, because simultaneously explaining the increase in weight observed, is that given in 1630 by the French physician Jean Rey, which will be dealt with later.

3. On August 1st, 1774, the English chemist John Priestley (1733—1804), to whom the science owes the invention of important methods for the collection and storing of gases, and who by substituting mercury for water as the liquid above which gases were collected, discovered gaseous ammonia, hydrochloric acid, etc., obtained another new gas of most startling properties. Priestley tells us himself how he believed that

(3) Priestley discovers a new gas which i⁵ a better supporter of combustion than air.

"More is owing to what we call chance,...than to any proper design or preconceived theory in this business" (*Experiments and Observations on Different Kinds of Air*, 1775),

and he proceeds to show how large a share this element of chance had in his discovery of this new gas, and how he was continually "surprised" by something perfectly unexpected happening. His results were not published till 1775, but when he was in Paris in the October of 1774 he told Lavoisier of his discovery and thereby supplied the French chemist with the most efficient weapon for attacking the phlogistic theory. The priority in the discovery of this new gas is no doubt due to Scheele who, two years earlier, had prepared it in a variety of ways including that first used by the English chemist, and who also investigated its properties, but did not publish his results till four years after his own discovery, and two years after Priestley's. The gas was obtained by heating with a burning-glass "mercurius calcinatus per se," the substance produced by heating mercury in air, and it had the remarkable property of supporting combustion much better than ordinary air. "Red precipitate," the substance obtained by precipitating the solution of a mercury salt by a mild or caustic alkali, was found to yield the same gas.

To account for the remarkable combustion-supporting property of the new gas, the additional hypothesis was framed that, whilst ordinary air already contained some phlogiston, the new gas, hence called "dephlogisticated air," did not. It could therefore take up more phlogiston, support combustion better than ordinary air, a gradation of property which we might compare with the water-absorbing power of a sponge. The dephlogisticated air—Priestley's new gas, our oxygen—would correspond to the dry sponge, the ordinary air to the wetted but not saturated sponge, and the phlogisticated air to the completely saturated sponge which can take up no more water. The fact that calcination and certain combustions when carried out in ordinary air always left a gaseous residue of phlogisticated air, but left no such residue when carried out in dephlogisticated air, was of course incompatible with the nature of this explanation, according to which ordinary air was a substance intermediate between dephlogisticated and phlogisticated air and which would require the final result of combustion to be the same in both cases.

Air is assumed to contain some phlogiston already, whilst the new gas termed dephlogisticated air does not.

4. The identification of phlogiston with hydrogen already referred to, which on the one hand gave valuable support to the phlogistic theory, bringing within its compass the phenomenon of the solution of metals in acids, on the other hand proved a great difficulty in its way.

(4) Phlogiston is identified with hydrogen.

As long as phlogiston was looked upon as a "principle" not isolated, any properties conveniently required from it—including that of negative gravity—could be assigned to it; it could be supposed to be adding itself to and removing itself from matter at will. But this became impossible when a definite substance, capable of having all its properties investigated, took its place. If the change of metal calx to metal simply consisted in the addition of phlogiston, and if phlogiston is inflammable air, why should water be formed when a metal calx is heated in hydrogen, and what becomes of this hydrogen when, in the process of calcination, it escapes from the metal? These and similar questions proved inconvenient, and no answer, at any rate no satisfactory answer, was forthcoming.

5. But what in the light of to-day should have proved the greatest blow to the already much discredited theory of phlogiston was the discovery that the relation

(5) Mercury calx changed into metallic mercury without the addition of phlogiston.

$$\text{metal calx} + \text{phlogiston} = \text{metal}$$

certainly was not universal, since the calx of mercury could be reduced to metallic mercury without the addition of any phlogiston whatever, *i.e.* without heating with carbon. The first observations on this subject are due to Bayen[1] who, in a paper entitled "Chemical Experiments made with some Precipitates of Mercury with the Object of discovering their Nature," gives the results of investigations undertaken primarily to account for the increase in weight on calcination. The results he obtained were unexpected:

"Such are the phenomena presented by the precipitates considered, when treated in the manner described, phenomena which are truly startling, and which demand long and laborious work from the chemist who wishes to prove their occurrence, and by experiment to find a cause for each."

So startling, because so contradictory to the fundamental tenets of the theory of phlogiston do his results appear to him, that he finds it necessary to offer at the outset a sort of apology, a declaration of good faith:

"Since I take no side but that of truth when I know it, it becomes my duty to retail simply and with perfect honesty the details and the results of my experiments; the first of these were imperfect and guided by prejudice, but they led me insensibly to others which made me turn from the error under which I had lain";

and further on

"In giving an account of the following experiments, I shall no longer use the language of the disciples of Stahl, who will be obliged to remodel their doctrine of phlogiston; or to admit that the precipitates of mercury, with which I deal, are not metal calxes, though they have been held to be so by some of their most celebrated chemists; or else to assume that there are calxes which can be reduced without the agency of phlogiston."

[1] *J. phys.* III. 1774 (pp. 135, 281).

Bayen's experiments were the following :

Weighed quantities of metallic mercury were dissolved in nitric acid, and then precipitated by fixed alkali (potassium carbonate), or caustic alkali (potash), or lime. The precipitates when well washed, dried and weighed, were invariably found heavier than the mercury taken. He assumes the identity of these precipitates with each other and with calcined mercury, though he admits that he has not absolutely proved it. For this assumption of identity he finds support in the fact that all these precipitates, when mixed with charcoal and heated, yielded metallic mercury and fixed air, just as does calcined mercury itself. But the "startling" part of the work consisted in the discovery that when the metal calx was heated by itself, without carbon to provide the phlogiston supposed necessary for reduction, metallic mercury was formed and a gas evolved. Priestley's discovery and investigation of this gas occurred later in the same year, but his results were not published till the year after. Bayen worked with weighed quantities of mercury calx in retorts of known capacity communicating with receivers, the volume of which was also known, and which were placed over water. The volume of gas evolved could thus be measured. This he did, and he also weighed the metallic mercury formed, and the unchanged residue. He did not investigate the nature of the gas evolved, nor did he even distinguish it from the fixed air obtained on reducing the calx with charcoal, but he noted that the volume of gas evolved was in both cases approximately the same. The result of this quantitative experiment is embodied in the following summary :

Marginal note: Bayen's experiments on mercury calx.

	onces	gros	grains
Weight of mercury calx prepared according to the description given (*i.e.* pp. by alkalis from the solution of mercury in nitric acid)............. =	1	0	0
This on reduction without the intermediary of carbon gave a few drops of water which deposited in the neck of the retort, and which were estimated........................ =	0	0	3
and revived mercury.................................. =	0	7	4
and earth remaining unchanged at the bottom of the retort............................... =	0	0	3
and an estimated loss of mercury.............. =	0	0	4
	0	**7**	**14**

"Hence the diminution in weight suffered by the mercury calx· on reduction to metallic mercury has been 58 grains[1]. I cannot state positively that these 58 grains are the true weight of the elastic fluid liberated from one ounce of this calx, but certainly everything leads to this belief."

Bayen then returns to the consideration of the increase in weight observed in the calcination of the mercury by his method, and he quite correctly puts it down to two causes:

Bayen's explanation of the increase in weight on the calcination of mercury. (i) contamination with the materials used in solution and in precipitation, the solutions adhering so tenaciously to the precipitated calx that their complete removal by washing becomes a matter of great difficulty, if not an impossibility[2]; (ii) combination with an elastic fluid:

"The experiments which I have performed and which may have suffered through not having been better presented to the Public, oblige me to conclude that in the mercurial calx to which I refer, the mercury owes its calcined state, not to any loss of phlogiston sustained, but to its intimate combination with the elastic fluid, the weight of which, in adding itself to that of the mercury, constitutes the second cause of the increase observed in the precipitates examined."

Bayen's work is more valuable for having brought to light a phenomenon inconsistent with the very fundamental conception of the phlogistic theory, than for having helped towards the substitution of a newer and better explanation for the one so discredited; and that, in spite of the fact, that the explanation given by him for the phenomenon investigated closely approximates to the correct one. But his proofs were not complete, and hence his own statement of his inference lacked definiteness and personal conviction.

Insufficiency of proofs, but certainly not want of self-confidence must also be charged against that explanation of the phenomena of calcination which is found in the work already

Rey's explanation of calcination. referred to (p. 39) of an investigator a century and a half earlier. In 1630 Jean Rey published a series of essays entitled: "Essays of Jean Rey, Doctor of Medicine, on the Researches of the Cause owing to which Tin and Lead increase in Weight when they are calcined[3]."

---

[1] 1 once = 8 gros ; 1 gros = 72 grains.
[2] Ordinary gravimetric analysis familiarises one with the difficulty of washing precipitated mercuric or cupric or argentic oxide.
[3] Alembic Club Reprints, No. 11.

"Now I have made the preparations, nay, laid the foundations for my answer to the question of the sieur Brun, which is, that having placed two pounds six ounces of fine English tin in an iron vessel and heated it strongly on an open furnace for the space of six hours with continual agitation and without adding anything to it, he recovered two pounds, thirteen ounces of a white calx; which filled him at first with amazement, and with a desire to know whence the seven ounces of surplus had come. And to increase the difficulty, I say that it is necessary to enquire not only whence these seven ounces have come, but besides them what has replaced the loss of weight which occurred necessarily from the increase of volume of the tin on its conversion into calx, and from the loss of the vapours and exhalations which were given off. To this question, then, I respond and sustain proudly, resting on the foundations already laid, 'That this increase in weight comes from the air, which in the vessel has been rendered denser, heavier, and in some measure adhesive, by the vehement and long-continued heat of the furnace: which air mixes with the calx [frequent agitation aiding] and becomes attached to its most minute particles: not otherwise than water makes heavier sand which you throw into it and agitate, by moistening it and adhering to the smallest of its grains'."

The manner in which Rey arrives at his answer is not by any direct experiments on calcination, but rather by experiments and the reference to experiments of a purely physical nature, such as the discussion of the causes, like change of volume, which can, and of those, like heat, which cannot, produce change of weight. Thus he lays a sound foundation for his method, which is one of elimination, showing that none of the causes to which it had been usual to ascribe the observed increase in weight could be considered legitimate; that it could not be due to the giving up of heat of negative gravity, nor to the absorption of fire matter of positive weight, not to an increase in density, not to the absorption of soot or of anything else from the materials of the containing vessels. And so none is left unchallenged of all the possible modes of explanation save that of the fixation of the air. Therefore his explanation, such as it is, does not partake of the nature of proof; it amounted to showing that every hypothesis made so far was inadmissible and to bringing forward another, which however was not tested deductively in any way. With regard to the part played by the air in producing the increase of weight, it is quite clear that Rey's view differed widely from that propounded later and now held. He did not look upon it as the cause of calcination, to him it was not a case of the combination of the metal with a definite

The insufficiency of Rey's explanation.

portion of the air, or even with a certain amount of the air as a whole, but it was formation of the calx—how this was produced he does not say—and subsequent condensation of air on the surface of this calx. So even if it had been more widely known than it was, Rey's work could not in itself have supplied a satisfactory theory of combustion.

It is usual to state that Hooke, who published his *Micrographia* in 1665, anticipated the modern view of the nature of combustion;

Hooke's explanation of combustion. but it will appear from the following extracts that whatever value may be assigned to his work, it cannot be claimed that he did more than recognise the part played by the air in the process, while still retaining the conception of a "sulphureous principle" which is lost by the body in the act of combustion. We are told that:

"From the experiment of charring of coals...we may learn...that the air in which we live, move, and breath, and which encompasses very many, and cherishes most bodies it encompasses, that this air is the menstruum, or universal dissolvent of all sulphureous bodies...that the dissolution of sulphureous bodies is made by a substance inherent, and mixt with the air, that is like, if not the very same, with that which is fixt in saltpeter, which by multitudes of experiments that may be made with saltpeter, will, I think, most evidently be demonstrated....The dissolving parts of the air are but few, that is, it seems of the nature of those saline menstruums, or spirits, that have very much flegme mixt with the spirits, and therefore a small parcel of it is quickly glutted, and will dissolve no more;...whereas saltpeter is a menstruum, when melted and red-hot, that abounds more with those dissolvent particles, and therefore as a small quantity of it will dissolve a great sulphureous body, so will the dissolution be very quick and violent....It is observable, that, as in other solutions, if a copious and quick supply of fresh menstruum, though but weak, be poured on, or applied to the dissoluble body, it quickly consumes it : so this menstruum of air, if by bellows, or any other such contrivance, it be copiously apply'd to the shining body, is found to dissolve it as soon, and as violently as the more strong menstruum of melted nitre."

Hooke finishes the exposition of his views concerning combustion with the remark:

"This hypothesis I have endeavoured to raise from an infinite of observations and experiments, the process of which would be much too long to be here inserted, and will perhaps another time afford matter copious enough for a much longer discourse."

This relegation of the definite statement of the facts on which his hypothesis depended to an uncertain future—the promise was never redeemed—is characteristic of the earlier attitude towards such problems and makes an exact estimate of the value of Hooke's contribution to a true theory of combustion impossible.

Towards 1775 something better was all that was required to bring about the final abandonment of the phlogistic theory which by then had become untenable. Encumbered with subsidiary hypotheses of all sorts, it could not even clumsily perform the work required from it, and something better was furnished by Lavoisier[1].

The nature of Lavoisier's combustion hypothesis and its development into a theory, the exposition of his method—most important for the purpose followed here,—these will best be appreciated by an account of his investigations on this subject in the original order of their presentation.

In 1772 he proved that phosphorus and sulphur, on being burnt, absorb a large volume of air and increase in weight[2]. The

Lavoisier on the increase of weight in the burning of phosphorus and sulphur.

phenomena themselves being of the same nature, the cause of this increase in weight should be the same as that to which is due the increase in weight of a metal on calcination. But to what is this due?

The phlogistic theory gave no answer, or rather a number of answers worse than none. Its success had always lain in its power of dealing satisfactorily, from one common point of view, with the qualitative aspect of phenomena. The quantitative relations it did not willingly take into account, and when forced to do so it could not cope with them. Lavoisier began his work with the consideration of these very relations.

In a memoir deposited in 1772 and published in 1774 entitled, "The Increase of the Weight of Metals on Calcination," he sets

Lavoisier on the calcination of tin.

himself the problem of accounting for this increase. He begins with a historical account of Stahl's views on combustion, and of the unsatisfactory explanation these supply for the generally known and accepted increase in weight suffered by metals on calcination. Much can be said, and so very much has been said, concerning Lavoisier's propensity to arrogate to himself the result of other people's

---

[1] Anton Laurent Lavoisier, born 1743; made a member of the French Academy at the early age of 25; in the service of the State as Fermier Général and director of the saltpetre industry; impeached under the Reign of Terror and executed 1794. Besides his brilliant work in pure chemistry, he introduced improvements in technology, and carried out most important researches on heat.

[2] *Œuvres*, II. (p. 103).

work, and so much research has been devoted to proving him utterly unscrupulous in this respect[1], that it a pleasure to find that if he failed in this way, he did not do so habitually. He refers to Rey's work on the calcination of lead and tin in the most appreciative manner possible and describes it as of the greatest importance. Just as in the 1770 research on the change of water to earth, he begins with a splendid piece of reasoning on the possible solutions of his problem. What is it that could possibly cause this increase in weight, and what would be the consequences of the action of each of these possible causes on the weight of the whole system and of each of its parts? Then he investigates what changes really do occur, and hence infers what the cause must have been.

"Thus then did I at the beginning reason with myself: if the increase in the weight of metals calcined in closed vessels is due, as Boyle had thought, to the addition of the matter of the flame and the fire which penetrate the pores of the glass and combine with the metals, then it follows that on introducing a known weight of metal into a glass vessel and sealing this hermetically, determining the weight exactly, and then proceeding to calcination by a charcoal fire—just as Boyle had done—and then finally after calcination, before opening it, again weighing the same vessel, this weight must be found augmented by that of the whole quantity of fire matter which had been introduced during calcination. But if, said I to myself, the increase in the weight of the metal calx is not due to the addition of fire matter nor of any other extraneous matter, but to the fixation of a portion of the air contained in the vessel, the whole vessel after calcination must be no heavier than before and must merely be partially void of air, and the increase in the weight of the vessel will not occur until after the air required has entered."

The calcination of lead and of tin in closed vessels is chosen for the experimental testing of these theoretical considerations. The account of the experimental operations and the results obtained, are for the sake of brevity, both embodied in the following tabular representation:

	EXPERIMENT I.			EXPERIMENT II.		
	ozs.	gros	grains	ozs.	gros	grains
(1) Weight of tin taken..................	8	0	0	8	0	0
(2) Weight of retort + air contained in it at ordinary temperature	5	2	2·50	12	6	51·75
(3) ∴ Weight of retort + tin + air at ordinary temperature = (1) + (2)	13	2	2·50	20	6	51·75

[1] Thorpe. *Priestley, Cavendish, Lavoisier, and La Révolution Chimique. Essays in Historical Chemistry*, 1894 (p. 110); *Nature*, London, 42, 1890 (pp. 313, 449).

	Experiment I. ozs. gros grains	Experiment II. ozs. gros grains

The neck of the retort was then drawn out and to prevent its being shattered in the subsequent heating, it was warmed to expel some of the air, the end of the neck fused whilst the retort was kept hot, the whole cooled and then weighed.

(4) Weight of retort + tin + air at higher temperature..........................

	Experiment I.	Experiment II.
(4) Weight of retort + tin + air at higher temperature	13 1 68·87	20 6 16·88

The sealed retort with its contents was then exposed to the heat of a coal fire; the tin after melting became coated with a black substance which sank in the tin. This process was continued until no further change seemed to occur. The retort was then cooled and weighed.

(5) Weight of retort + residual tin + burnt tin + residual air.........

	Experiment I.	Experiment II.
(5) Weight of retort + residual tin + burnt tin + residual air	13 1 68·60	20 6 15·88
(6) ∴ Change of weight in the retort and its contents due to heating. = (4) − (5)	0 0 0·27	0 0 1·00

These changes were commensurate with the limit of accuracy of the weighings. The retort was next opened, air rushed in and it was again weighed.

(7) Weight of retort + residual tin + burnt tin + air filling it at ordinary temperature..............

	Experiment I.	Experiment II.
(7) Weight of retort + residual tin + burnt tin + air filling it at ordinary temperature	13 2 5·63	20 6 61·81
(8) ∴ Change of weight due to the calcination of the tin and the replacement of the absorbed air by ordinary air = (7) − (3)	0 0 3·13	0 0 10·06

From the fact that (4) and (5) are almost identical, it follows that the gain in weight of the tin must have been equal to the loss in weight of the air. To actually determine the gain in weight of the tin, the unchanged tin + burnt tin were detached from the glass vessel and weighed. As a check, the now empty retort was also reweighed and compared with the original weight.

	EXPERIMENT I. ozs. gros grains			EXPERIMENT II. ozs. gros grains		
(9) Weight of the retort + air contained in it at ordinary temperature at end of experiment..................	5	2	2·50	12	6	51·62
(10) ∴ Change in weight of the retort = (2) − (9)...........................	0	0	0	0	0	0·13
Weight of the unchanged tin + burnt tin.........................	8	0	3·12	8	0	10·00
(11) ∴ Increase in weight of the tin on calcination = (10) − (1)	0	0	3·12	0	0	10·00

This increase in weight of the tin on calcination is therefore practically identical with (8), the weight of air which took the place of that absorbed in calcination, *i.e.* 3·12 and 3·13 grains ; 10·00 and 10·06 grains.

From the results of these two experiments—similar ones with lead were not successful—Lavoisier draws the following conclusions :

"Summing up the results of the two experiments on tin just described, it seems to me impossible not to draw the following conclusions :

First. In a given volume of air only a fixed quantity of tin can be calcined (much unchanged tin had remained behind)[1].

Secondly. This quantity is greater in a large retort than in a small one (compare the figures under 8 and 11 for the two experiments)[2].

Thirdly. The hermetically sealed retorts, weighed before and after the calcination of the tin contained in them, showed no difference of weight, which evidently proves that the increase in weight of the metal arises neither from the fire matter nor from any other matter extraneous to the vessel.

Fourthly. In all calcinations of the tin the increase in weight of the metal is sufficiently nearly equal to the weight of the air absorbed, to prove that the portion of the air which combines with the metal during calcination is of specific gravity approximately equal to that of atmospheric air."

Here then the problem he had set himself had been solved by Lavoisier. He had ascertained the cause of the increase in the weight of metals on calcination and had found it to be combination with a certain portion of the air. And having proved before that sulphur

Results of the experiments on the calcination of tin.

[1] The remarks in brackets are explanatory insertions.
[2] Knowing the density of air and the volume of his vessel, Lavoisier could calculate the weight of air contained in the retorts, and comparing this with the increase in weight of the burnt tin, could estimate what fraction of the whole air had been absorbed.

F.   4

and phosphorus on burning also increase in weight and absorb a large volume of air, Lavoisier must at that stage be supposed to have established that combustion consists in combination with a portion of the atmospheric air, whereby the increase in weight on combustion is accounted for. But he knew nothing as yet concerning the nature of the portion of air absorbed. In the time between the memoir on the calcination of tin and his next contribution to the subject of combustion, falls Priestley's discovery of a gas obtainable by the heating of red precipitate (burnt mercury), the investigation of the properties of this gas, and the recognition that it is a better supporter of combustion than ordinary air; Lavoisier hears of this new fact, and his next paper bears evidence of the manner in which he was helped thereby.

In 1775 he presented to the Académie des Sciences a memoir, entitled, "On the Principle which combines with Metals during Calcination and which augments their Weight[1]." He poses the question, what is it that unites with metals when they are calcined and increase in weight? To find the answer required, he must evidently try to regain the gas absorbed, and hence he is led to the special consideration of a case in which this reduction can be achieved without the addition of another substance like carbon. Such a substance he finds in the calx of mercury, then called red precipitate or mercurius praecipitatus per se, which, as had been shown by Bayen and by Priestley, yielded on heating the metal mercury and a gas. Lavoisier first investigates whether this substance may legitimately be termed a calx, a question which it will be remembered had somewhat disturbed Bayen, but which, though inclined to answer in the affirmative, he saw no way of settling conclusively. Lavoisier heats some of the red precipitate with charcoal in a small retort and obtains metallic mercury and a gas which he shows to be fixed air, whereby he proves the original substance to be a calx, as it is a class-characteristic of calxes that on being heated with charcoal they yield the metal and fixed air. He next heats the red precipitate by itself:

Lavoisier's experiments on burnt mercury.

> 1 ounce of red precipitate yields 7 gros and 18 grains of liquid mercury and 78 cubic inches[2] of a gas.

[1] *Œuvres*, II. (p. 122).
[2] 1 old French inch = ·02707 metre = 1·066 English inches; 1 old French cubic inch = 19·8 cubic centimetres.

This gas is not fixed air but is endowed with quite different and characteristic properties, chief amongst which is that of being a better supporter of combustion than ordinary air.

"Accordingly it seems proved that the principle which combines with metals during their calcination and which increases their weight is no other than the purest portion of the very air surrounding us, which

Results of Lavoisier's experiments on burnt mercury. we breathe, and which in this operation passes from the expansive to the fixed state. Hence in all the metallic reductions when charcoal is used, fixed air is obtained, which is due to the combination of the charcoal with the pure portion of the air; and it is very likely that all the metallic calxes would, like that of mercury, yield nothing but eminently respirable air, were it possible to reduce them all without any addition, as the mercury precipitated per se is reduced."

"Since in the reviving of the mercury calx all the charcoal disappears and nothing but mercury and fixed air are obtained, it must be concluded that the principle to which till now the name of fixed air had been given is but the result of the combination of the eminently respirable portion of the air with the charcoal."

Here then a second point has been established, namely, the nature of the portion of the air absorbed in calcination. But nothing short of unassailable proof satisfied Lavoisier, and in his *Elementary Treatise on Chemistry*, published in 1789, is found an experiment on the quantitative synthesis and analysis of the mercury calx, which is conclusive. No date is given as to when this experiment was first performed, but by its nature it joins itself to the one just described[1].

Lavoisier's quantitative synthesis and analysis of mercury calx.

Fig. 4. Lavoisier's apparatus for the Calcination of Mercury.

[1] *Œuvres*, I. III. (p. 35).

4—2

Synthesis of the Mercury calx.

4 ounces of very pure mercury were placed in a small retort communicating with a bell-jar containing a known volume of air confined over mercury in a trough (fig. 4). The retort with its contents was heated by means of a charcoal furnace for 12 days to just short of boiling. Beginning from the second day, red specks were formed on the surface of the heated mercury, and these increased for some days in quantity. By the 12th day it was seen that calcination did not proceed further. The fire was put out, the volume of residual air in the bell jar was measured, its nature investigated, and the red substance formed was collected and weighed, the quantitative results being :

Volume of air at the beginning
=50 cubic inches.

Volume of air at the end
=42 to 43 cubic inches.

∴ Volume of air absorbed in calcination
=7 to 8 cubic inches.

Weight of red substance
=45 grains.

"The air remaining behind after this operation, which owing to the calcination of the mercury had been reduced to ⅚ of its original volume, was no longer fit for respiration nor combustion ; animals introduced into it died in a few minutes, and lights were immediately extinguished."

Analysis of the Mercury calx.

The 45 grains of the red substance which had been collected were placed in a very small retort, arranged so that it could be heated, and the gaseous and liquid products of the decomposition collected. When the retort was heated, metallic mercury and a gas were formed, the quantitative results being :

Weight of burnt mercury taken
=45 grains.

Weight of metallic mercury formed
=41½ grains.

Volume of gas evolved
=7 to 8 cubic inches.

In a subsequent memoir Lavoisier states that the density of this gas is ½ grain per cubic inch, which makes the weight of the above volume of gas 3½ to 4 grains.
But 41½ + 3½ = 45 grains.

"There were formed 7 to 8 cubic inches of an elastic fluid much more adapted than ordinary air for supporting combustion and the respiration of animals."

When these two gases were mixed ordinary air was reformed. ·

" Reflection on the conditions of this experiment shows that the mercury on calcining absorbs the salubrious and respirable portion of the atmosphere —that the portion of the air which remains behind is a noxious kind of gas incapable of supporting combustion and respiration. Hence atmospheric air is composed of two elastic fluids of different and so to speak opposite nature." ·

Lavoisier realises that the results of this experiment do not lend themselves to determining the ratio in which these two gases are mixed in the atmosphere, since the affinity of the mercury is not sufficiently great to absorb to the last trace all that portion of the air which acts as a supporter of combustion. Such a substance he finds in phosphorus, but his investigations concerning the volumetric composition of air need not here be followed further. It is more pertinent to take note of some other experiments of his in which he again demonstrated that the increase in weight of a substance on calcination is equal to the weight of gas absorbed. This was proved in two ways : on the one hand, as in the case of the experiment on tin, he weighed a hermetically sealed glass balloon containing phosphorus and air, before and after burning, and found no change ; on the other hand he ascertained the actual increase in weight of a calcined metal, measuring the volume of gas absorbed, which when multiplied by the density of the gas gives its weight. The following data refer to such an experiment :

Other experiments of Lavoisier on combustion.

Weight of iron before calcination	= 145·6 grains
Weight of iron after calcination	= 192·0 ,,
Increase in weight due to calcination	= 46·4 ,,
Air absorbed	= 97 cubic inches
1 cubic inch of air weighs	= 0·47317 grains
∴ 97 cubic inches weigh	= 45·9 ,,
Difference between increase in weight observed and weight of gas absorbed	= 0·5 ,,

· So far Lavoisier's work had consisted in an experimental study of the actual phenomena of combustion, considered from the point of view of the burning body and of the air, without which the combustion could not occur. Cause and effect of everything observed were correlated, and each conclusion arrived at was confirmed experimentally. Not until this foundation had been firmly laid does he feel himself able to deal with the whole subject of combustion, and to propound his views on it. In the memoir,

" On Combustion[1]," published in 1778, he disposes of the view of a
special principle of combustibility and enunciates
his views on the nature of combustion, according to
which the burnt substance is more complex than
the combustible substance, and is formed by the
absorption of the gas, named by Priestley " dephlogisticated air,"
by Scheele " fire air," and by Lavoisier himself at first " eminently
pure air," and afterwards " oxygen."

Lavoisier's views on the nature of combustion.

The salient features of combustion and the hypothesis according
to which he can explain combustion, calcination, reduction, and
respiration, are:

(i) Combustion is accompanied by the disengagement of heat
and light.

(ii) Substances can burn only in Priestley's dephlogisticated air.

(iii) The gas in which the substance burns is absorbed, and
there is an exactly equivalent increase in the weight of the substance burnt.

(iv) The burnt substance is an acid (sulphuric or phosphoric,
etc.), or a metal calx (calcined mercury, litharge, etc., etc.).

Hence the mechanism of combustion according to Lavoisier is
represented by:

Lavoisier's and the phlogistic hypothesis contrasted.

$$\text{Metal (simple)} \quad + \quad \text{oxygen} \quad = \quad \text{metal calx (complex)}$$

$$\text{Metal calx} \quad - \quad \text{oxygen} \quad = \quad \text{metal}$$
Carbon withdraws oxygen, forming fixed air

in direct contradiction to the phlogistic scheme:

$$\text{Metal} - \text{phlogiston} = \text{metal calx}$$
$$\text{(complex)} \qquad\qquad \text{(simple)}$$

But even in 1778 Lavoisier is still wonderfully cautious, and
guards himself expressly against the pretension of having propounded an absolutely demonstrated theory.

" Further I repeat that in attacking the doctrine of Stahl, I do not aim
at substituting for it a theory rigorously demonstrated, but only a hypothesis
which to me seems more probable, more conformable with the laws of nature,
and one involving less of strained explanations and fewer contradictions."

[1] *Œuvres*, II. (p. 226).

But he soon speaks out more strongly. In 1783 appeared a memoir, entitled "Reflections concerning Phlogiston[1]." After giving his own explanation of combustion he proceeds to say:

"But if in chemistry everything can be satisfactorily explained without the aid of phlogiston, it thereby becomes eminently probable that such a principle does not exist, that it is a hypothetical being, a gratuitous assumption, and sound logic is opposed to unnecessary complication. Perhaps I might have confined myself to these negative proofs and remained content to show that the phenomena can be better explained without phlogiston than by means of it; but the time has come when I must speak out in a more definite and formal manner concerning a view which I consider an error fatal to chemistry, and which appears to me to have considerably retarded progress by the method of false reasoning it has engendered. All these reflections prove what I have advanced, what it has been my object to demonstrate, what I will repeat once more, namely that chemists have turned phlogiston into a vague principle, one not rigorously defined, and which consequently adapts itself to all the explanations for which it may be required. Sometimes this principle has weight and sometimes it has not; sometimes it is free fire and sometimes it is fire combined with the earthy element; sometimes it passes through the pores of vessels, sometimes these are impervious to it; it explains both causticity and non-causticity, transparency and opacity, colours and their absence; it is a veritable Proteus changing in form at each instant."

Lavoisier's criticism of the theory of phlogiston.

And surely the severity of this characterisation of the phlogistic theory was not undeserved. That which Lavoisier proposed to put in its place explained equally simply all those phenomena which it had been the merit of the older theory to collect and prove to be of the same type—combustion, calcination, reduction, respiration, solution of metals in acids. But everything which to the phlogistic theory had proved a difficulty (*ante*, p. 35), and which had led to its becoming ineffectually complicated by clumsy subsidiary hypotheses, finds a ready explanation, follows as a natural inference from Lavoisier's hypothesis; the increase in weight on burning is the consequence of combination with ponderable oxygen; the necessity for the presence of air and the remaining behind of a lesser volume of some different gas is due to the air containing the combustion-supporting oxygen mixed with nitrogen, a non-supporter of combustion; the better burning in oxygen and the complete absorption of this gas follow from the

Lavoisier's explanation solves the difficulties left unexplained by the phlogistic theory.

[1] *Œuvres*, II. (pp. 623, 640).

difference between oxygen and ordinary air; the evolution of hydrogen when metals are dissolved in acid is understood when water is recognised as a compound of oxygen and hydrogen, from which the metal on its change to calx—a change assumed to precede the solution of that calx in the acid—withdraws the oxygen and liberates the hydrogen; the reduction of some metal calxes without the intervention of carbon meets with the explanation that in these cases oxygen is liberated, whilst when carbon is used, fixed air, a compound of carbon with the oxygen withdrawn from the calx, is formed.

And so in spite of the opposition of eminent men like Priestley, Scheele and Cavendish who themselves, by the discovery of oxygen and the recognition of the composition of water, had supplied Lavoisier with the most valuable materials for the building up of his theory, but who yet to the end of their lives remained believers in phlogiston, Lavoisier's hypothesis was bound to carry the day, and did so before long. But this must not blind us to the very real merits of the Phlogistic Theory, and to the enormous services it has rendered in helping the development of chemistry along truly scientific lines.

The substitution of the oxygen hypothesis of combustion for the phlogistic.

"If our hypothesis renders a reason for the agreement of cases really similar, we may afterwards find this reason to be false, but we shall be able to translate it into the language of truth." (Whewell.)

All the change required by the substitution of Lavoisier's for the phlogistic hypothesis of combustion, consists in substituting the words "addition of oxygen" for "withdrawal of phlogiston," and the large number of facts correlated under the old theory, remain so under the new one. It seems fitting to close this chapter dealing with the rise and development of one theory and the subsequent replacement of it by another, with a well-deserved appreciation of the theory deposed.

"The Phlogistic Theory was deposed and succeeded by the Theory of Oxygen. But this circumstance must not lead us to overlook the really sound and permanent part of the opinions which the founders of the phlogistic theory taught. They brought together, as processes of the same kind, a number of changes which at first appeared to have nothing in common; as acidification, combustion, respiration. Now this classification is true; and its importance remains undiminished, whatever are the explanations which

Whewell's appreciation of the services of the phlogistic theory.

we adopt of the processes themselves. ...It has been said, that in the adoption of the phlogistic theory, that is, in supposing the processes [of acidification, combustion and respiration] to be subtractions rather than additions, '*of two possible roads the wrong one was chosen, as if to prove the perversity of the human mind.*' But we must not forget how natural it was to suppose that some part of a body was destroyed and removed by combustion; and we may observe, that the merit of Becher and Stahl did not consist in the selection of one road of two, but in advancing so far as to reach this point of separation. That, having done this, they went a little further on the wrong line, was an error which detracted little from the merit or value of the progress really made." (Whewell, *History of the Inductive Sciences.*)

CHAPTER II.

LAVOISIER AND THE LAW OF CONSERVATION OF MASS.

" Wrongly do the Greeks suppose that aught begins or ceases to be; for nothing comes into being or is destroyed; but all is an aggregation or secretion of pre-existing things; so that all becoming might more correctly be called becoming mixed, and all corruption, becoming separate."

ANAXAGORAS, cir. 450 B.C.

" Men should frequently call upon nature to render her account; that is, when they perceive that a body which was before manifest to the sense has escaped and disappeared, they should not admit or liquidate the account before it has been shown to them where the body has gone to, and into what .it has been received."

BACON, cir. 1623.

" We might as well attempt to introduce a new planet into the solar system, or to annihilate one already in existence as to create or destroy one particle of hydrogen."

DALTON, 1808.

SCIENTIFIC work of a very high order is always characterised by the great importance and value of what might be termed its side issues. In such investigations, apart from their main subject, there may be found discoveries of highly important new facts; indications of the existence of hitherto unsuspected laws; unexpected verifications and applications of principles already more or less clearly recognised. This is true in a high degree of Lavoisier's work; amongst the proofs and discoveries incidental to his researches on combustion is the experimental basis, by him supplied, for the law of conservation of mass.

The opening words of Wurtz's *History of Chemical Theory,*

Lavoisier's use of quantitative methods. "Chemistry is a French science founded by Lavoisier of immortal memory" have been made the subject of violent controversies, rendered unnecessarily bitter by national bias. The sweeping form of Wurtz's statement, and its consequent doubtful validity, are evident. But if, as is generally admitted, chemistry only took rank as a science when it made quantitative work its basis, then Lavoisier must be put before anyone else as having directed it into this road, and led it a considerable distance along it. It is not intended to convey the impression that before him no one had worked quantitatively. Black's[1] researches on magnesia alba (1752) furnish an admirable example of the successful application of quantitative methods to the solution of chemical problems. Black holds that the change from chalk to lime consists only in the withdrawal of fixed air, and he adduces in proof the changes in weight accompanying the change from chalk to lime and back again[2]:

Black's quantitative work.

"A piece of perfect quicklime, made from two drams of chalk, and which weighed one dram and eight grains, was reduced to a very fine powder, and thrown into a filtrated mixture of an ounce of a fixed alkaline salt[3] and two ounces of water. After a slight digestion, the powder, being well washed and dried, weighed one dram and fifty-eight grains[4]. It was similar in every trial to a fine powder of ordinary chalk."

But whilst some of Lavoisier's predecessors and contemporaries had worked quantitatively and whilst one of them, Cavendish[5], produced results of an accuracy which would rank high even now, and which for that time was almost miraculous, no one had like Lavoisier recognised the paramount importance of quantitative relations, nor used them to the same extent in the interpretation of the course of chemical action. Lavoisier's quantitative work

Comparison between Lavoisier's and Cavendish's quantitative work.

[1] Joseph Black (1728—1799), professor of chemistry in the Universities of Glasgow and Edinburgh. His researches on carbonic acid and its compounds with the alkalis and alkaline earths are classical, and are in conception and execution on the same lines as those followed by Lavoisier to whom they served as pattern. He was also the discoverer of latent heat.

[2] The changes referred to are:

$$\text{chalk heated} = \text{lime} + \text{fixed air}$$
$$CaCO_3 = CaO + CO_2$$
$$\text{lime} + \text{fixed alkali} = \text{chalk} + \text{caustic alkali}$$
$$CaO + K_2CO_3 = CaCO_3 + K_2O.$$

[3] Fixed alkaline salt = sodium or potassium carbonate.

[4] Since 1 dram = 60 grains, the powder weighed only 2 grains less than the original chalk.

[5] The Hon. Henry Cavendish (1731—1810), eminent alike as chemist and physicist. His contributions to chemistry comprise investigations on "inflammable air" (hydrogen), the discovery of the composition of water, and the recognition of air as a mixture of nitrogen and oxygen in constant proportions.

cannot, if accuracy alone is made the test, stand comparison with that of Cavendish, but yet in another sense how superior! Cavendish is not guided by the recognition of any general principle underlying the changes in weight which accompany chemical changes. He fires a mixture of inflammable air and common air in a closed vessel and obtains water, but does not succeed in what he admits would be "very extraordinary and curious," namely, the verification of the statement made by Priestley, that there is invariably loss of weight:

    "Though the experiment was repeated several times with different proportions of common and inflammable air, I could never perceive a loss of weight of more than one-fifth of a grain, and commonly none at all." (*Experiments on Air*, 1784.)

    Now if Lavoisier had performed this experiment, it is not unlikely that, considering the great experimental difficulties involved, and his own inferiority to Cavendish as an experimenter, he might have found a change in weight greater than $\frac{1}{5}$ grain; but it is also quite certain that he would promptly have put this loss down to experimental error, being guided in all his work and all his reasoning by the clear recognition and implicit acceptance of the principle of conservation of mass. From the very outset of his scientific career he fully believed, and acted on the belief, that in a chemical change there is neither creation nor destruction of matter; that an increase in the weight of one substance is counterbalanced by an exactly equal decrease in the weight of another, the total weight of the system comprising the two remaining absolutely unaltered; and that the weight of a compound is equal to the sum of the weights of its constituents. And here it must again be emphasized that Lavoisier was not the first to hold or promulgate these views.

Lavoisier is always guided by the recognition of the principle of the conservation of mass.

    "'Nothing is created,' said Lucretius, following Epicurus and the agnostic school ; even the alchemists never did pretend to create gold or metals, but only to transmute the fundamental and pre-existing metal." (Berthelot, *La Révolution Chimique*, 1890.)

    "It far exceeds the power of meerly naturall agents, and consequently of the fire, to produce anew so much as one atom of matter, which they can but modify and alter, not create ; which is so obvious a truth, that almost all sects of philosophers have denied the power of producing matter to second causes, and the Epicureans and some others have done the like, in reference to their gods themselves." (Boyle, *The Sceptical Chymist*, 1661.)

What Lavoisier did, was to assume this permanency of weight to apply to the substances with which chemists dealt, and to be independent of the effect of heat, till then supposed by many to be ponderable.

"Lavoisier established a radical difference between on the one hand ponderable matter,...matter of which the balance proved the invariability before, during, and after combustion ; and on the other hand,

Lavoisier recognises heat to be imponderable. the igneous fluid, of which the introduction from an outside source, or the withdrawal during combustion, neither increased nor diminished the weight of substances ; contrary to what the partisans of phlogiston had thought." (Berthelot, *La Révolution Chimique*.)

And Lavoisier's work, especially that on combustion, supplies the data which prove the validity of this view. It may be convenient to present in a tabular form (p. 62) such of

Experiments of Lavoisier in proof of the conservation of mass. these data as are derived from experiments already dealt with under another aspect, classifying them into: (A) those in which the constancy of the weight of the system comprising the reacting substances is ascertained, and (B) those in which the identity of the weight of a compound with the sum of the weights of the constituents is established. Agreement is of course only to be expected within the limits of experimental error, which itself depends on the accuracy of the measurements involved (*post*, chap. III.).

It must be realised that whilst from the very outset of his experimental career he was guided in all his reasoning by the recognition of the principle of the conservation of

Enunciation of the principle. mass, it was only when he had found this assumption amply verified that Lavoisier enunciated it formally (1785).

"Nothing is created, either in the operations of art or in those of nature, and it may be considered as a general principle that in every operation there exists an equal quantity of matter before and after the operation ; that the quality and the quantity of the constituents is the same, and that what happens is only changes, modifications. It is on this principle that is founded all the art of performing chemical experiments ; in all such must be assumed a true equality or equation between the constituents of the substances examined, and those resulting from their analysis." (*Œuvres*, I. 101.)

Here, then, is a precise statement of the principle of the conservation of mass, and the first suggestion for its deductive

Experiments of Lavoisier which show the conservation of mass in chemical change.

A. The reactions are made to occur in closed vessels which together with all their contents are weighed at the beginning and at the end of the experiments.

	Initial weight of the closed vessel and its contents				Final Weight of the closed vessel and its contents				Change
	lbs.	ozs.	gros	grains	lbs.	ozs.	gros	grains	grains
(i) The supposed change of water to earth (p. 9)	5	9	4	41·50	5	9	4	41·75	+ 0·25
(ii) The calcination of tin (1) (p. 47)		13	1	68·87		13	1	68·60	− 0·27
(2)		20	6	16·88		20	6	15·88	− 1·00

B. The weight of a compound compared with the sum of the weights of its constituents.

	Compound AB	Consti- tuent A	Consti- tuent B	Sum of the Con- stituents A + B	Difference between compounds and sum of con- stituents AB ∼ (A + B)	Difference calculated to 100 of AB
(i) The ana- lysis of mercury calx (p. 52)	45 grains red powder	41½ grains mercury	7 to 8 cubic inches of gas weigh- ing 3½ to 4 grains	45 to 45½ grains	0 to + ½ grains	0 to + 1
(ii) The syn- thesis of iron calx (p. 53)	192 grains iron calx	145·6 grains metallic iron	97 cubic inches of oxygen weighing 45·9 grains	191·5 grains	+ 0·5 grains	+ 0·25

application in the form of chemical equations. The manner and extent of the use of chemical equations is the best justification for the assertion that the whole of the modern science of chemistry is based on the principle of the conservation of mass. With every equation we write, we affirm our belief in its truth; in every analysis or synthesis by difference we assume its validity. When wishing to ascertain the composition of, say, copper oxide, that is, the ratio. between the quantities of copper and of oxygen, we find the weight of the copper and of the copper oxide[1], and we deduce the weight of the oxygen from the equation

Chemical equations.

$$\text{copper} + \text{oxygen} = \text{copper oxide}$$
$$A \quad + \quad X \quad = \quad B.$$

The establishment of the law of the conservation of mass has followed curious lines. Lavoisier did not arrive at it strictly inductively, by generalisation from a large number of cases in which the weights of the substances participating in a chemical reaction were compared with the weights of those resulting from it. The available data of chemical investigations did not supply him with the material for so doing. The belief growing amongst physicists of the imponderable nature of heat, together with the old view of the indestructibility of matter in general, must have supplied him with the basis for an assumption, from which he drew deductions that were verified by the result of experiment. And this is the method that has been adhered to. Except for some experiments on the subject of which those of Landolt described in the next chapter (P. 103) are typical, the world of science has been content to look upon the conservation of mass as an axiom, the verification of which lay in the fact that its deductions were found to agree with experiment. In all analytical work its validity is assumed, and it is proved by the fact that—within the limits of experimental error—the results of complete analysis always add up to 100 per cent. And even in the very few cases on record where *very accurate* complete

History of the establishment of the law of conservation of mass.

---

[1] This is the case whether we work synthetically, starting with a known weight of metallic copper, changing this by strong heating in a current of oxygen, or by some other method, into the oxide, and weighing the oxide so formed; or analytically, starting with a known weight of the oxide, reducing this in a current of hydrogen, and weighing the copper formed.

analyses or syntheses have been made, this has not been done to test the degree of accuracy to which the law could be proved to hold, but rather to make the degree of approximation between the actually found numbers and those required by the law a touchstone for the accuracy of the experimental work. But though not undertaken for this purpose, such complete syntheses and analyses can be used for the purpose of testing the law, and Stas[1] in his wonderful atomic weight determinations supplies us with the necessary material.

The law tested by the results of complete analyses and syntheses.

"I have considered it essential to radically alter the system of synthesis and analysis followed by all chemists. Hitherto, the syntheses and the analyses have been made by difference. This method presupposes in the case of synthesis that the weight of the element actually determined enters without loss into the compound, and further that the compound formed and weighed contains absolutely nothing but the other simple or compound substance that had been combined with the first. Similarly in analysis it presupposes that the difference actually represents the weight of the other simple or compound substance combined with the one actually weighed."

This passage makes us realise how, even in research work of a high order, certain assumptions are made, and how it may become necessary to test the validity of these in proportion as greater accuracy is required. Stas here challenges the assumption that nothing is lost, either of the constituent or of the compound actually weighed, and that nothing is added in the form of accidental impurities. He wishes to put these two assumptions to an experimental test by making complete analyses and syntheses, the test being therefore based on the implicitly made assumption of the absolute validity of the law of conservation of mass :

"Hence I came to the conclusion that in such syntheses and analyses... a method should be used in which, besides the weight of each separate element, we should actually determine the weight of the combined elements. Thus, in a synthesis from the substances A and B, we must determine the weight of A, the weight of B, and after their combination the weight of AB produced ; and likewise in analysis, to ascertain the ratio AB to C in a

[1] Stas, b. 1813 at Louvain, d. 1891 at Brussels, where most of his work was done in his private laboratory. He owes his place amongst the heroes of the science to his marvellously exact researches on chemical composition. The most important of his memoirs on this subject are: "Recherches sur les rapports réciproques des poids atomiques," and "Nouvelles recherches sur les lois des proportions chimiques." Bruxelles, *Bul. Acad. Belgique*, x. 1860; xxxv. 1865; *Œuvres complètes*, I. (pp. 308, 419).

compound ABC, we must weigh ABC, and also AB and C which are formed from it."

To Stas, on the assumption of the absolute validity of the law of conservation of mass, the smallness of the difference between the values AB and $(A + B)$, or between ABC and $(AB + C)$, becomes the measure of the purity of the material used and of the absence of mechanical loss in the process; but whenever these same numbers are quoted as a proof of the law of conservation of mass, the argument runs : " When the utmost care and skill are directed to purifying the materials employed, and to preventing any mechanical loss, as was done by Stas in a manner never attempted before and not surpassed since; and when therefore it may be supposed that no appreciable error arises from such causes : then it is found that the weight of a compound is practically that of the sum of the components from which it is formed (synthesis), or into which it is decomposed (analysis)." Thus interpreted the work becomes a direct experimental verification of the law of conservation of mass.

It remains now to show in what manner and with what results Stas carried out these experiments. Nothing short of the actual perusal of his own account in the *Nouvelles Recherches sur les Lois des Proportions Chimiques*, 1865, can convey an adequate idea of his method, of the care and ingenuity expended on the purification of all the materials used, of the elaborate precautions taken against mechanical loss, and of how at every step subsidiary investigations were undertaken to test the efficiency of all these measures, and to check the results obtained. So nothing will be attempted here beyond an account of the chemical processes involved in the methods used, and a statement of the numerical results. Nothing at all will be said about the purification of the materials, and no more than is indispensable concerning the elaborate details of the manipulations involved.

The complete syntheses made by Stas were those of silver iodide and of silver bromide. The principle of both these being identical, the synthesis of the iodide alone will be **Stas' complete** dealt with here. From previous experiments, it was **syntheses of** known that the ratio by weight in which iodine and **silver iodide.** silver combine is very nearly that of 1·17592 to 1. Quantities of most carefully and most elaborately purified iodine and silver were weighed out accurately in that ratio, in one special

F.   5

experiment 32·4665 grams of iodine and 27·6092 grams of s_{ilv}er[1]. The problem to be solved was how to bring about combination between these quantities of silver and iodine, under conditions which should minimise the possibility of any loss of either sub-stance or of the compound formed. With this object, the iodine was converted into a solution of ammonium iodide, and the silver into a solution of silver sulphate.

"[The method employed] for the conversion of a given weight of iodine into ammonium iodide consisted in the use of a titrated solution of ammonium sulphite, containing an amount of free ammonia equal to that used in the preparation of the neutral ammonium sulphite. The iodine dissolves in this liquid with evolution of heat, changing the neutral sulphite into the neutral sulphate of ammonium, and itself passing into ammonium iodide. The liquid is absolutely colourless, and in the presence of the least trace of excess of ammonium sulphite, it remains so indefinitely[2]."

[1] The numbers given here, and in all following accounts of Stas' work, are, unless otherwise specified, those of *weights in vacuo*, calculated from the numbers obtained by weighing, with platinum or brass weights, the substances contained in vessels which were either full of dry air or vacuous. Concerning the process of weighing and the application of the necessary corrections for obtaining the required weights *in vacuo*, Stas says: "Whenever the nature of the materials did not present an absolute obstacle, I carried out the weighings in air, using as counter-poise to the containing vessel another vessel of the same kind, of the same surface, and as far as possible of the same weight as that which it was intended to balance. ......I did not weigh the substances in evacuated vessels except as a check, or unless I was dealing with materials which were hygroscopic or capable of condensing air. When weighing under these conditions, I varied the procedure according to the form of the vessels which contained the substances and which had to be made vacuous······· I always counterpoised by an apparatus absolutely identical in external volume and in the nature of its surface,......and I took all the precautions ······required to carry out the vacuum weighings with certainty, and to obtain constant results······· In order to be able to reduce all my weighings to *vacuo*, I have determined the specific gravity of my platinum and brass weights, and have found that the platinum kilogram displaces 47 c.c. of air and the brass kilogram 125 c.c. These are the two data which have served for all my calculations, whether I weighed the bodies in air and applied the correction to the weights used and to the substance weighed, or whether I weighed the bodies *in vacuo* and applied the correction to the weights only."

[2] The reactions involved, and the equations representing them, are :

(i) Iodine + water + ammonium sulphite = hydriodic acid + ammonium sulphate.
$$I_2 + H_2O + (NH_4)_2SO_3 = 2HI + (NH_4)_2SO_4$$
.. The amount of hydriodic acid formed is exactly equivalent, as regards neutral-ising power, to the amount of sulphurous acid oxidised.

(ii) Hydriodic acid + ammonia = ammonium iodide.
$$2HI + 2NH_3 = 2NH_4I$$

This is the ammonia present in the ammonium sulphite solution, and which is equal in amount to that used in the formation of the neutral ammonium sulphite from sulphurous acid. It is required for the neutralisation of the hydriodic acid formed, since otherwise the reverse reaction might occur :

$$2HI + (NH_4)_2SO_4 = I_2 + H_2O + (NH_4)_2SO_3.$$

"To bring about the transformation of the silver into sulphate, I first dissolved it in dilute nitric acid, then evaporated the solution of silver nitrate to dryness, and converted the nitrate into acid sulphate, which latter was then heated with acid sulphate of ammonium in order to remove any traces of nitrogen compounds it might have retained."

The solution of ammonium iodide containing 32·4665 grams of iodine, and the solution of very acid silver sulphate containing 27·6092 grams of silver were mixed, and the silver iodide formed by their interaction was made to settle. The supernatant clear solution might contain slight excess of iodine in the form of hydriodic acid, or of silver in the form of silver sulphate. Previous experiments had shown that with iodine and silver mixed in the ratio of 1·17592 to 1, the element in excess must be iodine. To change the excess of iodine present as hydriodic acid into silver iodide, and to determine the amount of silver required for this purpose, formed the next step. A standard solution of silver sulphate, *i.e.*, a solution containing in a known volume a known weight of silver, was used, and for this purpose run into the clear liquid as long as any further turbidity was produced. In the experiment specially referred to, the volume of solution required to accomplish this contained 0·0131 grams of silver. The silver iodide resulting from the combination of all the iodine and all the silver present was then most carefully washed by decantation at a raised temperature, and after having been introduced into a suitable vessel, it was carefully dried and weighed. The weight found was 60·0860.

Stas explains why in these syntheses, contrary to his usual practice, he did not always work with quantities as large as those used in the last of the experiments summarised in the table on p. 68.

"Iodide of silver, produced in a liquid which contains much free sulphuric acid, retains this acid with extreme tenacity. But the complete removal of this acid by way of washing is an absolute necessity; otherwise, on drying the iodide, in consequence of the concentration of the sulphuric acid, iodine is liberated....I have therefore been obliged to do the washing at a high temperature, and because of the presence of traces of sulphurous acid, which, under the action of light, would slowly alter the silver iodide, this had to be done in a dark room. Further, to prevent the disintegration of the iodide and the production of a milky liquid, the clearing of which is slow and troublesome, I have been obliged to continue the washing day and night until completed, with the temperature slowly and steadily rising. But however rapidly one may perform this operation, as soon as the weight of the iodide exceeds 100 grams, forty-six to sixty hours of uninterrupted work is required in a dark room, the air of which becomes heated and damp from the vapours of the water bath. It is evident that such work soon exceeds human power, and that it is impossible to repeat it a large number of times."

It is a satisfaction that of such an experiment Stas himself could say:

"The synthesis has therefore been complete. I recovered with unexpected accuracy the weight of the elements I had made to react."

The following table embodies the results of four *complete* syntheses of silver iodide, carried out by Stas according to the above method.

Complete Syntheses of Silver Iodide.

Weight of iodine taken	Weight of silver taken	Weight of silver added for the precipitation of the iodine that had remained in the state of hydriodic acid	Sum of the weights of iodine and of silver used	Weight of the silver iodide produced	Difference between the weight of silver iodide produced, and the sum of the weights of iodine and of silver used	Difference between the weight of the compound and the sum of the weights of its constituents, calculated for 100 of the compound
32·4665	27·6092	0·0131	60·0888	60·0860	0·0028	− ·0046
46·8282	39·8223	0·0182	86·6687	86·6653	0·0034	− ·0039
44·7599	38·0620	0·0175	82·8394	82·8375	0·0019	− ·0023
160·2752	136·2952	0·0595	296·6300	296·6240	0·0060	− ·0020

The results of his complete analyses of silver iodate were no less wonderful:

"In the analysis of the iodate in terms of the weights of the oxygen and the iodide produced, I resorted to the action of heat and absorbed the oxygen liberated by means of copper heated to redness[1]. As I did not succeed in obtaining the silver salt free from traces of water, I collected and weighed the water obtained in the decomposition of the iodate analysed. In the reduction of the iodate by heat it always happens that traces of the salt itself, of silver iodide, and even of iodine, are carried away mechanically; I therefore took special precautions to protect myself from these sources of error. Pure

Stas' complete analyses of silver iodate.

---

[1] Silver iodate (heated) = Silver iodide + oxygen : $AgIO_3 = AgI + 3O$.
Copper (heated) + oxygen = copper oxide : $Cu + O = CuO$.

oxygen produces vivid incandescence in finely divided copper previously heated to dull redness; the oxide thus obtained adheres firmly to the tube, and may even cause its rupture on cooling.   To prevent such an accident I carried out the fixation of the oxygen by the copper in an atmosphere of nitrogen, which I had to lead in pure and dry.   Before the analysis I had to weigh the different pieces of apparatus empty of nitrogen, and I had to do so again after the decomposition of the iodate and the absorption of the oxygen by the copper....All these conditions combined have rendered the analysis of the iodate a most complicated and delicate operation, simple though it seems at first sight."

Apparatus used by Stas for the Complete Analysis of Silver Iodate.

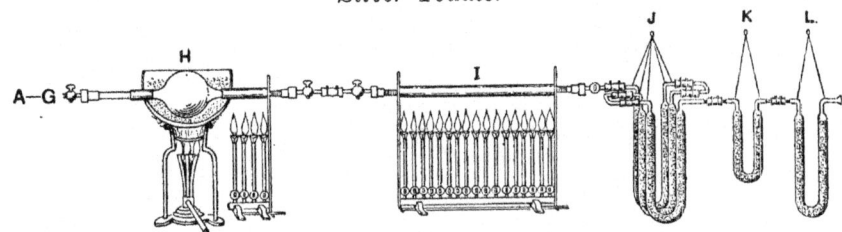

Fig. 5.

A—G (not reproduced here) comprises:

(i)   gasometer containing nitrogen,

(ii)   elaborate apparatus for freeing the nitrogen from oxygen, moisture and carbonic acid,

(iii) air-pump.

H = Two-necked bulb for the heating of the silver iodate.   The end nearest to *I* contains the different substances employed to retain the traces of silver iodate, silver iodide and iodine carried over mechanically.

I = Gas furnace in which lies the hard glass tube containing the finely divided copper destined for the absorption of the oxygen disengaged from the iodate of silver.

J = System for the condensation of the water; it is composed of 3 U-tubes filled with pumice soaked in sulphuric acid.

K = U-tube filled with pumice and sulphuric acid for testing the complete retention of the water by *J*.

L = U-tube filled with pumice and sulphuric acid to protect *K* from the moisture of the air.

The preceding reproduction of (part of) the plate accompanying Stas' memoir, together with the following tabular representation of his results, show what apparatus he employed, how he used it, and what results he obtained.

The results obtained in two experiments performed with iodate prepared from silver sulphate[1] and silver dithionate[2] respectively were:

	I	II
	Iodate prepared from silver sulphate and potassium iodate.	Iodate prepared from silver dithionate and potassium iodate.

Before the decomposition.

	grams.	grams.
(1) Weight of the bulb apparatus H, evacuated after having been filled with nitrogen ...	602·3015	928·5415
(2) „ of the same apparatus with the slightly moist iodate, evacuated after having been filled with nitrogen	700·6470	1085·4160
(3) „ of the tube I filled with copper, evacuated after having been filled with nitrogen	472·7535	472·8130
(4) „ of the system J, intended to collect the water contained in the iodate	832·5210	832·5930
(5) „ of the combined systems H, I, J ...	2005·9215	
(6) „ of the tube K	103·2068	103·2075

After the decomposition.

(7) Weight of the combined systems H, I, J ...	2005·9220	
(8) „ of the bulb apparatus H, evacuated from nitrogen	683·8945	1058·7245
(9) „ of the copper tube I, evacuated from nitrogen	489·4360	499·4230
(10) „ of the system J	832·5925	832·6720
(11) „ of the tube K	103·2072	103·2080

Hence:

	I	II
Weight of the slightly moist iodate in air $= (2) - (1)$	98·3455	156·8745
„ „ „ in vacuo [3] ...	98·3396	156·8649
„ of water contained in the iodate $=(10)-(4)$	0·0715	0·0790
„ of dry iodate in vacuo	**98·2681**	**156·7859**
„ of iodide in air $= (8) - (1)$	81·5930	130·1830
„ of iodide in vacuo	81·5880	130·1755
„ of oxygen in air $= (9) - (3)$	16·6825	26·6100
„ of oxygen in vacuo	16·6815	26·6084
Weight of silver iodide + weight of oxygen ...	**98·2695**	**156·7839**
Difference between the weight of iodate used, and the sum of the weights of its constituents ...	+0·0014	−0·0020
This difference calculated for 100 of iodate ...	**+0·0014**	**−0·0013**

---

[1] Silver sulphate + potassium iodate = silver iodate + potassium sulphate
(fairly sol. in water)   (sol.)   (insol.)   (sol.)

$$Ag_2SO_4 \;+\; 2KIO_3 \;=\; 2AgIO_3 \;+\; K_2SO_4$$

[2] Silver dithionate + potassium iodate = silver iodate + potassium dithionate
(very sol· in water)   (sol.)   (insol.)   (sol.)

$$Ag_2S_2O_6 \;+\; 2KIO_3 \;=\; 2AgIO_3 \;+\; K_2S_2O_6$$

[3] That the weight in vacuo is less than the weight in air, is accounted for by the fact that in the equation:

wt. in vacuo = wt. in air + wt. of air displaced by substance weighed − wt. of air displaced by weights used;

the second term (*i.e.* wt. of air displaced by substance weighed) is zero, the substance weighed having been contained in an exhausted tube.

A critical examination of the numbers set out in the table reveals in a variety of ways the extraordinary accuracy of Stas' work.

The differences between (11) and (6), which are ·0004 grams and ·0005 grams respectively, prove how well the system of drying tubes *J* had worked ; that is, how complete had been the retention of the water liberated from the moist iodate. The weighing of the combined systems *H, I, J*, before and after the decomposition, is an example of the manner in which Stas checked his results by different determinations of the same quantities.

"To check the accuracy of my weighings, I suspended from the balance, before and after the analysis, the three systems *H, I, J, en bloc.* Since the oxygen and the traces of water liberated in the decomposition [of the iodate contained in *H*] were retained, the one in the copper tube *I*, and the other in the tubes *J* provided for the condensation of the water, the total weight of the three systems should, after the experiment, within the limits of experimental error, be the same as before."

A counterpoise was used, consisting of a set of tubes, the external volume of which had been made identical with that of the component parts of the system *HIJ*, and it was ascertained by actual trial that when equilibrium had been established, it was maintained independent of changes in temperature and pressure of the surrounding air, and that no alteration in weight occurred consequent on the operation of filling *HIJ* with nitrogen and subsequently evacuating them.

"I left the system suspended from the balance during 120 hours, in the course of which time I made the temperature of the room vary between 5° and 28°, without any appreciable change in weight having occurred. During this interval, the barometric pressure varied between 761 and 769 mm."

The difference between 2005·9215, the weight of the whole apparatus before the decomposition, and 2005·9220, the weight after decomposition, is only ·0005, a brilliant proof of the skill of the experimenter. These numbers may be used directly for the purposes of the verification of the law of conservation of mass, and the experiment furnishing them is of the same type as some of those of Lavoisier described before (*ante,* p. 62). The proof consisted in showing that the weight of a system, within which a reaction had occurred, differed by not more than half a milligram from what its weight had been before the reaction.

But a more important demonstration of the validity of the principle of conservation of mass is supplied by the close agreement between the weights of the silver iodate and the sum of the weights of the silver iodide and oxygen; because for these it is possible to refer the differences found to the quantity of matter changed. The discrepancies between Stas' results and the requirements of the absolute validity of the principle of conservation of mass are only $+ \cdot 0014$ per cent. and $- \cdot 0013$ per cent. respectively.

Until a few years ago these results represented the highest perfection attained in measurements of this kind, and they stood out conspicuously, unapproached by any other work in the same field. But in 1895 E. W. Morley published in the *Smithsonian Contributions to Knowledge* the results of his many years' work on the gravimetric composition of water[1]. The

Morley's complete syntheses of water.

ingenuity, perseverance, and skill with which this investigation was planned, the enormous difficulties encountered and overcome, and the wonderful agreement between the results obtained by different methods, allow of its being classed with the very best of Stas' work. Amongst these experiments of Morley's is a set of complete syntheses of water, in which the hydrogen, the oxygen, and the water formed were each weighed. Of course for Morley, as before him for Stas, the object of the complete synthesis was to obtain a check for the purity of his oxygen and hydrogen, and for the completeness of the combination between them, the absolute validity of the law of conservation of mass being assumed. But, interchanging as before what is assumed and what is to be proved, if we assume sufficient purity of materials and absence of loss on combination,— assumptions which the quality of Morley's work fully justifies—we may use his results in proof of the law of conservation of mass. Here, again, it would be impossible in a small compass to give an adequate account of the details of the experiments, and of the many precautions taken and the corrections applied in the case of each of the measurements involved; the main points only can be dealt with.

Fig. 6 represents, reduced by one half the actual size, the apparatus used for the combination of the hydrogen and oxygen, and for the determination of the weight of water produced.

[1] Short abstract in: *Nature*, London, 53, 1896 (p. 428).

The hydrogen, which was led in through c, b, a, was obtained by the action of heat on palladium hydride contained in a tube. The difference between the weights of palladium hydride at the beginning and the end of the experiment gave the weight of hydrogen supplied (column 1 of table, p. 75).

Oxygen made from potassium chlorate was stored in two large glass globes, from which it was made to pass at a suitable rate through d, b, a, into the combustion chamber M. The difference between the weight of the globes at the beginning and the end of the experiment gave the weight of oxygen supplied (column 4 of the table).

The apparatus depicted was weighed vacuous, connection was then made at c and d with the hydrogen and oxygen reservoirs, one of these gases was allowed to flow in, sparks were passed at a a, and the other gas was introduced, whereby ignition was set up at a; this was continued until a suitable amount of water had been formed. The current of the gases was then stopped and combustion ceased, leaving in the apparatus the water formed, together with a residual mixture of hydrogen and oxygen, which were at too low a pressure for combustion to proceed. These gases were drawn off by a pump into a vessel suitable for their eudiometric analysis. In their exit from the combustion chamber they passed through the tubes bb, the phosphorus pentoxide in which re-

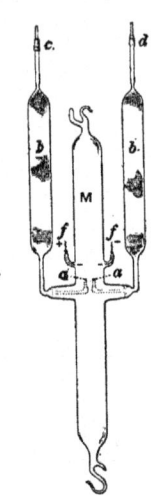

Fig. 6.

c, d, tubes through which the hydrogen and oxygen are led in.

b, b, tubes filled with phosphorus pentoxide.

a, a, small tubes in continuation of d, b, and c, b, by which the gases are led into the combustion chamber.

f, f, platinum wires for sparking and igniting the inflowing gases at either of the tubes a, a.

M, combustion chamber.

tained every trace of water. The weights of this hydrogen and oxygen which had escaped combination were determined by finding their respective volumes eudiometrically, and multiplying these by their densities. This determination involved therefore a knowledge of the ratio by volume in which hydrogen and oxygen combine, and of the densities of these gases, quantities which Morley himself had measured most accurately in the course of this investigation. Moreover, the amount of residual gas was so small that any slight error in the values used for the volume ratio and for the densities would have had but little influence on the final result. The weights of residual hydrogen and oxygen are given in columns 2 and 5. The apparatus from which the residual uncombined gases had been withdrawn was then weighed again, the difference between this and its original weight giving the weight of water formed (column 8).

The results of nine such complete syntheses are embodied in the following table. Morley's No. IV, in which experiment the apparatus broke before completion, does not contain sufficient data for the present purpose; Nos. VI and X are evidently disfigured by misprints, and therefore all three are omitted.

The last column gives the differences between the "experimental" numbers and those required according to the principle of conservation of mass. From the smallness of these numbers, and from the fact that they are sometimes positive and sometimes negative, we are justified in ascribing the discrepancy between experiment and theory to experimental error, rather than to limited validity of the principle involved.

Morley's Complete Syntheses of Water.

	1	2	3	4	5	6	7	8	9	10
	Weight of hydrogen introduced into the apparatus	Weight of hydrogen left uncombined	Hydrogen used	Weight of oxygen introduced into the apparatus	Weight of oxygen left uncombined	Oxygen used	Sum of weights of oxygen and hydrogen used	Weight of water formed	Difference between weight of water formed, and sum of weights of oxygen and hydrogen used	Difference calculated for 100 of water
I	3·2663	0·0018	3·2645	25·9299	0·0123	25·9176	29·1821	29·1788	- ·0033	- ·011
II	3·2563	0·0004	3·2559	25·8623	0·0092	25·8531	29·1090	29·1052	- ·0038	- ·013
III	3·8199	0·0006	3·8193	30·3294	0·0084	30·3210	34·1403	34·1389	- ·0014	- ·004
IV										
V	3·8392	0·0010	3·8382	30·4741	0·0041	30·4700	34·3082	34·3151	+ ·0069	+ ·020
VI										
VII	3·8299	0·0002	3·8297	30·4033	0·0020	30·4013	34·2310	34·2284	- ·0026	- ·008
VIII	3·8292	0·0006	3·8286	30·4085	0·0119	30·3966	34·2252	34·2261	+ ·0009	+ ·003
IX	3·8278	0·0053	3·8225	30·3692	0·0195	30·3497	34·1722	34·1742	+ ·0020	+ ·006
X										
XI	3·7642	0·0005	3·7637	29·8974	0·0109	29·8865	33·6502	33·6540	+ ·0038	+ ·011
XII	3·8223	0·0012	3·8211	30·3775	0·0346	30·3429	34·1640	34·1559	- ·0081	- ·024

CHAPTER III.

EXACT AND APPROXIMATE LAWS.

" Pendant longtemps les chimistes et les physiciens, dès l'instant qu'ils ont vu certains faits se reproduire avec une apparence de régularité, ont cru à l'existence d'une loi naturelle susceptible d'être exprimée par une relation mathématique simple ; de plus ils ont contracté l'habitude de considérer la loi comme démontrée du moment qu'ils avaient exécuté ou des pesées ou des mesures qui ne s'en écartaient pas trop."

STAS, 1865.

THE considerations of the last chapter should have directed our attention to the discrepancy between "experimental results" and the numbers calculated according to the quantitative law to which these results should conform. A critical examination of such discrepancies, and the referring of them to special causes, involves a previous discussion of the possible and probable accuracy of all experimental results. By the accuracy of an experimental result we mean the approximation of the value found to the true one, and we must from the outset realise clearly that the result of any quantitative experiment will never absolutely coincide with the real value of the effect measured. According to the conditions of the experiment this approximation may be more or less near.

Experimental results differ from the theoretical, being always affected by errors.

"Nothing is more certain in scientific method than that approximate coincidence alone can be expected. In the measurement of continuous quantity, perfect correspondence must be purely accidental, and should give rise to suspicion rather than to satisfaction." (Jevons, *Principles of Science.*)

The constant existence of various classes of error can be minimised, but never entirely removed.

"It is surprising to learn the number of causes of error which enter into even the simplest experiment, when we strive to attain the most rigid accuracy." (Jevons, *ibid.*)

But however large the number of possible causes of error may be; in chemical work, that is in the investigation of the properties of different kinds of matter, they can all be classified under the three heads of chemical, physical, and personal errors. Each of these classes requires separate consideration[1].

Classification of errors affecting chemical experiments according to their cause.

1. The errors arising from chemical causes, from the fact that we may not really be dealing with the amount and the kind of matter supposed to be present.

1. Errors due to chemical causes.

(a) The method employed may be faulty. It may assume a reaction to be completed when, as a matter of fact, one of the substances participating is present in excess. So, for instance, a colour change in a substance termed an indicator may be supposed to exactly mark a condition in which free acid and free alkali are absent, whilst in reality, according to the sensitiveness of the indicator, a greater or lesser amount of free acid or alkali may be required to produce the change. Or, when we estimate the quantity of an element or group of elements present by transformation into an insoluble compound of known composition, and make the amount of the compound formed the measure of the amount of the element or groups of elements present in it, the insoluble substance formed may, under the conditions of the experiment, not really have the composition assumed. For instance, barium sulphate precipitated from a solution containing much ferric iron always carries down with it some of that iron, and hence the weight of the precipitate formed is not a true measure of the amount of sulphuric acid to be estimated. An interesting example of error due to a faulty chemical method is found in Morley's criticism of Leduc's method for determining the ratio by volume in which hydrogen and oxygen unite to form water. Leduc had assumed that electrolytic gas (mixed oxygen and hydrogen obtained by the electrolysis of water) contained its constituents in

(a) The method is faulty.

---

[1] A more usual classification of errors is into those due to the method, the instruments, and the observer respectively.

the ratio in which these are combined in water, and hence that an accurate knowledge of the densities of the .mixture, the oxygen, and the hydrogen respectively, were all the data necessary for calculating the volume ratio required[1], and that the accuracy attained would depend only on that of the three density determinations. But Morley pointed out that, probably owing to secondary reactions in the electrolytic cell, electrolytic gas always contains excess of hydrogen.

No care bestowed on manipulation and on accuracy of measurement can reduce the error due to such faulty methods. Special investigations are required to detect these errors, and in order to remove them the method must be modified: *e.g.* if, previous to precipitation with barium chloride, all the ferric iron present in solution with the sulphate is reduced to the ferrous state, there will not be the same tendency for the iron to come down with the barium sulphate. Or the effect produced by the error must be measured and a suitable correction applied: *e.g.* the amount of free acid or alkali required to turn a certain amount of purple litmus to red or to blue must be ascertained; the amount of hydrogen left after explosion of the electrolytic gas must be found.

(*b*) The substances weighed may not be pure, and there may be mechanical loss in the process of the reaction investigated.

(*b*) Impurities and mechanical loss. For instance, in the quantitative synthesis by difference of silver iodide, that is, in the determination of the weight of silver iodide obtainable from a certain weight of silver, the experimental result will differ from the true one by an amount dependent on: firstly, the purity of the silver originally weighed out; secondly, the completeness with which this silver had entered into the composition of the silver iodide; and thirdly, the retention by the silver iodide of impurities derived from the substances it had come into contact with during its formation, such as sulphuric acid, water, etc. The influence on the final result of these various causes of error would of course be different. Impurities in the

[1] Let a vols. of hydrogen of density D_H combine with b vols. of oxygen of density D_o and let the density of the mixture of the two gases present in the ratio $a : b$ be D_M, then

$$a D_H + b D_o = (a + b) D_M,$$

from which it follows that

$$\frac{a}{b} = \frac{D_o - D_M}{D_M - D_H}.$$

silver would, according to their nature, make the final value higher or lower than the correct one; the presence of carbon, which under the conditions of the experiment does not form a stable iodide, would make it lower; the presence of copper, of which the combining ratio with iodine is less than that between silver and iodine, would lead to an increase in the weight of iodide formed. Mechanical loss of silver, or of silver iodide, would lower the final value; whilst the impurities in the silver iodide would raise it. The magnitude of the error due to loss of material, and to the presence of such impurities as are accumulated in the process of synthesis, depends upon the skill of the experimenter, and is likely to vary somewhat in the individual experiments. On the other hand, if the silver used is homogeneous, that is, if the impurities present in it are uniformly distributed, then the value of the error due to this cause will remain constant, however often the experiment may be repeated with the same material. Hence to detect a constant error due to impure material, the experiments must be repeated, not with the same material, but with material derived from different sources. The preparation of pure substances is a matter of extreme difficulty; and what is even more difficult is the determination of the degree of purity attained. Ordinary chemical methods fail at an early stage; very small quantities of substances, especially when very diluted, that is, mixed with a comparatively large amount of others, do not answer to their characteristic qualitative reactions. And hence the test of purity is looked for in the agreement between the results obtained with specimens of the same substance prepared by different methods. Thus, in the analysis of silver iodate described in the last chapter, Stas used two specimens of the salt, and found that:

Silver iodate from silver sulphate and potass. iodate contained 16·976 % oxygen.
    ,,   ,,   ,,   ,, dithionate   ,,   ,,   ,,   16·972 % ,,

The close agreement between these numbers leads to the inference, either that both samples of iodate had been free from an amount of impurity large enough to appreciably influence the final result, or else that both samples, though differently prepared, had contained exactly the same amount of impurities. The first alternative is the simpler, and hence the more likely to be true. And of course, the greater the number of methods used in the preparation of the materials, the greater is the probability that concordance in the results is due to the common absence of impurities, rather than to the presence of these in quantities which would produce the same amount of effect in each case.

2. The errors arising from physical causes, due to the fact that, owing to permanent or temporary physical conditions the values found for physical quantities such as weight, volume, pressure, etc. are either not the true ones, or not those actually sought.

2. Errors due to physical causes.

(*a*) The measuring instruments used may not be correct or not sufficiently sensitive. The arms of a balance may not be equal, the weights may not be true to the standard units nor bear to each other the required relations, the measuring vessels may be imperfectly graduated, the thermometer may have its fixed points marked wrong; or the mechanical construction of all these instruments may be such that small differences in the quantities to be measured escape detection: an overweight of several milligrams placed in one pan of the balance may fail to produce a shifting of the pointer, the neck of a graduated flask may be so wide that several drops of liquid added or taken away would still seem to leave the level of the liquid in the same position relatively to the mark, etc. The faults in the instruments which cause constant errors can only be detected by the special testing of these instruments, whereby the value of the error may be ascertained and used as a correction. A balance, the arms of which are of unequal length will, if sufficiently sensitive, lend itself perfectly to accurate weight determinations, provided that the ratio of the length of the arms is determined, or that the method of double weighing or of weighing by substitution is resorted to. Some extra work is thrown on to the experimenter who has to use an instrument affected by such an error, but no extra work would enable him to obtain a value correct to milligrams with a balance which will only turn with an overweight of ·01 gram. Again, if a narrow-necked measuring vessel marked 100 c.c. had on various trials been found to hold weights of water which correspond to volumes varying between 98·42 and 98·46 c.c., then it will be known and it may be permanently indicated, that its volume is 98·4 c.c. and that the process of filling it to the

(a) The instruments used are not correct, or not sufficiently sensitive.

mark will be attended with variations of probably not more than
·02 c.c.; whilst a similar very wide-necked vessel may be found
to hold from 99·71 to 100·32 c.c., which gives a mean value of
practically 100 c.c., but with the possibility of successive fillings
differing by as much as ·3 c.c. from this mean value and by as
much as ·6 c.c. from each other.

(*b*)   Changes in the physical conditions may produce temporary alterations in the value of the quantities measured.   Measuring
vessels and scales are correct for one temperature
only, and errors of several milligrams may be made
in weighing a body which is not at the exact
temperature of the balance case; moreover the influence of the
buoyancy of the air, which itself varies with the temperature,
pressure, and hygroscopic condition, is considerable.

(*b*) Influence
of conditions.

The accurately measured value of a length, or volume, or
weight in air, is true for certain physical conditions only, and
to make it serve for other occasions it is necessary to specify
the conditions to which it refers; for instance, in stating the
density of a solid or liquid substance the temperature must be
mentioned.   Hence, in making the measurements the physical
conditions must be carefully studied, and their influence on the
value sought considered.   If it is required to find a gaseous density
at a certain temperature and pressure, either the temperature and
pressure at the time of the experiment are made equal to that
required; or the density is determined at a certain other carefully
measured temperature and pressure, and the necessary corrections
are then calculated on the basis of known laws correlating change
in volume with change in temperature and pressure.

This holds quite generally: either the physical conditions are
adjusted to the standard value; or they are measured, and the
necessary corrections calculated from a knowledge of the numerical
relation between the quantity required and the conditions.   But
either course presupposes a complete knowledge of all the conditions which influence the quantity to be measured, and of the
manner of this influence, a requirement which is far from being
always satisfied.   The influence of pressure on the volume of a
gas finds its expression in Boyle's law, but it had been usual to
assume that pressure changes of not more than one atmosphere
had no appreciable influence on the volume of solids.   It was

F.   6

reserved for Lord Rayleigh[1] to point out that this assumption led to an appreciable constant error in the determination of gaseous densities by Regnault's method, in which a glass globe is weighed, first vacuous, and then full of gas, being both times counterpoised by another glass globe of exactly the same external volume as the full globe. But in the first case the external pressure of the atmosphere produces a shrinkage of the vacuous globe which can be found by direct experiment, and which amounts to between ·04 and ·016 per cent. of the volume of the globe. The effect of this shrinkage is to make the vacuous globe displace a smaller volume of air than assumed and hence to make its weight greater. Consequently, the weight of the gas, found from the difference between the weight of the globe when full and when vacuous, is smaller than the true value.

3. Errors due to the observer, and termed personal errors, which comprise two distinct kinds:

(*a*) The errors commonly termed mistakes, and

3. Errors due to the observer.

(*a*) Mistakes.

which consist in a wrong registration of the values measured. A weight may be omitted in the counting up, a scale misread. • Such errors are of course absolutely accidental and preventable, and repetition of the experiments or the checking of the results by another observer will lead to their detection, and to the rejection of any experiment in which such had occurred.

(*b*) The errors which arise from the imperfection of our sense organs, and which find their analogy in the errors due to the

(*b*) Limited sensitiveness of sense organs.

want of sensitiveness of the physical instruments used. In the recognition of colour changes, adjustment to marks, reading of scales, etc. etc. only a limited degree of certainty and accuracy can be attained. Training helps, and the perception can be assisted by the creation of special conditions. Practice will enable us to recognise colour changes so faint that at first they would have escaped detection; the change from purple to red in litmus will be much more sharp and sudden if viewed in the monochromatic light of a sodium flame, when the red solution appears perfectly colourless, while the blue or violet looks like a mixture of black ink and water.

[1] London, *Proc. R. Soc.* 43, 1888 (p. 361).

Certain of these personal errors arising from limitation in the perceptive power of our senses may, though varying slightly in amount, be in one direction only. In the fixing of the exact juncture at which a colour has appeared, the perception is sure to lag behind the effect, in the case of disappearance to forestall it; some people always read a scale persistently higher or persistently lower than do the majority. But with other personal errors it is not so; a scale reading which by accurate measurement with a telescope had been found to be 30·237, may, by ordinary reading in several repetitions be recorded as 30·23, 30·24, 30·25, that is, sometimes too high, sometimes too low.

The classification of errors is in itself a matter of no great importance, and in whatever way done, it must to a certain extent be arbitrary. The important point which should have been brought out by the classification attempted above and by the examples there given, is the fact that the number of possible errors is very great, and that they differ in their influence on the final result. Whilst some, termed *constant errors*, always affect the result in the same direction, make the value obtained always larger or always smaller than it should be, others, termed *accidental errors*, have no such constant effect, but are characterised by sometimes raising, sometimes depressing the true value.

Classification of errors according to their effect on the result, Constant and accidental errors.

Constant errors cannot be detected by simple repetition of the measurements; change of material, of method, of instrument, and even of observer, is required to accomplish this. The presence of a constant error may be proved without its cause being detected; but once the cause is known, steps can generally be taken to remove it. Of accidental errors, on the other hand, it is assumed that if the measurement made is repeated a sufficient number of times, it is as likely that the value obtained will be too great as that it will be too small; that therefore the number and magnitude of the deviations in both directions will be the same; and hence that the mean value will come nearer to the true one than the individual ones.

"If several results have been obtained which differ slightly owing to errors of observation, it is generally assumed that the most advantageous and most probable approximation to the true value can be found by taking a mean of the results. In scientific language then, the word *mean* signifies a value derived from a series

The arithmetical mean.

of observational results, and intermediate between them, which is believed to express the most probable and the most advantageous value of the quantity measured." (Lupton, *Notes on Observations*, XIV.)

If we assume that any one observation is of value equal to any other and that the number and magnitude of the positive errors is equal to that of the negative errors, we may add all the results and divide by their number, whereby we obtain a value termed the "arithmetical mean" or "*the* mean." With n measurements of the same quantity, giving the values $a_1, a_2, a_3, a_4 \ldots a_n$, the arithmetical mean m is given by

$$m = \frac{a_1 + a_2 + a_3 + a_4 + \ldots + a_n}{n} = \frac{\Sigma a}{n}.$$

It is essential to realise and to remember that when we look upon the arithmetical mean as representing the most probable and the most advantageous value of the quantity measured, we do so on the assumption of elimination or at any rate reduction of the accidental or variable errors, constant errors being of course in no way affected by this process of taking a mean. To find such a mean is so usual and simple an operation that a special example may be dispensed with.

"As we are not justified in assuming that the mean of a series of observational results is exactly equal to the actual quantity measured, we cannot in strictness call the difference between the mean adopted and any observational results the error of the observation ; the word residual is generally used to express this difference. Of course, if the mean is supposed to be equal to the value, the residuals become equal to the errors." (Lupton, *ibid.*)

Error of the observation.

So far any one of the observational results has been considered as reliable as any other, that is, equal values have been assigned to them. In the case of measurements made with specimens of the same material, by the same method, with the same instruments and by the same observer, this is a legitimate assumption ; but alter any one of the above conditions, and more likely than not, the accuracy of the measurement will have been influenced in one sense or another. And it becomes necessary to evaluate the relative reliability of the different measurements made, or as it is called, to assign a definite weight to each of them. The process of "weighting" the different results and of

Weighting experiments, and calculation of the general mean.

combining them into a final result, termed the "general mean," can be thus described:

"The opinion of the observer as to the value of different observations is expressed by multiplying each result by a number supposed to represent its relative weight; this is equivalent to assuming each observation to be repeated a number of times in proportion to its supposed accuracy. The fictitious results thus arrived at are dealt with just as though they were real and the 'general mean' is obtained by multiplying each observation by its weight and dividing the sum by the sum of these weights." (Lupton, *ibid.*)

$$M = \text{General Mean} = \frac{p_1 a_1 + p_2 a_2 + p_3 a_3 + \dots p_n a_n}{p_1 + p_2 + p_3 + \dots p_n},$$

where $p_1, p_2, p_3 \dots p_n$ are the weights assigned to the individual observations or to the arithmetical means of different series of measurements of the same quantity.

Empirical weighting, that is, weighting according to individual judgment, must always be a matter of great difficulty, and at best its results are open to some objections. Examples of a not uncommon method of weighting are found in the cases when the required ratio of the reacting quantities—whether volumes or weights—of two substances is not calculated for each separate observation and the results then combined into an arithmetical mean, but when the calculation is performed on the sums of the quantities of each of the two substances measured in the different experiments. For instance, let the case investigated be that of the interaction between solutions of an acid and of an alkali, and let the observational results be:

10 c.c. of the solution of the acid are neutralised by 12·50 c.c. of the solution of the alkali; ∴ 1 c.c. of the acid = 1·250 of the alkali.

25 c.c. of the solution of the acid are neutralised by 31·00 c.c. of the solution of the alkali; ∴ 1 c.c. of the acid = 1·240 of the alkali.

35 c.c. of the solution of the acid are neutralised by 43·35 c.c. of the solution of the alkali; ∴ 1 c.c. of the acid = 1·238 of the alkali.

$$\text{Arithmetical Mean} = 1\cdot2 + \frac{\cdot050 + \cdot040 + \cdot038}{3} = 1\cdot243$$

$$\text{General Mean} = \frac{12\cdot50 + 31\cdot00 + 43\cdot35}{10 + 25 + 35} = 1\cdot241.$$

The assumption made in such a plan of weighting is that an experiment performed with 25 c.c. is $2\frac{1}{2}$ times as good as one made with 10 c.c.; an experiment made with n grams, n times as good as one made with one gram of material. The discussion of the legitimacy of such weighting is beyond the scope of this book[1].

When all that is possible has been done to remove constant errors, and to reduce the accidental errors by letting them counteract one another, the result obtained will still be an approximate one only, differing by the algebraic sum of all the partial errors from the true value; and what that true value is we have no means of ascertaining.

The manner in which experimental results are influenced by the different kinds of errors that can be operative, and the recognition of the inevitably approximate nature of all experimental numbers should now enable us to separate that which is experimental error, that is, discrepancy between our measurements and the actual values of the quantities investigated, from actual anomalies in these values; to find answers to certain questions of the highest importance in scientific investigations, which are:

Firstly: Given the numerical results of the repeated measurement of a certain quantity, how can we interpret the discrepancies between the individual numbers or sets of numbers, and what inferences can we draw concerning the nature of the errors by which they are affected?

Secondly: Returning to what formed the starting point of all the considerations so far passed in review in this chapter, how are we to interpret the differences between experimental and theoretical numbers, which latter represent the requirements of the law supposed to describe accurately the relations between the quantities measured?

1. It would obviously be impossible to follow the process and the result of a critical examination of numbers without actual examples, and the experimental results obtained in some recent important investigations will be used to illustrate how an answer

I. The detection of the presence or absence of constant errors.

---

[1] The subject of "Means," "Weights of Observations," "Mean and Probable Errors," is dealt with to the amount required by the ordinary student of Chemistry in: Lupton, *Notes on Observations*, chaps. XIV. XV. XVI.—Kohlrausch, *Physical Measurements*, 1894. 1. Errors of Observations—mean and probable Error. 2. Influence of Errors of Observation on the Result.—Ostwald, *Physico-Chemical Measurements*, 1902, chap. I. Calculation.—Clarke, *A Recalculation of the Atomic Weights*, 1897, pp. 3–5, 7–8.

may be found to the first of the questions posed above. The importance and interest of the subject may, it is hoped, somewhat mitigate the inevitable dryness of strings of numbers.

(*a*) Morley, in his determination of the gravimetric composition of water already referred to, deduced the required ratio in one set of experiments from the densities of oxygen and of hydrogen, and from the ratio by volume in which these gases are combined in water. Each of these three quantities was measured most accurately, and in most cases by several different methods.

(*a*) In Morley's determination of the density of oxygen derived from different sources.

The density of oxygen was found in three different ways, and the results obtained by one of these methods (that described in the paper as the third) will now be examined. Details cannot here be given of all the elaborate precautions taken to ensure the greatest possible accuracy in each of the measurements made, and to remove possible sources of error. The principle of this density determination was the same as that of Regnault's classical work on the subject. A large glass globe holding from 8 to 21 litres, the volume of which had been determined most accurately, was filled with oxygen at the temperature of melting ice, and at a pressure P measured by a barometer. The weight W of the globe so filled was ascertained, and in quick succession to it W_0, that of the same globe made vacuous. $W - W_0$ will therefore represent the weight of a volume V of oxygen at 0° and pressure P, from which the normal density, *i.e.* the weight of 1 litre at 0° and 760 mm. pressure, was calculated. The oxygen was obtained by two different processes, (i) by the action of heat on potassium chlorate, and (ii) by the electrolysis of pure dilute sulphuric acid. The gas was in each case subjected to a most elaborate process of purification and drying. The results obtained are given in the following table:

Morley's Determinations of the Density of Oxygen.

Using Ice and a Barometer.

	Volume of the Globe used = V	Pressure = P	Weight of the Oxygen = $W - W_0$	Density of the Oxygen at 0° and 760 mm. in grams per litre D
I Oxygen from Potassium Chlorate	8793·9 c.cm.	727·04 mm.	12·0179 g.	1·42920 = Mean + ·00002
	8832·1 ,,	746·93 ,,	12·3951 ,,	1·42860 = ,, − ·00058
	8793·9 ,,	769·76 ,,	12·7227 ,,	1·42906 = ,, − ·00012
	8832·1 ,,	773·22 ,,	12·8400 ,,	1·42957 = ,, + ·00039
	8832·1 ,,	772·22 ,,	12·8192 ,,	1·42910 = ,, − ·00008
	8793·9 ,,	778·68 ,,	12·8745 ,,	1·42930 = ,, + ·00012
	8793·9 ,,	778·04 ,,	12·8628 ,,	1·42945 = ,, + ·00027
				1·42918 = Mean
II Electrolytic Oxygen	8832·1 c.cm.	774·39 mm.	12·8572 g.	1·42932 = Mean + ·00024
	8793·9 ,,	750·12 ,,	12·3983 ,,	1·42908 = ,, ± ·00000
	8832·1 ,,	769·83 ,,	12·7796 ,,	1·42910 = ,, + ·00002
	16517·2 ,,	765·35 ,,	23·7671 ,,	1·42951 = ,, + ·00043
	20057·6 ,,	761·52 ,,	28·7134 ,,	1·42933 = ,, + ·00025
	15081·7 ,,	774·98 ,,	21·9675 ,,	1·42905 = ,, − ·00003
	21557·8 ,,	772·55 ,,	31·3039 ,,	1·42914 = ,, + ·00006
	15383·4 ,,	747·88 ,,	21·6150 ,,	1·42849 = ,, − ·00059
	15383·4 ,,	754·99 ,,	21·8274 ,,	1·42894 = ,, − ·00014
	15383·4 ,,	763·80 ,,	22·0808 ,,	1·42886 = ,, − ·00022
				1·42908 = Mean

Difference between the mean values for "Potassium Chlorate" and "Electrolytic" Oxygen = ·00010 = ·007 per cent. (nearly).

An examination of this table shows: that the difference between the mean values of the two sets of data is ·00010; that in the chlorate set the individual values differ in one case by as much as ·00058 from the mean value for that set, and on an average by about ·00022; and that in the electrolysis set the greatest deviation from the mean value is ·00059, the average one ·00020. Hence the difference between the mean values obtained by an otherwise identical method for differently prepared specimens of gas, is only half that of the average difference between the mean value of a set and the individual determinations comprised in this mean. And the inference must therefore be drawn that the discrepancies observed are only due to experimental errors; that though both sets of data may be affected by constant equal errors due to the method, there cannot be present any constant error of appreciable magnitude due to impurity in the materials used, since except in the very unlikely event of this impurity causing in both cases the same amount of error, it would have produced a constant difference between the two sets of data, such as is not found.

(*b*) Lord Rayleigh's determinations of the density of nitrogen[1] have been already referred to in connection with the discovery of argon (*ante*, p. 14). The method was essentially the same as that used by Morley and just described. One particular globe was used throughout, and the numbers given in the table on p. 90 are not the weights of 1 litre of nitrogen but those of the volume of gas contained in the special globe at 0° C. and a special pressure defined by the manometer.

(b) In Lord Rayleigh's determination of the density of nitrogen derived from different sources.

An examination of this table shows good agreement between the individual values obtained with specimens of nitrogen prepared in the same way. This agreement is least good in the nitric oxide set, where the greatest deviation from the mean value is ·00192 and the average deviation ·00152. In the other five sets, which in this respect are comparable, the greatest deviation from the mean value of a set is ·00036, the average one ·00015. Turning to the comparison with each other of the different mean values obtained, it becomes apparent that the three means for chemically prepared nitrogen show very good agreement, the greatest difference being one of ·00139 between the nitric oxide and the ammonium nitrite series, which is quite comparable with the differences in the nitric oxide series itself. Or omitting the nitric oxide series, which, because of the much greater discrepancy between its individual values deserves less reliance, the two means differ from each other by not more than ·00035, a quantity just about the same as the average deviations from the means of the individual experiments constituting the sets. And similarly the three means for the nitrogen prepared in various ways from

[1] Rayleigh, "The Density of Nitrogen Gas," *Nature*, London, 50, 1894 (p. 157).

Lord Rayleigh's Determinations of the Density of Nitrogen.

I		
Nitrogen obtained from Chemical Compounds		
1 From Nitric Oxide by hot Iron	**2** From Nitrous Oxide by hot Iron	**3** From Ammonium Nitrite by hot Iron
2·30143 = Mean + ·00135	2·29869 = Mean − ·00035	2·29849 = Mean − ·00020
2·29890 = ,, − ·00118	2·29940 = ,, + ·00036	2·29889 = . ,, + ·00020
2·29816 = ,, − ·00192		
2·30182 = ,, + ·00174		
2·30008 = Mean	**2·29904 = Mean**	**2·29869 = Mean**
Mean of all the experiments under I = **2·29927**		
II		
Nitrogen obtained from the Air		
The Oxygen withdrawn by		
1 Hot Copper	**2** Hot Iron	**3** Ferrous Hydrate
2·31026	2·31017 = Mean + ·00014	2·31024 = Mean + ·00003
	2·30986 = ,, − ·00017	2·31010 = ,, − ·00011
	2·31003 = ,, ± ·00000	2·31028 = ,, + ·00007
	2·31007 = ,, + ·00004	
2·31026 = Mean	**2·31003 = Mean**	**2·31021 = Mean**
Mean of all the experiments under II = **2·31016**		
Difference between the mean values for "Chemical" and for "Atmospheric" Nitrogen = **·01089** = **·5** per cent. (nearly).		

atmospheric air show excellent agreement, the greatest difference being one of ·0023 between the hot copper and the hot iron experiment, a difference again comparable with that between the individual values in a set and the mean value of that set.

But a comparison between 2·29927, the mean value obtained by the combination of all chemical nitrogen data, and 2·31016, the mean value of all the atmospheric nitrogen data, reveals a difference of ·01089, which is about sixty times greater than ·00015 the average difference between the individual numbers in a set and the mean value for that set, or than ·00022 the average difference between the various sets belonging to the chemical or the atmospheric series respectively. Here then, differences of two entirely distinct orders of magnitude have to be accounted for. Experimental error is assumed to be the cause of the comparably small differences between the individual results in each set. Practical identity of the substances dealt with, and hence, for the reason already several times given, absence of any appreciable amount of accidentally present impurity, must be assumed in the case of the different samples of chemical and of atmospheric nitrogen respectively. But whilst all the differently prepared samples of chemical nitrogen must, because of the concordance in the value of their density, be looked upon as chemically identical, and whilst for the same reason all the samples of differently prepared atmospheric nitrogen must also be chemically identical; chemical and atmospheric nitrogen, the densities of which, when determined by the same method, differ from each other by an amount more than sixty times as great as that of the average difference due to experimental error, must be chemically different.

Here then the interpretation of the results of very accurate measurements leads to the recognition that two substances initially supposed to be the same are really chemically different. It has been described before in what manner the search for this difference was conducted, and how the discovery of Argon was its result.

2. In dealing with the interpretation of the differences between the experimental and the theoretical numbers which represent the

2. The interpretation of the differences between experimental and theoretical numbers.

relations between two quantities supposed to vary according to a certain law, we are confronted with the question : are these discrepancies due simply to experimental error, which makes the result obtained differ from the true value of the quantities measured, this true value being identical with the theoretical one ; or does the true value differ from the theoretical one ? The discussion of an actual case is required in order to illustrate how a decision is made between these two explanations.

The law which expresses the relation between the volume of a gas and its pressure, and which is variously known as Boyle's

(a) Boyle's law

$\frac{P_0 V_0}{P_1 V_1} = 1.$

law (1660) or Mariotte's law (1679) states that the volume varies inversely as the pressure, or that pressure × volume = constant.

(i) Dulong and Arago's results suggest, but do not prove, the existence of deviations.

The following table embodies measurements of Dulong and Arago, made in 1827 :

Dulong and Arago's Table of the Compressibility of Air.

Pressure in mm. mercury	Observed volume	Calculated volume	Difference
760·00	501·3		
3612·48	105·247	105·470	0·223
3757·18	101·216	101·412	0·196
4625·18	82·286	82·380	0·094
5000·78	76·095	76·193	0·098
5737·38	66·216	66·417	0·201
8596·24	44·308	44·325	0·017
9992·36	37·851	38·132	0·281
12620·00	30·119	30·192	0·073
13245·06	28·664	28·770	0·106
14667·36	25·885	25·978	0·093
16534·9	22·968	23·044	0·076
16584·4	22·879	22·972	0·093
18438·5	20·547	20·665	0·118
20236·66	18·833	18·872	0·039
20498·68	18·525	18·588	0·063

"On comparison, the observed and the calculated numbers are found to be very nearly the same, from which must be inferred that the real compressibility of air differs very little, if at all, from that calculated according to Boyle's law. But more than this must not be inferred, since the differences are not equal to zero, and since the observed volumes are always smaller than the calculated volumes. The cause for the discrepancy may be found in the law not being absolutely valid, or in want of accuracy in the measurements. But the kind of the deviation observed points rather to the first of these possibilities....It must be remembered that we are never able to make absolutely exact measurements; if the deviations between the observed values and those calculated according to the presumed law are very small, the assumption is justified that these differences would be equal to zero if the measurements were perfect, and hence the exact validity of the law may be assumed.   But in such a case the difference between observation and calculation will be sometimes positive, sometimes negative, since if the method is correct, it is equally probable that the inevitable errors in the observation would make the result sometimes higher, sometimes lower.   Constant deviations (*i.e.* in one direction only), however small these may be, lead to the supposition that the method employed is affected by a constant source of error, or that the law is not exact." (Wüllner, *Lehrbuch der Experimental Physik*, I.)

It remained therefore for future work to decide whether these constantly negative deviations from the law found by Dulong and Arago were due to some constant error or to a real discrepancy between the actual occurrences and the formula chosen to represent them.   Repetition of the experiments by different observers working under different conditions, employing different apparatus, and using different methods of measurement, together with a critical study of all circumstances which through the simultaneous action of some other law of nature could exert a disturbing influence on the quantity measured[1], constitutes the only way known to us of searching for a constant error.   No such error has been discovered in the experiments on the compressibility of air.

Regnault (1810—1878), a great French physicist and no mean chemist, specially famous for his experimental researches on heat, marvellously ingenious at devising experimental methods in which important sources of error were done away with and the accuracy of the measurements involved pushed far beyond its then limits, in 1846 took up the investigation of the compressibility of gases, and carried it out with all his characteristic skill.   Regnault's method consisted in accurately measuring a volume V_0 of a gas and the corresponding pressure P_0, then compressing the gas to a new volume V_1 (generally about half the original) and finding the new corresponding pressure P_1.   By the law $\frac{V_0}{V_1}$ should be equal to $\frac{P_1}{P_0}$.   The actually obtained numbers for air are given in the following table :

(ii) Regnault's results prove the existence of deviations for pressures below 14 atmospheres.

---

[1] The influence of change of temperature on the volume was known to be great, and hence the necessary care would have been bestowed on keeping the temperature constant.   But *a priori* it was not impossible that other physical conditions, such as, say, the nature of the walls of the containing vessel, might also have some influence on the volume of the gas.

Volumes V_0 and V_1	Corresponding pressures P_0 and P_1	Temperature °C.	$\dfrac{V_0}{V_1}$	$\dfrac{P_1}{P_0}$	$\left(\dfrac{V_0}{V_1}-\dfrac{P_1}{P_0}\right)$
1939·69 and 969·26	738·72 and 1476·25	4·44	2·001215	1·998389	·002826
1939·69 and 969·86	738·99 and 1475·82	4·40	1·999990	1·997076	·002914
1939·47 969·39	739·19 1476·80	4·43	2·000701	1·997863	·002838

In the last column are given the differences between the experimental and the theoretical numbers. This particular research enables us to judge of the accuracy attained. The results of the above three experiments which practically amount to a repetition of the same measurement, the quantities of air and the pressures employed having been nearly the same, show very good agreement. The differences between the individual measurements (the greatest of which is ·000088) are less than $\frac{1}{30}$ of the deviation from the law, which latter therefore certainly cannot be due to experimental error in the measurements. It must consequently be inferred that air is very slightly, but quite distinctly more compressible than it should be if it strictly obeyed Boyle's law. Regnault found the same for other gases, and the following table embodies results of some experiments in which different gases and greater pressures were used.

The data for each gas are arranged in two columns, the first of which contains the initial pressure P_0, the second the ratio $\left(\dfrac{V_0 . P_0}{V_1 . P_1}\right)$ which according to the requirements of the law should be unity. The deviations of the numbers of this column from unity represent therefore the magnitude of the deviations from the requirements of the law. A number greater than unity goes with a gas more compressible than it should be theoretically (*i.e.* V_1 is smaller than it should be, or what comes to the same, P_0 is greater), and a number smaller than unity goes with a gas not sufficiently compressible. V_1 was always made nearly exactly half of V_0, and P_1 is the pressure measured when the gas of initial volume V_0 and pressure P_0 was reduced to V_1.

Regnault's Values for the Compressibility of Gases.

Air		Nitrogen		Carbonic Acid		Hydrogen	
P_0	$\left(\dfrac{V_0 . P_0}{V_1 . P_1}\right)$	P_0	$\left(\dfrac{V_0 . P_0}{V_1 . P_1}\right)$	P_0	$\left(\dfrac{V_0 . P_0}{V_1 . P_1}\right)$	P_0	$\left(\dfrac{V_0 . P_0}{V_1 . P_1}\right)$
738·72	1·001414	753·62	1·000788	764·03	1·007597	—	—–
2112·53	1·002765	1159·26	1·000996	1414·77	1·012313	—	—–
4140·82	1·003090	2159·60	1·001381	2164·81	1·018973	2211·18	0·998584
4219·22	1·003495	3030·22	1·001955	3186·13	1·028494	3989·47	0·996961
6770·15	1·004286	4953·92	1·002860	4879·77	1·045625	5845·18	0·996121
9336·41	1·006366	5957·96	1·003277	6820·22	1·066137	7074·96	0·994697
		7297·06	1·003924	8393·68	1·084278	—	
		8628·54	1·004768	9620·06	1·099830	9175·25	0·993126
		9775·38	1·004881			10361·88	0·992327
		10981·42	1·006456				

Here we see the fact brought out, that air, nitrogen and carbonic acid are more compressible than they should be according to theory, whilst hydrogen is less so ; that the deviations from the law increase with increasing pressure; and that carbonic acid, of which we know that it can fairly easily be condensed to a liquid, shows much greater deviations than do the other gases.

Further experiments on the same subject, and the extension of the measurements into the domain of very high pressures, have confirmed and extended these results; extended, in so far as they have brought out the fact that hydrogen does not stand alone in having a compressibility lesser than the theoretical, *i.e.* a product PV greater than that required by the law, but that at high pressures all gases behave likewise. Hence in the case of gases for which PV is less than it should be by theory, this value must reach some minimum, after which it increases continuously. (Natterer, 1855 ; Cailletet, 1879 ; Amagat, 1880.)

(iii) **At high pressures, the compressibility of all gases is less than the theoretical.**

Cailletet's values for the compressibility of air at high pressures.

Pressure in metres of mercury = P	Volume = V	VP	Difference from minimum of VP
39·359	207·93	8184	284
44·264	184·20	8153	253
49·271	162·82	8022	122
59·462	132·86	7900	—
64·366	123·53	7951	51
69·367	115·50	8011	111
79·234	103·00	8162	262
84·388	97·97	8267	367
99·188	86·06	8536	636
114·119	76·69	8751	851
144·241	62·16	8966	1066
164·145	54·97	9023	1123
174·100	52·79	9191	1291
181·985	51·27	9330	1430

From these numbers it would appear that at a pressure of about 60 metres of mercury (79 atmospheres) air has reached a maximum of compressibility, that with further increasing pressure it behaves like hydrogen, and that up to the limit investigated the compressibility continually decreases. Similar experiments made with other gases show that their general behaviour in this respect is alike, and that the deviation from the minimum value of PV is specific and characteristic of each gas.

Evidently then the simple formula $PV = \text{constant}$ does not represent the actual occurrences. For gases like oxygen, nitrogen,

Boyle's law is therefore proved to be not an exact, but an approximate or ideal law.

and hydrogen, which are far removed from the temperature of condensation to a liquid, and at pressures below three atmospheres, it does so sufficiently closely to justify its being unhesitatingly used for practical calculations. But of course, when it may be so used and when not, depends altogether on the degree of accuracy aimed at, or rather on that attained in the rest of the work. So Morley, in his determinations of the densities of oxygen and hydrogen (*ante*, p. 87), in calculating the volume at one pressure from that at another pressure had to do so according to a more exact formula than that of Boyle. How is such a formula obtained? Regnault had already substituted for the simple formula $V_0 P_0 = V_1 P_1$ the more complex expression

$$V_0 P_0 = V_1 P_1 \{1 - A (P_1 - P_0) - B (P_1 - P_0)^2\}$$

which he had obtained in the usual manner by interpolation from his numerical results, and in which A and B were constants characteristic of the different gases. Within the limits of Regnault's measurements this formula represented the experimental results satisfactorily, but of course it would fail completely in the realm of the high pressures where the deviations from the law are reversed.

More success might be attained by approaching the subject from the theoretical side, and by investigating how the hypothesis which accounts for the simple law stands with regard to the observed deviations from theory. Obviously the hypothesis, if correct, should in its original or in a slightly modified form, account not only for a more complex and quite general empirical formula, but also for the fact that, under certain conditions, the formula representing the occurrences closely approximates to the simple form of Boyle's law. These requirements

are fulfilled by the kinetic hypothesis already referred to (Introd., p. 29), which gives a satisfactory explanation of all the simple gaseous laws. It was said then that the kinetic hypothesis assumed gaseous pressure to be due to the impact on the walls of the containing vessel of the constituent particles, which are supposed to be perfectly elastic, to be moving in straight lines subject to the laws of dynamics until they collide with each other or impinge against the walls of the containing vessel, to occupy a volume which when compared with that of the gas as a whole is negligible, and to be at such distance apart as to exert no attractive force on each other.

A simple argument, given so long ago as 1738 by Daniel Bernouilli, the originator of this kinetic hypothesis,

shows how the law of the direct proportionality between the density and the pressure of a gas follows as a necessary result from the hypothesis:

"If...the gas consists of a large number of moving particles, and the pressure exerted by it on the walls of the vessel arises from the impacts of the particles against these walls, then...if the gas is compressed and the volume diminished, the number of impacts of the now more closely packed particles against the walls increases, and for two reasons: first, there is a larger number of particles in the layer of gas immediately adjoining the walls; and, secondly, as the particles are more crowded together, they collide oftener, and, hurled back by the collision, are oftener flung against the walls. If, by the compression of the gas, the volume is diminished in the ratio $1 : s^3$, the distance between any pair of particles is diminished in the ratio $1 : s$; the number of particles, therefore, in the bounding layer, which is in contact with a given area of the walls, is increased in the ratio $s^2 : 1$; further, the number of collisions that take place between the molecules in a given time is increased in the ratio $s : 1$; and in this same ratio also is the number of impacts of any particle in the bounding layer against the walls increased. Since, then, the number of impinging particles is increased in the ratio $s^2 : 1$, and the number of impacts by each in the ratio $s : 1$, the number of impacts against a given part of the walls in a given time is increased in the ratio $s^3 : 1$, which is the inverse of the ratio in which the volume of the gas is diminished. *The pressure, therefore, of a gas varies inversely as its volume.*

Boyle's law is thus deduced from the hypothesis of molecular impacts." (O. E. Meyer, *Kinetic Theory of Gases*, 1899.)

This inference from the simple kinetic hypothesis, which takes the form $PV = $ constant, agrees, however, only approximately with the results of experiment. But some consideration shows that by its nature the kinetic hypothesis should lead to the expectation of such deviations.

Theoretical explanation of the observed deviations from Boyle's law.

"In the assumptions with which we started there are two different points which cannot be directly proved, and are therefore open to doubt. The first is the assumption that gases are made up of molecules of very small dimensions, and the second is the assumption that in gases there is no cohesion. Neither of these is exactly true, and therefore neither can be admitted except as an approximation to the truth; and in their inexactness lies ample ground for the departures from Boyle's law.

In the first place, if the dimensions of the molecules are not indefinitely small, the calculation which led to the law is not exact. For it is only if the space actually occupied by the molecules is absolutely negligible in respect of the volume which contains them that we may justifiably conclude that the frequency of collision is increased in the ratio $s : 1$ by a diminution of the volume in the ratio $1 : s^3$. If this condition is not fulfilled there is less actual distance between the molecules[1], which, therefore, collide the oftener with

[1] AC is the space supposed to be traversed in the simple hypothesis, whilst it really is only BC; and similarly for the return journey.

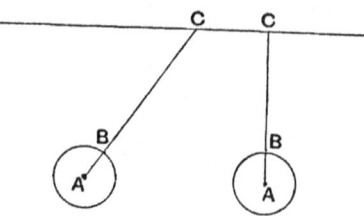

each other, and in the same ratio impinge the oftener against the walls of the vessel—in other words, the pressure is greater than according to the former calculation; and as this increment in the pressure is the more considerable the less the volume, the pressure must increase at a greater rate than the volume diminishes. The denominator $P_1 V_1$ of the ratio $[P_0 V_0/P_1 V_1]$, wherein P_1 denotes the higher pressure, is on this account greater than the numerator $P_0 V_0$, so that the ratio $P_0 V_0/P_1 V_1$ has, as actually happens with hydrogen, a value less than 1.

A deviation in the reverse direction occurs when the second hypothesis is sensibly in fault and the gas has marked cohesion. For such a property will tend to lessen the volume, which will, therefore, on this ground diminish more rapidly than the pressure increases; $P_1 V_1$ will thus be smaller than $P_0 V_0$ and the ratio $P_0 V_0/P_1 V_1$ greater than 1, as is the case with all the gases in [Regnault's] table (p. 95), except hydrogen.

Probably both influences occur in nature, and the numbers in Regnault's table seem to show that in the case of most gases the influence of cohesion is predominant so long as the pressure lies within certain limits. But when higher pressures are employed all gases exhibit, according to the observations of Natterer, Amagat, and Cailletet, the same behaviour as under lower pressure is noted with hydrogen. The product PV increases with the pressure P, because on account of the dimensions of the molecules the volume V cannot diminish so much as the law requires.

These considerations show that the departures from the strict law can also be explained by the theory." (O. E. Meyer, *ibid.*)

Van der Waals has deduced a formula in which for V, the volume occupied by the gas as a whole, he substitutes $(v - b)$,

Van der Waals' formula for the relation between volume and pressure in gases.

where b is a correction for the space occupied by the gaseous particles themselves; and for P, the pressure really exerted by the gas, he substitutes $\left(p + \dfrac{a}{v^2} \right)$, where $\dfrac{a}{v^2}$ is a correction for the counteracting effect of the attractive forces between the constituent particles. The formula thus corrected becomes $\left(p + \dfrac{a}{v^2} \right)(v - b) = \text{constant}$, where a and b are constants, the values of which depend on the special nature of the gas[1].

Van der Waals' formula well represents the relations between the volume and the pressure of a gas within the limits to which measurements have yet been pushed; it represents Regnault's

[1] Ostwald, *Lehrbuch*, I. p. 226, describes how these constants are found experimentally, and on page 223 an account is given of Van der Waals' reason for making the term expressing the molecular attraction inversely proportional to the square of the volume occupied by the gas.

numbers no less than Cailletet's, and it is easy to foretell from it the general course of the deviations from the simple formula of Boyle's law. When v is very large, $\frac{a}{v^2}$ will be very small and can be neglected, whilst $(v-b)$ will become equal to v, and the whole formula merges into $PV = $ constant, the expression for Boyle's law; Regnault's table shows that these conditions are realised in the case of the so-called permanent gases at pressures below three atmospheres. With decreasing volume $\frac{a}{v^2}$ becomes effective and generally does so sooner than b, with the result that with in-creasing pressure the compressibility increases; *e.g.* air and nitrogen in Regnault's tables. But at very high pressures, when p is large and v correspondingly small, the influence of $\frac{a}{v^2}$ on the value of the term $\left(p + \frac{a}{v^2}\right)$ becomes relatively less than that of b on the term $(v-b)$, leading to reduced and continually decreasing compressibility, *e.g.* hydrogen in Regnault's table, air in Cailletet's table.

Boyle's law is therefore only an ideal, or approximate law, explained by an approximate hypothesis. And this is the case with many other laws. The statement which at first

The charac-teristics of an approximate law.

seems to correctly embody the results of the measure-ments of two variable quantities, cannot stand the test of increased accuracy and extended investigation. Refinement of the experimental method, consisting in the elimina-tion of constant errors, and in greater and greater accuracy of measurement, leads to numerical results which show better agree-ment amongst themselves, and thereby emphasises the difference between the experimental mean value and that required by the law. Any law for which such a discrepancy, clearly not due to constant or experimental errors, has been established, must be classed as an approximate law; and if it represents a deduction from a certain hypothesis, that hypothesis must be such as to account for the discrepancy also.

But are all the empirical laws of this kind, or are there some

The charac-teristics of an exact law.

in the case of which increased refinement of mea-surement leads to closer and closer agreement with the theoretical values, instead of to more certain recognition of deviations? A critical examination of the data given in the last chapter in support of the law of conservation of mass will supply the answer to this question.

In Lavoisier's experiments (chap. II, p. 62) the deviations from the requirements of the law were found to be from ·25 to 2·0 per cent., quite comparable with the degree of accuracy attained in the individual measurements involved, those of gaseous volumes being very approximate only, as when we are told that "7 to 8 cubic inches" of a gas were evolved.

(b) Is the law of conservation of mass exact or approximate?

The difference in the accuracy aimed at and attained by Lavoisier on the one hand and by Morley and Stas on the other is enormous. And the magnitude of the deviations from the requirements of exact conservation of mass shows a corresponding change. In Morley's complete syntheses of water (p. 75), if we exclude experiments V and XII, in which the discrepancies are much larger than in any of the others and hence probably due to some accidental cause, the deviations are reduced to $\frac{1}{10000}$, ranging from − 0·01 to + 0·01 per cent.; and the fact that they are sometimes positive and sometimes negative, and about equal in amount in the two directions, characterises them as experimental errors. The enormous difficulties Morley had to overcome in the preparation of large quantities of pure oxygen and hydrogen, in the weighing of these gases, in the process of making them combine without loss at the high temperature at which this reaction occurs, and finally in the collection without loss of the water formed, were even greater than those met with by Stas in his complete syntheses of silver iodide, and complete analyses of silver iodate. Hence we should expect Stas' experimental results to approximate even more closely than Morley's to the theoretical, provided always that the law is really an exact one. And this is found to be the case. The syntheses of silver iodide gave numbers differing from the theoretical by not more than from − 0·002 to − 0·004 per cent. True, there was always loss, but this is readily accounted for by the fact that the greatest difficulty encountered lay in reducing the inevitable loss in the collection of the silver iodide, which had to be transferred from one vessel into another and washed by decantation for days (p. 67). The two analyses of silver iodate, in which no such transfer and no washings were required, gave results which differed from the theoretical by + 0·0014 and − 0·0013 per cent. respectively, discrepancies perfectly commensurable with the marvellously low experimental error. Stas' work therefore warrants the statement that the law of conservation of mass has been proved to an accuracy of about ·002 per cent., and that, considering the continued parallelism between experimental error and deviation from theory, the law is probably an absolutely exact one. More than this must not and cannot be inferred.

"Physicists often assume quantities to be equal provided that they fall within the limits of probable error of the processes employed. We cannot expect observations to agree with theory more closely than they agree with each other, as Newton remarked of his investigations concerning Halley's comet....Absolute equality is always a matter of assumption. We cannot even prove the indestructibility of matter; for were an exceedingly minute fraction of existing matter to vanish in any experiment, say 1 part in 10 millions, we could never detect the loss....The smallness of the quantities we can now observe is often very astonishing....We must nevertheless remember that effects of indefinitely less amount than these must exist...[and] when we pay proper regard to the imperfection of all measuring instruments and the possible minuteness of effects, we shall see much reason for interpreting with caution the negative results of experiments." (Jevons, *Principles of Science*, 1879.)

Laws can only provisionally be classed as absolutely exact.

The above quotation has been given in order to bring out with all possible emphasis our real position towards the principle of conservation of mass, to show what Stas' numbers have proved, and what they have not proved. They have proved that deviations from the law—if such do exist—cannot be greater than ·002 per cent., and they leave the existence or non-existence of smaller deviations an open question. All that was done and all that could be done lay in showing that the deviations actually observed were within the limits of the experimental error.

But neither Stas' nor Morley's experiments were undertaken with a view to investigating the validity of the law of the conservation of mass, and as a matter of fact they are particularly ill adapted to that purpose. The admiration called forth by the degree of approximation to the law attained, is mainly due to the fact that this was achieved in spite of all the conditions having been so unfavourable.

In experiments made by Landolt with the object of testing the law of conservation of mass, the conditions could be

specially adapted to this purpose, and hence a more rigorous
proof given[1]. The continuous advance made by
science, in which the practically impossible of to-day
becomes the actual achievement of to-morrow, is well
shown in the close approximation between what
Landolt established in 1893, and what in 1879 Jevons had given
as a safe example of a point we could not prove with the means at
our disposal. Landolt thus describes the object of his investigation,
and the method followed :

Investigations to test the law of conservation of mass.

"The problem is to accurately investigate by experiment how nearly the
weight of a chemical compound coincides with that of the sum of its
constituents. Stas' results only show that the difference cannot be other than
very small, and that conclusive results cannot be expected unless the work is
carried out in sealed vessels."

Two vessels *A* and *B* were used, which in order to obviate
all corrections for the buoyancy of air were made exactly equal
in volume by successive tentative additions of pieces of glass
to the smaller. The shape and size of these vessels are thus
described :

"The substances were contained in ∩-shaped glass vessels, of which the
two vertical limbs, closed below, were 18 cms. long and 5 cms. wide; whilst
the upper bent connection had a diameter of only 2 cms. Into this latter,
on each side, passed short open tubes, used for filling the vessel, and subsequently fused off."

· Approximately known quantities of the two substances that
it was intended to make interact were introduced separately into
the two limbs of each of the tubes *A* and *B*, previous to their
being sealed. The two tubes *A* and *B* were made of very nearly
equal weight, differing by at most a few milligrams. The amount
of this difference in weight was ascertained most accurately by
means of a specially constructed balance, which was sensitive to
within a few hundredths of a milligram. Suitable tilting of the
tube *A* made the contents of the two limbs mix and the reaction
occur, after which *A* was again weighed against *B*; then *B* was
treated likewise and a third weighing taken. All the reactions
used were quite simple ones, in which the heat evolution, and
hence the rise of temperature, was not great. The method
followed may be illustrated by the result of the measurements

[1] H. Landolt, *Zs. Physik. Chem.*, Leipzig, 12, 1893 (p. 1).

made in an experiment, when the reaction was that between silver sulphate and ferrous sulphate, resulting in the formation of metallic silver and ferric sulphate.

The sealed ∪-tubes, A and B, each weighed about 922·36 grams, and were practically identical in volume. A was about 1 mg. heavier than B.

The reacting masses of ferrous sulphate and silver sulphate weighed 114·22 grs.

(i) Before the reaction

$$\text{Weight of } A = B + 1·199 \text{ milligrams.}$$

(ii) After the reaction in A⎱
and before the reaction in B⎰

$$\text{Weight of } A = B + 1·032 \quad ,,$$

$$\therefore \text{Decrease} = 0·167 \quad ,,$$

$$\frac{\text{Change in weight}}{\text{Weight of reacting substances}} = -\frac{·000167}{114·22} = -\frac{1}{684,000} = -·000146 \text{ per cent.}$$

(ii) After the reaction in A⎱
and before the reaction in B⎰

$$\text{Weight of } B = A - 1·032 \text{ milligrams.}$$

(iii) After the reaction in B

$$\text{Weight of } B = A - 1·163[1] \quad ,,$$

$$\therefore \text{Decrease} = 0·131 \quad ,,$$

$$\frac{\text{Change in weight}}{\text{Weight of reacting substances}} = -\frac{·000131}{114·22} = -\frac{1}{870,000} = -·000115 \text{ per cent.}$$

Landolt embodies his final results in the table on p. 105.

His summary is:

"None of the reactions employed show a certain change of weight. If such changes should after all occur[2], they must be so small as to be of no practical importance to the chemist."

---

[1] The number given in the paper is 1·161, probably a misprint.

[2] A. Heydweiler (*Ann. d. Physik* (4), vol. v., 1901, p. 394), working by the same method as Landolt, has investigated a number of simple chemical changes, using about 200 grs. of the substances interacting. The error in weighing did not exceed ·04 mgs., and hence any change in weight exceeding ·05 mgs. could not be put down to experimental error. Whilst in some of the reactions investigated, the change in weight observed was within the limits of the experimental error, in others, such as that between iron and copper sulphate, there was always a loss of about ·2 mgs., that is, of about 1 in 1,000,000. No proportionality could be shown between the weight changes observed, and the masses of the interacting substances, nor any relation with changes in physical properties. These results require further confirmation before they can be accepted.

Reaction	Weight of the reacting mass	Change in weight observed	Probable error of weighing	Change in weight per cent. of reacting substance
Silver Sulphate and Ferrous Sulphate [1]	114·2 grams 114·2 ,, 171·3 ,,	− 0·167 mgs. − 0·131 ,, − 0·130 ,,	± 0·014 mgs. ± 0·020 ,, ± 0·012 ,,	− 0·000146 − 0·000115 − 0·000076
Hydriodic and Iodic Acid [2]	127·6 ,, 127·6 ,, 157·2 ,, 157·2 ,, 314·5 ,, 314·5 ,,	− 0·047 ,, − 0·114 ,, − 0·103 ,, − 0·102 ,, − 0·177 ,, − 0·011 ,,	± 0·015 ,, ± 0·009 ,, ± 0·009 ,, ± 0·011 ,, ± 0·008 ,, ± 0·009 ,,	− 0·000037 − 0·000089 − 0·000066 − 0·000065 − 0·000056 − 0·000003
Iodine and Sodium Sulphite [3]	157·0 ,, 157·0 ,, 192·0 ,, 192·0 ,,	+ 0·105 ,, − 0·031 ,, + 0·002 ,, − 0·127 ,,	± 0·006 ,, ± 0·012 ,, ± 0·014 ,, ± 0·012 ,,	+ 0·000067 − 0·000020 + 0·000001 − 0·000066
Chloral Hydrate and Potash [4]	201·0 ,, 201·0 ,,	+ 0·012 ,, + 0·007 ,,	± 0·016 ,,	+ 0·000006 + 0·000003

It might have been safer to say, to the chemist of to-day. If the chloral and potash experiment is taken by itself, Landolt may be said to have proved legitimately and conclusively that the change of weight in this reaction—if any—is less than one part in ten millions, that is, to have accomplished what Jevons had considered as beyond the limit of experimental investigation. But after all, the real position, as represented in the quotation given from Jevons, is not affected. All the modification required is the

[1] Silver sulphate + ferrous sulphate = silver + ferric sulphate.
$Ag_2SO_4 + 2FeSO_4 = Ag_2 + Fe_2(SO_4)_3$.

[2] Hydriodic + iodic acid = water + iodine.
$5HI + HIO_3 = 3H_2O + 3I_2$.

[3] Iodine + sodium sulphite + water = $\begin{cases} \text{sodium iodide + sodium dithionate} \\ or \\ \text{hydriodic acid + sodium sulphate.} \end{cases}$

$2I + 2Na_2SO_3 = 2NaI + Na_2S_2O_6$
or
$2I + Na_2SO_3 + H_2O = 2HI + Na_2SO_4$.

[4] Chloral hydrate + potash = chloroform + potassium formate + water.
$CCl_3.COH.H_2O + KOH = CCl_3H + H.COOK + H_2O$.

substitution of, say, " one part in one hundred millions " for " one part in ten millions." The limit has been pushed further, but our classification of any law as exact still retains its provisional nature.

The importance of the subjects dealt with in this chapter makes a short summary of its contents desirable. The main object was to find a standard whereby to interpret experimental results, so as to distinguish the differences due to the imperfection of our measurements from those due to real differences in the quantities measured. This necessitated some consideration of all the possible causes of error; these were classified, and what is more important, their influence on the final experimental result was investigated. It was recognised that whilst some errors exert a constant, others exert an accidental influence, having sometimes a positive, sometimes a negative value. The manner of detecting constant errors, by change of method, apparatus, etc. was next dealt with. The algebraic sum of all undetected constant errors and of all the accidental errors was given the name of the experimental error, which was found to be characterised by assuming in irregular succession positive and negative values. This knowledge of the effect of errors was then utilised for deciding to what cause we must assign certain observed differences in the experimental measurements of a quantity supposed to be the same. Using determinations of the density of oxygen and of nitrogen respectively as examples, it was shown how different types of results may be obtained by measurements of the same quantity in two sets of experiments, in which one condition only, viz. the source of the gases, was varied. In the one case, that of the oxygen, the mean values for the two sets of measurements differed by not more than did any two measurements in the same set; whilst in the other case, that of the nitrogen, the discrepancy between the mean values for the two sets was of an altogether different and higher order of magnitude. Hence it was inferred that in this latter case the change in condition had produced a real change in the quantity measured, whilst no such effect had been produced in the first case.

The same line of argument was next used in the interpretation of differences between experimental results and those calculated according to theory. The following scheme summarises the conclusions arrived at in this respect:

(Margin) Summary.

The differences between the experimental and the theoretical values are:

I	II	
Sometimes positive, sometimes negative.	*Always positive, or always negative.*	
	A	B
They are probably caused by experimental error, due to the combined, partly counteracting effects of small constant errors and of all the accidental errors.	*No constant error can be detected.*	*A constant error is present.*
The deviations are of the same order of magnitude as the experimental error. Improved methods of measurement decrease both proportionately.	Whilst improved methods of measurement diminish the experimental error, there is not a proportionate effect on the deviations from theory.	This is shown by the result of varying the conditions of the experiment.
The deviations from the law *are only apparent.*	The deviations from the law *are real.*	The deviations from the law *may be only apparent.*

This gave the basis for the classification into approximate and exact laws. A somewhat detailed consideration of Boyle's law served the purpose of showing how its approximate nature was established through the characteristics II A, and how this affected the relations with the hypothesis from which it follows deductively. An examination of the experimental foundation of the law of conservation of mass showed it to possess the characteristics of I, and hence to deserve the name of an exact law, realising that the term exact must always be used provisionally only, and with reference to the limit of the effect we are able to measure.

CHAPTER IV.

BERTHOLLET AND THE LAW OF MASS ACTION.

" Wo vier, bisher je zwei zu zwei verbundene Wesen, in Berührung gebracht, ihre bisherige Vereinigung verlassen und sich auf's neue verbinden; in diesem Fahrenlassen und Ergreifen, in diesem Fliehen und Suchen glaubt man wirklich eine höhere Bestimmung zu sehen; man traut solchen Wesen eine Art von Wollen und Wählen zu, und hält das Kunstwort Wahlverwandtschaften für vollkommen gerechtfertigt."

GOETHE, *Wahlverwandtschaften*, 1809.

LAVOISIER has introduced into Chemistry the recognition of the permanence in kind and in•amount of the matter in the universe. To him we owe the precise formulation of the principle which sets the limit to the possible changes of matter; many different substances may be produced by the combination in various ways of various constituents, but the amount and the ultimate properties of these constituents are unalterable: "the quality and the quantity of the constituents is the same, and modifications alone can take place." There is supplied for the use of Nature and of Man some building material, certain kinds and a certain amount of each kind; impossible to transform one kind into another[1], that is, change the quality; impossible to destroy or create any of it, that is, change the quantity. Hence it happens that the history of chemistry must chronicle as one of the greatest advances made in it at any time, the clear recognition, towards the end of the eighteenth century, of certain limitations to Nature's and Man's power of producing

The possible modifications in matter occur subject to the restriction that:

(1) The total quantity remains the same.

---

[1] The proved transformation of radium into helium and certain probable changes of one radio-active element into another form no exception to this statement, because these changes occur spontaneously and at rates unaffected by any known physical agency.

chemical changes. And the early years of the nineteenth century saw a further great advance in the proof then given of the existence of yet another such limitation. It was then established that

(2) Combination occurs only in certain fixed ratios. not only is it impossible to alter the quality and quantity of this building material, but also that combination can only be brought about between definite quantities of the constituents; it was proved that a certain amount of one kind of matter is only able to combine with a fixed amount of another kind of matter; if more is offered, it leaves some uncombined, if less is offered, some of itself remains uncombined. Put into the more precise form of scientific wording, this amounts to the statement, that, at the beginning of the nineteenth century, the fixity of the composition of chemical compounds was recognised and raised to the rank of a fundamental law of the science. The meaning to be attached to this term "fixity of composition" is, that, given two substances the same in their properties, they also have the same composition, that is, they contain the same constituents united in the same ratio. Quantitative analyses of compounds had been made long before the beginning of the nineteenth century, and of course there would

Fixity of composition was not proved until after the contrary had been maintained. have been no point in this, had not fixity of composition been at any rate tacitly assumed for certain cases. But fixity of composition was not definitely proved or formulated as a law, until doubt had been thrown on the validity of the relations to which it referred, and until it had been made the subject of a controversy for ever memorable in the history of science. In giving an account of this controversy, work must be considered which primarily deals with dynamical problems; an excursion must be made into the history of affinity.

Affinity is the name given to the cause which produces combination between different substances and which recalls the old

Affinity—Its cause is not known. view, long ago discredited, that similarity (*affinitas* = relationship) between the combining substances constitutes the cause of combination. What this cause really is, we know no more to-day than did the chemists of old. The problems of affinity represent a portion of the science of chemistry in which much work has been done of late, and still is being done, in which a huge number of facts have been investigated and laws formulated; but in which the step of devising a hypothesis which will satisfactorily account for these facts and laws has not yet been taken. All that we know concerns the effects, in kind and amount, of the unknown cause termed affinity. When copper combines with bromine to form copper bromide, this is said to occur owing to the affinity existing between copper and bromine; when copper bromide treated with chlorine gives copper chloride and bromine the reaction is said to be due to the fact that the affinity of chlorine for copper is relatively greater than that of bromine. And when, in the competition of sulphuric acid and of hydrochloric acid for an amount of potash insufficient for the neutralisation of both acids, the measurement of thermal and other physical effects (*post*, p. 125) points to a division of the base in which the hydrochloric acid takes three-fifths and the sulphuric acid two-fifths of the potash, we say that this happens because the relative affinity of sulphuric acid for potash is only two-thirds that of hydrochloric acid. In all these cases we simply state the nature of the occurrences, without any attempt to account for them theoretically.

The early study of affinity phenomena followed the course of all investigations in science. Individual phenomena were observed and then classified. Quantitative investigations proper, in which all the occurrences belonging to the same class of phenomena are measured in terms of some fixed unit, were preceded by a rough quantitative classification, in which one affinity effect was recognised as greater than another and the whole number of them arranged in the order of their magnitude relatively to each other.

Stahl, the founder of the phlogistic theory, has also contributed

Stahl arranges substances in the order of their affinities.

important work on affinity. He attempted (1720) to classify the effects of affinity in the manner above described, by arranging similar substances in a series in the order in which they expel one another from a compound:

"The following mechanical experiment may serve as an example. Dissolve silver in nitric acid, it will take up the silver and appear as a light liquor; into the clear and transparent liquor throw thin strips of copper foil, the nitric acid will dissolve these and will drop the silver in the form of a powder; pour this clear green solution on to lead foil, it will be attacked and the copper previously dissolved will be dropped; pour off the clear solution and pour it on to zinc, it will dissolve the zinc and allow the previously dissolved lead to drop; into this clear solution put chalk, it will be dissolved and the zinc dropped; then to this solution add spirits of urine, which will combine with it, releasing the chalk; and finally drop in lye, the solution will take it up and allow the volatile salt to go[1]."

S. F. Geoffroy (1672–1731) used such a classification as the basis of his tables of affinities, an arrangement destined to become

Geoffroy's tables of affinities.

very popular. The principle was to arrange similar substances so that the one following was always expelled by the one preceding from combination with the one heading the list. When thus represented Stahl's example just quoted becomes: *Nitric Acid*: potash, ammonia, lime, zinc, lead, copper, silver.

But yet another most important discovery concerning the action of affinity is due to Stahl. He recognised the fact that a

Stahl recognises the influence of temperature on the order of affinities.

reaction occurring at one temperature in one direction could be reversed at another temperature; that at ordinary temperatures calomel is decomposed by silver, whilst, under the influence of heat, silver chloride is decomposed by mercury:

"When *mercurium sublimatum dulcem*[2] is put into a *solutionem argenti* the silver falls down as a cornua[3] and the aqua fort.[4] attracts the mercury. But then I go and take fresh running mercury—it I mix with the lunam cornuam and give it due fire; there again rises for me a mercurius dulcis, such as had been used at first,—And yet in the former process the silver had taken over from the mercurio dulci its salty nature[5]...but let there be some intervention by another agent, heat, everything is reversed, and lo the silver throws back the acidum salis with which it had been saddled to the mercury, and the mercury with it must go its own way." (Kopp, *Geschichte der Chemie*.)

These two principles of Stahl, (i) the arrangement of substances in the sequence of their affinities towards one with which

[1] Ostwald, *Lehrbuch der Chemie*, II. part II. (p. 19).
The reactions referred to are :

Nitric acid + (solution)	silver (solid)	= silver nitrate (colourless solution)	+ water + oxides of nitrogen
Silver nitrate + (solution)	copper (solid)	= copper nitrate (green solution)	+ silver (solid)
Copper nitrate + (solution)	lead (solid)	= lead nitrate (colourless solution)	+ copper (solid)
Lead nitrate + (solution)	zinc (solid)	= zinc nitrate (colourless solution)	+ lead (solid)
Zinc nitrate + (solution)	calcium carbonate (solid)	= calcium nitrate (colourless solution)	+ zinc carbonate (solid)
Calcium nitrate + (solution)	ammonium carbonate (solid or solution)	= ammonium nitrate (colourless solution)	+ calcium carbonate (solid)
Ammonium nitrate + (solution)	potassium hydrate (solid or solution)	= potassium nitrate (solution)	+ ammonia + water. (gas)

[2] Mercurium sublimatum dulcem = calomel = mercurous chloride = HgCl.
[3] Lunam cornuam = horn silver = silver chloride = AgCl.
[4] Aqua fortis = nitric acid = HNO_3.   [5] The acid.

they all combine, and (ii) the recognition that this sequence may be different when the dry substances are heated together from what it is in the case of interaction between solutions at ordinary temperatures, form the basis of the work of Bergman, with whose name chiefly are associated the views on affinity current at the end of the eighteenth century.

Torbern Bergman (1735–1784), a great Swedish chemist, studied and taught at Upsala, where he first held the chair of mathematics and afterwards that of chemistry.   He did much to systematise analysis, and furthered the study of the chemical nature of minerals and rocks.

Bergman's work on affinity.

Here we are only concerned with his work on affinity, by him termed "elective attraction."[1]   He looks upon combination as due to the attraction between the smallest particles:

"It is found that all substances in nature, when left to themselves and placed at proper distances, have a natural tendency to come into contact with one another.   This tendency has long been distinguished by the name of 'attraction.'   I do not propose in this place to inquire into the cause of these phenomena, but in order that we may consider it as a determinate power, it will be useful to know the laws to which it is subject in its operations, though the mode of agency be as yet unknown."

Investigation of the effects, the cause of which is unknown.

This attraction is of different magnitude, and Bergman recognises the desirability of actual measurement, but in the absence of any means for realising this, he has to content himself with arranging affinities in the order of their magnitudes, for which purpose he uses the tabular form first introduced by Geoffroy.

Arrangement of affinities in the order of their magnitudes.

"In this dissertation I shall endeavour to determine the order of attractions according to their respective force; but a more accurate measure of each, which might be expressed in numbers and which would throw great light on the whole of this doctrine, is as yet a desideratum."

The meaning to be attached to the terms "order of attractions" is defined by him in the following manner:

"Suppose A to be a substance for which other heterogeneous substances a, b, c have an attraction; suppose further A combined with c to saturation (this union I shall call Ac) should upon the addition of b tend to unite with it to the exclusion of c, A is then said to attract b more strongly than c, or to have a stronger elective attraction for it; lastly let the union of Ab on the

[1] Bergman's work on Affinity was given to the world in a book entitled *Opuscula physica et chemica*.   Upsala, 1783.   An English translation of part of it appeared under the title: "*A dissertation on elective attractions*, by Torbern Bergman. Translated from the Latin by the Translator of Spalanzani's Dissertations, London, 1785."   The quotations given are taken from this translation.

addition of *a* be broken, let *b* be rejected, and *a* chosen in its place, it will follow that *a* exceeds *b* in attractive power, and we shall have a series *a, b, c* in respect of efficacy. What I here call attraction others denominate affinity; I shall employ both terms promiscuously in the sequel, though the latter being more metaphorical would seem less proper in philosophy."

The question is next posed whether this order of attraction is constant, and the answer is on the whole affirmative. But Bergman recognises the necessity of two sets of tables, one for solutions at ordinary temperatures, and another for the interaction between solids at high temperatures respectively:

"A difference in the degree of heat sometimes produces a difference in elective attractions."

"Since many operations cannot be carried on without the aid of heat, and since the power of this most subtle fluid is highly worthy of being observed, I think the table of elective attractions ought to be divided into two areas, of which the upper may exhibit the free attractions, that may take place in the moist way, as the expression is; and the lower those which are effected by the force of heat." ·

Tables of affinity "in the moist way" and "in the dry way."

The reproduction of two columns from Bergman's tables (p. 115) shows this division into "two areas."

Experimental determination· of the order of affinities.

But how are these single elective attractions to be determined? It is very important to obtain some definite idea of the experimental work done to show that *a* has a greater affinity for *A* than has *b*.

"Let *Aa* (*i.e. A* saturated with *a*) be dissolved in distilled water, and then add a small quantity of *c*, which may either be soluble in water by itself or not. First let it be soluble, then a concentrated solution ought to be employed, which when dropped into a solution of *Aa*, sometimes immediately affords a precipitate, which being collected and washed, either proves to be a new combination, *Ac*, with peculiar properties, or *a* is extruded, or sometimes both. It now remains to be examined, whether the whole of *a* can be dislodged by a sufficient quantity of *c* from its former union. It should be carefully noted in general, that there is occasion for twice, thrice, nay, sometimes six times the quantity of the decomponent *c*, than is necessary for saturating *A* when uncombined. If *c* effect no separation, not even in several hours, let the liquor stand to crystallise, or at least become dry by a spontaneous evaporation ; high degrees of heat must be avoided lest they disturb the affinities. Here the knowledge of the form, taste, solubility, tendency to efflorescence and other properties,—even those which in other respects appear of no consequence,—of the substance, is of great use in enabling us to judge safely, whether any and what decomposition has taken place.

F.

8

If only one of the compounds Aa and Ac be soluble in highly rectified spirit of wine, there is scarce any need for evaporating, for if the mixture be made, and left a few hours at rest, and then spirit of wine be added, that which cannot be dissolved in it is separated.

The smell also often indicates what is taking place. Thus vinegar, acid of ants, of salt, nitre, volatile alkali are easily distinguished when set free. The taste likewise also informs an experienced tongue.

Let Ad then be treated with b and a etc. separately in the same manner.

In like manner, let Ac, Ab, Aa be examined in their order. By such an examination properly conducted the order of affinities is discovered."

Bergman then passes in review the material to be thus dealt with, compares what has been done with what remains to be done. and, small wonder, is appalled by the magnitude of the task.

"This task however exercises all the patience and diligence and accuracy and knowledge and experience of the chemist. Let us suppose only 2 series of 5 terms, a, b, c, d and e to be examined with respect to A, 20 different experiments are required, of which each involves several others.

The tables which we have at present contain only a few substances, and each of these compared only with a few others. This is no reproach to the authors of them, for the task is laborious and long. Although therefore I have been employed upon it with all the diligence I could exert and as much as my many other engagements would permit, yet I am very far from venturing to assert that that which I offer is perfect, since I know with certainty that the slight sketch now proposed will require above 30,000 exact experiments before it can be brought to any degree of perfection. But when I reflected on the shortness of life, and the instability of health, I resolved to publish my observations, however defective, lest they should perish with my papers, and I shall relate them as briefly as possible. In itself it is of small consequence by whom science is enriched, whether the truths belonging to it are discovered by me or by another. Meanwhile if God shall grant me life, health and the necessary leisure I will persevere in the task which I have begun."

The tables as given by Bergman are published at the end of his book. At that time tables of affinity ranked with chemists in importance comparable to that now attached to

Bergman's tables. atomic weight tables. As soon as a new substance was discovered, it was at once investigated with a view to settling its position in the tables of elective attractions; and the value attached to these tables can be gathered from what the English translator of Bergman's dissertation says concerning them in his preface:

"Two sets of tables are subjoined, it was thought that many readers would be dissatisfied with the chemical characters alone, especially as the

SINGLE ELECTIVE ATTRACTIONS

Names now used	*Signs*	*Names*	*Names*	*Signs*	*Names now used*
		IN THE MOIST WAY			
Sulphuric acid	+ ⊕	Vitriolic Acid	Acid of Fluor	+𝓕	Hydrofluoric acid
Baryta	Ψ *p*	Pure ponderous earth	Lime	Ψ *p*	Lime
Potash	⊖√*vp*	Pure vegetable alkali	Pure ponderous earth	Ψ *p*	Baryta
Soda	⊖√*mp*	Pure fossil alkali	Pure magnesia	Ψ *p*	Magnesia
Lime	Ψ *p*	Lime	Pure vegetable alkali	⊖√*vp*	Potash
Magnesia	Ψ *p*	Pure magnesia	Pure fossil alkali	⊖√*mp*	Soda
Ammonia	⊖^*p*	Pure volatile alkali	Pure volatile alkali	⊖^*p*	Ammonia
Alumina	⚲▽*p*	Pure clay	Pure clay	⚲▽*p*	Alumina
Oxide of zinc	Ψ Ô	Calx of zinc	Calx of zinc	Ψ Ô	Oxide of zinc
etc.	Ψ ♂	Calx of iron	Calx of iron	Ψ ♂	etc.
	Ψ ⏀	Calx of manganese	Calx of manganese	Ψ ⏀	
	Ψ ♌	Calx of cobalt	Calx of cobalt	Ψ ♌	
	Ψ 8	Calx of nickle	Calx of nickle	Ψ 8	
	Ψ ♄	Calx of lead	Calx of lead	Ψ ♄	
	Ψ ♃	Calx of tin	Calx of tin	Ψ ♃	
	Ψ ♀	Calx of copper	Calx of copper	Ψ ♀	
	Ψ ♉	Calx of bismuth	Calx of bismuth	Ψ ♉	
	Ψ ♁	Calx of antimony	Calx of antimony	Ψ ♁	
	Ψ o-o	Calx of arsenic	Calx of arsenic	Ψ o-o	
	Ψ ☿	Calx of mercury	Calx of mercury	Ψ ☿	
	Ψ ☽	Calx of silver	Calx of silver	Ψ ☽	
	Ψ ☉	Calx of gold	Calx of gold	Ψ ☉	
	Ψ ☽☉	Calx of platina	Calx of platina	Ψ ☽☉	
Water	▽	Water	Water	▽	Water
Alcohol	▽ˢ	Spirit of wine	Spirit of wine	▽ˢ	Alcohol
Removal of oxygen	🜍	Phlogiston	Pure silicious earth	⩡	Silica
		IN THE DRY WAY			
Removal of oxygen	🜍	Phlogiston			
Baryta	Ψ *p*	Pure ponderous earth	Lime	Ψ *p*	Lime
Potash	⊖√*vp*	Pure vegetable alkali	Pure ponderous earth	Ψ *p*	Baryta
Soda	⊖√*mp*	Pure fossil alkali	Pure magnesia	Ψ *p*	Magnesia
Lime	Ψ *p*	Lime	Pure vegetable alkali	⊖√*vp*	Potash
Magnesia	Ψ *p*	Pure magnesia	Pure fossil alkali	⊖√*mp*	Soda
Metallic oxides	Ψ *m*	Metallic calces	Metallic calces	Ψ *m*	Metallic oxides
Ammonia	⊖^*p*	Pure volatile alkali	Pure volatile alkali	⊖^*p*	Ammonia
Alumina	⚲▽*p*	Pure clay	Pure clay	⚲▽*p*	Alumina

former edition of the tables has already been published in words. To suppress the signs entirely seemed improper for they are so convenient that every student of chemistry ought to make himself familiar with them. Besides as most chemists will wish for a set to stand always open for inspection, the two sets will scarce be thought superfluous by any."

In the above table, the columns from Bergman's two tables which refer to sulphuric acid and hydrofluoric acid are reproduced. It may here not be unnecessary to draw attention to the fact that this symbolic representation, which uses the old alchemical signs together with additional newer ones, is fundamentally different from the symbolic notation initiated by Dalton, developed by Berzelius and used at the present day. Bergman's was purely qualitative, ours is qualitative and quantitative. Between the columns containing the symbols are placed the corresponding columns from Bergman's other table giving the results in "words"; but since some of the old names are as puzzling as the old symbols, two columns are added in which, where necessary, the modern equivalents of the old names are given.

The following quotation, referring to the sulphuric acid

Bergman's explanation of his tables.

column, gives a specimen of Bergman's discussion of his results, the modern names, where necessary, being inserted in brackets.

"Amongst the substances hitherto tried, vitriolic acid adheres most tenaciously to *caustic terra ponderosa* (caustic baryta), which when added to a solution of vitriolated tartar (potassium sulphate) generates the ponderous spar (barium sulphate) which remains insoluble at the bottom. The liquor contains caustic vegetable alkali (caustic potash). *Caustic or pure vegetable alkali is incapable of decomposing ponderous spar.*

Next stands caustic vegetable alkali (caustic potash) which when added in sufficient quantity to a solution of Glauber's salt (sodium sulphate) yields vitriolated tartar (potassium sulphate) and uncombined mineral alkali (caustic soda) *which is unable to detach vitriolic acid from vegetable alkali.*

Caustic mineral alkali (caustic soda) precipitates the calcareous basis (lime) of gypsum (calcium sulphate) *but the inverse experiment does not succeed.*"

The salient points of Bergman's view as represented in the above quotations would seem to be:

(1) There is a sequence in the magnitude of the elective affinities of a series of substances towards one with which they all

Bergman's results summarised.

combine, and this is manifested by the fact that the one possessing the greater affinity, expels from the combination the one possessing the lesser affinity.

(2) This order is constant under each of the two different conditions of interaction in the moist and dry way respectively, but differs under these two distinct conditions.

(3) The substance of lesser affinity is completely expelled by that of greater affinity, subject however to the possibility that the mass of the expelling substance may have to be very much greater than that required for simply replacing the expelled substance in the combination.

(4) It is impossible to reverse such a reaction.

No hypotheses were involved in Bergman's work, it was simply a case of observation, a statement of what happens. His results and his general conclusions drawn from extensive experimental data, were attacked by Berthollet.

Claude Louis Berthollet (1748–1822) was one of the most brilliant of the many distinguished contemporaries and successors of Lavoisier. An early adherent of Lavoisier's oxygen theory of combustion, he did much to popularise it. Throughout the Revolution, the Empire, and the Restoration his high scientific achievements were recognised and honoured. His many-sidedness is striking. He began life as a medical man, held for many years a Professorship at the École Normale, served on Government commissions for the Republic, organised the scientific part of Napoleon's expedition to Egypt, and spent the last years of his life at Arcueil, where he formed the centre of a group of eminent scientists. He did more for the theory of chemistry than any one before him, Lavoisier excepted; he also enriched applied science by the most important discovery of the bleaching action of chlorine and of its technical application. Most clear-sighted in his detection of the fallacies of Bergman's affinity work, he displayed the most extraordinary tenacity in clinging to an erroneous view of his own devising; a striking figure in the history of chemistry, interesting no less for his positive achievements than for the error into which he fell.

Berthollet's work on affinity. Refutation of Bergman's results.

On the 9th of Messidor of the year VII. (July 1799) Berthollet, who had joined Napoleon's expedition to Egypt, read in Cairo a paper entitled " Recherches sur les lois de l'affinité," the subject matter of which was later on expanded into a book entitled " Essai de statique chimique." In this paper he attacks the validity of Bergman's results:

"Let us assume, says Bergman, that a substance A is completely saturated by another substance B, and let us call their combination AB. Then if a third substance C be added, and if it should expel the substance B from the compound and should in its place enter into combination with A, a new substance AC would result in place of AB. Therefore in order to determine the affinities of two substances B and C towards a third A, Bergman asks that we should investigate whether one of these substances expels the other from its combination with the third. He also recommends that the inverse experiment should be made, *i.e.*, that we should investigate whether the second substance can expel the first from its combination with the third. Hence he assumes the possibility of both experiments giving concordant results, and concludes that the first substance C has a greater affinity towards the third, A, than has the second, B. But he does add that it may sometimes be necessary to use a quantity of the decomposing substance 6 times as great as that required for the complete saturation of the substance with which it tends to combine[1]. The whole of Bergman's conception is based on the supposition that affinity is an unalterable force, of a kind such that one substance which has been expelled by another from one of its compounds, cannot itself again expel this substance. The idea of the un-alterableness of this force has been pushed so far, that celebrated chemists have attempted to represent the affinities of different substances by numbers, which when compared amongst themselves should indicate their magnitude, the quantities of the substances present exerting no influence.

I propose to prove in this paper that affinities do not act as absolute forces, so that one substance can be directly expelled by another from its compounds; but I hold that in all combinations and de-compositions which are the result of affinity, the substance on which two others act with opposing forces, always divides itself between these two substances, and that the ratio of the division does not only depend on the strength of the affinities, but also on the quantities of the active substances present; since to produce an equal division, the greater mass of one substance can compensate for what is lacking in its affinity.

Division of a substance between two others in the ratio of their affinities and their masses.

Hence I shall have to show that, in the case already quoted, when C is allowed to act on AB, the result is not AC alone (unless it so happens that other forces besides affinity come into play), but that A divides itself between B and C in the ratio of their affinities and of their masses."

Berthollet then proceeds to supply the experimental basis for his views. He chooses his examples very largely from the class of reactions in which alkalis and alkaline earths interact with acids. The experiments have a bearing on two distinct points:

[1] *ante*, p. 113.

Firstly: The proof that the result of a chemical action depends not only on affinity but also on the quantities interacting.

Secondly: The investigation of the conditions which modify the application of the above principle, and the properties of substances which are conducive or adverse to the display of their chemical activity.

1. (i) A solution of potash boiled for a long time with barium sulphate yielded a certain amount of soluble potassium sulphate, along with soluble baryta. The amount so produced was small, but if the barium sulphate was repeatedly treated with fresh amounts of potash, that is, if the potassium sulphate and baryta were always removed by treatment with water, the decomposition could be made all but complete, in contradiction to Bergman's result (p. 116).

Experimental evidence for: 1. The influence of mass.

(ii) Calcium phosphate and potash and water evaporated to dryness yielded on proper treatment a certain amount of potassium phosphate and lime.

(iii) Aqueous solutions of caustic soda and of potassium sulphate were evaporated to dryness, the alkaline hydroxides removed by treatment with alcohol—the sulphates being insoluble in alcohol,—the remaining solid sulphate was dissolved in water and the solution allowed to crystallise, when it yielded not only potassium sulphate but also sodium sulphate. This also contradicts Bergman's result (p. 116).

In interpreting these and a whole number of similar occurrences Berthollet points out in the clearest possible manner how the results of a chemical change depend not only on the affinities of the active substances but also on their relative quantities. This is shown by the fact that one substance does not completely expel another when all are present together in solution (exp. iii) and further by the actual reversal of a chemical change when the active masses are altered (exps. i and ii).

(a) Two substances are formed simultaneously.

(b) Reactions can be reversed.

2. Berthollet then proceeds to consider the conditions which can modify the general principle, just enunciated and proved, of the combined effect of affinity and mass action. He draws attention to the fact that the mass of a substance undergoing chemical

change can only make itself felt if each of its particles can act

2. The in-
fluence of
physical
conditions.
Decrease in
the active
mass of sub-
stances due to:
(a) Volatility,
(b) Insolu-
bility.

and react with each particle of the other substances participating in the change[1]. Hence any substance which owing to insolubility or volatility is removed from the solution in the form of a precipitate or gas cannot be considered to contribute to the "*active mass.*" Therefore the solubility and the volatility of the substances involved in a reaction have a profound influence on the final result of that reaction, as illustrated by the following experiments:

(i) Hydrochloric and nitric acids are volatile at temperatures at which sulphuric acid is not, and sulphuric acid therefore expels hydrochloric and nitric acids from their compounds, a reaction employed in the usual method of preparing these acids. Moreover hydrochloric acid gas is at once evolved when concentrated sulphuric acid acts on solid salt, whilst it is not evolved by the action of dilute acid. Again, at the boiling point of sulphuric acid (338°), phosphoric acid is non-volatile, and hence at that temperature phosphoric acid will expel sulphuric acid from its compounds. It will do this, not because it has greater affinity towards the various bases, but because the original partition of the base between the two acids which occurred according to their chemical masses (mass and affinity) is being continually upset in favour of the phosphoric acid. The active mass of the phosphoric acid increases relatively to that of the sulphuric acid, which latter, owing to its volatility, is always removed as soon as it is liberated by the other acid. The final result is complete replacement of the sulphuric by the phosphoric acid.

"It is a consequence of this effect of heat that at a sufficiently high temperature all fire-proof acids expel those that are volatile from their compounds, and since they differ much in volatility, certain acids must be looked upon as fire-proof when compared with some, and as volatile when compared with others. Thus at a sufficiently high temperature hydrochloric acid and nitric acid are completely expelled by sulphuric, whilst sulphuric acid itself is separated by phosphoric acid; and this occurs irrespective of affinity.

If therefore under the influence of heat one substance expels another from

---

[1] "Corpora non agunt nisi fluida," the old form of expressing this fact, is generally illustrated in chemistry lectures by a stock experiment: solid sodium bicarbonate and tartaric acid are ground up finely and mixed intimately, nothing happens; the solid mass is moistened with water, a copious evolution of carbonic acid gas sets in instantaneously.

its compounds, we must not conclude that it is at ordinary temperatures endowed with a greater affinity."

(ii) Berthollet gives many examples to illustrate the influence of solubility on the final result of a chemical change. Amongst these is the formation of barium sulphate. He says that, in comparing the affinity of baryta towards sulphuric acid with that of potash or soda towards this same acid, the influence of solubility is enormous. Owing to its almost complete insolubility the barium sulphate is always removed from the solution as soon as formed. Consequently, in the joint competition of the potash (or soda) and the baryta for the sulphuric acid, the alkali obtains only an imperceptibly small share of the acid; and an erroneous view would be gained were this made the basis for evaluating the relative affinities. The discussion of many similar cases leads to the recognition that the insolubility, or the lesser solubility of a substance is equivalent to reduction of its active mass, and hence that if amongst the possible combinations between the constituents participating in a change there is one which is insoluble or less soluble than the others, that one will be formed at the expense of all others. Thus the insoluble carbonates of calcium, barium and strontium are precipitated when soluble salts of these elements are mixed with soluble carbonates; and from aqueous solutions containing several salts, these will be precipitated, not according to the actual amounts of them present in the solution, but in the order of their solubilities. These relations give the law by which the exchange of bases can be foretold. But since the solubility of salts is differently affected by temperature, this circumstance must be taken into account. So from a mixture of potash and soda with hydrochloric and nitric acids (equivalent to a mixture of sodium chloride and potassium nitrate, or to one of sodium nitrate and potassium chloride) crystallisation in the cold results in a deposit of potassium nitrate, whilst crystallisation on evaporation yields sodium chloride[1].

And so all these experiments of Berthollet and their interpretation constitute a complete proof of the untenableness of Bergman's position:

[1] 100 parts of water dissolve

	at 20°	at 100°
Sodium nitrate	87·5	180
Potassium nitrate	31·2	247
Sodium chloride	35·6	40
Potassium chloride	34·7	56·6

"All tables of affinity must necessarily give a wrong impression of the
chemical action of the substances classified, because they are
based on the supposition that all substances are possessed of
forces of affinity of different strength, in virtue of which they
produce decomposition and combination irrespective of their
quantities and of all other conditions capable of exerting an
influence on the result."

It is important to note that the difference between Bergman's
and Berthollet's views of the action of affinity is not one due to
different *à priori* reasoning or to different hypotheses used in the
interpretation of the experimental results; but that it is one of facts,
of differences in the experimental results themselves.
In Bergman's work we note the curious influence
of results "expected" on the nature of results "obtained." Firmly convinced that the substance of
greater affinity always completely expels from its
compounds the one of lesser affinity, he arranges the
conditions of his experiments, if need be, so as to produce a result
somewhat approximating to this expectation, using large excess of
the substituting substance and thereby greatly increasing its effect.
He realises that the assumption •

$$Ab + d = Ad + b \ does \ occur,$$

and $$Ad + b = Ab + d \ does \ not \ occur,$$

should be put to a double experimental test, but if he tried the
reverse action, his observation was certainly incorrect, stating as
he did that no effect was produced, as for instance
that caustic or pure vegetable alkali is incapable of
decomposing ponderous spar, etc., etc. (p. 116).

But Berthollet's correct experiments and their
correct interpretation swept away, not only the tables
of affinity, but also the method whereby these had
been obtained, and he had no other to put in its place. He says:

"To find the affinity of two substances towards a third, in accordance
with the conception we have now gained of affinity, can mean nothing other
than to determine the ratio in which this third substance divides itself between
the two first."

Suppose we deal with a solution of sulphuric acid, soda and
potash, the acid present being insufficient to saturate all the
alkali—or with what has been experimentally proved to be equiva_

lent, namely, potassium sulphate and soda, or sodium sulphate and potash—then the problem is to find the ratio in which the sulphuric acid divides itself between the two alkalis, *i.e.* how much sodium

Chemical methods are not available for finding the distribution of the acids and bases amongst the salts present in a solution. sulphate and potassium sulphate are present in the solution. The resources of quantitative analysis are not equal to this. In the case of all salts in solution we can only determine the total quantity of the acid and of the base, but not the manner of their distribution amongst the salts. This is a limitation of the utmost importance in connection with the solution of affinity problems. In the above case, the addition of a soluble barium salt will produce a precipitate of insoluble barium sulphate the weight of which serves as measure of the total amount of sulphuric acid present, but it obviously tells us nothing whatever concerning the way in which this was distributed between the two alkalis, whether equally, or whether a larger part had gone to the soda or to the potash. And so it is with the other constituents also. Their total amount and not their distribution is all we can ascertain. To obtain the results that he required, Bergman allowed such a solution to crystallise, and potassium sulphate being the result, he inferred that in the solution likewise the sulphuric acid had been combined with the potash only. But Berthollet's work had gone to show that such a crystallisation indicates neither the complete nor the preponderating saturation of the sulphuric acid by potash, but that it only indicates the lesser solubility of potassium sulphate as compared with sodium sulphate:

"To find the degree of saturation to which each substance had been able to attain, a separation is required, and this can only be achieved either by crystallisation or precipitation, or by the action of a solvent. But we have seen how all such processes must be looked upon as foreign forces which alter the results and which themselves determine what compounds shall be formed. And as we are unable to measure the influence of these forces, we cannot deduce how much is due to affinity."

Taking the special case of the two sulphates above discussed, Berthollet points out that the solution containing the two sulphates and the two free alkalis might be evaporated to dryness, and that, by the action of alcohol, the alkalis which are soluble in alcohol might be separated from the sulphates which are not. The relative quantities of the sulphates could then of course be determined by the estimation of the potassium and sodium, but "difference in the action of the spirit of wine on potash and soda would in itself already influence the amount of these two substances, and through them the final result." He concludes:

"It is therefore clear that the affinity of two substances towards a third cannot be determined by direct experiment, not even when the substances are all soluble...because, in order to determine the degree of their saturation, other forces must be brought into play."

The position then is that analysis gives us no information of how the sulphuric acid is distributed between the two alkalis present in excess; neither does crystallisation, because the special distribution represented by the crystallised substance may owe its existence to the crystallising process itself. The absolutely clear realisation of this is of such importance that one more quotation bearing on the subject shall be given:

"I have examined whether it is possible to determine the ratio of the affinities of two substances towards a third; I have shown that to do this we should have to determine the ratio in which this substance divides itself between the two competing ones. But I have also shown that the determination of this ratio is attended by unsurmountable obstacles originating in the very means which must be used for the purpose."

Summarising the salient points of Berthollet's views on affinity, as has been done in the case of Bergman's, these appear to be:

Berthollet's results summarised.

(1) The result of a chemical change depends on the nature and on the relative masses of the substances participating in it.

(2) These masses are themselves dependent on the states of aggregation of the substances originally present and of those formed in the reaction.

From these two principles it follows that:

(a) By virtue of affinity only, one substance can never completely expel another from its compounds.

(b) By suitably varying the masses, reactions can be reversed; volatility and solubility through their continuous influence on the active masses have a greater effect on the final result than affinity itself, and may lead to the complete expulsion of one substance by another.

(c) The relative affinities of two substances towards a third cannot be measured.

This impossibility of subjecting affinities as understood by Berthollet to any process of measurement may in part have been the cause of the long continued neglect of the striking results above enumerated. The two reasons generally given for this neglect, and no doubt the chief ones are, that for a long time after the promulgation of these views, all the working power available in the department of chemistry was concentrated on the development of the atomic theory and of organic chemistry respectively; and that the permanently true in Berthollet's affinity work had to suffer from the discredit brought on it by the errors into which it led him concerning chemical composition.

Causes of the long neglect of Berthollet's work.

A period of complete neglect was followed by one when facts accumulated which could only be explained on the basis of Berthollet's views on affinity, and then came the time when new methods were developed for ascertaining the distribution of acids and bases in solution. These methods are of a physical nature, and are all based on the same principle. A short description of a special case must suffice for illustration. Iron sulphocyanide dissolves in water to a red solution, the depth of colour produced depending on the strength of the solution. The colour corresponding to a certain amount of salt present in a given volume of solution, or the amount of salt corresponding to a certain colour produced can therefore be ascertained. Solutions of potassium sulphocyanide and ferric chloride are mixed in equivalent quantities, that is, a certain amount of ferric chloride is taken and an amount of sulphocyanide containing as much sulphocyanic acid as would be required to replace all the hydrochloric acid in the ferric chloride; a red colour appears, showing that at any rate some hydrochloric acid has been replaced by the sulphocyanic acid. Has it all been replaced, and if not, how much? An investigation of the depth of colour produced gives the means of evaluating the total amount of iron sulphocyanide present, which together with a knowledge of the total amount of iron used, gives the ratio in which the iron is divided between the two acids. It is found that only about one-eighth of the total possible amount of iron sulphocyanide is formed. Addition of either sulphocyanic acid in the form of potassium sulphocyanide, or of iron in the form of ferric chloride leads to an increase in the iron sulphocyanide, made evident by

Physical methods for finding the distribution of the components of substances present in solution.

the deepening of the colour, and in perfect agreement with the requirements of Berthollet's doctrine. And many other physical properties besides colour, such as density, fluorescence, refractive index, etc., etc., lend themselves to ascertaining in a similar manner the ratio in which bases and acids are distributed amongst the salts present[1]. "Relative affinity" or "relative avidity" is the name given, in the case of the neutralisation of acids and bases, to the ratio in which one substance divides itself between two others.

And since 1864, when Guldberg and Waage, two Norwegian chemists, clearly formulated the law of the influence of mass on the results of chemical changes in the statement:

Berthollet's work recognised in Guldberg and Waage's Law of Mass Action.

"chemical effect is proportional to the active masses," and supplied the necessary experimental evidence, the development of affinity measurements on the basis of this law, that is according to Berthollet's doctrine, has been continuous and enormous.

It was in order not to break off abruptly at a point where quantitative application seemed impossible, that these necessarily quite inadequate indications concerning the gradual recognition and application of Berthollet's work have been given. These problems really are beyond the purpose of this chapter, which is merely to give an account of Berthollet's work on affinity, sufficiently full to prepare for a proper understanding of how it influenced his views on chemical combination.

[1] Ostwald, *Outlines of General Chemistry*, 1890 (p. 325), gives the principles of such physical methods, as well as examples of their application.

CHAPTER V.

PROUST AND THE LAW OF FIXED RATIOS.

"Es ist das Schicksal unserer Wissenschaft, dass die wichtigsten Ent-deckungen einer Epoche zuerst zu extremen Hypothesen führen."
MENDELEEFF.

BERTHOLLET'S work on affinity in the form of the law of mass action is the basis of all modern work on chemical change. The recognition that chemical effect, *i.e.* the amount of change produced, depends on two factors, of which one is the active mass of the substance considered, and the other a constant conditioned by the nature of the substance and termed the coefficient of affinity, is as fundamental in the province of chemical dynamics as are the laws of combination in that of chemical statics. From the perfectly correct premise of the influence of mass on the chemical effect produced, Berthollet drew the erroneous inference that mass had an influence, not only on the amount of change but also on the kind, producing a continuous variation in the ratio in which the constituents are united in the compound. He affirmed the variability of the composition of chemical compounds, the possibility of combination between constituents in all sorts of continuously varying ratios.

His argument took the form : since chemical effect depends on the masses of the reacting substances, an increase in the relative mass of one of the constituents of a compound must necessarily lead to the formation of a compound richer in that constituent. Hence a gradual increase in the relative amount of that constituent present at the time of the formation of the compound will—all other conditions being the same—give rise to the formation of a whole series of compounds in which

[marginal note:] Variability of composition asserted by Berthollet.

[marginal note:] The principle of variable composition theoretically deduced from the law of mass action.

the percentage of that constituent will show a continuous increase. But as the quantity of the one constituent increases, the attractive force of the other diminishes, and hence combination with further quantities becomes gradually more and more difficult[1].

"When a compound has been formed, its two constituents are retained in virtue of their mutual affinities and of their respective masses, so that in accordance with the general law of chemical action, if one of the two constituents preponderates, the quantity of it present in excess is held by the other constituent the less firmly, the greater that excess; but in a condition such as that of neutrality the power of action of each constituent on the other is yet far from being exhausted....Hence chemical action, the strongest as well as the weakest, is exerted in consequence of the mutual affinity of the substances and of the quantities present within the sphere of action, and the action diminishes with the saturation; but there is no stage at which it gives rise to fixed ratios[2]."

The experimental refutation of Berthollet's assumption of variable composition will be left till after the exposition of his reasoning and his evidence, but it should be noticed from the outset that the function here assigned to the influence of mass is different from that in the portions of his work dealt with in the last chapter.

Now we have *one* substance combining with a *second*, and the assumption is made that the combining ratio varies in some way

Mass is assumed to determine, besides the kind and amount of chemical effect produced, also the composition of the substance formed.

directly as the quantities of these substances present at the time of combination, thus:

$$mA + Ma \text{ produces } m'A \cdot M'a.$$

For instance: m sulphuric acid $+ M$ potash produces potassium sulphate containing the acid and alkali in the ratio $\dfrac{m'}{M'}$, which is some function of $\dfrac{m}{M}$, and which may assume any value between 0 and ∞.

Before we had *two* substances competing for combination with a *third*, and dividing this third in the ratio of their affinities and of their active masses, thus:

$$c + Ma + mA = pca + (1-p)\,cA + \text{excess of } a + \text{excess of } A.$$

[1] Berthollet always uses the term "combinaison" which will be translated by either "combination" or "compound."

[2] All the quotations from Berthollet given in this chapter are taken from the *Essai de Statique Chimique*, 1803, in which his views on variable composition find their fullest expression.

For instance:

Sulphuric acid $+ m$ potash $+ M$ soda,

$$= \begin{matrix} p \text{ sulphuric acid} \\ \text{combined with potash} \end{matrix} + \begin{matrix} (1-p) \text{ sulphuric acid} \\ \text{combined with soda} \end{matrix} + \begin{matrix} \text{excess} \\ \text{of potash} \end{matrix} + \begin{matrix} \text{excess} \\ \text{of soda} \end{matrix}$$

where the ratio $\dfrac{p}{1-p}$ depends (i) on the relative affinities of potash and soda towards sulphuric acid, and (ii) on $\dfrac{m}{M}$, the ratio of the masses of potash and soda. Here the possibility of variation in the composition of the sulphates is not considered. Berthollet's experiment (p. 119, 1, iii.) in which from a system such as this he isolated the two sulphates by the removal of the excess of alkali, allows us to infer that in this case at any rate he had recognised the possibility of excess of alkali remaining uncombined; but the vague and contradictory nature of his statements whenever these touch on the quantitative aspect of neutralisation makes it difficult to correctly appreciate his attitude in this matter.

How then did Berthollet proceed to harmonise his theoretical inferences concerning the necessary existence of variable composition with the actual facts? He knew that in the case of certain combinations the percentage of the one constituent was never less than a certain fixed value, or never more than another fixed value, or that it was thus limited in both directions; moreover a large number of substances were known, for which it had been definitely proved that they always contained their constituents in the same ratio. How did he explain the occurrence of such limiting values, and the existence of compounds of fixed composition, and what were the cases of variable composition he could adduce? This amounts to an inquiry as to how he dealt with the facts contradictory to his views, and as to the nature of his positive evidence.

Experimental evidence for variable composition.

The natural course would be to give the positive evidence first, were it not that all through the greater number of his most telling examples of variable composition there runs the recognition of the existence of some limiting value, which makes it necessary to begin by showing how he accounted for the existence of such limitations to the supposed general law. But since the principle of his explanation is the same as that used for the cases of fixed composition, it will be natural to deal with these next, leaving the positive evidence to the last.

Whilst posing in theory quite generally that in the ratio $\frac{m}{M}$ in which two substances combine, M can vary continuously from 0 to ∞, he had to admit from the outset that such

1. Explanation of apparent exceptions.

a series of combinations is not always found. He knew that no lead oxide could be formed containing a greater percentage of oxygen than that of the brown oxide; that no mercury oxide was known higher than the one which corresponds[1] to corrosive sublimate, and none lower than the one which corresponds to calomel, etc., etc.

Influence of physical conditions made to account for:

(i) Cases of limited variability.

"It is in the opposing forces that we must look for causes of limitation in the combining ratios." The influence of the physical conditions of the substances participating in, and resulting from a chemical change, which had been used so successfully to account for the preponderating effect of one affinity over another, is again made to serve. The argument is that the composition representing the limiting value is that of the most insoluble, or the most volatile, or the least volatile of all the possible combinations, which therefore is formed in preference to all others.

"A metal like zinc which volatilises as soon as it liquefies, is from the outset in the condition most favourable to combination. Hence it must unite directly with a fixed quantity of oxygen, with that amount of it in which the mutual action produces the greatest condensation; and this condensation then itself becomes the cause which sets a limit to the amount of oxygen entering into combination; beyond this point the action of the gaseous oxygen cannot overcome the obstacle offered to it by the condensation."

The case of the oxide of mercury of fixed composition obtained by calcination of the metal is dealt with in the same manner.

The identical reasoning is resorted to when dealing with sub-

(ii) Cases of fixed composition.

stances of fixed composition; these are considered as limiting cases which owe their existence to the operation of special physical conditions.

"An interesting problem remains to be solved: namely to ascertain the predisposing influences and the conditions which in the case of certain compounds produce fixed ratios, whilst other combinations occur in all

---

[1] The oxide "corresponding" to a salt produces that salt when acted on by the acid, and may itself be obtained from that salt by the action of an alkali.

ratios. But in the matter of these compounds of fixed composition, it behoves us to examine first what is true and what exaggerated in this property attributed to them.

Of all chemical problems, there is none that has been more neglected than the investigation why fixed ratios are met with in some cases, whilst in others combination occurs in all ratios. Having found that a certain number of compounds have a more or less fixed composition, no further attempts have been made to ascertain how far this constancy of ratios is real, how far it extends, and wherein the chemical action of the substances which exhibit this property differs from that of others which do not."

Neutralisation phenomena supply the example needed in the exposition of Berthollet's views on the problem stated above. The fact under consideration is, that if to an alkali which exerts a certain effect on a substance termed an indicator (purple litmus turns blue) an acid is added gradually, a point is reached at which the substance formed is termed neutral (purple litmus remains unchanged), and that after this it shows some of the properties of the free acid (purple litmus turns red).

"We see therefore that acidity and alkalinity saturate one another and may in turn predominate, according to the ratio in which the combination occurs; but there is no check, no interruption whatever in the progress of the combination and of the saturation[1] which accompanies it, unless it so happens that the forces of cohesion and elasticity produce the separation[2] of a substance in which the ratios are fixed in consequence of the effect of one of these conditions. Chemists, struck by the fact that in many compounds they had found fixity of composition, have often assumed it to be a general property of compounds to be thus constituted. They hold that when a neutral salt receives an excess of acid or of alkali, the resulting homogeneous substance is a solution of the neutral salt in a portion of the free acid or free alkali. This is a hypothesis which is founded on nothing else than the distinction between solution and combination."

The reason why the substance crystallising out from any such solution is neutral and of fixed composition, he finds in the least solubility of this particular combination; and he warns against drawing from the composition of such a substance any inferences as to the ratio in which the constituents are combined in the solution itself.

"The cause of the separation of substances in the solid state is that of their special composition; the ratios characteristic of these combinations are those

[1] Berthollet seems to distinguish "saturation" (neutralisation) from "combination."
[2] Formation of insoluble or of volatile substances.

of compounds possessed of sufficient cohesive force...for producing separation ; these ratios must be constant because the conditions [of formation] are the same....Hence it so happens that salts separate out by crystallisation in a neutral state, because in the neutral state the insolubility is greatest....It is generally supposed that the salts present in a solution are the same as those obtained from it by crystallisation ; but in such an assumption no attention is paid to what happens in the liquid, and nothing is taken into account but the final result."

To Berthollet substances formed with the greatest condensation—decrease of volume—are necessarily the most insoluble, and there is a curious argument concerning the crystallisation of such neutral salts. He knows that the amount of heat liberated when acids and alkalis combine is greatest when these substances are mixed in the exact ratio required for producing neutral salts. Heat evolution he thinks must always mean condensation, and hence the greatest heat effect would correspond to the greatest condensation, which in its turn would cause the greatest insolubility; but this special degree of cohesion is itself the proximate cause of the formation of a compound of fixed composition. For gases his argument runs on the same lines. Gaseous ammonia, which is formed from nitrogen and hydrogen with considerable condensation (the volume of the compound is half that of the constituents), has a fixed composition because of the cohesive force being a maximum ; nitric oxide, in the formation of which there is very little contraction (correct determinations show none), possesses the power of combining with further amounts of oxygen, which according to Berthollet increase gradually and continuously.

So the explanation which had served to show why in the competition of two substances for a third, physical conditions may lead to exclusive formation of one of the two possible combinations, now serves also to explain why the substance so formed has a fixed composition. When potash and baryta compete for sulphuric acid, the baryta gets all the sulphuric acid, because, owing to its insolubility and consequent precipitation, the active mass of barium sulphate present in the solution is being continually decreased; and the fact that the barium sulphate so formed, or formed in any way, has a fixed composition, independent of the relative amounts of baryta and sulphuric acid available, is also put down to the circumstance that this particular combination is much less soluble than all the other possible ones between the same constituents.

Hence, according to Berthollet's argument, the constant compo_ sition of the substances separating by precipitation or crystallisation from solutions still leaves open the possibility of the existence in these solutions of combinations in all ratios; and the inability at that time to determine the distribution of the components of salts in solutions made this reasoning fairly plausible, and the existence of such combinations of fixed composition not *à priori* incompatible with a general principle of variability.

Berthollet fared worse when it came to producing direct positive evidence, to finding cases of variable composition. Solu-

2. Examples of cases of variable composition:

tions, alloys and glasses, metallic oxides, and what we now call basic salts supply him with the material required.

(i) Solutions, alloys, and glasses.

"A solid, a salt for example, dissolves in water in all ratios, up to that extreme value which corresponds to saturation, a condition in which the tendency to solution has itself become weaker than the opposing tendency to cohesion; but the degree of saturation varies with the temperature, which diminishes the resistance due to cohesion....[Metals which alloy dissolve each other in all ratios unless the] difference in density or in fusibility should interrupt this mutual solution....Substances which vitrify combine in all ratios up to a point at which the insolubility of some of them presents an obstacle to the formation of this uniform transparent solution, which has all the characteristics of a chemical combination, in that all the properties have become common."

The composition of certain metallic oxides and of basic salts offers to Berthollet what he considers excellent proof of his proposition. Herein he was considerably helped by the fact that since faulty and hence non-concordant analyses gave results of the kind he required, there was no lack of such data in the then state of analytical chemistry.

It has already been said that Berthollet recognised and explained the existence of fixed limiting values in the composition of certain oxides. With regard to these he says:

"A number of chemists, struck by the fixed values to which certain oxidations are limited, suppose that there always exist fixed ratios to which the combination of the oxygen must conform; they assign to nature a balance which subject to certain decrees regulates the composition of compounds; and they pay no attention to the conditions which may be the causes that set a limit to the ratio of the combining substances....I must therefore show that the quantities of oxygen in the oxides depend on the same conditions which also effect other combinations, that these ratios can vary continuously from the point at which combination becomes possible, to the point at which it attains its highest value, and that if such variation is not found, it is only because the conditions which I have named constitute an obstacle to such progressive action."

According to Berthollet, a volatile metal, such as mercury or antimony, cannot form an oxide containing less than a certain percentage of oxygen, but this is not the case with non-volatile metals:

"It is not so with metals entering into quiet fusion, like tin and lead; their oxidation proceeds from the slightest amount up to one which however does not always represent the maximum of oxidation they
(ii) The oxides can be made to undergo under other conditions[1]; and these
of non-volatile
metals.   oxides show a gradation in colour and other properties, characteristic of all the different stages of oxidation. So lead forms an oxide which to begin with is grey, then passes through various shades of yellow, and finishes by being red....Iron also passes, as the oxidation proceeds, through different shades of colour, and assumes different properties.... The red oxide of mercury when rubbed with mercury divides its oxygen with an indefinite quantity of this latter, and forms an oxide which differs according to this ratio and which assumes various shades of greyish yellow.
    Vauquelin, by heating together equal quantities of red oxide of iron and of iron sheet, has obtained, without evolution of any gas, a black oxide containing not more than 0·25 parts of oxygen, while the red oxide had contained 0·40 to 0·49; and no doubt on varying the quantities, oxides could be thus produced in which the oxygen would be present in a ratio very different from that of the black oxide.
    Chevenix by fusing an oxide containing 20 per cent. of oxygen with more of the metal, has made an oxide of copper containing only 11½ per cent. of oxygen and approximating in colour to copper."

In the same manner Berthollet deals with the precipitates obtained by the action of varying amounts of alkali—soda or potash—on the salts of copper, mercury and bismuth. The precipitates produced by using comparatively small amounts of the alkalis are not the oxide, but carry down with them some acid; they are what are now termed basic salts, substances the composition of which is equivalent to that of hydrate plus neutral salt[2]. Further, cold solutions of potash or soda precipitate from copper, not the black oxide, but a green substance, soon after recognised as a compound differing from the oxide in that it contains water. To Berthollet the colour of this substance seemed to indicate the

---

[1] Heating lead in air leads to red lead at most; treatment of the red lead with dilute nitric acid, produces the higher, the brown oxide.
[2] Copper sulphate...$CuSO_4$.   Copper nitrate...$Cu(NO_3)_2$.
Basic copper sulphates...$CuSO_4.3Cu(OH)_2$ and $CuSO_4.2Cu(OH)_2$.
Basic copper nitrate...$Cu(NO_3)_2.3Cu(OH)_2$.

presence of some acid, which analysis showed to be very small in amount, and which we now know must have been due to insufficient washing from the adhering solution of the undecomposed salt.

This then was Berthollet's case in his attack on the principle of constant composition till then accepted, his theoretical foundation based on the principle of the influence of mass, and his experimental evidence for the doctrine of variable composition which he wished to establish.

The old view of constant composition, which had never been definitely proved, found a vigorous and finally victorious champion in Proust.

Fixity of composition asserted by Proust. Joseph Louis Proust (1755–1826), a Frenchman by birth and at the outset of his scientific career attached to the pharmaceutical department of the Salpêtrière Hospital in Paris, carried out his most celebrated researches in Madrid, where he was professor for some time. The war in 1808 put an end to his work there.

Proust's many papers in his controversy with Berthollet appeared in the *Journal de Physique* between the years 1802 and 1808. The clearness of the arguments used, the variety of the experimental work described, and the liveliness and keenness of the style make these papers which deal exclusively with the distinctly dry subject of quantitative analysis most interesting reading even now.

Proust proves that substances formed under the most diverse conditions have a fixed composition, and he shows that Berthollet's examples of variable composition were all cases of mixtures. This involves him in the necessity of differentiating between mixtures and compounds, an undertaking the difficulties of which he fully realised, and with which he dealt in a manner very much like that still resorted to for the same purpose.

As far back as 1799 Proust had proved the identity in composition of native and artificial carbonate of copper, and had enunciated the general principle of which this formed an instance. He found that the hot solution of 100 parts by weight of copper in sulphuric or in nitric acid, when precipitated by either potassium or sodium carbonate invariably gave 180 parts of green carbonate, which when distilled yielded 10 parts

1. Examples of cases of fixed composition. (i) Native and artificial substances have the same composition.

of water, and left, after the expulsion of the carbonic acid, 125 parts of black oxide of copper; and 100 of copper gave very concordantly 125 of black oxide. Hence the composition of artificial carbonate of copper is:—

Copper...	...	...	=100	= 55·6	} 69·4	
Oxygen...	...	...	= 25	= 13·8		
Carbonic acid	...	= 45	= 25		
Water ...	...	...	= 10	5·6		
Copper carbonate	=180	=100			

"The malachites[1] of Arragon when dissolved in nitric acid lose carbonic acid and leave 1 per cent. of an earthy residue; precipitation reproduces 99 parts of artificial carbonate. 100 grains of the same malachite calcined in a crucible at a moderate heat leave 71 grains of black oxide. If 2 per cent. are deducted for the foreign matter contained in it, there remains 69 grains for the oxide contained in the malachite, but to within a small fraction these 69 of copper oxide correspond also to 99 of artificial carbonate."

His argument concerning native and artificial carbonate is:

"Since 100 parts of this carbonate dissolved in nitric acid and thrown down by alkaline carbonates give us 100 parts of artificial carbonate, and since the base of these two compounds is the [same] black oxide, we must recognise an invisible hand which holds the balance in the formation of compounds; we must conclude that nature acts not differently in the depths of the earth than on its surface and through the agency of man. These ratios always the same, these constant properties which characterise the true compounds of art or of nature, in one word this *pondus naturae* so well realised by Stahl is, I say, no more left to the power of chemists, than is the law of election which governs all these combinations."

Proust recognises the existence of more than one compound between the same elements, but knows that the number of such different combinations is small—often not more than two—that the composition of each is perfectly fixed, and the change in the combining ratios sudden and considerable. Thus he finds that iron can be artificially made to combine with sulphur in two ratios, one of the two substances so produced being in properties and composition identical with native iron pyrites.

(ii) The number of possible combinations between the same constituents is small, and the change in composition sudden.

---

[1] Malachite = basic carbonate of copper = $CuCO_3 . Cu(OH)_2$.

The results of his experiments on these two sulphides, when summarised, are:

<table>
<tr><td>1. Compound containing minimum of sulphur.</td><td>2. Compound containing maximum of sulphur.</td></tr>
<tr><td>*Production:* (i) Heating iron and sulphur to fusion.</td><td>(i) Native.</td></tr>
<tr><td>. (ii) Ignition of pyrites, or the higher artificially prepared sulphide.</td><td>(ii) Heating iron or the lower sulphide with sulphur to a temperature at which pyrites is not yet decomposed.</td></tr>
</table>

Composition: For every hundred parts by weight of iron:

60 of sulphur.	90 of sulphur[1].

"From this it follows that at a sufficiently high temperature one hundred parts by weight of iron can bind 60 of sulphur, which gives the sulphide in the minimum. But at lower temperatures this amount of iron can fix an additional quantity of sulphur, equal to half the above weight, and the result is a sulphide in the maximum, containing 90 of sulphur. If however the second compound is exposed to the temperature at which the first is produced, it passes into this; that is to say, on depriving it of all the sulphur that it had been able to bind above 60 parts per hundred of iron, it is brought back to the minimum. Iron sulphide in the maximum is simply pyrites; it has very nearly the same density, and shares all other properties with pyrites."

Not only have artificial and native compounds the same composition, but native compounds, whatever parts of the earth they come from, have the same composition also.

(iii) Minerals of different origin are identical in composition.

"According to our principles a compound is a substance such as the sulphide of silver, of antimony, of mercury, of copper, such as an oxidised metal, an acidified combustible; it is a privileged product to which nature assigns fixed ratios, it is, in short, a being which nature never creates even when through the agency of man, otherwise than balance in hand, *pondere et mesura.* Let us recognise, therefore, that the properties of true compounds are invariable as is the ratio of their constituents. Between pole and pole, they are found identical in these two respects; their appearance may vary owing to the manner of aggregation, but their properties never. No differences have yet been observed between the oxides of iron from the South and those from the North. The cinnabar [mercuric sulphide] of Japan is constituted according to the same ratio as that of Almaden. Silver is not differently oxidised or muriated in the muriate [chloride] of Peru than in that of Siberia. In all the known parts of the world you will not find two muriates of soda

---

[1] These results are not accurate. Iron pyrites, FeS_2, contains twice as much sulphur as FeS, the sulphide obtained by strongly heating sulphur with excess of iron (*post*, p. 166).

[rock salt], two muriates of ammonia, two saltpetres, two sulphates of lime, of potash, of soda, of magnesia, of baryta, etc., differing from each other. ...The native oxides follow the same relations of composition as the artificial. This is a fact which analysis confirms at every step. We find in the bosom of the earth copper oxide containing 25 per cent. of oxygen, arsenic with 33, lead with 9, antimony with 30, iron with 28 and 48, and others still[1]."

Berthollet accounted for the fixed composition of certain of the substances artificially produced by the chemist by saying that the special physical conditions cause the formation of a compound of a particular composition, and hence that sameness of the conditions of formation is the cause of constancy of composition. The identity of the composition of native and artificial oxides etc. deprives this argument of its force, because, however ignorant we may be of the conditions under which the different minerals have been formed, it is certain that these conditions have been very varied and very different from those of the laboratory methods:

"But if it is found that in our hands metals cannot bind oxygen beyond the fixed ratios known to us, because the progress of their oxidation is suddenly checked by the action of the opposing forces specified by Berthollet, shall we also be obliged to believe that when nature does not proceed beyond these very same ratios in the oxides that she offers us, this is because her resources for oxidation are limited by the very same opposing forces as those that prevail in our laboratories? And yet this is what would have to be granted in order to account for the constant agreement we find between the composition and properties of nature's oxides and ours. This, I hold, involves an identity of causes which it will not be found easy to admit; on the contrary, we shall rather concur in the belief that if the combinations which we make every day in our laboratories have a perfect resemblance to those of nature, this is due to the fact that the powers of nature hold invisible sway over all the operations of our arts. If we find it

Alteration in the conditions of formation does not produce a change in composition.

impossible to make an ounce of nitric acid, an oxide, a sulphide, a drop of water, in ratios other than those which nature had assigned to them from all eternity, we must again recognise that *there is a balance which, subject to the decrees of nature, regulates even in our laboratories the ratios of compounds.* And even if some day we should succeed in clearly recognising the causes which retard or accelerate the action of substances tending to combine, we could only flatter ourselves with knowing one more thing, namely, the means which nature uses to restrict compounds to the ratios in which we find them combined. But such knowledge, would it invalidate the principle I have proved? I think not, because the principle is

---

[1] These are also the percentages of the artificially formed oxides.

only the corollary of the facts which we discover every day; there is nothing hypothetical about it; facts have led to it, facts alone could overthrow it."

Proust's diligence in supplying and multiplying the facts on which the principle is based was prodigious. He investigates the compounds of antimony, of tin, of cobalt, of nickel, of copper; throughout he finds that, vary the conditions and the relative masses as you will, the oxides and sulphides produced always have a definite composition:

"100 parts of antimony and as many of sulphur heated in a glass retort until all is well fused, and the excess of the sulphur expelled, leave 135 parts of sulphide. This experiment, repeated as many times as desired, always gives the same result. 100 parts of antimony heated with 300 parts of cinnabar give from 135 to 136 of sulphide[1]. 100 parts of these artificial sulphides kept in fusion for one hour lose nothing; 100 parts of these sulphides heated with as much sulphur lose nothing in weight. Hence antimony follows the law of all the metals which can unite with sulphur. It attaches to itself an invariably fixed quantity of it, which we have no power to increase or diminish."

These are some striking cases chosen from amongst the large number of investigations whereby Proust adduced positive evidence

2. Refutation of examples of cases of varied composition. (i) Glasses etc. classed as mixtures.

for the constancy of the composition of compounds; and he was none the less successful in refuting Berthollet's examples of variable composition. Solutions, alloys and glasses he looked upon as mixtures, and differentiated them from compounds proper. How he characterises a mixture as opposed to a compound will be dealt with later.

It has already been shown how Proust recognised the existence of two or more different oxides of certain metals. Berthollet's many

(ii) Oxides of supposed variable composition proved to be mixtures of two oxides of fixed composition, or of one oxide and the metal.

oxides, supposed to contain the metal and oxygen in ratios which vary continuously between certain limits, are shown to be mixtures in varying ratios of two oxides of different but individually fixed composition, or of one oxide and the metal.

So with regard to the oxides of lead and of bismuth, said to contain quantities of oxygen varying from zero upwards (p. 134), Proust showed that on treating them

---

[1] Antimony + mercuric sulphide
      (cinnabar)
  = antimony sulphide + mercury + excess of mercury sulphide.
    (not volatile)   (volatile)   (volatile)

with an acid so dilute as not to attack the metal itself, whilst acting on the ordinary oxide, the nature of these substances is revealed as a mixture of the metal and the oxide corresponding to the ordinary salts of the metal.

"If the grey or greenish oxide is treated with water acidified with nitric acid, an immediate separation results ; finely granulated metal is separated, and the solution contains nothing but ordinary nitrate of lead,— vinegar acts on such oxides in the same manner and with the same results. ... The greenish or incomplete oxide of bismuth shows itself to be of the same nature as the calcined oxide of lead; weak acids detect in it nothing but finely divided metal and yellow 12 per cent. oxide of bismuth. It is the admixture of metallic powder with the oxide, of black with yellow, which gives us these light or dark greenish tints, through which the lead and bismuth pass before assuming the clear yellow colour, which is the sign of complete oxidation. It is the same with all other calcinations by fire. At whatever stage they may be, it is always a case of a mixture between molecules that are metallic and others that are oxidised in the various degrees fixed by their nature."

Berthollet believed that when mercury is dissolved in nitric acid under varying conditions of concentration and temperature, a variety of nitrates corresponding to a large number of different oxides is produced; but he had to explain why these nitrates when treated with hydrochloric acid always gave a mixture of only two chlorides of perfectly fixed composition, the insoluble calomel and the soluble corrosive sublimate[1].

"To remove this truly embarrassing difficulty, Berthollet says that the mercury passes into two compounds of constant composition, only at the moment when it can separate in these two combinations ; and hence it would have to be assumed that in mixing the hydrochloric acid with the nitrate, these two combinations are only formed at the precise moment when the separation of the soluble and insoluble salts is initiated by the action of this acid. That is to say, the numerous oxides which in Berthollet's opinion give rise to as many different nitrates all present in the same solution, urged by the affinity of the hydrochloric acid as much as by the insolubility of one of the chlorides about to be formed, these oxides, I say, suddenly and simultaneously abandon

[1] Mercury + nitric acid = x mercuric nitrate (containing about 33·3 per cent.
of nitric acid)
    + y mercurous nitrate (containing about 20·6 per cent.
of nitric acid)
∴ the percentage of nitric acid will vary continuously from 20·6 when $x = 0$ to 33·3 when $y = 0$.

Mercuric nitrate + hydrochloric acid = mercuric chloride
                                        (corrosive sublimate)
Mercurous nitrate + hydrochloric acid = mercurous chloride.
                                        (calomel, insoluble in water)

the positions they had occupied in the series, to rush to its two extreme values, to place themselves just where are found the corrosive sublimate and the calomel, the only two compounds remaining after the operation. It must be admitted that this is a case of oxides behaving with much intelligence.... If by raising the temperature we reduce the weight of an oxide which had reached the highest degree of oxidation, and which does not suffer from the inconvenience of being volatile; or on the other hand, if by long-continued heating we raise a metal to this highest stage of oxidation: will it be permissible to believe with good reason that all the ascending and descending terms of oxidation, which by these means we can introduce between the extremes, represent as many different oxides? Certainly not! I will not recognise in this the ordinary course of nature, and I do believe that in such cases we produce nothing but mixtures in all possible ratios, of the oxide at the minimum with the oxide at the maximum. The following experiments, if they do not lend themselves to actually proving this view, at least make it very plausible.

Having had occasion to analyse some of these oxides, described by Rinmann and said to be of all proportions, I found that on calcining steel, or wrought iron, or cast iron, I met with none but the two known oxides of this metal[1] mixed in various ratios."

The recognition that lower percentage oxides are mixtures between an oxide of fixed composition and the metal, readily explains why Berthollet had found that the calcination of a volatile metal like mercury and antimony did *not* give him oxides of a composition below a certain limiting value, the simple reason being that heat removes from the mixture the excess of the volatile metal. Hence also, whilst calcination did not give a lower oxide of mercury, rubbing the ordinary oxide with metallic mercury did.

In his investigations on antimony compounds Proust finds that he can get the same ruby coloured substances either by fusing together varying quantities of the oxide and the sulphide of antimony, when no gas is evolved, or by fusing the oxide and sulphur, when sulphurous acid is evolved; thus clearly indicating that in the second case oxygen is eliminated from some of the oxide, sulphide being formed in its place. His general inference is:

"These results furnish us with the proof that the livers, crocuses, and glasses, all the hepatiques[2] transmitted to us by ancient chemistry

---

[1] The oxides meant are those to which correspond the ferrous and ferric salts.

[2] Names given to different antimony compounds: *Livers of antimony* are brown or black substances, more or less soluble in water, formed by the combination between antimony sulphide and alkaline sulphides. *Glass of antimony* or *vitrum antimonii* is a transparent dark red ruby mass formed from the oxide and sulphide.

are nothing but the oxide at the minimum holding in solution not, as had been thought, sulphur, but varying quantities of the sulphide of antimony; in short, they are mixtures, which can be represented by the formula, oxide of antimony $+1+2+3+4$ etc. of sulphide of antimony. This clears the oxide of this metal from the suspicion cast upon it, of being able to unite with sulphur in all ratios, regardless of the unalterable laws of combination."

(iii) Livers, crocuses, etc. are mixtures of the oxide and sulphide of antimony.

And just as with these antimony compounds, which really are mixtures of different combinations in various ratios, so it is with many minerals found in nature, in which oxides may be associated with sulphides, or one oxide with another; and mineralogists are advised to distinguish between "combinations of elements" and "associations of combinations," the latter of which would on analysis give variable results.

"But what difference, it will be asked, do you recognise between your chemical *combinations* and the *unions of combinations*, which latter you tell us nature restricts to no fixed ratios?

Proust's distinction between compounds and mixtures.

Is the power which makes a metal dissolve in sulphur different from that which makes one metallic sulphide dissolve in another? I shall be in no hurry to answer this question, legitimate though it be, for fear of losing myself in a region not yet sufficiently lighted up by the science of facts; but my distinctions will, I hope, be appreciated all the same when I say: The attraction which makes sugar dissolve in water may or may not be the same as that which makes a fixed quantity of carbon and of hydrogen dissolve in another quantity of oxygen to form the sugar of our plants, but what we do see clearly is that these two kinds of attractions are so different in their results that it is impossible to confound them."

He quotes as examples of the two kinds of actions:

(i)   The solution of nitre in water, compared with that of nitrogen in oxygen which gives nitric acid, or of nitric acid in potash which gives saltpetre.

(ii)   The solution of ammonia in water, compared with that of hydrogen in nitrogen which gives ammonia.

(iii)   The solution of antimony sulphide in antimony oxide, compared with that of antimony in sulphur.

Proust's criteria are that solutions of the first kind can occur in any ratios, those of the second kind only in fixed ratios. This

Antimonial saffron or *crocus antimonii*, a brownish powder melting to a yellow glass, is obtained by the deflagration of stibnite (antimony sulphide) with nitre (Roscoe and Schorlemmer, *Treatise on Chemistry*, II.).

is of course to argue in a circle, but all the same the appeal is to facts true and easily apprehended.

"Sulphide of antimony can dissolve in the lower oxide in an infinite number of ratios which give rise to the livers, the glasses, the crocuses, and all the intermediate shades. But is it so with antimony itself in its relations to sulphur? Do we know of two solutions of the one substance in the other, of two sulphides of antimony? In the unions termed 'compounds,' nature imposes laws on herself and on us, so that no chemist can make compounds in new proportions. Chemistry no longer confounds these two types of union, but needs names to distinguish them."

For the sake of completeness in the refutation of Berthollet's cases of supposed variable composition, it may be mentioned that

(iv) Copper hydrate identified as a fixed compound.

Proust showed successfully that the coloured substance precipitated in the cold from copper salts by an alkali differs from the black oxide only in that it contains water. The assumption of the presence of acid because the substance has the colour of the salts, is not justified, since the presence or absence of water accounts also for the difference in colour between white anhydrous copper sulphate and the blue crystalline salt.

And so all along the line Proust had proved his point, which may be summed up in his own words:

"Election and proportion [that is, affinity and fixity of composition] are the two poles about which revolves immutably the whole system of true compounds, whether produced by Nature or by Man."

Fixity of composition, recognised by Proust as a perfectly general characteristic of compounds, is one of the fundamental principles of chemistry. The law of fixed ratios, or as it is more commonly called, the law of fixed proportions, may be expressed in one or other of the following forms:

"The relative weight of combining bodies is always fixed in every combination" (Wurtz, *The Atomic Theory*).

The Law of Fixity of Composition formulated.

"All true chemical changes take place between definite volumes or weights of the substances" (Lothar Meyer, *Outlines of Theoretical Chemistry*).

"Definite chemical compounds always contain their constituents in fixed and invariable proportions" (Lothar Meyer, *ibid.*).

"If one substance is transformed into another, then the masses of these two substances always bear a fixed ratio to each other.

If several substances react together then their masses as well as those of the new bodies formed always bear fixed proportions to each other" (Ostwald, *Outlines of General Chemistry*).

"The masses of the constituents of homogeneous substances bear a constant ratio to each other as well as to the mass of the compounds" (Ostwald, *Lehrbuch der Chemie*).

Does this law describe absolutely exact or only approximate relations? Must it be classed with the law of the conservation of mass as an exact law, or with that of gaseous compressibility as an approximate law? Proust's work with experimental errors of at least 1 to 2 per cent. (*e.g.* "give from 135 to 136 of sulphide"—"if 2 per cent. are deducted for the foreign matter contained" pp. 139, 136), and with its intentionally rounded off

Is it an exact or an approximate law? Proust's results do not supply an answer.

figures does not, of course, lend itself to the necessary test. But soon after 1810 an enormous improvement in analytical work set in, and the accuracy aimed at and attained grew marvellously. Greater and greater became the concordance between the results representing the composition of different specimens of the same substance when determined by different methods and by different observers. Berzelius had led the way[1], others had followed, and the development of the methods of accurate analysis culminated in the work of Stas.

The discrepancies decrease with increase in the accuracy of the measurements, and are of the same order as the experimental errors.

The law of fixity of composition was beginning to earn the title of "exact law," in that the values for the discrepancies between experimental results and theoretical requirements, and those for errors of observation, kept of the same order of magnitude, and together grew pleasantly lesser and lesser. For instance, the analyses of, say a specimen of potassium chloride made by the neutralisation of potash by hydrochloric acid, and of another specimen made by heating potassium chlorate, did not differ from each other more than two separate analyses of portions of the same potash-hydrochloric acid specimen. Theory requires that the analyses of different specimens should give

---

[1] "Most of the chemical experiments have been made in the laboratory of M. Berzelius, where I enjoyed the advantage of witnessing and of acquiring the most exact methods to which chemical analysis has yet attained." "Six months have elapsed since the reading of this Memoir; in this time I have learned in the laboratory of M. Berzelius methods of working of which I had had no idea before" (Mitscherlich, 1821 and 1819).

identical results. But it has been shown before that the inevitable presence of experimental error makes the realisation of such agreement quite impossible; and hence that if the discrepancies between experiment and theory are only of the magnitude of the experimental error, they may be accounted for by experimental error alone, or by experimental error together with actual deviations from the law, which latter must however in such case be smaller than the experimental error. So for the law of fixed composition, as for that of any other quantitative law, all that can be done at any time is, after reducing experimental error to a minimum: to prove that if deviations from the law exist at all, these cannot be greater than a certain value, which is that of the experimental error; and not to forget that whilst the magnitude of such deviations would prevent the possibility of their being detected by the experimental means available at the time, this might not be so in the future. The final conclusion in such a case is, that since from a purely empirical point of view, and leaving out of consideration any philosophical bias against inexactness in the laws of nature, the probability of the law being an exact one is as great as the opposite, and since for all practical purposes it may be considered so, it should *provisionally* be classed as an exact law.

What then is the degree of accuracy to which the law of fixity of composition has been proved?

Degree of accuracy to which the law has been proved.

Marignac on theoretical grounds considers that the law is possibly only approximate.

J. C. G. de Marignac (1817—1894), Professor at Geneva, and most famous for his accurate determinations of the combining ratios of the constituents of compounds, guided thereto by zealous support of the hypothesis of the English doctor Prout (*post*, chap. xvi) in 1860 threw out the suggestion that the composition of compounds might vary within very small limits. The experimental data already available at the time, and amongst which his own ranked in point of accuracy with the highest, excluded the possibility of any but very slight variations. He says:

"I do not consider it as absolutely demonstrated that a number of compounds may not constantly, in the normal course of things, contain an excess of one of their constituents, very minute no doubt, but still perceptible in very delicate experiments."

This passage occurred in an article in which Stas' determinations[1] of the combining ratios of certain elements were discussed and criticised with special reference to the discrepancies between his results and the requirements of Prout's hypothesis which Marignac championed. Stas, by nature always averse to assuming anything that could be investigated and proved, made this hypothesis the subject of one of his now classical researches.

Stas' experimental investigation of the question raised by Marignac. "I am fully aware that amongst the fundamental principles of chemistry, as in all other sciences, there are a number which have been accepted as proved, whilst this is far from being the case. What is the position, in this respect, of the law of fixed ratios? This law rests on the analyses and the syntheses made in the course of nearly a century. Therefore it seems to me that, even in the most exacting mind, there should be no room for doubt concerning the generally admitted fact of the constancy of composition.

Although in the realm of well-established facts I have not met with a single one calculated to support the view of the celebrated Geneva chemist, I have all the same felt bound, even if it were only from deference to his opinion, to submit this question to a new examination. Amongst the conditions which might conduce to make the composition of stable compounds variable are temperature and pressure."

Syntheses of silver chloride under different conditions. The influence of temperature found nil. As far back as 1860 Stas had published the results of syntheses of silver chloride made in a variety of ways. The table on page 147 summarises his methods and results.

Concerning this work Stas, in the 1865 discussion on fixity of composition, says:

"An examination of the synthesis of silver chloride contained in the work published by me in 1860 will show that this substance, though produced under very different conditions, has a composition which I am forced to regard as constant. In fact I dare not attribute the insignificant differences in the results to other causes than to the inevitable errors of observation. And yet in some of the experiments the chloride of silver had been produced by the combustion of the silver in chlorine at red heat, whilst in others it had been obtained at the ordinary temperature by precipitating with hydrochloric acid the silver dissolved in nitric acid. The influence of temperature in these instances seems to me, therefore, to have been *nil*."

---

[1] *Recherches sur les rapports réciproques des poids atomiques*, 1860.

Stas' Syntheses of Silver Chloride.

Preparation	Weight of silver	Weight of silver chloride	Weight of silver chloride derived either from the silver carried away in the process of solution or from the evaporation of the washings	Total weight of silver chloride	Weight of silver chloride produced by 100 of silver
Burning of the metal in chlorine and subsequent heating in a current of carbonic acid to remove the chlorine retained by the fused chloride	91·462	121·4993		121·4993	132·841 = Mean − ·004
	69·86735	92·8145		92·8145	132·843 = ,, − ·002
	101·519	134·861		134·861	132·843 = ,, − ·002
Solution of the silver in nitric acid; precipitation of the nitrate formed by gaseous hydrochloric acid led in near the surface of the solution; evaporation of the whole mass in the same vessel; fusion of the chloride produced in an atmosphere of hydrochloric acid afterwards replaced by air	108·549	144·1725	0·0345	144·207	132·849 = ,, + ·004
	399·651	530·826	0·0940	530·920	132·846 = ,, + ·001
Solution of the silver in nitric acid ; precipitation by a solution of hydrochloric acid; washing of the precipitate by water acidulated with nitric acid; fusion of the chloride in an atmosphere of hydrochloric acid, etc.; evaporation out of contact with air of all the liquids, in order to recover the chloride carried away or dissolved	99·9925	132·8379	0·00035	132·8382	132·848 = ,, + ·003
Solution of the silver in nitric acid ; precipitation of the nitrate by solution of pure ammonium chloride; washing of the precipitate by water acidulated with nitric acid; fusion of the chloride in an atmosphere of hydrochloric acid, etc.; destruction of all the ammoniacal salts in the washings by gaseous chlorine, and subsequent evaporation out of contact with air, in order to recover the silver chloride carried away or dissolved	98·3140	130·566	0·0360	130·602	132·842 = ,, − ·003
					132·845 = Mean

The following quotations are from the 1865 memoir:

"With the object of settling by new researches whether the temperature and pressure during the formation of a compound exert an influence on the ratio in which the constituents unite, I again took up the determination of the proportional number between silver and ammonium chloride, which had once before formed the subject of a very long investigation on my part. I chose this particular reaction for two reasons: firstly, because it allowed me to make the influence of temperature and pressure intervene in the conditions of formation; and secondly, because the process could be carried out with an accuracy approaching to the mathematical. The experimentally determined proportional number between silver and chloride of ammonium depends simultaneously on the purity of the silver and on the composition of the silver chloride and the ammonium chloride; whilst the composition of the ammonium chloride itself depends on its purity. And since the determination can be made at ordinary temperature as well as at 100°, it becomes possible to ascertain with great exactness whether the composition of silver chloride can within these limits experience any change. On the other hand, the ammonium chloride can be produced at the ordinary or at a higher temperature; and in the latter case it can be volatilised at the ordinary pressure or *in vacuo*.

Analyses of ammonium chloride produced in different ways.

Conditions, the influence of which was investigated: (i) source of preparation, (ii) state of aggregation, (iii) temperature, (iv) pressure.

And finally the ammonia intended for the preparation of the chloride may be derived from different sources, which allows of a reciprocal control of the results. With this object I have prepared the ammonium chloride by three different methods, namely:

(a) From commercial ammonium chloride suitably purified.

(b) From purified commercial ammonium sulphate[1].

(c) From ammonia derived from the reduction of potassium nitrite[2].

I first investigated whether under the same conditions the results were the same for the different chlorides (Table, p. 150, Exps. I., III., V., VI.) and whether they were the same as those I had obtained on a former occasion (Exps. X., XI., XII.)."

In the preparation of the pure silver required Stas displayed

[1]
$$\left.\begin{matrix}\text{Ammonium chloride}\\ \text{or}\\ \text{ammonium sulphate}\end{matrix}\right\} + \text{calcium h drate}\atop(\text{lime}) = \left.\begin{matrix}\text{calcium chloride}\\ \text{or}\\ \text{calcium sulphate}\end{matrix}\right\} + \text{water} + \text{ammonia}$$

$$\left\{\begin{matrix}2NH_4Cl\\ \text{or}\\ (NH_4)_2SO_4\end{matrix}\right\} + \underset{(\text{non-volatile})}{Ca(OH)_2} = \left\{\begin{matrix}CaCl_2\\ \text{or}\\ CaSO_4\end{matrix}\right\} + 2H_2O + \underset{(\text{gas})}{2NH_3}$$

[2]
$$\underset{KNO_2}{\text{Potassium nitrite}} + \underset{3H_2}{\text{nascent hydrogen}} = \underset{KOH}{\text{potash}} + \text{water} + \text{ammonia} \atop + H_2O + NH_3$$

and:

$$\underset{2KOH}{\text{potash}} + \underset{Zn\ (\text{and Fe})}{\text{metallic zinc}} = \underset{K_2ZnO_2}{\text{potassium zincate}} + \underset{H_2}{\text{hydrogen}}$$

all the extraordinary carefulness and ingenuity of which he was capable. Silver prepared in a variety of ways was compared by its action with pure sodium chloride. 10·000 grams of the different specimens of silver were dissolved with all possible precautions in nitric acid, and were mixed with a solution of 5·420 grams of sodium chloride, a quantity almost exactly sufficient for the complete precipitation of the silver as silver chloride; the slight excess of silver remaining in the solution was then estimated by the addition of standard solution of sodium chloride. The all but identical results obtained for the different specimens of silver could be taken as proof of the absence of impurities. The procedure for finding the proportional number between the ammonium chloride and the silver prepared in the manner just described was as follows:

Experimental procedure in the determination of the ratio silver : ammonium chloride.

Quantities of silver and of ammonium chloride in the ratio of 2·0187 : 1 were weighed out most accurately, it being known from previous experiments that if mixed in this ratio there remains after the action a very slight excess of silver in the solution above the precipitated silver chloride. The silver was dissolved with all possible precautions, and to prevent any chance of loss of hydrochloric acid when, in the subsequent determinations at 100°, the ammonium chloride is added to a hot solution of the silver nitrate containing much free nitric acid, this acid was neutralised by the gradual addition of pure ammonia. The neutralised silver nitrate solution was either kept at the ordinary temperature or heated to 100°, and the weighed quantities of solid ammonium chloride dropped into it. The vessel with its contents was shaken in order to make the silver chloride settle, and the excess of silver in the clear solution was then estimated by means of sodium chloride solution of known silver value. To this end, the sodium chloride solution was run drop by drop into the liquid through which a strong beam of monochromatic yellow light passed, until no further turbidity was produced. This method of illumination devised by Stas allowed him to determine the end point of the reaction with great accuracy, and enabled him to measure with ease $\frac{1}{20}$ of a milligram of silver in 1 litre of solution, and to detect with certainty $\frac{1}{100}$ of a milligram.

The following table embodies his results concerning the proportional numbers between ammonium chloride and silver:

Stas' Analyses of Ammonium Chloride.

Preparation		Physical conditions		Temperature at precipitation of the silver chloride	Weight of ammonium chloride	Weight of silver added	Weight of the excess of silver left after the double decomposition	Weight of ammonium chloride interacting with 100·000 of silver
		Pressure at sublimation of the ammonium chloride	State of aggregation of the ammonium chloride					
Ammonia from ammonium chloride Neutralisation by pure hydrochloric acid gas of the ammonia solution obtained on dissolving in water the ammonia liberated by lime from ammonium chloride, which had previously been purified by nitric acid.	I	Ordinary		Ordinary	11·7964	23·8133	0·0290	49·598 = Mean + ·001
The ammonium chloride was sublimed in a current of dry ammonia gas	II	Ordinary		100°	39·6213	79·9831	0·0970	49·597 = ,, ± ·000
Ammonia from ammonium sulphate The commercial sulphate was heated with concentrated sulphuric acid, and then	III	Ordinary		Ordinary	11·8084	23·8376	0·0290	49·597 = ,, ± ·000
treated with nitric acid to destroy organic impurities; liberation of the ammonia, and subsequent treatment as above	IV	Ordinary		100°	13·4063	27·0632	0·0355	49·602 = ,, + ·005
Ammonia from potassium nitrite Potassium nitrate was by fusion with lead reduced to the nitrite; the nitrite was reduced to ammonia by the nascent hydrogen resulting from the action of a mixture of zinc and fine iron wire on potash. Subsequent treatment of the ammonia as above. Some of this ammonium chloride was resublimed *in vacuo*	V	Ordinary	crystalline mass	Ordinary	6·2521	12·6212	0·0140	49·593 = ,, - ·004
	VI	Ordinary	fine powder	Ordinary	10·7176	21·6355	0·0262	49·597 = ,, ± ·000
	VII	Ordinary	crystalline mass	100°	7·6011	15·3442	0·0187	49·597 = ,, ± ·000
	VIII	Vacuo	crystalline mass	Ordinary	13·5129	27·2784	0·0355	49·598 = ,, + ·001
	IX	Vacuo	fine powder	Ordinary	6·2250	12·5663	0·0140	49·592 = ,, - ·005
Experiments belonging to an earlier date. A mean value of 49·5944 was obtained in 19 determinations, of which three are given here	X	Ordinary		Ordinary	11·0088	22·2236	0·0300	49·600 = ,, + ·003
	XI	Ordinary		Ordinary	10·9290	22·0673	0·0280	49·599 = ,, + ·002
	XII	Ordinary		Ordinary	12·2604	24·7499	0·0305	49·598 = ,, + ·001

49·597 = Mean

An examination of the table[1] shows that in most cases the deviations from the mean value were not above 1 milligram per 100 grams of ammonium chloride used, that is equal to the probable errors of the weighings. And these small differences are distributed quite at random over the individual experiments. No parallelism can be detected between the deviations of the values from the mean and variations in temperature, or pressure, or mode of preparation of the ammonium chloride, or aggregation of the salt. Experiments II. IV. and VII. made at 100°, show no constant difference from I. III. V. and VI. made at ordinary temperature; and the same was found concerning the influence of all the other variables investigated. Hence in these ammonium chloride analyses, just as in the silver chloride syntheses, the differences observed can all be accounted for by experimental error, and the law of fixed composition can accordingly be classed as an exact law. But the statement that it is a law the exactness of which has been proved to ·004 per cent. for the composition of silver chloride, and to ·008 per cent.[2] for the composition of ammonium chloride, represents the truth more strictly. Stas' own *résumé* is:

Stas' results justify the classification of the law as exact.

"The results given prove that within the limits I had been obliged to impose to render the experiment possible, temperature has no influence on the composition of the ammonium chloride and of the silver chloride; they prove further that pressure exerts no influence whatsoever on the composition of the ammonium chloride.

Stas' résumé of the evidence for the exact nature of the law of fixed ratios.

If the recognised constancy of stable chemical combinations has needed further demonstration, I consider that the all but absolute identity of ... [my results] has now completely proved it."

[1] In Stas' own table the results are arranged in four series: *First.* Old experiments comprising Nos. X. XI. XII. *Second.* Experiments at ordinary pressure and ordinary temperature, Nos. I. III. V. VI. *Third.* Experiments at ordinary pressure and at 100° C., Nos. II. IV. VII. *Fourth.* Experiments *in vacuo* and at ordinary temperature, Nos. VIII. and IX.

[2] These are rough evaluations, in which twice the *average* deviation from the mean has been calculated to 100 of the compound.

CHAPTER VI.

DALTON AND THE LAW OF MULTIPLE RATIOS.

*" What have the ratios of small whole numbers to do with concord?
This is an old riddle propounded by Pythagoras, and hitherto
unsolved."*

HELMHOLTZ, 1857.

PROUST had recognised that in many cases the same constituents may form *two* combinations, each of fixed composition,
and had proved that a mixture in varying ratios
of two such combinations accounted for the existence
of a number of substances containing the same
constituents in many different, continuously varying
ratios. But most scrupulous about not asserting
more than what the facts observed warranted, he
carefully guarded against his argument being interpreted to mean that only *two* compounds between
the same constituents could exist:

*Proust though
recognising
the combina-
tions of the
same constitu-
ents in differ-
ent ratios, had
failed to see
the simple re-
lation between
these various
ratios.*

"In considering metals from this point of view I have never intended to
set a limit to the number of oxidations which they are capable of undergoing.
Who, in fact, would dare to assert that the progress of chemistry may not
soon bring to our notice others, which already exist in nature, hidden from
us as yet, but ready for discovery?"

The orthodox treatment of the subject of the quantitative
composition of compounds containing the same constituents, would
here demand some remark to the effect that Proust was within
measurable distance of the discovery, that in such a case the
combining ratios change according to a definite law. It would
be correct to say that it is strange Proust should not, instead of
always representing his results as percentages, have calculated in
addition the amount of one substance uniting in each of the
different compounds with the *same* amount of the other; and how
if he had done so, he could not have failed to notice that the
numbers, inaccurate though they were, bore some simple relation

to each other.   And it is further usual to add that when just about the time of Proust's work on fixity of composition, the existence of such a simple relation was discovered in England . (1802—1808), this was done on the basis of experimental data not any more accurate than Proust's.   But is it after all strange that a man who was an experimenter from first to last, one averse to any speculations and hypotheses not absolutely necessitated by the facts of the science, should not have seen much of simple regularity in numbers which, for all we know, he may have represented in this manner?

		Oxides of Tin.				*Oxides of Copper.*		
I	Tin	78·4	100	I	Copper	80	100
	Oxygen	21·6	27·5		Oxygen	20	25
	Tin Oxide	100·0	127·5		Copper Oxide	100	125
II	Tin	87	100	II	Copper	86·2	100
	Oxygen	13	14·9		Oxygen	13·8	16
	Tin Oxide	100	114·9		Copper Oxide	100	116

$$27·5 = \mathbf{1·84} \times 14·9 \qquad\qquad 25 = \mathbf{1·57} \times 16$$

Oxides of Antimony.

I	Antimony		77	100
	Oxygen		23	29·8
	Antimony Oxide		100	129·8
II	Antimony		81·5	100
	Oxygen		18·5	22·7
	Antimony Oxide		100	122·7

$$29·8 = \mathbf{1·31} \times 22·7$$

And if, as did happen, Dalton saw the regularity where Proust did not, this is not strange either, considering the fundamental difference between their ways of looking at nature.

Dalton calculates the different quantities of A which combine with B to the same amount of B, which reveals the law of multiple ratios.

Dalton, a theorist before everything, full of speculations concerning the hidden nature of phenomena, was the man from whom, with the material available, the discovery of such a regularity could be expected; Proust was not.   And this is borne out by the view recently propounded on good evidence[1], and likely to become the generally accepted one, that it was not the available data which led to an empirical law, but that in the experimental numbers Dalton found a verification of his theoretical speculations.

[1] Roscoe and Harden, *A New View of the Origin of Dalton's Atomic Theory*, 1896.

But to turn from the history of what might have happened or should have happened, to the history of what did happen: to Dalton, the famous Quaker chemist of Manchester, is usually assigned all the credit of having referred the quantities of one substance A which can combine in different ratios with another substance B, to a fixed amount of B; and of having thus been able to show that the quantities of A which combine in the various compounds with the same amount of B bear a simple whole ratio to each other.

John Dalton (1766—1844) was born in a Cumberland village, the son of a poor weaver; endowed with natural aptitude and an indomitable will, he utilised all possible opportunities for the study of mathematics and natural philosophy. From 1781 to 1793 he kept school, taught and lectured at Kendal, devoting all the time and energy he could spare to scientific investigations, chiefly meteorological.

In 1793 he went to Manchester as tutor of mathematics and natural philosophy at a Presbyterian College. Though he resigned this post six years later, he remained in Manchester to the end of his life, earning his living as a private teacher, and devoting himself uninterruptedly and whole-heartedly to scientific research.

It is characteristic of Dalton's work that speculations con-cerning the properties of matter, starting from the most diverse phenomena and considerations, and leading him on to others more or less related, follow in quick succession, and that in his publications he does not always give the connecting links. Much has to be surmised concerning the genesis, the development and the final form of his views, with the inevitable result that there is some difference of opinion and a great deal of uncertainty as to the exact history of his chemical dis-coveries and theories. This renders the task of giving any complete and correct account of his discovery of the above specified numerical relation, termed the law of multiple ratios, a difficult matter. At this point we are, however, more con-cerned with the actual facts which Dalton presented to the scientific world in support of the existence of the law, than with the question whether he arrived at the law purely inductively, or whether the facts

Difficulty of settling the origin and sequence of Dalton's discoveries.

Facts given by Dalton in sup-port of the existence of a simple nume-rical relation between the different quan-tities of A com-bined with the same quantity of B.

AP.

ned
: to

one
her
ble
ous
itio

ge,
an
he
to
all
ns,

nd
ed
nd
ng
h.
u-
se
on
ek
ot
oe
nt
le
at
s-
of
is
n,
lt
n-
ed
of
ed
ts

VI]   *Dalton's Evidence for Multiple Ratios*   155

observed were interpreted as a welcome verification of prior theoretical ideas. Some of the facts were derived from experiments of his own, others from those of contemporaries.

(i) In a paper read in 1802 and published in 1805[1] is found the first example of the law. The combination of nitrous air (nitric oxide), a colourless gas insoluble in water,

(i) Combination of nitrous air with oxygen. with a portion of the atmospheric air to form a red gas soluble in water, was well known at that time, and had been used by Cavendish and others for determining the volumetric composition of air. This is what Dalton says concerning its use for this purpose :

"If 100 measures of common air be put to 36 of pure nitrous gas in a tube 3-10ths of an inch wide and 5 inches long, after a few minutes the whole will be reduced to 79 or 80 measures, and exhibit no signs of either oxygenous or nitrous gas. If 100 measures of common air be admitted to 72 of nitrous gas in a wide vessel over water, such as to form a thin stratum of air, and an immediate momentary agitation be used, there will, as before, be found 79 or 80 measures of pure azotic gas [nitrogen] for a residuum. If, in the last experiment, *less* than 72 measures of nitrous gas be used, there will be a residuum containing oxygenous gas ; if *more*, then some residuary nitrous gas will be found. These facts clearly point out the theory of the process : the elements of oxygen may combine with a certain portion of nitrous gas, or with twice that portion, but with no intermediate quantity. In the former case *nitric* acid is the result ; in the latter *nitrous* acid : but as both these may be formed at the same time, one part of the oxygen going to *one* of nitrous gas, and another to *two*, the quantity of nitrous gas absorbed should be variable ; from 36 to 72 per cent. for common air."

(ii) Also based on his own work were the results of the analyses of two hydrides of carbon. He says, "It was in the summer of 1804 that I collected at various times

(ii) Heavy and light carburetted hydrogen. and in various places the inflammable gas [marsh gas] obtained from ponds." He found that marsh gas, like olefiant gas (ethylene) contains nothing but carbon and hydrogen, and that these two substances, termed light and heavy carburetted hydrogen[2] respectively, showed a

---

[1] Alembic Club Reprints, No. 2, p. 8.

[2] Light carburetted hydrogen = marsh gas, is found in the gases given off by ponds and as a constituent of coal-gas ; it does not combine directly with chlorine, and is not absorbed by concentrated sulphuric acid.

Heavy carburetted hydrogen = ethylene, olefiant gas, is a constituent of coal-gas ; it combines directly with chlorine, and is absorbed by concentrated sulphuric acid.

The composition by weight according to present standard values, is :

2 of hydrogen and 6 of carbon.   1 of hydrogen and 6 of carbon.

simple multiple ratio between the weights of the constituent elements.

In carburetted hydrogen⎱ 4·3 of carbon were combined with 2 of hydrogen.
from stagnant water ⎰

In olefiant gas   4·3   ,,   ,,   1   ::

Obviously the numbers expected by theory, and not the experimental results, are given.

The actual method of analysis used and the results arrived at by Dalton are the following[1] :—

	Carburetted hydrogen.	Olefiant gas.
Explosion with oxygen whereby water of negligible volume and carbonic acid are formed :	"If 100 measures of carburetted hydrogen be put to upwards of 200 of oxygen, and fired over mercury, the result will be a diminution of near 200 measures, and the residuary 100 will be found to be carbonic acid.... I think it proper to observe, that according to my most careful experiments, 100 measures of this gas require rather more than 200 measures of oxygen, and give rather more than 100 carbonic acid; but the difference is not more than 5 per cent. and may in general be neglected."	"Olefiant gas...explodes with uncommon violence when mixed with oxygen,... my results have always given that 100 measures of the gas require less than 300 of oxygen, but more than 270; the [vol. of the carbonic] acid formed should be about 185 or 190 [measures]."
Decomposition by sparking whereby the hydrogen present is liberated and the carbon deposited :	"When a portion of carburetted hydrogen is electrified for some time, it increases in volume, in the end almost exactly doubling itself; at the same time a quantity of charcoal is deposited. The whole of the gas is then found to be pure hydrogen."	"When olefiant gas is electrified charcoal is deposited. According to most careful experiments made by Dr Henry and myself... 100 measures of olefiant gas will contain 195 of hydrogen."
Hence vol. of hydrogen is to vol. of carbonic acid :	:: 200 : 100	:: 195 : 190 or :: 200 : 195 approx.

(iii) In a note-book table dated September 6, 1803 he gives numbers which show the occurrence of multiple ratios in the case of the oxides of nitrogen. Sir Humphry Davy's analyses of nitrous oxide and nitrous gas gave as the mean of three determinations the results :

(iii) Nitrous and nitric oxide.

---

[1] Dalton, *A New System of Chemical Philosophy*, 1810.

In nitrous oxide (laughing gas)—1·648 nitrogen combine with 1 oxygen[1].

In nitrous gas (nitric oxide)—0·798 „ „ 1 „

∴ with *equal* quantities of oxygen are combined quantities of nitrogen which are in the ratio of........ 0·798 : 1·648

$$= 0·798 : 0·798 \times 2·06$$
$$= 1 : 2 \text{ (nearly).}$$

(iv) The same table also shows that Dalton considered the difference in the two known oxides of carbon to be

(iv) Carbonic acid and carbonic oxide. due to the fact that for the same amount of carbon the one contains twice as much oxygen as the other.

"...experiment confirmed the truth of Lavoisier's conclusion that 28 parts of charcoal + 72 parts of oxygen constitute carbonic acid, and also that carbonic oxide contained just half the oxygen that carbonic acid does[2], which indeed had been determined by Clément and Desormes, two French chemists, who had not however taken notice of this remarkable result."

Dalton's discovery of the law of multiple ratios as a part of his new theory of chemical combination reached a wider public in 1807. Thomas Thomson (1773—1852), Professor

Dalton's discovery published by Thomson in 1807. of Chemistry in Glasgow, remembered as the author of an important text-book and of a history of chemistry, as a very prolific if not quite reliable contributor to the determinations of the quantitative composition of compounds, and chiefly as the first disciple of Dalton and the populariser of the atomic theory, published in 1807, in the third part of the third edition of his *System of Chemistry*,

Thomson on the two oxalates of strontian. what in 1804 he had learnt from Dalton himself concerning his atomic hypothesis. In 1808 Thomson could supply an observation of his own in support of the law of multiple ratios[3].

---

[1] Nitrous oxide = laughing gas, a colourless gas, does not turn red in air, is fairly soluble in water and an excellent supporter of combustion.

The composition by weight, according to present standard values, is: 12·25 of nitrogen and 7 of oxygen.

[2] Carbonic oxide, a gas less dense than carbonic acid, is not absorbed by lime water, but is combustible.

The composition by weight, according to present standard values, is: 12 of carbon and 15·88 of oxygen.

Nitrous air, or nitrous gas = nitric oxide, gives red fumes with air, is not soluble in water and a not very good supporter of combustion.

6·12 of nitrogen and 7 of oxygen.

Carbonic acid is absorbed by lime water, is not combustible and does not support combustion.

12 of carbon and 31·76 of oxygen.

[3] Alembic Club Reprints, No. 2, p. 41.

"It appears that there are two oxalates of strontian, the first obtained by saturating oxalic acid with strontian water, the second by mixing together oxalate of ammonia and muriate [chloride] of strontian. It is remarkable that the first contains just double the proportion of base contained in the second."

The data which Thomson had obtained, and from which he drew this inference, were:

(1)   A solution containing 7 grains of real oxalic acid[1] was neutralised by ammonia, and the oxalic acid precipitated by muriate of strontian. The salt obtained weighed 12·3 grains, therefore 7 parts of acid had combined with 5·3 parts of base.

(2)   To a solution of oxalic acid containing 7 grains of real oxalic acid, strontian water was added till it ceased to produce any change. The liquid was evaporated to dryness and the residue weighed. The weight obtained was 17·6 grains, and therefore the weight of base added 10·6 grains; and 7 parts of acid had combined with 10·6 parts of base. This consequence was so surprising that the experiment was repeated, but the result remained the same.

∴ 7 parts of oxalic acid combine with..............5·3 parts strontian.
or with $10·6 = 2 \times 5·3$ „   „

These numbers have evidently been rounded off, a usual practice of Thomson.

Every science can furnish examples in support of the proposition that certain times are sure to bring forth certain discoveries as the natural and inevitable sequence to what has gone before, and that important discoveries are therefore often made practically simultaneously by different investigators; it seems almost as if within a certain number of the truly great, it were a chance to whom the priority falls. And so at the time when the attention of chemists began to concentrate on the investigation of the quantitative aspect of chemical phenomena, two men simultaneously discovered the simple relation in composition by weight, which is the law of multiple ratios.

---

[1] Real oxalic acid = oxalic acid less the elements of water is the hypothetical anhydride related to oxalic acid as sulphur trioxide is to sulphuric acid; it is that portion of the acid which in salt formation can be supposed to be adding itself to the base.—$H_2C_2O_4$ = present formula of oxalic acid, according to which C_2O_3 would have represented Thomson's "real oxalic acid."

W. H. Wollaston (1766—1828), an ingenious and careful experimenter, published in 1808 the results of experiments on super-acid and sub-acid salts[1] which he had carried out before he heard of Dalton's work, and which he did not pursue further after getting acquainted with the theory of his great contemporary.

Wollaston's work on acid salts, and his independent discovery of the law of multiple ratios.

"Dr Thomson has remarked, that oxalic acid unites to strontian as well as to potash in two different proportions, and that the quantity of acid combined with each of these bases in their super-oxalates, is just double of that which is saturated by the same quantity of base in their neutral compounds. As I had observed the same law to prevail in various other instances of super-acid and sub-acid salts, I thought it not unlikely that this law might obtain generally in such compounds, and it was my design to have pursued the subject with the hope of discovering the cause to which so regular a relation might be ascribed. But since the publication of Mr Dalton's theory of chemical combination, as explained and illustrated by Dr Thomson, the inquiry which I had designed appears to be superfluous, as all the facts that I had observed are but particular instances of the more general observation of Mr Dalton.......Since some persons may imagine that the results of former experiments on such bodies do not accord sufficiently to authorise the adoption of a new hypothesis, it may be worth while to describe a few experiments, each of which may be performed with the utmost facility, and each of which affords the most direct proof of the proportional redundance or deficiency of acid in the several salts employed."

It was known at that time that the same acid and alkali could form two or three different salts, and the names of sub-acid and super-acid salts expressed the fact that the observed differences between these were due to the presence of a relatively lesser or greater amount of acid. Taking the two compounds termed *subcarbonate of potash* and *carbonate of potash*, our knowledge of the qualitative properties of these two salts may be summarised as represented in the table on p. 160.

(i) The two carbonates of potash.

That the difference between these two substances is simply due to the presence of a relatively greater or lesser amount of carbonic acid is proved by the facts (iii and iv in the table) that removal by heat of some of the acid changes the carbonate into subcarbonate, and that the latter absorbs carbonic acid, passing into the carbonate. Wollaston's experiment settled the relative quantities of carbonic acid combined with the *same* amount of potash in the two salts.

[1] London, *Phil. Trans. R. Soc.* 98, 1808 (p. 96). Alembic Club Reprints, No. 2.

Carbonates of Potash.

	I Carbonate or Normal Carbonate. Wollaston's Sub-carbonate	II Acid Carbonate or Bicarbonate. Wollaston's Carbonate
(1) Similarities: (i) Both salts are compounds of the same constituents, as proved by: (a) Synthesis	Made directly by leading carbonic acid gas into potash solution	
(b) Analysis	Treated with dilute acid they evolve carbonic acid gas; they give the flame coloration and all other reactions characteristic of potassium compounds	
(ii) Behaviour towards indicators	Red litmus is turned blue	
(2) Differences: (i) Crystalline form	Crystallises with water	Crystallises without water
(ii) Solubility in water. Parts by weight of salt dissolved by 100 parts by weight of water at 20° and at 60°	Very soluble 112 127	Easily crystallised from hot water 26·9 41·3
(iii) Action of heat	Solid fuses and then remains unchanged; solution not changed	Solid and solution lose carbonic acid gas and change into the normal salt
(iv) Carbonic acid	Is absorbed and the salt changed into the bicarbonate	No action
(v) Magnesium sulphate solution added to solutions	Immediate precipitate	No precipitate until heated
(vi) Mercuric chloride solution added to solutions	Red precipitate	White precipitate

"Sub-carbonate of potash recently prepared, is one instance of an alkali having one-half the quantity of acid necessary for its saturation, as may thus be satisfactorily proved. Let two grains of fully saturated and well crystallised carbonate of potash be wrapped in a piece of thin paper, and passed up into an inverted tube filled with mercury, and let the gas be extricated from it by a sufficient quantity of muriatic acid, so that the space it occupies may be marked upon the tube. Next, let 4 grains of the same carbonate be exposed for a short time to a red heat; and it will be found to have parted with exactly half its gas; for the gas extricated from it in the same apparatus will be found to occupy exactly the same space, as the quantity before obtained from two grains of fully saturated carbonate."

The two distinct substances obtained by the interaction between potash and sulphuric acid which now are usually dis-

(ii) The two sulphates of potash. tinguished by the names *normal sulphate* or *neutral sulphate,* and *bisulphate* or *acid sulphate* were known to Wollaston as *sulphate* and *super-sulphate* respectively. The super-sulphate is produced by the action of excess of the acid on the alkali; it turns blue litmus red and neutralises potash or potassium carbonate, forming the sulphate which has no action on litmus.

"By an experiment equally simple, super-sulphate of potash may be shown to contain exactly twice as much acid as is necessary for the mere saturation of the alkali present. Let 20 grains of carbonate of potash (which would be more than neutralised by 10 grains of sulphuric acid) be mixed with about 25 grains of that acid in a covered crucible of platina, or in a glass tube ¾ of an inch diameter, and 5 or 6 inches long. By heating this mixture till it ceases to boil, and begins to appear slightly red hot, a part of the redundant acid will be expelled, and there will remain a determinate quantity forming super-sulphate of potash, which when dissolved in water will be very nearly neutralised by an addition of 20 grains more of the same carbonate of potash; but it is generally found very slightly acid, in consequence of the small quantity of sulphuric acid which remains in the vessel in a gaseous state at a red heat."

Thomson in his paper already quoted had dealt with the two oxalates of potash which he describes as follows:

(iii) The two oxalates of potash. "Oxalate of potash readily crystallises in flat rhomboids, commonly terminated by dihedral summits. The lateral edges of the prism are usually bevelled. At the temperature of 60° it dissolves in thrice its weight of water....This salt combines with an excess of acid, and forms a super-oxalate, long known by the name of *salt of sorrel.* It is very sparingly soluble in water....It occurs in commerce in beautiful four-sided prisms attached to each other. The acid contained in this salt is very nearly double of what is contained in oxalate of potash. Suppose 100 parts of potash; if the weight of acid necessary to convert this quantity into oxalate be x, then $2x$ will convert it into super-oxalate."

Wollaston also had investigated these two oxalates.

"The common super-oxalate of potash is a salt that contains alkali sufficient to saturate exactly half of the acid present. Hence, if two equal quantities of salt of sorrel be taken, and if one of them be exposed to a red heat, the alkali which remains will be found exactly to saturate the redundant acid of the other portion[1]."

[1] Salt of sorrel which is acid to litmus interacts with alkaline potassium carbonate forming the neutral potassium oxalate.

$$2KHC_2O_4 + K_2CO_3 = 2K_2C_2O_4 + H_2O + CO_2.$$

Oxalates are decomposed by heat yielding carbonates which contain *all* the base originally present in the oxalate.

$$K_2C_2O_4 = K_2CO_3 + CO, \quad or \quad 2KHC_2O_4 = K_2CO_3 + H_2O + 2CO + CO_2.$$

These are Wollaston's contributions to the experimental establishment of the law of multiple ratios. The importance of the numerical relation thus made evident was fully appreciated by the chemists of the day, but to Berzelius it seemed that the number and accuracy of Dalton's experiments did not warrant an immediate and unqualified acceptance of the law and of the vast theoretical speculations dependent on it. This want he himself proceeded to supply by a most comprehensive and most admirably executed study of the quantitative composition of a large number of substances, including practically all the more important of the compounds then known.

Berzelius draws attention to the insufficiency of Dalton's experiments, and supplies this want.

Johann Jakob Berzelius (1779—1848), a Swede by birth and education, spent all his life in his native country where he studied medicine and chemistry in Upsala and where, from 1807 onwards, he held a professorship in Stockholm. There is no domain of chemistry which he did not enrich by valuable discoveries. The amount of work he accomplished and its uniform excellence are alike wonderful; but the department in which he obtained his most important results was that of the investigation of the quantitative composition of compounds. His earliest work in this direction consists in a series of papers published between 1811 and 1812 under the name "Essay to ascertain the Fixed and Simple Ratios in which the Constituents of Inorganic Nature are combined[1]." He prefaces these researches by the following considerations:

"Berthollet, who is one of the most celebrated chemists of our age, in the course of his ingenious investigations into the laws of affinity, has attempted to demonstrate that substances could combine in an infinite number of continuous ratios. But Proust, another authority in chemical science, has proved on the contrary, that no such infinite variations occur in Nature, but that all complex, definite substances contain their fundamental constituents in a fixed ratio. So that if for instance a suboxide, by addition of oxygen changes to the oxide, this leads *per salto* to the formation of another compound with a fixed amount of oxygen; and hence the existence of a series of compounds continuous between the first and the last cannot be admitted. The truth of Proust's view cannot have failed to strike the experienced chemist; but what so far had not been known was, whether these sudden changes in composition occurred according to one and the same law for all substances, or in some indeterminate manner peculiar to each substance.

(1) General discussion on the subject of composition by weight.

[1] Reprinted in Ostwald's *Klassiker der Exacten Wissenschaften*, No. 35.

The experiments which I propose to communicate will, as a matter of fact, lead to some general laws for these combinations.

I have been attracted to these investigations...through finding that in the basic chloride of lead and in the basic chloride of copper the acid is saturated by four times as much of the base as in the neutral salts[1].

I had hoped to discover the cause of so remarkable a relation by accurately investigating the result of mixing different substances of this type. Whilst engaged in this work I came across Nicholson's Journal for November 1808 and found in it the experiments of Wollaston on acid salts which had been suggested[2] by Dalton's hypothesis. This hypothesis affirms that if substances can be made to combine in different ratios, these ratios are always produced by simple multiplication of the weight of the one substance by 1, 2, 3, 4 etc. Wollaston's experiments seem to support this hypothesis. But such a doctrine of the composition of compounds would so illuminate the province of affinity, that supposing Dalton's hypothesis be found correct, we should have to look upon it as the greatest advance that chemistry has ever yet made in its development into a science. I have no knowledge whatever of how Dalton developed his law and on what experiments he has based it, and hence I cannot judge whether my own experiments confirm his hypothesis in its full extent, or whether they modify it to a greater or lesser degree[3].

The experiments about to be described will show that when two substances A and B combine with each other in different ratios, it is always in the following fixed proportions: $1A$ with $1B$ (which is the composition in the minimum); $1A$ with $1\frac{1}{2}B$ or perhaps more correctly $2A$ with $3B$; $1A$ with $2B$; $1A$ with $4B$. But in my experiments there is not a single instance of $1A$ with $3B$."

Berzelius then proceeds to give the results of his experiments, investigations which are marked by the high degree of accuracy attained, the ingenuity displayed in the devising of the experimental methods, and the critical examination and interpretation of the results obtained. The relation stated above was found to hold in the

(2) Experimental results in support of the law of multiple ratios.

[1] A third independent discovery of the law of multiple ratios.

[2] This is not the case; Wollaston's work had been done before he knew of Dalton's.

[3] A verdict passed years later by Berzelius on the method followed by Dalton in the discovery of this law is of interest because showing clearly the difference in the attitude of these two great men towards the problems of Nature. "It appears...as if in this investigation the illustrious scientist had not at the outset been provided with a sufficiently firm experimental basis; and it may be doubted whether he displayed sufficient caution in applying the new hypothesis to the system of chemistry. To me it has seemed as if in the smallness of the number of analyses given, one could sometimes perceive the desire of the experimenter to obtain a certain result; but this is just the attitude to be avoided when seeking proofs for or against a preconceived theory. Notwithstanding all this, it is to Dalton that belongs the honour of the discovery of that part of the doctrine of chemical composition termed the law of multiple ratios, which none of his predecessors had observed."

following cases amongst others: (i) The two oxides of lead; equal quantities of lead are combined in the yellow litharge and the brown oxide respectively with quantities of oxygen which are in the ratio 1:2. (ii) The two oxides of copper; to change the black oxide into the red oxide, it is necessary to add an amount of copper equal to that already contained in it. (iii) The two oxides of sulphur; the same weight of sulphur is combined in sulphuric and sulphurous acid respectively with quantities of oxygen which are in the ratio 3:2. (iv) The two oxides of iron; the quantity of oxygen combined with a certain amount of iron in the red oxide is $1\frac{1}{2}$ times that combined with the same amount of iron in the lower oxide.

The following table gives the numerical results from which Berzelius deduced these relations, and allows us to judge of the approximation of the experimental values to simple whole numbers.

			Ratio between the quantities of the element B combined with equal amounts of A
	Oxides of Lead		
Lead Oxygen	Yellow Litharge 100 7·8	Brown 100 15·6	7·8 : 15·6 = 1 : 2·00 = 1 : 2
	Oxides of Copper		
Copper Oxygen	Red 100 12·3	Black 100 25	12·3 : 25 = 1 : 2·03 = 1 : 2 (nearly)
	Oxides of Sulphur		
Sulphur Oxygen	Sulphurous 100 97·83	Sulphuric 100 146·427	97·83 : 146·427 = 2 : 2·993 = 2 : 3 (nearly)
	Oxides of Iron		
Iron Oxygen	Ferrous 100 29·6	Ferric (Red) 100 44·25	29·6 : 44·25 = 2 : 2·99 = 2 : 3 (nearly)

It may be well to give for one special case a detailed account of the method followed by Berzelius in obtaining data such as the above. The compounds selected for this purpose are the two sulphides of iron because of the interest there is in comparing Berzelius's method and results with those of Proust, whose work

on these substances has been dealt with before (p. 137). The
two sulphides are the native pyrites and the artificial
sulphide, which latter is obtained by fusing together
excess of iron and sulphur and detaching the compound formed from the unchanged iron.

(3) Determin-
ation of the
composition of
two iron sul-
phides.

The amount of iron present in a given weight of either of the
sulphides was found by changing it, by strong heating in air, or
otherwise, into red oxide and weighing the oxide so formed; but
according to a previous determination of Berzelius 100 of red oxide
contain 69·34 of iron (see table, p. 164), and hence the amount of
iron present could be calculated.

Oxidation by nitric acid changes all the sulphur present in
the sulphides into sulphuric acid which can be precipitated and
weighed as barium sulphate; and since Berzelius had found by a
previous synthesis from sulphur, that 100 of barium sulphate contain 34 of sulphuric acid in which there are 13·795 of sulphur, all
the necessary data are supplied for calculating the sulphur in the
sulphides.

The results thus obtained for the composition of the two
sulphides of iron are given in the table on p. 166.

This then constitutes the experimental evidence for the law of
multiple ratios which may be thus formulated: "If
two substances A and B unite in more than one
ratio, the various masses of A which combine with
a fixed mass of B bear a simple ratio to each other."

Enunciation
of the law
of multiple
ratios.

It still remains to investigate the degree of accuracy to which
the law has been proved and its consequent classification as an
exact or an approximate law. None of Dalton's data
are suitable for this purpose; it has been shown before
how he obviously always gave the theoretically expected and not the experimental numbers. And
Wollaston's experiments do not lend themselves any better to this
end, as we have no means of judging of the degree of accuracy of
the measurements involved, which depended chiefly on the detection
of neutrality by means of an indicator. Berzelius proved the law
to within ½ per cent.; but great as was his experimental skill,
none of the cases investigated were suitable for anything beyond
his immediate purpose, which was the establishment of the general
relation. An inspection of the two results—that obtained directly
and that from difference—for the composition of the artificial
sulphide of iron (see table, p. 166) reveals differences of about ⅓

Classification
of the law
as exact or
approximate.

Composition of the two Sulphides of Iron. (Berzelius.)

Artificial Sulphide	Percentage	Native Sulphide (Pyrites) containing a small amount of admixture of silica	Percentage
Sulphur: 2 grams of the sulphide prepared as before described were digested with *aqua regia* and the solution then precipitated with barium chloride giving 5·38 grams of barium sulphate which contain $\frac{5\cdot38 \times 13\cdot795}{100} = 0\cdot742$ of sulphur ∴ 100 sulphide contain $\frac{0\cdot742 \times 100}{2} =$	37·1	*Sulphur*: determined by difference, calculated from the equations: Sulphide = pyrites − silica Sulphur = sulphide − iron Substituting the values for these quantities from the determinations given below: Sulphide = 10·00 − 0·07 = 9·93 Sulphur in 9·93 sulphide = 9·93 − 4·5775 = 5·3525 ∴ 100 sulphide contain $\frac{5\cdot3525 \times 100}{9\cdot93} =$	53·92
But by difference from the directly determined iron the percentage of sulphur is 100 − 63 = ∴ Mean =	37·0 37·05		
Iron: after precipitation of the sulphuric acid the excess of barium chloride was removed by sulphuric acid, ammonia added, the precipitate ignited and weighed and found = 1·82 grams 1·82 grams of red oxide contain $\frac{1\cdot82 \times 69\cdot34}{100} = 1\cdot26$ iron ∴ 100 sulphide contain $\frac{1\cdot26 \times 100}{2} =$	63·00	*Iron*: 10 grams of pyrites ignited in a platinum crucible in a muffle furnace gave 6·67 grams of a mixture of red oxide and silica. Removal of the red oxide by solution in strong hydrochloric acid left ·07 gram of silica ∴ (10·00 − ·07) = 9·93 sulphide contained (6·67 − ·07) = 6·60 red oxide = $\frac{6\cdot60 \times 69\cdot34}{100} = 4\cdot5775$ iron ∴ 100 sulphide contain $\frac{4\cdot5775 \times 100}{9\cdot93} =$	46·08
But by difference from the directly determined sulphur the percentage of iron is 100 − 37·1 = ∴ Mean =	62·9 62·95		
∴ *Sulphur combined with* 100 *iron* = 58·80		∴ *Sulphur combined with* 100 *iron* = 117	

58·80 : 117 = 1 : 1·99

per cent., which makes it of course impossible to investigate the exact or approximate nature of the law within a narrower limit than the value of this difference.

There seems to be no record of any experiments made with the object, and specially devised for the end, of testing the accuracy of the law of multiple ratios, as is the case for the law of fixity of composition so splendidly investigated by Stas. But in searching amongst the data accumulated in that province of analytical work where accuracy has to be pushed to the utmost possible limit, that is amongst atomic weight determinations, a set is found which may be utilised for this purpose. Two investigations combined supply the numbers wanted, and Stas' name is associated with each of them. In 1840 Dumas and Stas found the ratio in which carbon and oxygen are united in carbonic acid, and in 1849 Stas determined the ratio in which carbonic oxide and oxygen combine to form carbonic acid. Thus we get the data required for comparing the quantities of oxygen combined with the *same* weight of carbon in the two oxides.

The results obtained are given in the following two tables :

No investigation has been undertaken to test the accuracy of the law, but experiments on composition of two oxides of carbon serve the purpose.

I. *Stas' syntheses of carbonic acid gas from carbonic oxide and oxygen.* 1849.

Carefully purified carbonic oxide was passed over a known weight of copper oxide at red heat, and the loss of weight of the oxide as well as the weight of carbonic acid gas formed were determined. The method was in all respects similar to that employed by Dumas in the gravimetric synthesis of water.

Weight of oxygen given up by the copper oxide	Weight of carbonic acid gas formed	Weight of carbonic oxide yielding 100 carbonic acid		
9·265	25·483	63·641 = Mean	+	·001
8·327	22·900	63·637 =	,,	− ·003
13·9438	38·351	63·643 =	,,	+ ·003
11·6124	31·935	63·637 =	,,	− ·003
18·763	51·6055	63·641 =	,,	+ ·001
19·581	53·8465	63·636 =	,,	− ·004
22·515	61·926	63·641 =	,,	+ ·001
24·360	67·003	63·642 =	,,	+ ·002
		63·640 = Mean		

II. *Dumas' and Stas' syntheses of carbonic acid gas from carbon and oxygen.* 1840.

Weighed quantities of pure carbon were burned in excess of oxygen; the carbonic acid gas was collected in the usual manner by absorption in potash and then weighed.

	Weight of carbon taken	Weight of carbonic acid gas produced	Carbon contained in 100 of carbonic acid gas		
Natural Graphite	1·000 ·998 ·994 1·216 1·471	3·671 3·660 3·645 4·461 5·395	27·241 = Mean 27·268 = 27·270 = 27·258 = 27·248 =	,, ,, ,, ,,	− ·025 + ·002 + ·004 − ·008 − ·018
Artificial Graphite	·992 ·998 1·660 1·465	3·642 3·662 6·085 5·365	27·237 = 27·253 = 27·281 = 27·307 =	,, ,, ,, ,,	− ·029 − ·013 + ·015 + ·041
Diamond	·708 ·864 1·219 1·232 1·375	2·598 3·1675 4·465 4·517 5·041	27·251 = 27·276 = 27·301 = 27·263 = 27·275 =	,, ,, ,, ,, ,,	− ·015 + ·010 + ·035 − ·003 + ·009
			27·266 = Mean		

The "General Mean" (*post*, chap. VIII, Appendix) for the percentage of carbon in carbonic acid gas from all the best syntheses (Dumas and Stas, 1840; Erdmann and Marchand, 1841; Roscoe, 1882; Friedel, 1884; Van der Plaats, 1885) is:—

27·278

But on the supposition of the fixity of composition of carbonic acid gas, that is on the assumption that the amount of carbon in 100 of carbonic acid gas is the same whether it had been formed by the combustion of carbon or of carbonic oxide, 63·640 of carbonic oxide must contain as much carbon as 100 of carbonic acid gas.

100 carbonic acid are produced from:

I. 63·640 carbonic oxide and 36·360 oxygen (Stas).

II. (i) 27·266 carbon and 72·734 oxygen (Dumas and Stas).

(ii) 27·278 „   „   72·722   „   (General Mean).

Therefore 63·640 carbonic oxide contain:

(i) 27·266 carbon and 36·374 oxygen (Dumas and Stas, and Stas).

(ii) 27·278 „   „   36·362   „   (General Mean, and Stas).

Therefore the quantities of oxygen combined in the two oxides with the *same* amount of carbon are:

(i) 27·266 carbon combined with 36·374 and 72·734 oxygen (D. and S., and S.).

(ii) 27·278   „   „   „   36·362 „ 72·722   „   (G. M., and S.).

But   (i) 36·374 : 72·734 = 1 : 1·9996.

(ii) 36·362 : 72·722 = 1 : 1·99995.

The deviations from the requirements of the exact law are ·02 per cent. and ·003 per cent. respectively, values of about the same order of magnitude as those found in the case of the law of fixed ratios, and perfectly commensurable with the experimental error of the measurements involved.

The law may be classed as exact.

It is interesting to compare the two tables, that of 1840 containing the joint work of Dumas and Stas, and that of 1849 which gives the work of Stas alone. In the quantities used and in the much greater concordance of the results obtained it may be seen that in the intervening years Stas had found himself. And this is why in the above calculations, in order to make the two sets of data employed of about the same accuracy, a second value has been used for the percentage of carbon in carbonic acid gas, which is called the "General Mean" (*post*, chap. VIII, Appendix) and which embodies a great many more recent and more concordant determinations.

Stas on the evidence for the law of multiple ratios.

Whether he considered the 1840 piece of work affected by too great an experimental error, or whatever else may have been his reason, Stas in 1865 when discussing the foundation of the law of multiple ratios makes no use of these data and ignores their bearing on the question.

"A careful study of all the available facts dealing with the relation between the weights of the elements forming chemical compounds, has convinced me that chemists have let themselves be more influenced by constancy of composition than by rigorous demonstration of the law of Wenzel[1] and of the hypothesis of Dalton known under the name of the law of multiple ratios.... The experiments of Wollaston on the ratio between oxalic acid and potash in the neutral and acid oxalates have been executed on so small a scale, that it is impossible to deduce from them whether the law of multiple ratios is a mathematical [exact], or limited [approximate] law. Besides, even supposing that the quantities had been sufficient, the principle on which the celebrated English chemist has based his work, that of *neutrality measured by means of colouring matter*, is nothing but a hypothesis, the validity of which would require to be itself proved first."

---

[1] The reference is to Richter's law of equivalent ratios dealt with in chapter VII.

CHAPTER VII.

RICHTER AND THE LAW OF EQUIVALENT RATIOS.

"By measure and number and weight thou didst order all things."
Wisdom of Solomon.

THE discovery of the law of multiple ratios, which involves the validity of the law of constant ratios, at a time (1802—1808) when fixity of composition was being assailed by Berthollet, shows that the views of the great French chemist on this matter cannot have been very seriously considered outside his own country. It is to an even earlier time that we must go back for the discovery of another of the fundamental laws of chemical com-

Cavendish in 1767 first uses the term "equivalent." bination. According to Kopp[1] Cavendish had as far back as 1767 designated a definite amount of "fixed alkali" (potassium carbonate) as *equivalent* to a certain amount of "lime," explaining this to mean that these quantities neutralised the same weight of the same acid; and in several experimental investigations in 1788 he applied the principle that the equivalency of bases is independent of the acid neutralised.

Bergman observes and explains main- tenance of neutrality in precipitation of one metal by another. Bergman knew that when one metal is precipitated by another from its neutral salt, neutrality is maintained; and he interpreted this in terms of the phlogistic theory by saying, that the quantities of the two metals, which in the salts were united with the same amount of acid, must contain the same amount of phlogiston[2]. The translation into the

---

[1] Hermann Kopp, *Die Entwickelung der Chemie in der neueren Zeit*, has been followed in this account of the early history of the law of equivalent ratios.

[2] This follows directly from the phlogistic view of the solution of metals in acids and of the nature of salts:

(i)   Acid   + calx of metal A = salt of A.
(ii)   Acid   +   metal A = salt of A + phlogiston.
                (calx+phlogiston)
(iii) Salt of A +   metal B = salt of B + metal A.
       (neutral)   (neutral)

Equation (iii), when expressed in terms of (i) and (ii), becomes

(Acid + calx A) + (calx B + phlogiston) = (acid + calx B) + (calx A + phlogiston).

These two quantities of phlogiston must be equal, since the solution had remained neutral.

terms of the oxygen theory is simple, and asserts that the quantities of two metals which in the salts are combined with the same amount of acid, are combined in their oxides with the same amount of oxygen.

Lavoisier states it as a problem requiring solution, to ascertain whether, when silver nitrate is precipitated by an alkaline chloride, the metal in the exchange of acids takes as much of the new acid as the amount of the other previously combined with it; or quite generally, whether the two acids in the interchange of the bases enter completely into the new compounds, and if not, what becomes of the excess of either[1].

But all this was but a preliminary skirting of the edges of the problem, which met with its solution in the researches of Richter.

[1] Since neutralisation and the composition of the substances termed *salts* furnished most of the data on which the law of equivalent ratios is based, it is desirable to understand the views held at the time of the discovery and establishment of this law concerning the nature of salts, and the terminology used. The view originated by Lavoisier, strongly supported by Berzelius, and generally accepted was, that a salt is a substance made up of two components, both of them oxides: one the base, the oxide of a metal; the other the acid, the oxide of a combustible substance. Sulphuric and sulphurous acids were therefore two oxides of sulphur which we now term acidic oxides or acid anhydrides. What we now call sulphuric acid, was then "hydrated sulphuric acid." The name of "real oxalic acid" for $H_2C_2O_4$ less H_2O has already been referred to (p. 158). That the substance then termed "the acid" could not always be isolated, was as little of an obstacle as our present inability to isolate carbonic, sulphurous, arsenious, etc. acids. In terms of our formulae these substances were:

Acids	Hydrated Acids
CO_2 =carbonic	
SO_3 =sulphuric	$H_2SO_4 = H_2O . SO_3$
SO_2 =sulphurous	
C_2O_3 =oxalic	$H_2C_2O_4 = H_2O . C_2O_3$
$H_6C_4O_3$ =acetic	$2H_4C_2O_2 = H_2O . H_6C_4O_3$
N_2O_5 =nitric	$2HNO_3 = H_2O . N_2O_5$.

The corresponding formulae and equations for the salts and their formation are:

$$\begin{cases} \text{Base} + \text{acid} = \text{salt} \\ CaO + SO_3 = CaO . SO_3. \end{cases}$$

$$\begin{cases} \text{Base} + \text{hydrated acid} = \quad \text{salt} \quad + \text{water} \\ CaO + \quad H_2O . SO_3 \quad = CaO . SO_3 + H_2O. \end{cases}$$

This conception of salts led to a curious result for hydrochloric acid:

$$\text{Chloride} = \text{Base} + \text{Acid}$$

$$\text{Silver chloride} = \text{Silver oxide} + \text{hydrochloric acid}$$

Composition of hydrochloric acid then assumed = (chlorine − oxygen).

The recognition of the elementary nature of chlorine by Sir Humphry Davy made this view untenable and marks the beginning of the change in the conception of salts.

Benjamin Richter (1762—1807) a native of Silesia, held official posts in connection with applied chemistry, first in the department of mines in Silesia and later at the Berlin porcelain manufactory. He is the inventor of the term "Stoichiometry" (from στοιχεῖα the fundamental constituents and μέτρον measure)[1] for the quantitative relations between chemically interacting substances, and the years 1792—1799 roughly mark out the time during which he made the observations and quantitative measurements which, though not accurate, though used by him for a fantastic interpretation in his determination to find numerical relations where Nature had provided none, still contain enough of solid truth for his name to be coupled with the discovery of one of the fundamental laws of chemical combination[2]. Richter, who looked upon chemistry as a branch of applied mathematics, showed all the distressing qualities of a person possessed by a fixed idea; he spent his life in looking for arithmetical regularities in the weights of acids and alkalis neutralising each other, and in finding them in spite of their non-existence.

Richter considers chemistry a branch of applied mathematics.

But all the same he managed to make discoveries of the highest importance. He not only noticed, but also correctly interpreted the fact, that when two neutral salts decompose one another, the resulting salts are still neutral:

Richter's discoveries in the quantitative study of salt formation.
(1) Maintenance of neutrality after double decomposition.

"...concerning that very common experience that two neutral salts on decomposition again produce neutral compounds, I could draw no direct inference other than that fixed quantitative relations must exist between the constituents of the neutral salts. If a solution of two components is so constituted that neither of them, as long as it remains in the solution, exhibits the peculiar characteristics it had before solution (*e.g.* the reactions of an acid or of an alkali), then such a solution is called saturated or neutral, or also a neutral compound....When two neutral solutions are mixed and a decomposition ensues, the newly formed products are also, almost without exception, neutral...."

1 The word Stoichiometry has until lately not been much used in England, and was rarely met with outside translations of German books on chemistry. It is however a very convenient term, comprising as it does a large and well defined set of phenomena.
2 *Anfangsgründe der Stöchiometrie*, 1792—4; *Ueber die neueren Gegenstände der Chemie*, 1792—1807.

If therefore the weights or masses of two neutral compounds be A and B,

and the weight of one constituent in A is a
and that of one constituent in B is b
then the masses of the constituents in A are $(A - a)$ and a
and those in B are $(B - b)$ and b.

The relative masses of the two constituents before the decomposition are $(A - a) : a$ and $(B - b) : b$; whilst after the decomposition the masses of the newly produced substances are $a + (B - b)$ and $b + (A - a)$, and the relative masses of their constituents are $(B - b) : a$ and $(A - a) : b$. Hence it follows that if the combining ratios in the original compounds be known, those in the newly formed compounds are known also."

Richter's style was extremely involved and marred by his clinging to phlogistic phraseology, but the passage just quoted is quite clear and straightforward. All the same the importance of the argument contained in it warrants an example.

A of barium nitrate consisting of a of baryta and $(A - a)$ of nitric acid interact with B, a weight of potassium sulphate consisting of b of potash and $(B - b)$ of sulphuric acid, and containing the amount of sulphuric acid requisite for combination with a of baryta.

But since after reaction the product is still neutral, the $(A - a)$ of nitric acid originally combined with a of baryta must have been the exact amount required for the neutralisation of the b of potash originally combined with sulphuric acid, and we have the relation:

$\{a + (A - a)\}$ of barium nitrate $+ \{b + (B - b)\}$ of potassium sulphate
$\qquad = \{a + (B - b)\}$ of barium sulphate $+ \{b + (A - a)\}$ of potassium nitrate

$\{a \text{ baryta} + (A - a) \text{ nitric acid}\} + \{b \text{ potash} + (B - b) \text{ sulphuric acid}\}$
$\qquad = \{a \text{ baryta} + (B - b) \text{ sulphuric acid}\} + \{b \text{ potash} + (A - a) \text{ nitric acid}\}$

from which we get:

$\left\{ \begin{array}{l} a \text{ baryta neutralises } (A - a) \text{ nitric acid } \textbf{or} \ (B - b) \text{ sulphuric acid} \\ b \text{ potash } \quad\quad\;\; \text{,,} \quad\quad (A - a) \quad\; \text{,,} \quad\; \text{,,} \quad (B - b) \quad\; \text{,,} \quad\quad \text{,,} \end{array} \right.$

and

$\left\{ \begin{array}{l} \quad (A - a) \text{ nitric acid neutralises } a \text{ baryta } \textbf{or} \ b \text{ potash} \\ (B - b) \text{ sulphuric acid} \quad\quad \text{,,} \quad\quad a \;\; \text{,,} \;\; \text{,,} \;\; b \;\; \text{,,} \end{array} \right.$

These relations can be expressed by saying that the ratio between the amounts of two acids which neutralise the same amount of any one base is independent of the base; and the ratio between the amounts of two bases which neutralise the same amount of any one acid is independent of the acid. A more terse formulation states that the quantities of acids and bases equivalent in one neutralisation are so in all.

Richter determined the quantities of the different acids and bases which neutralise one another, and spent much time and no

(2) Determination of the neutralisation equivalents of acids and bases. Recognition of the proportionality of the different columns.

doubt doctored the numbers obtained, in an attempt to show that whilst the weights of the bases increased in an arithmetical progression, the weights of the acids increased in a geometrical progression. But at the same time he enunciated with all possible definiteness the great truth, that the various amounts of the different bases which severally form neutral salts with the same amount (1000 parts) of anhydrous muriatic acid, do so also with the same amount (1394 parts) of anhydrous sulphuric acid; and that given the composition of a neutral chloride, that of the corresponding neutral sulphate could be calculated. And he uses the constancy of the ratio of the quantities of the two acids neutralising the same amount of base for checking his determination of the composition of the chlorides and the sulphates of the alkalis.

"The masses of alkalis or alkaline earths which maintain neutrality with a given mass of either of the three other volatile acids will always bear to each other the same ratio."

	1000 Sulphuric Acid	1000 Muriatic Acid	1000 Nitric Acid
Potash	1606	2239	1143
	$\frac{1606}{1218}=1{\cdot}319$	$\frac{2239}{1699}=1{\cdot}318$	$\frac{1143}{867}=1{\cdot}318$
Soda	1218	1699	867
	$\frac{1218}{638}=1{\cdot}909$	$\frac{1699}{889}=1{\cdot}911$	$\frac{867}{453}=1{\cdot}914$
Volatile Alkali	638	889	453
	$\frac{638}{2224}=0{\cdot}287$	$\frac{889}{3099}=0{\cdot}287$	$\frac{453}{1581}=0{\cdot}287$
Baryta	2224	3099	1581
Lime	796	1107	565
Magnesia	616	858	438
Alumina	526	734	374

The above table gives Richter's results concerning the neutralisation of sulphuric, muriatic, and nitric acids by the most common bases. Kopp says that Richter had evidently altered the numbers in the sense required by theory—a common practice in those days—but that notwithstanding they are very incorrect, though correlated by the just conception of the proportionality of the different columns.

Richter had also deduced from the maintenance of neutrality when one metal precipitates another from a neutral solution, that the quantities of two metals which dissolve in the same amount of acid also unite in their oxides with the same amount of oxygen.

(3) Maintenance of neutrality in precipitation of one metal by another.

Fischer in 1802 reduces Richter's many tables to one of two columns.

Thus, then, he had established that the quantities of two substances—be these compounds like acids and bases, or elements like metals—which are equivalent in one reaction are also equivalent in others. And yet he gives a separate table of neutralisation equivalents for *each* acid and for *each* base. Richter's contemporary Fischer in 1802 translated into German Berthollet's *Recherches sur les lois de l'affinité*, and in a note he gave an account of Richter's work. After stating the general principles established by Richter he proceeds to say:

"He has taken the trouble to examine each acid in its relations towards the bases, both by experiment and by calculation, and to give his results in the form of tables. It seems that Richter paid no attention to the fact that all his tables could be reduced to a single one containing 21 numbers divided into two columns. I give the one which I have calculated from Richter's newest data:

Bases		Acids		
Alumina	525	Hydrofluoric Acid		427
Magnesia	615	Carbonic	„	577
Ammonia	672	Sebacic	„	706
Lime	793	Muriatic	„	712
Soda	859	Oxalic	.	755
Strontia	1329	Phosphoric	„	979
Potash	1605	Formic		988
Baryta	2222	Sulphuric	„	**1000**
		Succinic	„	1209
		Nitric	„	1405
		Acetic		1480
		Citric		1583
		Tartaric	„	1694

The meaning of this table is that if a substance is taken from one of the two columns, say potash from the first, to which corresponds the number 1605, the numbers in the other column indicate the quantity required of each acid to neutralise these 1605 parts of potash; there will in this case be required 427 parts of hydrochloric acid, 577 parts of carbonic acid, etc. If a substance is taken from the second column, the first column will be used to ascertain how much of an earth or of an alkali is required to neutralise it."

By this means $(m+n)$ numbers for n bases and m acids, of which numbers 1000, the arbitrarily chosen unit for sulphuric acid, is one, enable us to foretell the composition of $(m \times n)$ salts.

Berthollet accepted the law of proportionality and gave an account of it in his *Essai de Statique Chimique* in which he reprinted Fischer's note. In this manner Richter's

Neglect of Richter's work. work, which at the time of its publication had been almost completely ignored, got more widely known and duly appreciated. Berzelius is responsible for having originated the error of assigning to C. F. Wenzel, a German chemist older than Richter, the discovery and the correct interpretation of the law of neutrality, an error which passed current till after 1850; but in spite of this he did full justice to Richter, whom he names as the discoverer of the law of proportionality.

In his *Lehrbuch* he speaks of Richter as being

Berzelius' estimate of Richter's discoveries. the chemist to whom above all others we owe a just, positive and comprehensive explanation of chemical ratios. He enumerates the three discoveries made by Richter, namely: (1) The observation and correct interpretation of the maintenance of neutrality after double decomposition, in which he represents him as having been forestalled by Wenzel, (2) the determination of the neutralisation equivalents of acids and bases, (3) the observation and correct interpretation of the maintenance of neutrality after the precipitation of one metal by another[1].

"On reading Richter's work on chemical ratios, we are amazed that the study of this subject could ever have been neglected."

But Berzelius proceeds to specify the causes which may be held responsible for this want of appreciation, namely: the inaccuracy

[1] This is the relation which Bergman observed and interpreted quite correctly, though in terms of another theory of combination.

F.   12

of Richter's numbers; the peculiarities of his style, which attempted to compromise between the phlogistic and oxygen terminology, thereby displeasing both parties; and the complete concentration just then of the attention of chemists on the phlogiston-oxygen controversy.

Wollaston in 1814, in his important paper entitled "A Synoptic Scale of Chemical Equivalents," says:

Wollaston
and Thomson
on Richter's
work.

"It is to Richter we are originally indebted for the possibility of representing the proportions in which the different substances unite with each other in such terms that the same substance shall always be represented by the same number. He discovered the law of permanent proportions."

In 1827 Thomson published the results of a large number of (not very accurate) analyses in a book entitled "An Attempt to establish the First Principles of Chemistry by Experiment." The introduction is interesting, being a short account of the history of the discovery of the laws of chemical combination, and of the genesis of the atomic theory from the point of view of a contemporary. He speaks of Richter as "having endeavoured to give chemistry a mathematical form," enumerates his discoveries in stoichiometry, and comments on the small effect produced by his work at the time of its publication. He quotes Berzelius' explanation on this point. But for England Thomson adduces an additional reason, going into details concerning contemporary political history which would not find a place in a scientific work of to-day.

"But in Great Britain, where the importance of Richter's opinions was likely to have been early appreciated, another cause operated to prevent them from gaining ground. The French Revolution was at its acme when Richter began his stoichiometrical investigations, and all Europe was plunged into the bloodiest and most inveterate war that has almost ever afflicted this part of the world. Great Britain soon became involved in the dispute and gradually not only bore the brunt of the war, but was by degrees deserted by all her allies, and at last left alone to wage war against all the world. Hence the intercourse between the men of science in Great Britain and on the Continent was gradually interrupted. This, together with the little attention paid to German literature in this country, prevented us from being aware of the labours of Richter. My first knowledge of them was derived from a notice in Berthollet's *Statique Chimique* published in 1803, and I found it impossible to procure a copy till after the battle of Waterloo. The notice in Berthollet's *Statique* would probably have speedily drawn the attention of our countrymen to this most important department of chemistry, had not the genius of Dalton led him to a much more general view of the subject."

The investigation of Richter's law and its proof form an important part of the 1811—1812 memoir of Berzelius already mentioned in connection with the law of multiple ratios. This is what he says concerning its general bearing:

Berzelius' work on Richter's law.

"I made the discovery that in all chlorides the quantities of bases saturated by the same amount of hydrochloric acid contained the same amount of oxygen. This was also found to be the case with sulphates.... This discovery really belongs to the meritorious investigator J. B. Richter, who, in the 6th part, page 113, of his "Treatise on the Newer Subjects of Chemistry" (Breslau, 1796), has tried to substantiate it by ingenious, though not always sufficiently accurate experiments. It is true his numerical determinations are nearly without exception incorrect, but since his errors arise from common sources there is perhaps after all more truth in his calculations than is commonly believed....

When two substances A and B have affinity for two others C and D, the ratio of the quantities of C and D which saturate the same amount of A is the same as that between the quantities of C and D which saturate the same amount of B. If for instance 100 parts of lead combine in the lowest oxide with 7·8 of oxygen and in the sulphide with 15·6 of sulphur, and further if, according to an analysis I shall have occasion to quote later, 100 of iron combine in the lowest sulphide with 58·8 of sulphur, it becomes possible to calculate the composition of the oxide of iron from the simple proportion :

Enunciation of the law of proportionality.

$$15 \cdot 6 : 7 \cdot 8 = 58 \cdot 8 : 29 \cdot 4,$$

and in the oxide of iron 100 of iron must be combined with 29·4 of oxygen. The experiments which I am going to describe confirm this result. In this manner we can calculate the composition of all binary compounds. It is the merit of Richter to have shown long ago that the composition of salts could be ascertained by a calculation such as this.

It must be evident that when the data involved are firmly established, the result of such calculations must be much more reliable than ordinary analyses. With this object, I have endeavoured to make the analyses about to be described as accurate as possible, and I have repeated the most important ones several times, before I ventured to trust to them. The error in these latter is certainly not more than 1 or 2 in 1000, and in the others not more than $\frac{1}{2}$ per cent.; but nevertheless they are not yet sufficiently accurate to give by calculation more than approximations."

Deductive application to calculating the composition of some substances from that of others.

The experiments contained in Berzelius' paper which supply a proof for these relations may be divided into three classes.

(1) The ratio between the quantities A and B which combine with the same amount of C in the compounds AC and BC is the same as that between the quantities which combine with the same amount of D in the compounds AD and BD.

Berzelius found that :

100 lead combine { in the sulphide with 15·42 sulphur / in the oxide with 7·7 oxygen $\therefore \dfrac{\text{sulphur}}{\text{oxygen}} = \dfrac{15\cdot42}{7\cdot7} = 2\cdot02.$

100 copper combine in the sulphide with 25·6 sulphur / in the oxide with 12·3 oxygen $\therefore \dfrac{\text{sulphur}}{\text{oxygen}} = \dfrac{25\cdot6}{12\cdot3} = 2\cdot08.$

100 iron combine { in the sulphide with 58·73 sulphur / in the oxide with 29·6 oxygen $\therefore \dfrac{\text{sulphur}}{\text{oxygen}} = \dfrac{58\cdot73}{29\cdot6} = 1\cdot98.$

Another set of data show that the ratio between the quantity of any one acid interacting with a certain amount of an oxide, and the amount of oxygen contained in that weight of oxide, remains the same for different oxides.

100 of sulphuric acid are combined in :

(i) Lead sulphate, with 279 of lead oxide containing 21·48 of oxygen.

(ii) Copper sulphate, with 103·66 of copper oxide „ 20·73 „ .

(iii) Ferrous sulphate, with 65·5 of ferrous oxide „ 19·39 „

(2) The ratio in which A and B are contained in the binary compound AB is the same as that in which they are contained in the ternary compound ABC.

The amount of sulphur[1] combined with :

100 of iron is   58·73 in the sulphide and 58·9 in the sulphate
100 of copper is 25·6 „   „   „ 50 = 2 × 25 „   „   [2].

---

[1] The sulphur in the sulphides was found by direct synthesis: the sulphur in the sulphates was estimated by precipitating and weighing as barium sulphate, the percentage of sulphur in barium sulphate being known.

[2] All these relations are of course subject to the operation of the law of multiple ratios.

And here Berzelius used an extremely ingenious method to prove the constancy of the ratio $A : B$ without the intervention of any analytical processes and the consequent errors.

Binary compound changed to ternary. Lead sulphide is oxidised by strong nitric acid to lead sulphate which is insoluble. Any excess of sulphur would, under the conditions of the experiment, be present as sulphuric acid, whilst any excess of lead would be in solution as lead nitrate. Lead and sulphur present in this form can be tested for in the manner below described, and their absence proved. The result is therefore that *all* the lead and sulphur originally present in the sulphide are also contained in the sulphate, and that the ratio *lead : sulphur* is the same in the binary compound lead sulphide as in the ternary compound lead sulphate.

"10 grams of finely powdered lead sulphide contained in a weighed glass flask were digested with *aqua regia* as long as oxidation could be perceived to occur, and the resulting mass was then evaporated and ignited in the flask. It weighed 12·65 grams. After complete cooling and subsequent weighing it was covered with water mixed with a little strong vinegar and warmed. The liquid did not acquire the sweet taste of sugar of lead (lead acetate), and therefore contained no lead. The sulphur in the lead sulphide had therefore been sufficient to form the quantity of sulphuric acid required for the neutralisation of the lead oxide."

The sulphur had been sufficient, but had it been more than sufficient?

"The experiment was repeated in a glass retort with a recipient joined to it, and all the acid that had distilled over was at the end poured back into the retort and re-distilled. The portion passing over at the end [sulphuric acid which is much less volatile than nitric acid would pass over last] was collected separately, and gave no reaction for appreciable traces of sulphuric acid. Hence the lead oxide had been sufficient to saturate all the acid which had been formed from the sulphur of the lead sulphide.

From these results I infer that lead sulphide contains its two constituents in the exact ratio required for the formation of lead sulphate."

(3) The ratio of the quantities A and B of two substances which combine with the same amount of C in the compounds AC and BC is the same as that in which A and B combine with each other in the compound AB.

(3) Constancy of the ratio $A : B$ in the compounds AC, BC, AB.

"I have shown in former experiments that lead sulphide contains the lead and the sulphur in the same ratio as does lead sulphate. If 100 parts by weight of lead give 146·33 of lead sulphate, then by former experiments 7·7 of oxygen must belong to the lead oxide, and 38·63 parts are left for the sulphuric acid which contains 15·42 of sulphur, *i.e.* the same quantity as is combined with 100 of lead in the sulphide. Hence the lead oxide which saturates a certain amount of sulphuric acid contains just one-third as much oxygen as does the sulphuric acid."

This argument somewhat expanded may be put into the following form.

From the data:

(i) Lead sulphate is a compound of lead oxide and sulphuric acid;

(ii) The ratio between lead and sulphur is the same in lead sulphide as in lead sulphate;

(iii) 100 lead yield 107·7 lead oxide
   „   „   115·42 lead sulphide
   „   „   146·33 lead sulphate;

it follows that:

(i)   146·33 lead sulphate derived from 100 lead consist of

107·7 lead oxide containing   38·63 sulphuric acid containing

100 lead   7·7 oxygen   15·42 sulphur   23·21 = 3 × 7·74 oxygen.

(ii) 15·42 sulphur and 7·7 oxygen are the quantities of these two elements combined with the same amount, *i.e.* with 100 of lead.

(iii) 15·42 sulphur and 3 × 7·74 oxygen are the quantities of these two elements combined with each other in sulphuric acid.

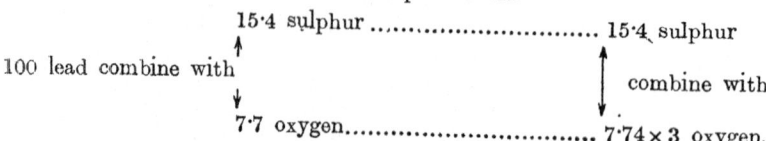

15·4 sulphur .... 15·4 sulphur

100 lead combine with   combine with

7·7 oxygen................................ 7·74 × 3 oxygen.

The quantitative relation discovered by Richter and put on a firm experimental basis by Berzelius is known under the various names of the law of "*equivalent ratios*," "*reciprocal* Designation and formulation of the law. *ratios*," "*permanent ratios*," "*definite ratios*." The first two of these designations refer to parts of the law only, and therefore one of the two last should be used by preference. The law may be cast into the form: The masses of substances, elementary or compound, which are equivalent in any one reaction, *i.e.* which combine or interact with the same amount of a third substance, are identical with or bear a simple numerical relation to the quantities equivalent in any other reaction, including that of combination with each other.

It is the operation of this law which reduces the study of chemical composition to something manageable. Just as in the table for the equivalents of acids and bases (p. 176) the numbers, in terms of the arbitrarily chosen unit 1000 of sulphuric acid, directly represent the quantities which unite to form all the salts that can be produced by the combination of any one base with any one acid, numbers can be found equally characteristic of the elements. If any quantity of any one element is taken as the arbitrary standard, and the amounts of all the other elements are found which combine with this standard or with the amount of any one other element itself combining with the standard, a table of numbers will be obtained giving for all the different elements the quantities (or simple whole multiples or submultiples of the quantities) which combine to form all the possible compounds.

Effect of the law on the relation between the composition of all possible compounds.

So if the standard amount of the standard element is A,

if the quantities of other elements combined with A are B, C, D, etc.,

if the quantities of other elements combined with D are E, F, G, etc.,

then any combination between these elements will be represented by the general formula

$$mA + nB + pC + qD + rE + sF + tG...,$$

where m, n, p, q, r, s, t ... are small whole numbers.

As in the case of the laws of chemical combination already dealt with, a question must be put concerning the degree of accuracy to which the law of definite ratios has been proved. An examination of Berzelius' numbers shows very considerable discrepancies between the empirical and the theoretical composition of the compounds investigated, the differences ranging from $\frac{1}{3}$ to 10 per cent. But in the passage quoted (p. 179) Berzelius discusses

Is the law exact or approximate? Berzelius' results not decisive.

and explains how under the circumstances closer agreement could not be expected; and the fact that the experiments on the compounds of lead which were based on the data determined with the relatively greatest accuracy, showed the best agreement with theory, is in favour of the law being exact.

In the experiment on the oxidation of lead sulphide to lead sulphate we are told that no excess of lead nor of sulphur could be detected, but unfortunately we have no means of evaluating the accuracy of the test, since we do not know what were the least quantities of lead and sulphuric acid that could under the circumstances have been detected. That the solution did not taste of sugar of lead can certainly not be considered a very rigorous proof of the absence of small quantities of lead. A special investigation concerning the exactness of the law was undertaken by Stas and carried out with most brilliant success[1].

Stas investigates accuracy of the law.

"The constancy of the composition of compounds does not prove that the ratio between the weights of the constituent elements is exactly the same in their combinations with other substances. Thus the composition of the sulphide and of the sulphate of barium may each be constant, and yet the ratio between the weights of sulphur and barium in 'the sulphide need not be absolutely identical with the ratio between these same elements 'in the sulphate.... All the analyses and syntheses performed during the last century are equally inadequate for demonstrating the mathematical exactness of the law of definite ratios. In fact, however great may be the skill of a chemist, it is impossible that he should perform an analytical or synthetical operation without incurring errors in the observations. And so far there is nothing to prove that the differences found in certain analyses between experiment, and calculation in accordance with a given hypothesis, must be attributed *entirely* to the error committed in the mechanical process; there is nothing to show that a certain portion of the error is not due to the inexactness of the law of definite proportions, considered as a mathematical law.... Constancy of composition being admitted, what is wanted to decide this question? It may be proved, for instance, that in binary and ternary substances which have two elements the same, these common elements are present in invariably constant ratios. Thus in the two substances AB and ABC the relation of weight between A and B must be exactly the same in AB as in ABC.

It is evident that the solution of the problem when thus stated becomes independent of analysis proper. In fact, all that is needed to decide the

[1] "De l'invariabilité des rapports en poids des éléments formant les combinaisons chimiques," *Nouvelles Recherches.*

question, is to ascertain whether the ternary substances can be brought to the state of the binary compounds without a fraction, however minute, of the common elements being liberated; or conversely, whether the binary substances can be transformed into ternary compounds without a fraction of the elements of the binary compounds being excluded from the ternary compound formed[1].

Ratio A : B in AB and ABC. Ternary compound changed to binary.

It is known that sulphurous anhydride transforms an iodate suspended in water into the iodide, being itself converted into sulphuric acid. I have ascertained that by the action of the same reagent the bromate passes into bromide and the chlorate into chloride. The absolute insolubility of the iodide, bromide and chloride of silver in water acidulated with sulphuric acid, and the possibility of detecting in a liquid one ten-millionth of the silver, or iodine, or bromine, or chlorine present, constitute quite exceptionally favourable conditions for subjecting the law of definite proportions to a decisive test."

Procedure in reduction by sulphurous acid of silver iodate to silver iodide.

The manner in which these investigations were carried out is illustrated by the following example:

"32·819 grams of iodate[2], which had been heated to 130°, were suspended in 100 c.c. of boiled water, to which 15 c.c. of sulphuric acid had been added. The air in the vessel containing this mixture having been replaced by pure carbonic anhydride, I poured in slowly 400 c.c. of a solution of sulphurous anhydride made that very instant, and containing only 1 per cent. more of the sulphurous anhydride than the exact quantity required for the transformation of the iodate into iodide. Whilst the reduction was proceeding the flask was plunged into a mixture of water and ice. When the reaction was completed, I withdrew half of the supernatant liquid, and after enclosing it in a stoppered bottle, I left it in complete darkness in order that it might clear spontaneously. Eleven days were required to accomplish this.

The flask containing the other half of the liquid together with the iodide produced, was immediately placed on the water bath and heated with continuous shaking until the liquid had cleared, which occurred at 53° C. The limpid liquid was divided into two parts; one, tested by a normal solution of iodide, proved to be completely free from silver; the other, tested by a normal solution of acid sulphate of silver, was found completely free from iodine[3]. The iodide of silver formed was washed three or four times with distilled water, and then digested at 60° with pure nitric acid mixed with five times its own volume of water. After digesting for a whole day, the liquid

---

[1] This was the method first used by Berzelius in his transformation of lead sulphide to lead sulphate.

[2] The iodate had been prepared from iodic acid and silver sulphate.

[3] Iodine present in any form other than that of silver iodide would in the presence of sulphurous acid be changed to hydriodic acid, thus:

$$I_2 + 2H_2O + SO_2 = 2HI + H_2SO_4$$
$$HIO_3 + 3H_2O + 3SO_2 = HI + 3H_2SO_4.$$

was absolutely free from silver, proof that not a trace of sulphide or sulphate of silver had been precipitated.

Finally the iodide of silver was washed with pure boiling water to remove the nitric acid as completely as possible, and after cooling it was moistened with a solution of cyanide of potassium mixed with hydrocyanic acid, in which it dissolved completely, producing a colourless limpid liquid.

The liquid referred to above which had been left to clear spontaneously, proved to be completely free from silver or hydriodic acid."

The results were the same for iodate prepared by other methods, and for bromate and chlorate prepared in several different ways. In one experiment nearly 260 grams of chlorate were reduced, and since $\frac{1}{100}$th of a milligram of silver or of chlorine could still be detected, Stas could assert that the absence of any reaction for silver or for chlorine proved that if either of these elements had been present in the solution above the precipitated silver chloride, their amount must have been less than one ten-millionth of the quantity of the silver and chlorine that had entered into reaction. Stas sums up by saying:

All the silver and halogen of the ternary compound are retained in the binary compound.

"Hence by the action of sulphurous acid the iodate, the bromate, and the chlorate of silver can be changed to the iodide, bromide and chloride without a fraction, however minute, of iodine, bromine, chlorine or of silver being liberated. The concordance of the results observed in the transformation of these three ternary compounds to the state of binary combinations demonstrates the invariability of the relative weights of the elements constituting them. I have also proved, if proof was required, the constancy of the composition of one of these binary compounds[1]. From the combination of these two sets of facts it follows that substances unite in absolutely fixed and invariable ratios, that these ratios are true *constants*, and that the laws of chemical combination are mathematical [exact] laws, as they have been regarded by chemists for nearly half a century."

Stas declares the laws of chemical combination to be exact.

[1] Silver chloride (p. 146).

CHAPTER VIII.

COMBINING OR EQUIVALENT WEIGHTS.
SYMBOL WEIGHTS.

" Wenn das [in 1808] aufgegangene Licht sich über die ganze Wissenschaft verbreiten sollte, mussten zuerst die Atomgewichte einer möglichst grossen Anzahl von Grundstoffen mit möglichster Genauigkeit ausgemittelt werden. Ohne eine solche Arbeit konnte auf diese Morgenröthe kein Tag folgen."

BERZELIUS, 1845.

THE law of fixed ratios affirms that for a compound AB, the value $A : B$ is a true constant; the law of permanent ratios extends this constancy from one compound AB, to any other compounds into the composition of which A and B may enter, but does so subject to the operations of the law of multiple ratios according to which $A : B$ may become $mA : nB$, where m and n are small whole numbers.

Generalisation from the laws of chemical combination leads to the conception of the combining weight.

Hence as has been said already: if a certain arbitrarily chosen amount of an arbitrarily chosen element be taken as standard, and the amounts of all elements which unite with this standard—or with the amount of any one other element itself uniting with the standard—be determined, numbers will be obtained which are true constants. These constants, termed *combining weights* or *proportional numbers*, represent the quantities of all the different elements which, either directly or after multiplication by a simple whole number, unite to form all the known and all the possible combinations between the elements. It is therefore possible to embody the three laws of chemical combination in the one wider generalisation: *Elements combine with each other in the ratio of their combining weights, or of simple whole multiples of these.*

The combining weight, the direct outcome of the empirical laws of chemical combination, a value absolutely non-hypothetical, may be defined as: "Any one of the quantities of an element which unites with the standard amount of the standard element, or with the combining weight of any other element." The combining weights being the numbers which regulate the quantitative aspect of all chemical change, the most accurate determination of these constants must constitute the fundamentally most important problem of the science. What is there to be gathered from the definition concerning the nature and the experimental determination of these magnitudes?

The combining weight defined.

(1) The combining weight may be determined either by direct reference to the standard element; or, when the element considered forms no compounds with the standard element, or when the compounds formed are not suitable for accurate analysis, it may be determined indirectly through an element the accurate combining weight of which is known. On the basis of the law of permanent ratios it can be affirmed that:

(1) The comb. wt. may be determined by direct or by indirect reference to the standard element.

comb. wt. C (standard A) = comb. wt. C (standard B) × comb. wt. B (standard A).

Thus, taking 8·00 of oxygen to be the standard, and having found that 35·45 of chlorine are equivalent to 8·00 of oxygen, then the combining weight of any other element may be ascertained from the composition of either the oxide or the chloride by determining the amount of that element combined with 8·00 of oxygen in the oxide, or with 35·45 of chlorine in the chloride.

(2) The standard must be chosen, which is an arbitrary process. There is no theoretical *a priori* preference for any one standard against any other; it is altogether a practical question to find what standard will lead to the most accurate and most permanent, as well as most simple numbers. Two points are left for decision, namely, the element, and the amount of it. Of these two it is the choice of the element which is the matter of real practical importance.

(2) The choice of the standard.

It is evident that if the combining weight of a certain element

M cannot be determined by direct reference to the standard, the accuracy of the value obtained must depend on that of each of the two factors in the equation given under (1), and that if B is the element through which a great many others are referred to A, uncertainty concerning the value: combining weight $B_{(\text{standard } A)}$ should disqualify A from being chosen as standard. Every redetermination of $B : A$, undertaken with the object of attaining to greater accuracy, will necessitate a recalculation and consequent change in all the values $M : A$ which are expressed in terms of the ratios $B : A$ and $M : B$, even though no new measurement had been made of $M : B$.

(i) The standard element and its necessary qualifications.

It must therefore be recognised as an essential qualification of the standard element that it should be one which forms compounds with a large number of other elements; that these compounds should be susceptible of very accurate analysis; and that when indirect determinations must be resorted to, the ratio $B : A$ should be known with a degree of accuracy which represents the utmost attainable in these measurements.

On the other hand, the quantity of the standard element taken is a mere matter of convenience in the handling of the numbers obtained. It is desirable that these numbers should not be unnecessarily large, and preference has been shown for a system in which they are all greater than unity.

(ii) The quantity of the standard element.

Hydrogen and oxygen have in succession, and unfortunately also at the same time, been used as the standard.

Hydrogen $= 1{\cdot}000$, is the number which has been championed chiefly for the reason that it is the element whose combining weight is the smallest of any; and that unit weight of the element of smallest combining weight seems the natural standard, leading for the other elements to the smallest possible numbers, when all are greater than unity. But against these more or less senti- mental recommendations have to be set the very real disadvantages, that very few elements combine with hydrogen to form hydrides, and that when they do, these compounds do not lend themselves to the purposes of very accurate analyses; that as a matter of fact practically all elements can be referred directly to oxygen, and that the ratio *oxygen : hydrogen* was, until a very few years ago,

The case for and against the hydrogen and the oxy- gen standard respectively.

not known with a degree of accuracy equal to that of the most exact determinations of other combining ratios.

The claims of *oxygen* to the position of the standard element are based on the fact that it forms compounds with nearly all elements; that the composition of a large number of oxides may be determined very accurately; that when oxygen compounds are not available chlorides (or bromides) generally are, and that the value *chlorine : oxygen* or *bromine : oxygen* is known with a degree of accuracy not surpassed by that of any other such ratios.

The history of the succession of standards deserves to be told very shortly. Thomson gives us the early part of it[1]. *Combining weight* may be substituted wherever he says *atomic weight* without altering the exact meaning, or the force of the argument:

"Mr Dalton made choice of the atom of hydrogen for his unity, and in this he has been followed by Dr Henry of Manchester, and by one or two chemical gentlemen in London. But this method has been rejected by almost all the British Chemists, and by all the Chemists without exception in Europe and America. The choice was unhappy for very obvious reasons. Because the atom of hydrogen is the most difficult of all to determine.... Now if we reckon the atomic weight of Hydrogen as unity and commit an error respecting its relation to that of other bodies, this error will affect the atomic weight of all other bodies and will make them all either too heavy or too light; whereas if we make choice of oxygen for our unity, an error respecting the atom of hydrogen will be confined to that atom and will not affect the accuracy of the atomic weights of other bodies.... Hydrogen as far as we know at present combines with but few of the other simple bodies, while oxygen unites with them all, and often in various proportions. Consequently very little advantage is gained by representing the atom of hydrogen by unity; but a very great one by representing the atom of oxygen by unity."

The standards actually used and advocated in succession: Hydrogen = 1; oxygen = 1, = 10, = 100.

Whilst Thomson used *1 of oxygen*, Wollaston was in favour of *10 of oxygen*. Both these numbers were rejected by Berzelius as too small, and he used *100 of oxygen*, which for a long time remained the generally accepted standard.

About the middle of the century a return was made to the Daltonian unit, and it was used for some time without giving rise to practical difficulties. Dumas had just (1842) redetermined the ratio *oxygen : hydrogen* from the composition of water, had found

[1] Thomas Thomson, *An Attempt to establish the First Principles of Chemistry by Experiment*, 1827.

it to be 15·96 : 2, and had expressed his belief that the true value
was probably 16 : 2. His number, which in the same year was
corroborated by the results of Erdmann and Marchand, was for
a long time considered as extremely exact, and in fact was so by
comparison with the accuracy attained for the other combining
ratios.

But Stas' work soon after supplied numbers for the combining
weights of a large number of elements which had all been determined by direct reference to 16 of oxygen, and

Stas recommends oxygen=16·00.   the accuracy of which was far superior to that of the
ratio *oxygen : hydrogen*. Indirectly he had obtained
for hydrogen in terms of *oxygen = 16·00*, the value
2·02. Though not attaching excessive importance to this result,
Stas was emphatic in pointing out that the composition of water
was not known with sufficient accuracy for hydrogen to be a
suitable standard, and he expressed himself in favour of 16 of
oxygen. Since, however, 16·00 was then widely accepted as the
exact combining weight of oxygen in terms of the hydrogen
standard, Stas' recommendation involved no recalculation; that is
no change in practice, only a change in theory desirable from the
point of view of future possibilities. The adoption of the oxygen
standard advocated would have meant that if a redetermination of
the composition of water should lead to a change in the value of the
ratio *oxygen : hydrogen*, the effect of this would only be an alteration
of the value used for the combining weight of hydrogen, and not
the necessity of a recalculation and a consequent change of all the
combining weight values determined directly in terms of oxygen.
Stas' suggestion, which it must be admitted was not pressed
strongly, did not receive much support.

From 1888 onwards, there appeared in quick succession redeter-
minations of the ratio *oxygen : hydrogen*, all bringing out the fact
that Dumas' experimental number 15·96 was too high; but in the
meantime Lothar Meyer and Seubert, strong champions of the
hydrogen standard, had made this value the basis of very popular
atomic weight tables. The new data did not provide a sufficiently
reliable number to make recalculations of all the other values
desirable, and the obvious way out of the difficulty, the use of the
tables of Ostwald in which the *16·000 oxygen* standard was used,
was not generally accepted. The uncertainties and inconveniences
arising from the fact that of the different text-books some give

the one table, and some the other, are a matter of present day practical experience. A very small counterbalancing advantage, one not likely however to be made much of in the voluminous controversy on the subject of the relative merits of the oxygen and the hydrogen standard is, that no theoretical discussion could so efficiently impress on those beginning the study of the science the fact that combining ratios are referred to an *arbitrary* standard, as does the actual simultaneous existence of two such standards.

When in 1898 a Committee appointed by the German Chemical Society unanimously reported[1] in favour of making "*16·000 oxygen*" the universal and sole standard, there seemed good hope of uniformity being attained before long. But to

International Committee decides for the concurrent use of the standards hydrogen and oxygen.

the great regret of a large section of chemists[2], a permanent International Committee has in 1902 decided to retain both standards concurrently[3]. This Committee has begun the annual publication of tables containing in two columns the standard values of these constants[4] referred to both *hydrogen = 1·000*, and *oxygen = 16·00*, and the differences between the two sets of numbers are very nearly 1 per cent.

Morley, by his wonderfully exact determination of the gravimetric composition of water (pp. 72, 87), solved in 1895 a long standing and most difficult problem, and supplied the required accurate value for the ratio *oxygen : hydrogen*. His number 15·879 : 2 is certain to within one or two units in the last place and therefore comparable with the most accurate determinations obtained by Stas for certain other elements. Hence the most serious practical difficulty in the way of the hydrogen standard is for the time removed. But the theoretical superiority of direct determinations retains its full force, and there is the further practical advantage that with *oxygen = 16·00*, a number of the most commonly used combining weights, such as those of carbon (12·00), nitrogen (14·0), sulphur (32·0), phosphorus (31·0), sodium (23·0), are represented by whole numbers, whilst a hydrogen unit (oxygen = 15·879)

[1] London, *J. Chem. Soc.* 76, 2, 1899 (p. 86).
[2] Ostwald, *Zs. physik. Chem.*, Leipzig. 42, 1903 (p. 634).
[3] *Chem. News*, London, 87, 1903 (p. 78).
[4] *Ibid.* 89, 1904 (p. 68).

would in these cases lead to the less convenient numbers 11·9, 13·9, 31·8, 30·8, 22·9.

(3)   More than one value may be obtained for the combining weight of an element; but all these values must be simple whole multiples and sub-multiples of each other, and hence the *accurate* determination of any *one* of them is all that is required. Any one accurate value for the combining weight of an element, together with the approximately correct determination of the composition of a compound into which this element enters in a *different* combining ratio, will be sufficient to settle the exact value of its combining weight in

(3) Different values may be obtained for the combining weight, but these bear a simple numerical relation to each other.

that compound. Two compounds of hydrogen and oxygen are known, viz. water and hydrogen peroxide, and experiment shows that in the latter compound there is *about* twice as much oxygen combined with the same amount of hydrogen as in the former; an *accurate* determination of the combining ratio in water will at the same time supply an equally accurate one for the peroxide, which will be exactly twice as great.

An extremely simple and beautifully conceived symbolic notation allows us to show the rule of the combining weights in all the quantitative relations of chemical combination and decomposition.

The beginning of all symbolic notation goes back to the alchemists, who used it with a qualitative meaning only. ☉ represented the substance gold, a metal endowed with certain special properties, without indication of any special amount of it; ☽ represented silver in the same sense; and the number of such signs grew with the increasing number of substances discovered and investigated. Inspection of Bergman's table (p. 115),

Symbolic notation. The alchemists' and Bergman's signs have only qualitative meaning.

in which the symbolic notation of the time is used, shows that no very systematic attempt was made to represent even the qualitative composition of compounds by a joining together of the symbols for the constituents; and in the then very imperfect state of knowledge concerning chemical composition, it would have been scarcely possible to accomplish this.

Dalton in 1803 devised a symbolic notation which was both qualitative and quantitative (*post*, chap. x). ☉ stood for 1 part by weight of the substance endowed with the properties of hydrogen, ◯ for the weight of oxygen which combines with one

Dalton's nota-
tion is qualita-
tive and
quantitative.

of hydrogen, which according to Dalton was 5·66; whilst ◯ ⊙ represents the qualitative and quantitative composition of water, which therefore in 6·66 parts by weight contains 1 of hydrogen and 5·66 of oxygen. ● stood for 4·5 parts by weight of carbon, the quantity which according to Dalton combined with 1 of hydrogen in olefiant gas, or with 5·66 of oxygen in carbonic oxide; and whilst ◗ ◯ represented the qualitative and quantitative composition of carbonic oxide, ●◯ ◯ represented that of carbonic acid. This was a tremendous step in advance, and though we no longer use the practically somewhat awkward symbols of Dalton, we have retained the principle unaltered.

Berzelius substituted for Dalton's geometrical symbols a more convenient representation by letters; the first letter, or the first together with one other letter of the Latin name of the element being chosen to represent that quantity which combines or interacts with simple whole numbers of the symbol weights of the other elements. A small number placed below indicates the number of times the quantity represented by this symbol weight is contained in the compound[1].

$H = 6·24$ of hydrogen ; $O = 100$ of oxygen ; $C = 75$ of carbon.

$H_2O = 112·48$ of water, containing $2 \times 6·24$ of hydrogen and 100 of oxygen.

$CO = 175$ of carbonic oxide, containing 75 of carbon and 100 of oxygen.

$CO_2 = 275$ of carbonic acid, containing 75 of carbon and 2×100 of oxygen.

These *symbol weights* are combining weights, but it is obvious that a definite convention must be adhered to as to which of the

The symbol for
an element re-
presents some
one of its
combining
weights.

different possible combining weights, all of which are in a simple numerical relation to each other, should be chosen to be represented by the symbol. It would be impossible to put this more simply and more clearly than was done by Laurent.

Auguste Laurent (1807—1853), a pupil of Dumas', at one time of his life Professor of Chemistry at Bordeaux, later on Master of

Laurent on
combining
weights.

the Mint in Paris, is most conspicuous by his work in organic chemistry. He was endowed with a great power for seeing the contradictions and weak points

[1] Dalton's criticism of Berzelius' modification of his symbols is quaint: "Berzelius' symbols are horrifying; a young student in chemistry might as soon learn Hebrew as make himself acquainted with them."

in the views then most commonly held concerning the fundamental nature of chemical combination. His exposition of the subject of combining weights, taken from his *Méthode de Chimie*, published after his death in 1854, calls for reproduction *in extenso*, lengthy though it is:

"Chemists have in turn used 'atoms,' 'proportional numbers,' or 'equivalents,' to represent the composition of compounds. But recognising the uncertainties and the changes to which an atomic notation is subject, the majority have given it up and have adopted a notation of equivalents which simply represents the result of experiment, and is not, like the former, subject to changes arising from individual opinion, as well as from the progress of the science[1]. The importance of this subject is such that we must dwell on it at some length. We will proceed to examine it, and will establish a difference between 'proportional numbers' and 'equivalents,' a difference not shown by our system of notation.

The empirical "proportional numbers" made the basis of notation in preference to the hypothetical "atomic weights."

We know from experiment that oxygen combines with simple substances in the following ratios :

100 of oxygen combine with :

12·50 and	6·25	.....................parts of hydrogen.
442·00 and	221·00 and 88·4	..............parts of chlorine.
200·00 and	100·00 and 66·6	..............parts of sulphur.
175·00 and	87·50 and 58·3 and 43·7	parts of nitrogen.
75·0 and	37·5	.............................parts of carbon.
350·00 and	233·3	...........................parts of iron.
2600·00 and 1300·0 and 866·0 and 650·0 parts of lead.

And experiment further demonstrates that 6·25 parts of hydrogen can combine with 221 parts of chlorine, with 100 and 200 parts of sulphur, with 75 and 37·5 parts of carbon ; that 221 parts of chlorine can combine with 100, 200 and 66·6 parts of sulphur, with 75 and 37·5 parts of carbon, with 350 and 233 parts of iron. That is to say, whenever two elements unite with each other, combination always takes place in 1, 2, 3, 4 or 5 different ratios, and that these ratios are represented by precisely the numbers found in the table of the oxygen compounds, or by simple multiples or sub-multiples of these numbers.

Let us then *arbitrarily* choose, for each element, *one* of the numbers representing the different ratios in which it combines with 100 parts of oxygen, let us call this number the *proportional number* for this element, and let us represent it by the initial of its name ; let us, for instance, choose the largest numbers, that is those inscribed in the first column. In that case we would represent the combination formed by 100 parts of oxygen and

[1] A criticism deserved at the then stage of development of the atomic theory, but which ceased to be applicable soon after Laurent's death.

12·5 parts of hydrogen, by the formula OH, and consequently the one containing 100 of oxygen and 6·25 of hydrogen by $OH_{\frac{1}{2}}$ or by O_2H. In like manner, we would represent the compound containing 442 parts of chlorine and 200 parts of sulphur by ClS, and the one containing 442 parts of chlorine and 66·6 parts of sulphur by $ClS_{\frac{1}{3}}$ or by Cl_3S and so on."

Laurent proceeds to point out that such a system would be intelligible, but that if a language real or symbolical is to be perfect, it is not sufficient that to each idea, as to each substance, there should correspond a fixed word or a fixed symbol. In order to assist the memory and to supply a basis for future reasoning, analogies between different ideas and different substances should be accompanied by corresponding analogies in the words or the symbols chosen to represent them.

Arbitrary selection of a combining ratio for the proportional number leads to notation which though correct is not scientific.

"But if, as was supposed, we arbitrarily select for the proportional number any one of the numbers from the table of the oxygen compounds, this choice might be such that the formulae assigned to the sulphate, the selenate, and the tellurate of potassium became :

$$SO_4K, \quad SeO_8K_2, \quad Te_2O_{12}K_3\,;$$

that those for the chlorides, bromides, and iodides, of potassium and of lead, became :

$$ClK, \quad Br_2K, \quad I_3K,$$
$$ClPb, \quad Br_4Pb, \quad I_6Pb.$$

Such a notation would be intelligible....But we perceive at once that it would singularly embarrass the memory, and that it would disguise from us a number of relations which we might otherwise realise at once.

Let us therefore suppose that instead of selecting the proportional numbers arbitrarily, we specially choose those that comply with the two following conditions :

Considerations which should guide the selection of proportional numbers.

(1) to represent the different series of compounds by the simplest possible formulae ;

(2) to assign to analogous compounds analogous formulae."

He goes on to show that these two conditions are not always compatible; also that in a growing science the analogies to be expressed change with the increase of knowledge, and that therefore the formulae representing compounds and the proportional numbers represented by the symbols may require occasional revision.

"But it would be absurd to wrangle over the invariability of a number accepted only by convention. Let us add that the proportional numbers by no means serve the sole purpose of representing in the simplest possible manner the composition of substances; striking relations often exist between the formulae[1] and the properties of certain substances, relations which would be completely hidden, if in the determination of their composition the only point kept in view were that of the representation of the composition. To illustrate this, I will employ three different notations to represent the composition of certain substances allied to bicarburetted hydrogen, compounds between which there are very remarkable relations of composition, of properties, of volume and of derivation."

Table illustrating Laurent's method in assigning formulae. Laurent's table is reproduced, p. 198; the values used in all three columns for the proportional numbers are: $O = 100$, $C = 75.0$, and $H = 6.25$.

" Half the formulae in column I are simpler than those in II and III.

Looking at the first two columns, who could perceive the reason why the first five substances are not attacked by potash[2], and why the boiling point rises from the 1st to the 5th; why the following five substances **Discussion of the Table.** can be decomposed by potash, and why their boiling point also rises continuously?... In the sequence of the compounds as represented in columns I and II, why is it that we pass from C to C_2, C_3, C_4 and back again to C or C_2; and why do the compounds when in the gaseous state occupy volumes which are in the ratio $1 : 2 : 4$?

But when we turn to the third column, we find that: (i) The weight of carbon remains constant. (ii) The volume is the same for all the compounds. (iii) The rise in the boiling point from the 1st substance to the 5th, and again from the 6th to the 10th, is due to the fact that in the first five compounds the quantity of hydrogen decreases regularly, whilst that of chlorine increases in the same manner, and that the same relation holds for the next five substances. (iv) The first five compounds, which are not attacked by potash, contain besides the carbon a constant number of *four* other equivalents, whilst the five following substances contain a different number of equivalents, namely *six*."

Laurent pointed out how the notation takes no account of the difference between *combining weights* or proportional numbers, and *equivalent weights* (*ante*, p. 195). In what then does this difference consist?

The last chapter, which dealt with the law of equivalent ratios, should have brought out the meaning of chemical equivalency, of the power possessed by certain quantities of different substances

---

[1] Laurent here uses the term "nombres," but the sense of the argument seems to require the translation "formulae."

[2] Alcoholic potash really acts on both classes of substances, but does so more readily in the one case than in the other. The effect always consists in the withdrawal of hydrogen and chlorine in the ratio in which they are contained in hydrochloric acid, producing from the saturated compounds (*e.g.* 6 to 10 in the table) substances of the type of ethylene (*e.g.* 1 to 5 in the table), and from ethylene derivatives, substances of the type of acetylene C_2H_2.

Names	I Simplest Formulae	II Berzelius' Formulae	III Gerhardt and Laurent's Formulae	Column added to Laurent's table, giving the boiling point (B.P.), the mode of formation, and some characteristic relations of the substances considered. B.P.
		All the quantities represented by the formulae of column III occupy the same volume when gasified at the same temperature and pressure. Call this volume V; then the volumes occupied by the quantities represented by the formulae of columns I and II vary between $\frac{1}{2}V$ and $2V$		
1. Ethylene, or Olefiant gas, or Heavy carburetted hydrogen	CH_2 $\frac{1}{2}V$	CH_2 $\frac{1}{2}V$	C_2H_4 V	$-103°$ C.	Gas, formed by dehydrating alcohol with sulphuric acid. Combines directly with chlorine to form Dutch Liquid; not acted on by potash
2. Monochlor ethylene, or Vinyl chloride	C_2H_3Cl V	$C_4H_6Cl_2$ $2V$	C_2H_3Cl ,,	$-18°$	Gas, formed by the action of alcoholic potash on ethylene chloride, $C_2H_4Cl_2$. Combines directly with chlorine, forming $C_2H_3Cl_3$
3. Dichlor ethylene	$CHCl$ $\frac{1}{2}V$	$C_2H_2Cl_2$ V	$C_2H_2Cl_2$,,	$37°$	Liquid, formed by the action of alcoholic potash and chlorinated ethylene chloride, $C_2H_3Cl_3$. Combines directly with chlorine, forming $C_2H_2Cl_4$
4. Trichlor ethylene	C_2HCl_3 V	$C_4H_2Cl_6$ $2V$	C_2HCl_3 ,,	$88°$	Liquid, formed by the action of alcoholic potash on $C_2H_2Cl_4$ (dichlor ethylene). Combines directly with chlorine, forming C_2HCl_5
5. Tetrachlor ethylene, or Protochloride of carbon	CCl_2 $\frac{1}{2}V$	CCl_2 $\frac{1}{2}V$	C_2Cl_4 ,,	$121°$	Liquid, obtained by strongly heating perchloride of carbon, which is reformed by action of dry chlorine in the sunlight
6. Dutch liquid, or Ethylene chloride, or Dichlor ethane	CH_2Cl $\frac{1}{2}V$	$C_2H_4Cl_2$ V	$C_2H_4Cl_2$,,	$84°1$	Liquid, formed by direct combination of chlorine with ethylene. Action of alcoholic potash solution gives vinyl chloride. Chlorine leads to substitution, not addition products
7. Chlor ethylene chloride, or Trichlor ethane	$C_2H_3Cl_3$ V	$C_4H_6Cl_6$ $2V$	$C_2H_3Cl_3$,,	$74·5°$	Liquid, formed by the substituting action of chlorine on ethyl chloride, C_2H_5Cl
8. Dichlorethylene chloride, or Tetrachlor ethane	$CHCl_2$ $\frac{1}{2}V$	$C_2H_2Cl_4$ V	$C_2H_2Cl_4$,,	$135°$	Liquid, obtained by the substituting action of chlorine on ethylene chloride
9. Trichlor ethylene chloride, or Pentachlor ethane	$CHCl_5$ V	$C_4H_2Cl_{10}$ $2V$	C_2HCl_5 ,,	$159°$	Liquid, formed by substitution of chlorine in C_2H_5Cl, or addition to C_2HCl_3. With alcoholic potash gives C_2Cl_4
10. Tetrachlor ethylene chloride, or Hexachlor ethane, or Perchloride of carbon	CCl_3 $\frac{1}{2}V$	C_2Cl_6 V	C_2Cl_6 ,,	$185°$ (M.P.)	Crystalline solid, obtained as the final product of the chlorination of ethylene chloride or ethyl chloride. Decomposed by alcoholic potash

[1] A substance of the same composition as Dutch Liquid is known which boils at $59°$.

to produce the same chemical effect. Those quantities of all the different bases which neutralise 10000 or any other amount —provided it is the same—of an acid, are equivalent as regards neutralising power; the quantities of oxygen and sulphur, which combine with the same amount of lead, are equivalent as regards combination with that metal; the weights of zinc and aluminium which evolve the same volume of hydrogen from dilute acid are equivalent as regards their action on the acid, etc. The conception of equivalency, which, through Richter's neutralisation work, arose from a study of compounds, was soon extended to elements when Richter himself recognised that the quantities of two metals equivalent in their action on acids were also equivalent in their power of combining with oxygen. Equivalency always implies equality in the effect produced by two substances on a third. This is what Laurent, in continuation of his discussion of proportional numbers, says on the subject of equivalents:

Equivalent weights. Meaning of chemical equivalency.

"If we take a quantity of the nitrate or the sulphate of silver containing 1350 parts by weight of the metal, and if copper, or lead, or iron, or potassium is added, we shall find that:

1350 parts of silver are displaced by 1300 parts of lead,
  ,,   ,,   ,,   ,,   400 parts of copper,
  ,,   ,,   ,,   ,,   350 parts of iron,
  ,,   ,,   ,,   ,,   490 parts of potassium.

And since the nitrates of lead, copper, iron, and potassium thus formed have properties analogous to those of nitrate of silver, we may say that 1300 parts of lead, 400 of copper, 350 of iron, and 490 of potassium play the same part, fulfil the same functions as 1350 parts of silver, or shortly, that they are its equivalents....Hence, to find the equivalents of two substances, it is necessary that they should exhibit analogies in their properties."

He goes on to show how one and the same substance may have different equivalents, according to the reaction involved. Thus perchlorate of potash, which in 700 parts contains 442 of chlorine, exhibits many important analogies with permanganate of potash, which differs from it in composition, only by containing 700 of manganese in place of the 442 of chlorine; but analogies exist also between sulphates and manganates in which 200 of sulphur are equivalent to 350 of manganese; and finally, when deduced from another state of oxidation, the equivalent of manganese becomes 233.

Laurent on equivalents. The same substance may have different equivalents.

"Hence by the method of equivalency we should arrive at the inference that manganese has three equivalents at least, viz., 700, 350, and 233. I will add that to me this conclusion seems correct. As I have said already, the equivalent depends on function, and hence if a substance performs different functions, it may have different equivalents. We must therefore say: the equivalent of manganese is 700 when it fulfils the function of chlorine, 350 when it plays the part of sulphur in manganates, and that in the manganic salts it is 233. To the question: what is the equivalent of such and such a substance? there should therefore always be added the words: 'when it performs such and such a function.'"

Accordingly, in any statement concerning equivalency, we should always expect to find the names and quantities of two substances, and the reaction in which these participate. But if the two substances considered are elements, and the one is the standard amount of the standard element; and if we recognise the possibility of the existence of more than one value, we arrive at the conception of the equivalent weight of an element which is: The amount which is of equal chemical value to the standard, that is, which will replace the standard in its compounds. Thus, 32 of sulphur are replaced by 16 of oxygen, when a sulphide is heated in air and thereby changed to an oxide, etc., etc.

The equivalent weight of elements.

Obviously if the standard is the same there must be a close connection between the combining weight and the equivalent weight of an element, in consequence of the law of equivalent ratios according to which the quantities which are equivalent to each other are also the quantities which will combine with each other.

$$\left.\begin{array}{l}\text{.................................}m \text{ of element } A \\ \text{standard amount of standard element}\end{array}\right\} \text{combines with } x \text{ of element } M.$$

∴ m, or $2m$, or $3m$, etc. of A will, if combination is possible, combine with 1, or 2, or 3, etc. times the standard amount of the standard element.

Hence, if referred to the same standard, the numerical values for the combining weight and the equivalent weight are the same, or simple whole multiples of one another, and the two terms may without error be considered synonymous, as is usually done. But all the same it would be well to remember that there is a certain difference in the conception of the two quantities: equivalent weight referring to elements and compounds alike, combining weight to elements only;

If the standard is the same, the numerical value for the equivalent and the combining weight is the same.

the equivalent, that which is equal to something else in its relations to a third substance, refers to equality of function, whilst the combining weight refers simply to composition, to the quantity of one substance combined with a certain amount of another.

And so the two terms may be included in the definition: *The combining or equivalent weight of an element is that amount of it which will combine with, or which will replace the standard amount of the standard element, or the quantities of any other element equivalent to this.*

Definition comprising combining and equivalent weights.

Laurent's treatment of the subject of combining weights or proportional numbers has shown how it is possible by means of these to represent, not only the quantitative composition of compounds, but also their mutual relations as exhibited in the graduation of their physical properties and in their reactions of formation and decomposition. Two things are required for this purpose: (i) a very accurate knowledge of any one of the combining weights of each element concerned; (ii) an extensive

The symbolic representation of compounds involves: (i) Accurate knowledge of the combining weights of the constituent elements. (ii) Extensive study of the properties of the compounds.

study of the chemical and physical properties of the substances into the composition of which these elements enter, with the object of ascertaining which simple multiple or sub-multiple of the one accurately determined combining weight shall be chosen for the symbol weight, and which multiples of the simplest formulae will best represent the compounds. It is proposed to treat here of the first of these requirements only, the second one being, in the present condition of the science, most readily met by the application of the molecular theory, in a manner which will be dealt with in subsequent chapters.

The object of all combining weight determinations is to ascertain, directly or indirectly, with the utmost possible accuracy, the quantities of the different elements which combine with or replace 16·000 parts by weight of oxygen. The general method for achieving this consists in a determination of the ratio between the element of unknown combining weight and one or more other elements with which it combines or interacts, and the combining weight of which is known. Obviously, the more direct the method, the greater is the probable accuracy. The following may serve as examples of

The determination of the combining weights. The general principle. Examples of the application of different methods.

methods frequently employed in combining weight determinations.

(1) *Determination of the composition of the oxide.* This of course is an absolutely direct method.

(1) Ratio of the weights of the constituents of an oxide or other binary compound.

(i) Carbonic acid is synthesised by burning a known weight of pure carbon in excess of pure oxygen, and the oxide formed is collected and weighed.

Roscoe in burning transparent Cape diamonds used in six experiments 6·4406 grams and obtained 23·6114 grams of carbonic acid.

∴ 17·1708 of oxygen combine with 6·4406 of carbon
16·000   „   „   „   **6·002**   „

(ii) The combining weight of hydrogen is deduced from the composition of the oxide water.

The data given before (p. 75) relating to Morley's complete syntheses of water show that, in nine experiments, a total quantity of 33·2435 grams of hydrogen combined with 263·9387 grams of oxygen, to form 297·1766 grams of water.

∴ 16 oxygen combine with **2·0152** hydrogen (from the ratio ox. : hyd.)
„   „   „   **2·0149** hydrogen (from the ratio water : ox.).

(iii) The composition of the oxide of copper, of lead, of zinc, of cadmium, of mercury, of phosphorus, etc. has been determined either by synthesising or by analysing the oxide. With the principle the same, the experimental method may take different forms.

Harden dissolved a known weight of pure mercuric oxide in potassium cyanide, and weighed the mercury separated on electrolysis.—Schrötter burnt a known weight of pure amorphous phosphorus in dry oxygen and weighed the oxide formed.—Berzelius and several successive experimenters found the composition of copper oxide by weighing the metallic copper left after a known weight of the oxide had been reduced in a current of hydrogen, etc., etc.

The analysis or synthesis of any binary compound containing one element of accurately known combining weight is in principle so closely allied to the above, that mention may here be made of it. Chlorides, bromides, and sulphides are the compounds that have been utilised for the purpose.

Richards and Cushmann reduced a known weight of pure nickel bromide in a stream of hydrogen and weighed the residual metal.

They found in eight experiments that a total of 24·28947 grams of nickel bromide gave 6·52285 grams of nickel.

∴ 17·76662 of bromine combine with 6·52285 of nickel

and 79·955, *i.e.* the comb. wt.   „   „   **29·355**   „

(2) *Analysis of salts which under the action of heat break up into two oxides, a non-volatile basic oxide, and a volatile acidic oxide.* A great many carbonates, nitrates, and sulphates and salts of organic acids decompose in this manner, a certain weight of the pure salt leaving behind a certain weight of the oxide. Independent experiment has made us acquainted with the constant ratio between the quantities of oxygen in the acidic oxide and that combined with the metal in the basic oxide[1]; and if we also know accurately the combining weight of the carbon, the nitrogen, or the sulphur, we have all the data required for the calculation, the object of which always is to ascertain the amount of metal combined with 16 of oxygen in the residual oxide. Since it has been found that in carbonates twice as much oxygen is combined with the carbon as with the metal, in nitrates five times as much with the nitrogen as with the metal, and in sulphates three times as much with the sulphur as with the metal, we have the general relation :

(2) Ratio between the weight of a salt and that of the basic oxide it contains.

$$(\text{Comb. wt. of metal} + 16 \text{ oxygen}) : \begin{cases} (12{\cdot}01 \text{ carbon} \ + 32 \text{ oxygen}) \\ (28{\cdot}08 \text{ nitrogen} + 80 \text{ oxygen}) \\ (32{\cdot}07 \text{ sulphur} \ + 48 \text{ oxygen}) \end{cases}$$

$$= \text{wt. residual oxide} : (\text{wt. salt} - \text{wt. residual oxide}).$$

(i) Erdmann and Marchand found in six experiments that a total of 158·2712 grams of calcium carbonate left 88·6887 grams of lime.

∴ $(X + 16) : (44{\cdot}01) = 88{\cdot}6887 : 69{\cdot}5825$,

∴ X = combining weight of calcium = **40·10**.

(ii) Svanberg found in four experiments that a total of 33·79035 grams of lead nitrate left on ignition 22·7754 grams of oxide.

∴ $(X + 16) : (108{\cdot}08) = 22{\cdot}7754 : 11{\cdot}01495$,

∴ X = combining weight of lead = **207·47**.

---

[1] The data given on p. 182 referring to Berzelius' synthesis of lead sulphate are an example of the manner in which these ratios have been found.

(iii) The calcination of the sulphate is a method which has been much used, notably in the case of magnesium, copper, zinc, aluminium.

Baubigny found in two experiments that a total of

6·2135 grams of anhydrous aluminium sulphate gave 1·8537 of alumina.

$$\therefore (X + 16) : (80\cdot07) = 1\cdot8537 : 4\cdot3598,$$

$$\therefore X = \text{combining weight of aluminium} = \textbf{18·045}.$$

(3) *Analysis of the chloride or bromide by ascertaining the amount of silver required for interaction with all the chlorine or bromine contained in the weight of halide taken, or by weighing the silver halide resulting from this reaction.* The antecedent data required are the combining weights of the halogen and of silver.

(3) Ratio between the weight of a halide and that of the silver interacting with it.

(comb. wt. X + comb. wt. halogen) : (comb. wt. silver)

= wt. of halide taken : wt. of silver interacting with it ;

or

(comb. wt. X + comb. wt. halogen) : (comb. wt. silver

+ comb. wt. halogen)

= wt. of halide taken : wt. of silver halide obtained.

This is a method of very general application and one which lends itself to very exact measurements, owing to the extreme sensitiveness of the reaction and the great accuracy with which the combining weights of silver and of the halogens are known.

Richards and Baxter found in six experiments that a total of 13·27104 gms. of nickel bromide interacted with 13·10434 gms. of silver giving 22·81273 gms. of silver bromide.

$$\therefore (X + 79\cdot955) : 107\cdot93 = 13\cdot27104 : 13\cdot10434,$$

$$\therefore X = \text{comb. wt. of nickel} = \textbf{29·348},$$

and

$$(X + 79\cdot955) : (107\cdot93 + 79\cdot955) = 13\cdot27104 : 22\cdot81273,$$

$$\therefore X = \text{comb. wt. of nickel} = \textbf{29·345}.$$

(4) *A compound of the element of unknown combining weight with elements of known combining weights is changed into another compound containing besides the element of unknown combining weight others of known combining weight.*

(4) Ratio between the weights of two salts containing the same weight of the metal of unknown combining weight.

The change of chloride to nitrate or sulphate is a case in point. Referring back to (2) we can put directly :

$$(X + 16 + 108\cdot08) : (X + 16 + 80\cdot07)$$

= wt. of nitrate : wt. of sulphate.

Turner determined the ratio between barium nitrate and barium sulphate, and found as the mean of three experiments that:

112·028 of nitrate correspond to 100·00 of sulphate.

$$\therefore (X+124\cdot08):(X+96\cdot07)=112\cdot028:100.$$

$$X = \text{combining weight of barium} = \mathbf{136\cdot8}.$$

Instead of adding more examples of general methods, this chapter will be brought to a close by a short account of Stas' determinations of certain of the combining weights used as antecedent data in the most usual types of indirect determinations. His work supplies material for the calculation (in terms of the oxygen standard) of the combining weights of silver, chlorine, bromine, iodine, potassium, sodium, lithium, nitrogen, sulphur and lead. For several of these elements—silver notably—a number of independent values are obtained, and the admirable concordance of these is a proof alike of the wonderful accuracy of Stas' work and of the exactness of the laws of chemical combination. A set of experiments from which the combining ratios of silver, chlorine and potassium, in terms of oxygen, can be deduced, will be dealt with first. Obviously to find these three values in terms of oxygen, three measurements relatively to compounds containing the four elements are required. The substances investigated were: (i) potassium chlorate containing potassium, chlorine and oxygen, and which readily gives up the oxygen, passing to potassium chloride, (ii) potassium chloride, and (iii) silver chloride.

Stas' determination of the combining weights of silver, chlorine and potassium.

The compounds investigated were: potassium chlorate and chloride, and silver chloride.

Since elements combine with each other in ratios which are those of their combining weights or of simple whole multiples of these, it follows that if we can ascertain the quantity of potassium chloride obtained from the chlorate by the loss of 1 or 2, or 3, etc. combining weights of oxygen, this will be an amount of chloride containing integral numbers of combining weights of chlorine and potassium; that the amount of silver required to precipitate all the chlorine in this amount of chloride must be a combining weight of silver; that the amount of chlorine found to combine with this amount of silver in silver chloride is the combining weight of chlorine; and that the difference between the above amount of potassium chloride and the combining weight

The combining wts. of silver, chlorine, and potassium, expressed in terms of the ratios:

of chlorine so found is a combining weight of potassium. The manner of finding each of these ratios will become apparent from the following.

1. The ratio in potassium chlorate of

$$potassium\ chlorate : potassium\ chloride = a : b.$$

∴ oxygen combined in a chlorate with b (chlorine + potassium)................ $= (a - b)$

∴ (chlorine + potassium) combined with $m \times 16$, that is, with an integral number of combining weights of oxygen, and therefore itself containing an integral number of combining weights of chlorine and potassium.. $= \dfrac{m \times 16 \cdot b}{(a - b)}$.

The knowledge that another compound between potassium, chlorine and oxygen, which is called potassium perchlorate, contains, for the same amount of chlorine and potassium, $\frac{4}{3}$ the amount of oxygen contained in the chlorate, indicates that in these substances there are present 3 and 4 combining weights of oxygen respectively and that m should be put equal to 3. There is nothing to indicate the number of combining weights of chlorine and potassium in the chloride, and the simplest provisional assumption of 1 combining weight of each may therefore be made. The correctness of these inferences is borne out by the fact that the combining weight of chlorine deduced on this supposition, that potassium chlorate is $KClO_3$, gives for hydrochloric acid the simple formula HCl, which perfectly represents the chemical properties and chemical analogies of that substance.

2. The ratio in the interaction between solutions of potassium chloride and silver nitrate of

$$potassium\ chloride : silver\ =\ c : d.$$

∴ silver interacting with $\dfrac{48b}{(a - b)}$ chloride, that is with the amount containing a combining weight of chlorine, and which therefore must be the *combining weight of silver*...... $= \dfrac{\dfrac{48b \cdot d}{(a - b)}}{c}$.

3. The ratio in silver chloride of

$$silver : silver\ chloride\ =\ e : f.$$

$$\therefore \text{ chlorine uniting with } \frac{\frac{48b}{(a-b)} \cdot d}{c},$$

the above found combining weight of silver, and which therefore must be the *combining weight of chlorine* .... $= \dfrac{\frac{48b}{(a-b)} \cdot d \cdot (f-e)}{c \cdot e}$.

From this the *combining weight of potassium* follows as

$$\frac{48b}{(a-b)} - \frac{\left(\frac{48b}{a-b}\right) \cdot d \cdot (f-e)}{c \cdot e} = \frac{48b}{(a-b)}\left(1 - \frac{d \cdot (f-e)}{c \cdot e}\right).$$

The experimental methods employed for the determination of these ratios and the results obtained have in part been given already.

1.   *The ratio potassium chlorate : potassium chloride $= a : b$.*

Two methods were used for the reduction of the chlorate: (i) Heating the salt, when potassium chloride and oxygen are formed. (ii) Evaporation of the salt with solution of hydrochloric acid, when potassium chloride, water, and chlorine result[1]; the excess of hydrochloric acid and all the products of the reactions except the potassium chloride are volatile and can therefore be removed. Elaborate precautions were taken to prevent incomplete reduction and mechanical loss by spurting.

The experimental determination of the above three ratios.

I. Potassium chlorate changed to potassium chloride.

The results were :

	Wt. of chlorate taken $= a$.	Wt. of chloride left $= b$.	Oxygen $= a - b$.
	69·8730	42·5094	27·3636
	82·1260	49·9648	32·1612
(i) Ignition of the chlorate.	86·5010	52·6305	33·8705
	132·9230	80·8800	52·0430
	127·2125	77·4023	49·8102
	498·6355	303·3870	195·2485
(ii) Evaporation of the chlorate with hydrochloric acid.	59·727	36·3440	23·3830
	95·7975	58·2955	37·5020
	147·318	89·6340	57·6840
	302·8425	184·2735	118·5690

---

[1] $KClO_3 \text{ heated} = KCl + 3\overset{\nearrow}{O}$

$KClO_3 + m \cdot HCl = KCl + 3H_2O + 6\overset{\nearrow}{Cl} + (m-6) HCl.$

Substituting in the equations the values for the total quantities of chlorate used and of chloride obtained in each set of determinations, we get :

$$(\text{comb. wt. potassium} + \text{comb. wt. chlorine}) = \frac{48 \cdot b}{a - b} \cdot$$

$$= \frac{48 \cdot 303 \cdot 3870}{195 \cdot 2485} = 74 \cdot 585$$

and

$$= \frac{48 \cdot 184 \cdot 2735}{118 \cdot 5690} = 74 \cdot 599$$

Mean $= \mathbf{74 \cdot 592}$.

2. *The ratio potassium chloride : silver $= c : d$.*

The method is in principle and in detail identical with that described (p. 149) for ascertaining the ratio ammonium chloride : silver. Nearly equivalent quantities of the chloride and silver were weighed out, the silver was dissolved in nitric acid, the chloride added and the slight excess of silver determined by titration with standard sodium chloride solution. It will be remembered that in the case of the mixing of the ammonium chloride with the silver solution, which in some cases was done at 100°, the excess of nitric acid in the silver solution was first neutralised to prevent any loss of volatile hydrochloric acid; in the case under consideration this neutralisation was not required.

2. Precipitation of the chlorine in potassium chloride by silver in solution.

The results were :

Weight of chloride taken $= c$.	Weight of silver required to interact with the chloride $= d$.
7·450	10·7807
7·450	10·7810
7·450	10·78094
7·450	10·7809
7·450	10·7811
2·0945	3·03086
1·98685	2·87528
4·4786	6·48090
4·7041	6·80720
7·09352	10·26470
8·88805	12·8617
9·66160	13·98165
8·10100	11·72353
22·3500	32·3428
4·12706	5·97225
3·26516	4·72508
5·88845	8·52137
5·17232	7·48502
3·83415	5·54853
3·84461	5·5635
4·19350	6·0684
5·18237	7·49967
3·59191	5·19780
145·70775	210·85488

Substituting the values for the total quantities in the equation we get:

$$\text{Combining weight of silver} = \frac{\dfrac{48b}{(a-b)} \cdot d}{c}$$

$$= \frac{74 \cdot 592 \cdot 210 \cdot 8549}{145 \cdot 7078} = \mathbf{107 \cdot 943}.$$

3. *The ratio silver : silver chloride = e : f.*

Stas' synthesis of silver chloride has already been dealt with in the chapter on fixity of composition (p. 146), and it has been shown there how the following numbers have been arrived at.

3. Synthesis of
silver chloride.

Weight of silver $= e$	Weight of silver chloride $= f$
91·462	121·4993
69·86735	92·8145
101·519	134·861
108·549	144·207
399·651	530·920
99·9925	132·8382
98·3140	130·602
969·35485	1287·7420

Substituting the values for the total quantities in the equation we get:

$$\text{Combining weight of chlorine} = \frac{\dfrac{48b}{(a-b)} \, d \cdot (f-e)}{c \, . \, e}$$

$$= \frac{107 \cdot 943 \cdot 318 \cdot 3871}{969 \cdot 3549} = \mathbf{35 \cdot 454}.$$

∴ Combining weight of potassium $= 74 \cdot 592 - 35 \cdot 454$ $= \mathbf{39 \cdot 138}.$

Another value for the combining weight of silver can be deduced from Stas' complete analyses of silver iodate and complete syntheses of silver iodide, dealt with in the chapter on conservation of mass (pp. 65 *et seq.*); a value for the combining weight of iodine is obtained at the same time. The equations are the same as those above deduced for the relations between potassium chlorate, potassium chloride, and silver; only that two equations suffice, the number of elements of unknown combining weight being not more than two.

The comb.wts.
of silver and
iodine calcula-
ted from Stas'
complete ana-
lyses of silver
iodate and
syntheses of
silver iodide.

1.   The ratio in silver iodate of

$$(silver + iodine) : oxygen = b : r.$$

For the same reasons as those given before (p. 206)

$$(\text{comb. wt. silver} + \text{comb. wt. iodine}) = \frac{48b}{r},$$

but for r we have two independent values:

   (i) the directly determined weight of oxygen obtained by heating a of iodate $= s$

   (ii) the difference between a the weight of iodate taken and b the weight of iodide formed $= (a - b)$

$$\therefore (\text{comb. wt. silver} + \text{comb. wt. iodine}) = \frac{48b}{s} \ldots\ldots\ldots\ldots(i)$$

$$= \frac{48b}{(a - b)} \ldots\ldots(ii).$$

2.   The ratio in silver iodide of

$$silver : silver\ iodide = d : p$$

$$\therefore \text{comb. wt. silver} = \text{silver in } \frac{48b}{r} \text{ silver iodide} = \frac{48b}{r} \cdot \frac{d}{p},$$

but for p also we have two independent values:

   (i) the directly determined weight of iodide produced from d of silver $= c$

   (ii) the sum of d, the weight of silver, and q, the weight of iodine which had combined $= (d + q)$

$$\therefore \text{comb. wt. silver} = \frac{48b}{r} \cdot \frac{d}{c} \ldots\ldots(i)$$

$$= \frac{48b}{r} \cdot \frac{d}{d + q} \ldots(ii).$$

Experimental determination of the above two ratios.   The values for the quantities a, b, etc., have been given on pp. 70, 68, and are:

1.

Iodate $= a$	Iodide $= b$	Oxygen $= s$	(Iodate − Iodide) $= a - b$
98·2681	81·5880	16·6815	16·6801
156·7859	130·1755	26·6084	26·6104
255·0540	211·7635	43·2899	43·2905

Substituting in the equations the sums obtained by adding the results of the individual determinations gives :

$$\text{Comb. wt. silver} + \text{comb. wt. iodine} = \frac{48b}{s} = \frac{48 \cdot 211 \cdot 7635}{43 \cdot 2899} = 234 \cdot 805$$

$$= \frac{48b}{a+c} = \frac{48 \cdot 211 \cdot 7633}{43 \cdot 2905} = 234 \cdot 801.$$

$$\text{Mean} = 234 \cdot 803.$$

2.

Silver $= d$	Iodine $= q$	Silver iodide $= c$	Silver + iodine $= d+q$
27·6223	32·4665	60·0860	60·0888
39·8405	46·8282	86·6653	86·6687
38·0795	44·7599	82·8375	82·8394
136·3547	160·2752	296·6240	296·6299
82·3601	96·7964	179·1590	179·1565
324·2571	381·1262	705·3718	705·3833

Substitution of these values in the equation gives :

$$\text{Combining weight of silver} = \frac{48b}{r} \cdot \frac{d}{c} = 234 \cdot 803 \cdot \frac{324 \cdot 2571}{705 \cdot 3718} = 107 \cdot 938$$

$$= \frac{48b}{r} \cdot \frac{d}{d+q} = 234 \cdot 803 \cdot \frac{324 \cdot 2571}{705 \cdot 3833} = 107 \cdot 937.$$

and hence :   Mean $= 107 \cdot 9375$.

Combining weight of iodine $= (234 \cdot 803 - 107 \cdot 937) = 126 \cdot 866$.

The numbers above quoted and made the basis of the combining weight calculations do not comprise all the work done by Stas on the determination of the composition of silver iodide and silver iodate; there are a great many other analyses and syntheses by difference.   The concordance of the results throughout is excellent.

The following are therefore the values for the combining weights determined in the two sets of experiments of Stas just considered :

Results for the comb. weights of silver, chlorine, iodine, and potassium.

Silver	$=$	**107·943** and **107·937**
Chlorine	$=$	**35·454**
Iodine	$=$	**126·866**
Potassium	$=$	**39·138**.

The two values for silver differ from each other only by ·006 per cent.   A number of other methods were also used for finding the combining weight of silver.   Determinations were made of the composition of silver chlorate and chloride, of silver bromate and bromide, quite analogous to those of silver iodate and iodide given above; of silver sulphate and silver sulphide, in which the sulphide was synthesised and the amount of silver in the sulphate found by heating the salt in a current of hydrogen, when metallic silver remains behind.

The values obtained for the combining weight of silver were:

From the ratios $\dfrac{\text{potass. chlorate}}{\text{potass. chloride}}$ and $\dfrac{\text{potass. chloride}}{\text{silver}} = 107\cdot943.$

,, ,, ,, $\dfrac{\text{silver iodate}}{\text{silver iodide}}$ and $\dfrac{\text{silver}}{\text{silver iodide}} = 107\cdot937.$

,, ,, , $\dfrac{\text{silver chlorate}}{\text{silver chloride}}$ and $\dfrac{\text{silver}}{\text{silver chloride}} = 107\cdot9406.$

,, ,, ,, $\dfrac{\text{silver bromate}}{\text{silver bromide}}$ and $\dfrac{\text{silver}}{\text{silver bromide}} = 107\cdot9233.$

,, ,, ,, $\dfrac{\text{silver sulphate}}{\text{silver sulphide}}$ and $\dfrac{\text{silver}}{\text{silver sulphide}} = 107\cdot9270.$

Stas' determinations of combining ratios have served and are still serving chemists as the model, perhaps more truly as the ideal of what such measurements should be, and the faith he himself had in them has been amply justified. He concludes the *Nouvelles Recherches* by saying:

"Having reached the end of this long research, I venture to express the wish that some chemist of sufficiently established scientific authority would take the trouble to check any one of my fundamental results, and to publish the numbers obtained in such an investigation. Without any reservation whatever will I submit to his verdict."

APPENDIX.

A SELECTION OF COMBINING WEIGHT VALUES.

An accurate knowledge of the combining ratios of the elements being of paramount importance in chemistry, it is proposed to supplement the preceding examples of how certain of these quantities have been determined by a tabular representation of the results obtained for some of the more commonly occurring elements. This must be prefaced by a short theoretical consideration of the principles guiding us in our selection of the value termed the "general mean value," and in the attendant evaluation of the quantity termed the "probable error."

For the combining weight of nearly every element we have a number of independent values, the result of the work of different observers who had employed different methods. With few exceptions, such as Stas' determinations for silver and Morley's for hydrogen, the values obtained by different methods, and even those obtained by the same method when used by different observers, show less agreement than the individual values in a set of measurements made by the same observer working by the same method.

Thus, in the determination of the combining weight of arsenic from the composition of the chloride and bromide respectively, the following results were obtained:

Dumas' determination of the ratio between arsenic chloride and the silver required to interact with all the chlorine contained in the chloride		Wallace's determination of the ratio between arsenic bromide and the silver required to interact with all the bromine contained in the bromide	
100 of silver react with the following weights of arsenic chloride $= a$	Comb. wt. arsenic $= \dfrac{\text{comb. wt. silver}.a}{100} - \text{comb. wt. chlorine}$ $= \dfrac{107\cdot93 \cdot a}{100} - 35\cdot45$	100 of silver react with the following weights of arsenic bromide $= b$	Comb. wt. arsenic $= \dfrac{\text{comb. wt. silver}.b}{100} - \text{comb. wt. bromine}$ $= \dfrac{107\cdot93 \cdot b}{100} - 79\cdot96$
56·015	25·01 = Mean + ·02	97·023	24·76 = Mean + ·02
56·022	25·02 = ,, + ·03	97·022	24·76 = ,, + ·02
55·970	24·96 = ,, − ·03	96·970	24·71 = ,, − ·03
55·993	24·98 = ,, − ·01		24·74 = Mean
	24·99 = Mean		

The mean value for the combining weight of arsenic is 24·99 from the chloride and 24·74 from the bromide series, the difference between the two numbers being ·25, which is more than eight times as great as ·03, the greatest deviation of the individual determinations in each series from the mean of the series. Therefore, if the combining weight is regarded as absolutely constant, one or both values must be affected by some unknown constant error.

The favourite and most important of the methods used for ascertaining the combining weight of bismuth is that in which the composition of the oxide is made the basis of the calculation. Two investigations, in each of which the oxide was synthesised from metal purified with all possible care, gave the following results :

Syntheses of Bismuth Oxide.

CLASSEN'S.   SCHNEIDER'S.

Percentage of bismuth in the oxide	Percentage of bismuth in the oxide
89·703 = Mean + ·007	
89·7035 = „ + ·0075	
89·693 = „ − ·003	89·661 = Mean + ·004
89·700 = „ + ·004	89·648 = „ − ·009
89·6944 = „ − ·0016	89·659 = „ + ·002
89·692 = „ − ·004	89·662 = „ + ·005
89·694 = „ − ·002	89·653 = „ − ·004
89·693 = „ − ·003	89·660 = „ + ·003
89·695 = „ − ·001	**89·657** = Mean
89·696 = Mean	
$\therefore$ Comb. wt. bismuth $=\dfrac{89\cdot696.16}{10\cdot304}=139\cdot28$	$\therefore$ Comb. wt. bismuth $=\dfrac{89\cdot657.16}{10\cdot343}=138\cdot69$

Here the difference between the two mean values for the percentage is very much greater than the average difference between the individual and the mean values in each series, and again we attribute the cause of this to some unknown constant error vitiating one or both values.

True, these examples represent somewhat extreme cases, but to a greater or lesser degree the same occurs in nearly all the combining weight determinations made according to different methods by different observers.

We are therefore confronted with the problem of how to combine the different results of repeated measurements of the same quantity into one final value. Several eminent chemists[1] have accomplished the task of calculating from all the experimental data available the *general mean* for each combining weight. The formula used is that given before (p. 85),

$$M = \text{General Mean} = \frac{p_1 m_1 + p_2 m_2 + p_3 m_3 \ldots + p_n m_n}{p_1 + p_2 + p_3 \ldots + p_n},$$

where p_1, p_2, $p_3 \ldots p_n$ are the *weights* assigned to m_1, m_2, $m_3 \ldots m_n$, the *arithmetical means* of the different series of measurements. The same formula applies of course in the case of the combination of several general means M_1, M_2, $M_3 \ldots M_n$ to a final value.

The final results of these calculations differ somewhat according to the weight assigned to the different series of experiments as expressed in the values given to p_1, p_2, etc. This weighting may be purely empirical, each set of determinations being considered only from a chemical point of view, and assigned a value according to the number and the importance of the constant errors probably involved, a process which is to a great extent arbitrary, depending as it does on individual judgment; or the weighting may be purely mathematical, based on the *probable* error of each value; or it may consist in a combination of the two methods.

If in any one series of experiments which is made up of n independent measurements of the same quantity, the deviations from m, the arithmetical mean for the series, are represented by d_1, d_2, d_3, $\ldots d_n$, and their sum by d, then the formulae which the calculus of probability gives for the quantities designated *probable errors* are :

$$\text{Probable Error of each Observation} = \frac{2}{3} \sqrt{\frac{\Sigma d^2}{(n-1)}},$$

$$r = \text{Probable Error of the Arithmetical Mean } m = \frac{2}{3} \sqrt{\frac{\Sigma d^2}{n(n-1)}}.$$

In mathematical weighting, the measure of p_1, p_2, $p_3 \ldots p_n$, that is, of the accuracy of the different arithmetical means

---

[1] Clarke, *A Recalculation of the Atomic Weights*, 1897. Meyer and Seubert, *Die Atomgewichte der Elemente*, 1883. Ostwald, *Lehrbuch der allgemeinen Chemie*, I, 1891. Richards, "A Table of Atomic Weights," Baltimore Md., *Amer. Chem. J.*, 20, 1898 (p. 543).

m_1, m_2, $m_3 \ldots m_n$ (or of the different general means M_1, M_2, $M_3 \ldots M_n$), is taken in each case as inversely proportional to the square of the probable error r_1, r_2, $r_3 \ldots r_n$, giving for the general mean

$$M = \frac{\dfrac{m_1}{r_1^2} + \dfrac{m_2}{r_2^2} + \dfrac{m_3}{r_3^2} + \cdots \dfrac{m_n}{r_n^2}}{\dfrac{1}{r_1^2} + \dfrac{1}{r_2^2} + \dfrac{1}{r_3^2} + \cdots \dfrac{1}{r_n^2}},$$

and for the probable error of this general mean

$$r = \frac{1}{\sqrt{\dfrac{1}{r_1^2} + \dfrac{1}{r_2^2} + \dfrac{1}{r_3^2} + \cdots \dfrac{1}{r_n^2}}}.$$

These formulae for M and r are used at two distinct stages in the course of the determinations under consideration.

(i) *In the calculation of a combining weight from sets of data obtained for the* **same** *ratio by one or more observers, with or without variations in the experimental procedure.* The first step is the computation of a General Mean from the various arithmetical means obtained for the ratio measured. Stas' determinations of the ratio potassium chlorate : oxygen by the two processes referred to on page 207, Classen's and Schneider's numbers for the ratio bismuth oxide : bismuth on page 214, are cases in point, to which may be added another example. The ratio between q, the weight of pure anhydrous barium chloride, and p, the weight of silver required to interact with it, has been determined by different observers with the following results for the value of p when $q = 100$.

	Arithmetical Mean	Probable Error
1845 Pelouze (3 experiments)	96·459	± ·0036
1848 Marignac (11 exps. in 4 series)	96·360	± ·0024
1860 Dumas (16 exps. in 3 series)	96·316	± ·0055
1893 Richards (14 exps. in 4 series)	96·520	± ·0025
General Mean	96·434	± ·0015

But what we require to know is not the probable error of the general mean for $\dfrac{p}{q}$ the ratio actually measured, but that of the combining weight deduced from this ratio by a calculation which, except in the comparatively rare case of direct reference to oxygen only, involves a greater or lesser number of antecedent data of which each carries its own probable error, and which for the above example takes the form

$$\text{Comb. wt. barium} = \frac{p}{q} \cdot \frac{1}{2 \text{ comb. wt. silver}} - 2 \text{ comb. wt. chlorine}$$

$$= \frac{96 \cdot 434 \pm \cdot 0015}{2 (107 \cdot 9376 \pm \cdot 0037)} - 2 \cdot (35 \cdot 4529 \pm \cdot 0037)$$

and quite generally[1]

$$X : (A + B + \ldots) = p : q$$
$$(X + A + B + \ldots) : (M + N + \ldots) = p : q$$
$$(X + A + B + \ldots) : (X + M + N + \ldots) = p : q$$

where X is the combining weight required, and $A, B, M, N \ldots$ are the combining weights used as antecedent data.

Hence a further mathematical operation has to be performed, in which by means of appropriate formulae[2] the probable error of X is calculated in terms of that of $A, B, M, N, \ldots$ and of $\frac{p}{q}$. The fourth column in the table on p. 220 contains the results of such calculations from the data given in the second and third.

(ii) *In the calculation of a general mean from the values obtained for an element by different methods, that is, by the measurement of* **different** *ratios*. The five independent values obtained by Stas for silver (p. 212) when so treated give, according to Ostwald's calculation, the final value $107 \cdot 9376 \pm 0 \cdot 0037$; whilst Clarke, by summing up the work of all investigators, gets $107 \cdot 924 + \cdot 0031$.

A quotation from the Introduction to Clarke's " A Recalculation of the Atomic Weights" may serve as a concise statement of the principles just expounded.

"The mode of discussion and combination of results was briefly as follows....Each series of experiments was taken by itself, its arithmetical mean was found, and the probable error of that mean was computed. Then the several means were combined according to the appropriate formula, each receiving a weight dependent upon its probable error. The general mean thus established was taken as the most probable value...for the atomic weight, and at the same time its probable error was mathematically assigned."

"When several independent values have been calculated for an atomic weight, they are treated like means and combined according to the formula [given on p. 215]. Each final result is therefore to be regarded as the general or weighted mean of all trustworthy determinations."

[1] *Ante*, pp. 202—205.　　　　[2] Clarke, *loc. cit.* (pp. 7, 8).

"But although the discussion of combining ratios is ostensibly mathematical, it cannot be purely so. Chemical considerations are necessarily involved at every turn. In assigning weights to mean values I have been, for the most part, rigidly guided by mathematical rules; but in some cases I have been compelled to reject altogether series of data which were mathematically excellent, but chemically worthless because of constant errors. ...Concerning the subject of constant and accidental errors...my own method of discussion eliminates the latter, which are removable by ordinary averaging; but the constant errors, vicious and untractable, remain at least partially. Still, where many ratios are considered, even the systematic errors may in part compensate each other, and do less harm than might be expected."

To illustrate the process described, a summary is given in the following table of Clarke's evaluation of the combining weight of one special element.

The Combining Weight of Lithium.

$\dfrac{p}{q} = \dfrac{100}{q}$	m Arithmetical mean and probable error for q	M General mean and probable error for q	Antecedent data $H = 1\cdot000$	Comb. wt. and its probable error, calculated from M
(1) $\dfrac{Silver\ chloride}{Lithium\ chloride}$ 1856. Mallet (2 exps.) 1862. Troost (2 exps.)	29·581 ± ·0087 29·5925 ± ·0145	29·584 ± ·0075		
(2) $\dfrac{Silver}{Lithium\ chloride}$ 1865. Stas (3 exps.)	39·358 ± ·001	39·358 ± ·001	AgCl = 142·287 ± ·0037 Cl = 35·179 ± ·0048 Ag = 107·108 ± ·0031	6·9752 ± ·0051
(3) $\dfrac{Lithium\ carbonate}{Carbonic\ anhydride}$ 1862. Diehl (4 exps.) 1862. Troost (2 exps.) Dittmar (10 exps.)	59·417 ± ·0060 59·456 ± ·0200 59·638 ± ·0173	59·442 ± ·0054	C = 11·920 ± ·0004 O = 15·879 ± ·0003	6·9628 ± ·0077
(4) $\dfrac{Lithium\ chloride}{Lithium\ nitrate}$ 1865. Stas (3 exps.)	162·5953 ± ·0025	162·5953 ± ·0025	Cl = 35·179 ± ·0048 O = 15·879 ± ·0003 N = 13·935 ± ·0015	6·9855 ± ·0129

General Mean (H = 1) = 6·9729 ± ·0040
General Mean (O = 16) = 7·026

The formulae used are for (1) $Li = \dfrac{q \cdot AgCl}{p} - Cl$..................(p. 204. 3).

   ,,   ,,   ,,   (2) $Li = \dfrac{q \cdot Ag}{p} - Cl$....................(p. 204. 3).

   ,,   ,,   ,,   (3) $Li = \dfrac{1}{2}\left(\dfrac{p \cdot CO_2}{q} - CO_3\right)$ ....(p. 203. 2).

   ,,   ,,   ,,   (4) $Li = \dfrac{p \cdot NO_3 - q \cdot Cl}{q - p}$ ....(p. 204. 4).

The following table contains the *results* of a number of combining weight determinations, most of them selected from Clarke's compilation, and hence found before 1897, a few of later date. This summary is not intended to be complete in any sense whatever; but it is thought that the provision of further examples may tend to make clearer and to emphasise the principles expounded in this chapter. Only the more common and more important of the elements are considered, and their selection, as well as that of the special determination given, has been somewhat arbitrary. Since it was a distinct object to include examples of a great number of different methods, it was not possible to give for each element the special determination considered to have yielded the most reliable results.

The numbers are those of the combining ratio multiples which, on Laurent's principles, would be chosen for the symbol weights, and which, according to the theory of chemical constitution now held, represent relative atomic weights in terms of the weight of the oxygen atom taken as 16·000. Though always referred to as atomic weights, the only object of these determinations is that of the *accurate combining weight*; and the decision of the special multiple selected for the atomic weight is a problem of fundamentally different nature, which is therefore always kept separate, and the results of which are tacitly assumed. But if for instance it should be decided for good reasons that the properties of the compounds of beryllium are better represented by formulae corresponding to the oxide Be_2O_3 than by those now used, which are derived from BeO, the atomic weight would become $\frac{3}{2}$ of its present value; this would however have no influence on Clarke's critical consideration of the data dealt with by him, and on the calculation to which he has subjected these. His final general mean 9·08 would simply be multiplied by $\frac{3}{2}$.

Element. Authority and Date	Method. Reaction used. Ratio measured	Antecedent Data. $O=16\cdot000$	Mean Value and Probable Error calculated from the two preceding columns	Probable Value from a combination of all independent Mean Values, according to:		
				Clarke[1] 1897	Richards[2] 1898	Internat. Com., 1904
Hydrogen. *Morley*, 1895	1. Complete . . sis of water (i) H_2 : O $= 2 \cdot 15\cdot8792 \pm \cdot00032$ (ii) H_2 : $H_2O = 2 : 15\cdot8785 \pm \cdot00066$ 2. Determination of densities of hydrogen and oxygen, and of their volume ratio in water $D_O = 1\cdot42900 \pm \cdot000034$ $D_H = \cdot089873 \pm \cdot00027$ $vol_O : vol_H = 1 : 2\cdot00269$	$O=16\cdot000$	$1\cdot00761 \pm \cdot00016$ $1\cdot00765 \pm \cdot00033$ $1\cdot00762 \pm \cdot0007$	$1\cdot008$	$1\cdot0075$	$1\cdot008$
Silver. *Stas*[3], 1865	*1. IO_3 : KCl = 100 : 60·846 ± ·0009 Ag : KCl = 100 : 69·1122 ± ·00014 2. IO_3 : AgCl = 100 : 74·9205 ± ·0010 Ag : AgCl = 100 : 132·8445 ± ·0008 3. $AgBrO_3$: AgBr = 100 : 79·651 ± ·0014 Ag : AgBr = 100 : 174·081 ± ·0006 4. $AgIO_3$: AgI = 100 : 83·0253 ± ·0009 Ag : AgI = 100 : 217·5344 ± ·0008 5. $AgSO_4$: Ag = 100 : 69·203 ± ·0012 Ag : Ag_2S = 100 : 114·8522 ± ·0007	$O=16\cdot000$	$107\cdot9401 \pm \cdot0058$ $107\cdot9406 \pm \cdot0049$ $107\cdot9233 \pm \cdot0140$ $107\cdot9371 \pm \cdot0045$ $107\cdot9270 \pm \cdot0090$ $107\cdot9376 \pm \cdot0037$ = General Mean	$107\cdot92$	$107\cdot93$	$107\cdot93$
Halogens, Stas, 1865. Chlorine Bromine Iodine	Synthesis of silver halides Ag : AgCl = 100 : 132·8445 ± ·0008 Ag : AgBr = 100 : 174·081 ± ·0006 Ag : AgI = 100 : 217·5344 ± ·0008	$Ag=107\cdot9376 \pm \cdot0037$	$35\cdot4524 \pm \cdot0040$ $79\cdot9632 \pm \cdot0033$ $126\cdot8654 \pm \cdot0037$	$35\cdot45$ $79\cdot95$ $126\cdot85$	$35\cdot455$ $79\cdot955$ $126\cdot85$	$35\cdot45$ $79\cdot96$ $126\cdot85$[5]

Alkalis, Stas, 1865. **Potassium** **Sodium** **Lithium**	Analysis of the chlorides Ag : KCl = 100 : 69·1122 ± ·00014 Ag : NaCl = 100 : 54·2062 ± ·0002 Ag : LiCl = 100 : 39·358 ± ·001	Ag = 107·9376 ± ·0037 Cl = 35·4529 ± ·0037	39·1355 ± ·0037 23·0575 ± ·0041 7·0303 ± ·0042	39·11 23·05 7·03	39·140 23·050 7·03	**39·15** **23·05** **7·03**
Sulphur. Stas, 1865	1. Analysis of silver sulphate Ag_2SO_4 : Ag = 100 : 69·203 ± ·0012 2. Synthesis of silver sulphide Ag : Ag_2S = 100 : 114·8522 ± ·0007	O = 16·000 Ag = 107·9376 ± ·0037	32·0676 ± ·0151 32·0622 ± ·0044	32·07	32·065	**32·06**
Nitrogen. Stas, 1865	Change of chlorides to nitrates by evaporation with nitric acid 1. KCl : KNO_3 = 100 : 135·6423 ± ·0014 2. NaCl : $NaNO_3$ = 100 : 145·4526 ± ·0030 3. LiCl : $LiNO_3$ = 100 : 162·5953 ± ·0025	O = 16·000 Cl = 35·4529 ± ·0037 K = 39·1361 ± ·0032 Na = 23·0575 ± ·0041 Li = 7·0303 ± ·0042	14·0368 ± ·0055 14·0489 ± ·0052 14·0458 ± ·0053	14·04	14·045	**14·04**
Lead. Stas, 1865	1. Synthesis of the nitrate Pb : $Pb(NO_3)_2$ = 100 : 159·9704 ± ·0010 2. Synthesis of the sulphate Pb : $PbSO_4$ = 100 : 146·4275 ± ·0024	{O = 16·000 {N = 14·0410 ± ·0037 {O = 16·000 {S = 32·0626 ± ·0042	206·919 ± ·014 206·911 ± ·011	206·92 206·92	206·92	**206·9**
Aluminium. Mallet, 1880	Measurement of hydrogen evolved by action of aluminium on pure soda Al : 3H = 26·890 ± ·0034 : 1	H = 1·0076 ± ·00015 $Density_H$ = ·089872 ± ·0000028	27·094 ± ·003[6]	27·11	27·1	**27·1**
Antimony[7]. Cooke, 1877	1. Synthesis of the sulphide Sb : Sb_2S_3 = 71·4818 ± ·0120 : 100 2. Analysis of the bromide $SbBr_3$: 3AgBr = 63·830 ± ·008 : 100 $SbBr_3$: 3Ag = 111·114 ± ·0014 : 100 3. Analysis of the iodide SbI_3 : 3AgI = 71·060 ± ·023 : 100	S = 32·070 ± ·0015 Ag = 107·924 ± ·0031 Br = 79·949 ± ·0062 AgBr = 187·873 ± ·0054 AgI = 234·771 ± ·0062	120·535 ± ·01 119·918 ± ·00 119·997 ± ·04 120·205 ± ·01	120·43	120·0	**120·2**
Arsenic. Kessler, 1861	Oxidation by potassium chlorate of arsenious to arsenic oxide $3As_2O_3$: $2KClO_3$ = 100 : 41·172 ± ·009	O = 16·000 Cl = 35·447 ± ·0048 K = 39·112 ± ·0051	75·23 ± ·022	75·01	75·0	**75·0**

Element. Authority and Date	Method. Reaction used. Ratio measured	Antecedent Data. O=16·000	Mean Value and Probable Error calculated from the two preceding columns	Probable Value from a combination of all independent Mean Values, according to:		
				Clarke[1] 1897	Richards[2] 1898	Internat. Com., 1904
Beryllium. *Kriiss and Moraht,* 1891	Ignition of the crystallised sulphate $BeSO_4.4H_2O : BeO =100 : 14\cdot144 \pm \cdot0017$	$O=16\cdot000$ $H=1\cdot0076 \pm \cdot00015$ $S=32\cdot070 \pm \cdot0015$	$9\cdot047 \pm \cdot003$	9·08	9·1	9·1
Bismuth. *Marignac,* 1883	Conversion of oxide into sulphate $Bi_2O_3 : Bi_2(SO_4)_3=100 : 151\cdot728 \pm \cdot0099$	$O=16\cdot000$ $S=32\cdot070 \pm \cdot0015$	$208\cdot183 \pm \cdot044$	208·11	208·0	208·5
Boron. *Ramsay and Aston,* 1893	Conversion of anhydrous borax into sodium chloride by evaporation with hydrochloric acid and methyl alcohol $Na_2B_4O_7 : 2NaCl=100 : 57\cdot930 \pm \cdot0081$	$O=16\cdot000$ $Cl=35\cdot447 \pm \cdot0048$ $Na=23\cdot048 \pm \cdot0046$	$10\cdot966 \pm \cdot005$	10·95	10·95	11
Cadmium. *Morse and Arbuckle,* 1898	Synthesis of cadmium oxide[8] $Cd : CdO=87\cdot536 \pm \cdot0021 : 100$	$O=16\cdot000$	$112\cdot377 \pm \cdot001$	111·95	112·3	112·4
Calcium. *Hinrichsen,* 1902	Ignition of native carbonate $CaCO_3 : CaO=100 : 56\cdot062 \pm \cdot0003$	$O=16\cdot000$ $C=12\cdot003 \pm \cdot0006$	$40\cdot142 \pm \cdot001$	40·07	40·0	40·1
Carbon. *Van der Plaats,* 1884	Burning of different varieties of carbon in oxygen $C : CO_2=1 : 3\cdot6660 \pm \cdot0001$	$O=16\cdot000$	$12\cdot003 \pm \cdot001$	12·01	12·001	12·00
Chromium. *Rawson,* 1889 *Meineke,* 1891	Reduction of ammonium bichromate to chromic oxide $(NH_4)_2Cr_2O_7 : Cr_2O_3=100 : 60\cdot337 \pm \cdot0029$	$O=16\cdot000$ $H=1\cdot0076 \pm \cdot00015$ $N=14\cdot041 \pm \cdot0021$	$52\cdot168 \pm \cdot006$	52·14	52·14	52·1

Element / Authority	Method	Standard	Value			
Cobalt. *Richards and Baxter, 1899* **Nickel.** *Richards and Cushman, 1899*	Reduction of bromide in hydrogen $CoBr_2$: Co=100 : 26·951 ± ·0013 $NiBr_2$: Ni=100 : 26·855 ± ·0006	Br = 79·949 ± ·0062	58·998 ± ·006 58·709 ± ·005	58·93 58·69	59·00 58·70	59·0 58·7
Copper. *Richards 1886 and 1887*	Silver deposited from silver nitrate solution by weighed quantities of copper Cu : 2Ag = 100 : 339·402 ± ·0040	Ag = 107·924 ± ·0031	63·593 ± ·002	63·60	68·60	63·6
Iron. *Richards and Baxter, 1900*	Reduction ot oxide to metal in current of electrolytically prepared hydrogen Fe_2O_3 : 2Fe = 100 : 69·958 ± ·0013	O = 16·000	55·883 ± ·003	56·02	56·0	55·9
Magnesium. *Richards and Parker, 1896*	Analysis of the chloride $MgCl_2$: 2Ag = 44·137 ± ·0003 : 100 $MgCl_2$: 2AgCl = 33·226 ± ·0013 : 100	Ag = 107·924 ± ·0031 Cl = 35·447 ± ·0048 AgCl = 143·371 ± ·0037	24·363 ± ·012 24·369 ± ·017	24·28	24·36	24·36
Manganese. *Weeren, 1890*	Reduction of sulphate to sulphide by ignition in stream of sulphuretted hydrogen $MnSO_4$: MnS = 100 : 57·633 ± ·0004	O = 16·000 S = 32·070 ± ·0015	54·991 ± ·002	54·99	55·02	55·0
Mercury. *Hardin, 1896*	Simultaneous electrolytic deposition of silver and mercury from solution of their double cyanides Hg : 2Ag = 92·660 ± ·0051 : 100	Ag = 107·924 ± ·0031	200·001 ± ·012	200·00	200·0	200·0
Phosphorus. *Van der Plaats, 1885*	Silver precipitated from the sulphate by weighed quantities of phosphorus P : 5Ag = 5·7322 ± ·0045 : 100	Ag = 107·924 ± ·0031	30·930 ± ·024	31·02	31·0	31·0
Silicon. *Thorpe and Young, 1887*	Change of bromide to oxide by evaporation with water $SiBr_4$: SiO_2 = 100 : 17·347 ± ·0027	O = 16·000 Br = 79·949 ± ·0062	28·400 ± ·012	28·40	28·4	28·4

Element. Authority and Date	Method. Reaction used. Ratio measured	Antecedent Data. $O=16\cdot000$	Mean Value and Probable Error calculated from the two preceding columns	Probable Value from a combination of all independent Mean Values, according to:		
				Clarke[1] 1897	Richards[2] 1898	Internat. Com., 1904
Tellurium. *Gooch and Howland,* 1894	Oxidation of dioxide to trioxide by standard potassium permanganate $TeO_2 : O = 100 : 10\cdot068 \pm \cdot0100$	$O = 16\cdot000$	$126\cdot916 \pm \cdot157$	127·49	127·5	127·6⁹
Tin. *Bongartz and Classen,* 1888	Analysis of bromide by electrolytic deposition of metal contained in a certain weight of bromide $SnBr_4 : Sn = 100 : 27\cdot123 \pm \cdot0020$	$Br = 79\cdot949 \pm \cdot0062$	$119\cdot017 \pm \cdot013$	119·05	119·0	119·0
Vanadium. *Roscoe,* 1868	Reduction of a higher to a lower oxide by ignition in hydrogen $V_2O_5 : V_2O_3 = 100 : 82\cdot491 \pm \cdot005$	$O = 16\cdot000$	$51\cdot379 \pm \cdot022$	51·38	51·4	51·2
Zinc. *Richards and Rogers,* 1895	Analysis of the bromide $ZnBr_2 : 2Ag = 104\cdot38 \pm \cdot0007 : 100$ $ZnBr_2 : 2AgBr = 59\cdot962 \pm \cdot0004 : 100$	$Ag = 107\cdot924 \pm \cdot0031$ $Br = 79\cdot949 \pm \cdot0062$ $AgBr = 187\cdot873 \pm \cdot0054$	$65\cdot402 \pm \cdot013$ $65\cdot406 \pm \cdot013$	65·41	65·40	65·4

[1] "The figures are given only to the second decimal, the third being rarely, if ever significant. In most cases even the first decimal is uncertain, and in some instances whole units may be in doubt."
(Clarke, *loc. cit.*, p. 364.)

[2] "The last figure of each number...cannot be considered in any case certain. It is often probably not much more than 1 unit in error, although the uncertainty may amount to as much as 6 or 8 units in some cases."
(Richards, Baltimore Md., *Amer. Chem. J.*, 20, 1898, p. 545.)

[3] Stas' determinations are given first and by themselves. Their great accuracy entitles them to this special treatment. The values in the fourth column are those calculated by Ostwald (*Lehrbuch*, I, pp. 30—42) from Stas' data. The slightly different values used subsequently as antecedent data are those calculated by Clarke from all reliable determinations.

[4] For the reactions used see *ante*, pp. 205—212.

[5] "Ladenburg has shown that the accepted number for iodine is probably too low, but other investigations upon the subject are known to be in progress, and until they have been completed it would be unwise to propose any alteration."

(*Report*, 1904.)

A preliminary notice by Köthner and Aeuer states as the result of a number of determinations carried out by different methods that "the atomic weight of iodine cannot be less than 126·963."

(Berlin, *Ber. D. chem. Ges.*, 37, 1904, p. 2536.)

[6] This is one of the comparatively few cases in which, the determination being by direct reference to hydrogen, the recent change in the value accepted for the ratio $O:H=16:1·0032$, and of Regnault's value for the density of hydrogen subsequently corrected for shrinkage of the exhausted glass globe) is given as 27·067.

[7] Given as instance of a case in which the discrepancy in the results indicates the presence of constant errors.

[8] The oxide obtained in this and similar syntheses, by solution of the metal in nitric acid and subsequent ignition, always retains some oxygen and nitrogen. The quantity of occluded gas was actually determined and subtracted from the weight of the oxide formed.

[9] The results are not yet conclusive; the discrepancies between the values obtained by different methods indicate the presence of some constant error not yet detected. The many values determined in the course of the last few years are in this respect not much better than the older ones, thus: Metzner in 1898 found $Te = 127·9$, Heberlein in the same year, 126·98; in 1901 Steiner found 126·4, Pellini 127·6, and Köthner 127·3 and 127·6; Guthier in 1902 found 127·5. The point is one of exceptionally great interest, because of the place of tellurium in the periodic law, which requires an atomic weight value less than that of iodine (*post*, chap. xvi).

CHAPTER IX.

THE ULTIMATE CONSTITUTION OF MATTER.
HYPOTHESES PRIOR TO 1800.

" If then after so many men have said diverse things concerning the generation of the Universe, we should not prove able to render an account everywhere and in all respects consistent and accurate, let no one be surprised; but if we can produce one as probable as any other, we must be content."

PLATO, *Timaeus.*

THERE is inherent in the human mind a desire to find an explanation—or as some would prefer to have it called, a description—of the phenomena of nature, by means of
speculations concerning the ultimate constitution of matter.

Hypotheses concerning the ultimate nature of matter go back to antiquity.

All the different types of civilisation have made their contribution to the solution of the problem. The early ones are distinguished by the audacity with which the scope of the phenomena to be explained was settled, including as it did not only what we should now call the physics of matter, but also the phenomena of life and of thought, and a code of conduct; and all that at a time when there was but little accurate knowledge of the individual phenomena, and practically no formulation of laws. The modern ones are characterised by the fact that much less is attempted and consequently more achieved. Modern physical science has set itself the task of devising and applying a system by which the aggregate of the phenomena associated with the conception of matter shall find an easy and satisfactory explanation or "description" by the assumption of some simple properties inherent in matter.

Character- istics of the modern method: (1) Simpli- fication is the object.

The object and process is one of simplification:

"To reduce the number of laws as far as possible, by showing that laws at first separated, may be merged into one; to reduce the number of the chapters in the book of science, by showing that some are truly mere subsections of chapters already written." (Poynting, *Opening address Section A, British Association*, 1899.)

And where mere observation and experiment do not lead to further simplification, where our ordinary method of explanation fails, we *imagine* a constitution of matter such that the apparently isolated laws appear but as the necessary outcome of the funda.. mental properties of this matter:

"We are no longer content to describe what we actually see or feel, but we describe what we imagine we should see or feel if our senses were on quite another scale of magnitude and sensibility." (Poynting, *ibid.*)

To frame such hypotheses is nothing new, but what characterises the present method is the rigorous testing of the adequacy of the tool thus devised for the work expected from it; the explanations in terms of the hypothesis are compared with the phenomena themselves, and the value of the hypothesis is measured by the indications it affords towards the further study of nature.

The modern method also keeps in mind that hypotheses are simply instruments, tools; that they are essentially temporary.

(2) The hypotheses and theories are admittedly temporary. "While the building of nature is growing spontaneously from within, the model of it we seek to construct in our descriptive science, can only be constructed by means of scaffolding from without, a scaffolding of hypotheses. While in the real building all is continuous, in our model there are detached parts, which must be connected with the rest by temporary ladders and passages, or which must be supported till we can see how to fill in the under-structure. To give the hypotheses equal validity with facts is to confuse the temporary scaffolding with the building itself." (Poynting, *ibid.*)

And what the influence of a hypothesis devised and applied in this spirit may be, is shown• by the development of chemistry from the time when Dalton introduced into it the **The influence of the modern atomic hypothesis on the development of chemistry.** atomic hypothesis, a development whose results are to be presented in the succeeding chapters. Of the atomic hypothesis it has aptly been said that it

"arose so early in the history of science as to almost tempt one to suppose that it is a necessity of thought, and that it has warrants of some higher order than any other hypothesis which could be imagined." (Poynting, *ibid.*)

15—2

Be this as it may, it must be of considerable interest to the chemist to follow the early history of this hypothesis; but to

Consideration of other hypotheses concerning the ultimate nature of matter. properly appreciate its nature and merits, its scope and importance at different times, it should not be presented alone, not be detached from its proper setting, which is that of the general history of the views concerning the ultimate constitution of matter. To give such a history at the same time shortly and faithfully is not possible, and that for several reasons: Distortion becomes inevitable, when in the different schemes of philosophy we sever the part dealing with physical phenomena from that relating to thought, life, and conduct; the sources of our knowledge of early hypotheses are often scant and indirect; and commentators and critics—old and modern—are apt to be misleading, personal bias

Different estimates of the relative value of the ancient and the modern atomic hypotheses. influencing their interpretation and evaluation in the usual manner. The different estimates of the value of the atomic hypothesis as framed in Greece will exemplify this last point. The classical scholar is apt to consider it in aims and achievements almost identical with the atomic theory of to-day; whilst the scientist—in this instance evidently better able to judge and compare—denies any real value to that of the ancients.

"The modern atomic doctrine is not by fortuitous coincidence identical with that of Leucippus and Democritus; but is its direct descendant, flesh of its flesh, and bone of its bone....Democritus is in complete agreement...with the actual results of the scientific research of the last three centuries. It borders on the marvellous how, lifting the veil which obscures our ordinary perceptions, he caught a glimpse of what the telescope and the spectroscope have but recently revealed as actual truths. When Democritus tells of an infinitely large number of cosmic systems different in size, some with many moons and others without sun or moon, of some in the process of formation whilst others, through collision, are being destroyed, and of some that are devoid of water; we seem to hear the voice of a modern astronomer, who has seen the moons of Jupiter, who has recognised the absence of aqueous vapour round the moon, and who has observed the nebulae and burnt out stars, phenomena to him revealed by the highly developed appliances of modern times." (Gomperz, *Greek Thinkers*, 1896.)

"From the earliest times that men began to form any coherent idea of [the world] at all, they began to guess in some way or other how it was that it all began, and how it was all going to end....Modern speculations are attempts to find out how things began and how they are to end, by consideration of the way in which they are going on now....A great number of people appear to have been led to the conclusion that [the modern theory of the molecular constitution of matter] is very similar to the guesses which we find in ancient writers—Democritus and Lucretius....It so happens that these ancient writers did hold a view of the constitution of things which in many striking respects agrees with the view which we hold in modern times.... The difference between the [ancient and modern views] is mainly this : the atomic theory of Democritus was a guess, and no more than a guess. Everybody around him was guessing about the origin of things, and they guessed in a great number of ways ; but he happened to make a guess which was more near the right thing than any of the others." (W. H. Clifford, *Lectures and Essays*, 1879.)

What follows does not and cannot aim at completeness in the account of the history of the hypotheses concerning the ultimate constitution of matter, nor at critical estimation and comparison. All that will be attempted is to give the salient features of some of the old and some of the newer views in their purely physical bearings, in order to convey some impression of the kinds of speculation of which the human mind seems capable, and of the fundamental differences dividing these.

Early Indian Philosophy[1] presents us with a view of matter which may be called atomistic. Kanada, whom some authorities, *Indian speculations: Kanada's atomistic philosophy.* without assigning him a definite time, place anterior to 1200 B.C., and whose philosophy is embodied in short aphorisms (Soutras), held that:

"...substance is that in which qualities abide and in which action takes place. Earth, water, light, air, ether, time, place, soul, mind, such are the substances."

Detaching from this saying that which refers to matter as we understand it now, we see that according to Kanada, there are five kinds of matter. With each of them is associated a definite characteristic property. Earth besides other properties possesses the distinguishing one of smell; water has all the properties of earth except that of smell, but in addition that of cold; light is coloured and hot; air is temperate; ether is the carrier of sound. These five elements are apprehended by, and each belong .to' a definite sense organ, to smell, taste, sight, touch and hearing respectively. The existence of an ethereal element is not deduced from perception, but is inferred, owing to the necessity of something to act as the vehicle of sound, something to be apprehended by

---

1 The material for what is said on this subject has been taken from Colebrooke's *Essays on the Religion and Philosophy of the Hindus*, 1858; and from Mabilleau's *Histoire de la Philosophie Atomistique*, 1895.

the sense of hearing, none of the other substances lending themselves to this. Concerning the ultimate constitution of these five primary substances, Kanada assumes that they are composed of what we should term atoms, and that there are as many kinds of atoms as there are elementary substances, that is five; earth being made up of smelling atoms, water of cold atoms, etc. The atom is simple, for matter is not capable of infinite division, otherwise

"there would be no difference of magnitude between a mustard seed and a mountain, a gnat and an elephant, each alike containing an infinity of particles."

The atom is reckoned to be the sixth part of a mote in a sunbeam and is supposed to have no extension. Two primary atoms combine, and by the combination of three such binary ones is formed the particle, possessed not only of the quality of extension, but also of all the other qualities which characterise the different elements. The further combination of these extended particles leads to substances such as we apprehend. The atoms themselves are indestructible and eternal, their combinations are transient. If the component particles are all of one kind, such combination leads to the elementary substances before enumerated; if they are of different kinds, the resulting substances are possessed of intermediate properties; the hard substances wood and stone are made up of earthy atoms only; flowers, wool, etc. which are soft are composed of the atoms of earth mixed with those of water or air; gold is made up of atoms of earth and atoms of light.

Kanada's conception of the nature of atomistic combination.

"But some may object that gold is not earth because it is without odour, which is held to be the characteristic of earth, nor is it water because it is devoid of viscidity and of natural fluidity, nor is it fire because of its weight— and for the same reason neither is it air or ether (both of which are held to be devoid of weight), therefore it is different from all the nine; if you say this, then I the Commentator say, your first two reasons are valid (against gold being earth or being water) but your third is a case of the fallacy termed unreality of the alleged nature, for the followers of this doctrine hold that gold is not really heavy in itself, but it appears to be heavy through the admixture with earthy particles and the gold itself is composed of fire or light."

Such an atomic doctrine was maintained not only by the followers of Kanada, but also by the sect of Buddha and others, heterodox and orthodox.

The Indian speculations on the ultimate structure of matter have had no direct result scientifically; some scholars hold that they influenced Greek thought on the same subject, but there seems to be no conclusive evidence to support such a view.

It was in Miletus, the chief Ionic city in Asia Minor, that a school of philosophy flourished all through the 6th century B.C., which set itself the problem of finding an answer to the question of what the world is and of how it had arisen[1]. Thales (640—546) reputed for his mathematical and astronomical knowledge and counted one of the seven sages of Greece, Anaximander (611—547) his junior by 28 years, Anaximenes (*circ.* 560—500) represent a school which explained all the phenomena observed by assuming one single kind of matter capable of suffering certain changes whereby it could be transformed into all the other substances which constitute the world. The power of undergoing these changes was supposed inherent in matter, which therefore must be looked upon as endowed with life, and hence the names of "Hylozoism" and "Hylozoists" ($ὕλη$ = stuff, $ζóη$ = life) given to the tenets of this school and its followers.

The Ionian philosophers assume one primitive matter.

"And some say that it (the soul) is mixed up with the whole, whence likewise Thales considered all things to be full of gods." (Aristotle, *de Anima.*)

We know that Anaximander and Anaximenes wrote works entitled "On Nature" ($περὶ φύσεως$), but we possess practically nothing of the works of this school, and are indebted for a knowledge concerning them to Aristotle (384—322 B.C.) and to the commentary of Simplicius who flourished *circ.* 500—550 A.D.

"...the great majority of the earliest philosophers thought the only foundation was a material first principle, in one or other of its forms, because that out of which everything springs—in other words, that from which as a primal element everything comes at its birth, and into which at its death it is resolved again—the substance remaining permanent throughout all its changes of conditions—this they say is the primal element and this the foundation and basis of all Existences. For this reason they hold that nothing fresh comes into Life, nothing passes out of Life; their idea being

[1] For this, as for all the succeeding accounts of the different Greek views of the nature and the formation of the material universe, the following books amongst others have been used: Mabilleau, *Histoire de la Philosophie Atomistique*, 1895; G. H. Lewes, *History of Philosophy*, 1880; Erdmann, *Grundriss der Geschichte der Philosophie*, 1878; Windelband, *A History of Philosophy*, 1893.

that a natural body, such as described, keeps itself permanently intact...for there must exist some natural body,—either one or more than one—from and out of which everything else is produced, whilst it, however, keeps itself unchanged.... According to these philosophers, then, one

Thales assumes water to be the one primal element, Anaximenes air.

would fancy that that cause which...we classed under the head of matter was the sole cause of all things....What however all are *not* agreed on is the exact kind of this first principle, and whether there be one or more than one primal element....Thales...pronounces it to be water (showing therefrom among other things how that the earth rested on a basis of water), having got the idea perhaps from seeing that the nutriment of all things was moist, and that from it heat itself was generated and by it was kept alive (and that out of which everything is generated obviously *is* the universal first principle)....But according to Anaximenes...air is prior to water, and has more claims than any other simple element to be the first foundation of all things." (Aristotle, *Metaphysics*[1].)

The cosmogony of Anaximander seems to have been elaborated in greater detail—at any rate we know more about it. The

The cosmogony of Anaximander.

fundamental principle by him first named "the Beginning" ($\dot{a}\rho\chi\dot{\eta}$) is postulated to possess the property of infinite extension, and hence its name "the Infinite" ($\tau\grave{o}$ $\ddot{a}\pi\epsilon\iota\rho\rho\nu$). It is an all-pervading, eternal, indestructible mass, itself devoid of qualities, but containing potentially all the various substances which are formed from it by separation. From the $\ddot{a}\pi\epsilon\iota\rho\rho\nu$ everything proceeds, to it everything returns:

"Certain people[2] starting from the hypothesis of a single definite matter, let all things proceed from this by a process of rarefaction and condensation.... Others like Anaximander...let the opposites evolve from it by separation." (Aristotle, *Physics*.)

That which modern science still hopes for as the consummation of its labours, the reduction of all known kinds of matter to variations, according to fixed law, of one kind of primitive matter, the Ionian Philosophy assumed *à priori* as a self-evident fact; but it broke down in the attempt to show how this one cosmic matter changed into all the different substances such as we perceive them, how the homogeneous changed into the heterogeneous.

The search after the unchangeable underlying all the changes

---

[1] The translation followed in quotations from *Metaphysics*, Book I, is that by *A Cambridge Graduate*, 1881.

[2] The reference is to Thales and Anaximenes.

observed led to very different results in the case of other schools of Greek philosophy. For the Pythagoreans, a religious-political association which appeared towards the end of the 6th century B.C., in Magna-Graecia (the south Italian colonies of the Greeks) and took its name from its founder Pythagoras (about 580—500), number was the permanent thing underneath all the changing; things have not arisen out of numbers, but are formed or evolved according to them; they are in fact an imitation of the evolution of successive numbers from each other. The discovery of the numerical relations leading to harmony in music, and of the orderly motion of the heavenly bodies formed the empirical basis for this conception. Its deductive result was the attempt to show complete correspondence between the system of numbers and the actual order of things in the universe. Besides the different elements supposed to correspond to the individual numbers, there entered into the Pythagorean scheme of the world as a reality the unlimited void corresponding to the endlessness of the number series. And it is characteristic of the Greek attitude of mind towards natural phenomena, that where the facts observed were not sufficient for the establishment of complete analogies, imagination was called in to supply what was required:

The Pythagoreans take the elements of numbers to be the elements of all things.

"The so-called Pythagoreans, who applying themselves to mathematics, were the first to bring these studies into prominence, and by being trained exclusively therein, came to think that what was the foundation of mathematics was the foundation of all things whatsoever. And as numbers are naturally what mathematics begin with, and as they fancied they could discern many more points of resemblance between these numbers and the facts and processes of Life and Mind than they could in fire, in earth, or in water... and observing moreover that the properties and laws of the different Harmonies were all dependent upon numbers—since, I say all things else in all their qualities seemed to be modelled upon numbers, and numbers were prior to anything else in Nature, they hence formed the conception that the first elements of numbers were the first elements of all things whatsoever, and that the whole heaven was but an instance of harmony, in other words (the outcome of) numbers. And so all the parallelisms they could point to between either numbers or the harmonies on the one hand, and the various phases and parts of the heaven or the general disposition of the universe on the other, these they would collect and piece carefully together. And if there were still any gap visible, they clung eagerly to the attempt to make their system show a continuous thread throughout. For instance, as the number ten is thought by them to be a perfect thing,...they say that the heavenly bodies are ten in number too, but since there are only nine that can be seen, they are, fain to invent a tenth (which they call) the Antichthon." (Aristotle, *Metaphysics*.)

"The Pythagoreans too held the existence of void...as that which separates all natural objects, as if the void were a kind of separation and limitation of successive things. And they thought that this was so primarily in numbers; for the void, they said, determines their individual natures." (Aristotle, *Physics*.)

For the Eleatics, named after Elea, the town in South Italy which was the birthplace of Xenophanes the founder of the school,

The Eleatics take a fundamental idea the Unity of all Being, from which follows:
1. eternity of existence,
2. absence of Void and motion,
3. unchangeability of all that is.

and who flourished in the 5th century B.C., the *à priori* assumption of the unity of all being supplied the fundamental principle sought.

This conception of the absolute unity of all being led to certain necessary logical inferences concerning the properties and the origin of matter and motion, viz.:

1. The eternity of the existence of all that is, because creation and destruction would involve passage from or to non-being, and hence an antithesis to being, incompatible with unity.

2. The absence of void and hence also that of motion, because motion without void is impossible. But a separation of space into the full and the empty would again be a violation of the principle of unity:

"Some of the ancients held that existence was of necessity one and motionless. For they say that void is non-existent and there could not be motion without the existence of void as a separate thing." (Aristotle, *Generation and Corruption*.)

"Melissus attempts to show that the universe is without motion; for if it is to have motion, there must he says be void; but void is not included among existing things." (Aristotle, *Physics*.)

3. The unchangeability of all that is, due primarily to the impossibility of motion and secondarily to the fact that thereby would be produced something that had not existed before, in contradiction to the all-embracing principle of unity.

Here then we have a theory of the universe which, starting from purely metaphysical considerations without regard to physical facts, leads by the application of vigorous unflinching logic, such as the Greek mind excelled in, to inferences concerning the condition of the world, which are found to be in direct opposition to sensual perception. The contradiction must be removed; and the Eleatic method of doing so marks in a striking manner the fundamental

The Eleatic philosophy denies the validity of sense perception. difference between the attitude towards nature of the Greek Philosopher and the Modern Scientist respectively; the Eleatic Zeno denies the validity of sense testimony. We can but echo antiquity's own criticism of such a method:

" Some, passing over and disregarding sense perception, say that the universe is one and motionless and infinite, for if there were a limit, next to that limit would follow void....This seems true logically, but practically it seems like madness to hold such views; for no madman is so beside himself as to believe fire and ice to be one and the same." (Aristotle, *Generation and Corruption*.)

Whilst the Ionian, Pythagorean and Eleatic schools agreed in their assumption of the permanency of that which they made the

Heraclitus: Change is the essence of all being, and Fire is the emblem of change. foundation of the universe, the very opposite is at the basis of the system of Heraclitus of Ephesus (about 536—470), concerning whom Lucretius says:

" For they who have held fire to be the matter of things and the sum to be formed out of fire alone, are seen to have strayed most widely from true reason. At the head of whom enters Heraclitus to do battle, famous for obscurity more among the frivolous than the earnest Greeks who seek the truth. For fools admire and like all things the more which they perceive to be concealed under involved language." (Lucretius, *De Rerum Natura*[1].)

To Heraclitus and his school all persistence is an illusion; change is the abiding, is that which pertains to the essence of substances:

" ...the opinions of Heraclitus and his school,—how that the things of sense one and all together are in a perpetual flux, and that no precise knowledge about them is possible." (Aristotle, *Metaphysics*.)

All is in a continuous state of flow, and this conception takes the concrete form of making fire the basis of everything—not fire such as we apprehend it and such as in early Hindu and later Greek philosophy ranked as an element, but the attribute of fire, its extreme mobility and continual change. Fire changes into everything else, and everything changes back again into fire; movement in space—and this is where the originality of this conception of change comes in—is the operative cause. Fire in

---

[1] The translation followed in all quotations from Lucretius is that of Munro.

descending becomes water, and water descends further still and thus becomes earth; and the water descends again and there is reproduced the fire for the feeding of the sun and the fiery stars, returning to them that which they had lost.

Empedocles of Agrigentum (490—430) divided matter into the four elements retained by Aristotle, through him made dominant in science till the end of the 17th century, and even now still met with in popular thought and speech. Fire, earth, air, and water are called the roots of all things (ῥιζώματα), and are endowed with the properties of being without beginning, indestructible, homogeneous, and unchangeable; they are divisible into parts and these parts are capable of change of place; from the mixture of these elements arise the different substances such as we perceive them, and whose properties are the result of the kind of mixture made. The cause of the combination and separation of the elements is due to the forces "love" (φιλία), and "hate" (νεῖκος), which are not inherent in the elements but constitute an outside active principle. Hence Empedocles is sometimes said to have assumed six elements:

Empedocles assumes the existence of four elements, earth, air, fire, water, which are eternal, homogeneous, and unchangeable.

"[To the simple elements air, water, fire, Empedocles adds] earth as a fourth; for these four elements, he maintains, remain throughout all changes constant and never come into Being, except in the way of becoming greater or smaller, according as they are collected together into one mass, or broken up and dispersed out of one mass into several." (Aristotle, *Metaphysics.*)

"Empedocles did indeed call the corporeal elements four, but altogether including the motive powers (Love and Strife) he made them six in number." (Aristotle, *Gen. and Cor.*)

"They appear to have strayed exceedingly wide of the truth, who believe that all things grow out of four things, fire, earth, air and water. Chief of whom is Agrigentine Empedocles." (Lucretius.)

Empedocles denies the existence of void:

"Some of those who deny the existence of void have not defined light and heavy at all, for example Anaxagoras and Empedocles." (Aristotle, *De Caelo.*)

The Ionian conception, so impressive in its simplicity, of one kind of primitive matter had been departed from by Empedocles, and the same was done, but in a more rational manner, by Anaxagoras (500—427), a native of Klazomene, who settled towards the middle of the fifth century in Athens where he enjoyed the friendship of Pericles. He also chose the title of " On Nature " (περὶ φύσεως) for the work in which he embodied his natural philosophy, fragments of which have come down to us. Be.. sides this his tenets are much referred to and criticised by Aristotle.

Anaxagoras assumes as many " Elements " as there are simple substances, that is kinds of matter which by repeated division always yield parts having the properties of the whole. These ultimate parts were in accordance with the definition given of them termed " Homoeo-meriae " (ὅμοιος = like, μέρος = a part).

Anaxagoras' Homoeo-meriae. Great number of substances supposed elementary.

At a time when mechanical division and change of temperature were the only means of separation, the majority of the substances met with had to be classed as simple, and hence the large number of the homoeomeriae.

" Empedocles said that the corporeal elements were four in number, but that altogether, including the motive powers (Love and Strife) they were six ; on the other hand, Anaxagoras, as well as Leucippus and Democritus, held that their number was infinite. Indeed Anaxagoras calls the homoeomeriae elements, for example bone and flesh and marrow and other substances in each of which the part bears the same name as the whole substance." (Aristotle, *Gen. and Cor.*)

" Let us now examine also the homoeomeriae of Anaxagoras....First of all then, when he speaks of the homoeomeriae of things, you must know he supposes bones to be formed out of very small and minute bones and flesh out of very small and minute fleshes and blood by the coming together of many drops of blood, and gold he thinks can be composed of grains of gold and earth be a concretion of small earths and fires can come from fires and water from waters, and everything else he fancies and supposes to be produced on a like principle." (Lucretius.)

These elements were supposed to be present in a very finely divided state throughout the universe, and their coming together constitutes the production, their separation the destruction of things such as we perceive them. Every individual thing contains something of every element, but it exhibits mainly the properties of that substance of which it contains most and which therefore preponderates.

Every sub-stance con-tains some-thing of every one of the Homoeo-meriae, but exhibits the properties of the one that preponderates.

" ...if all the bodies which grow out of the earth, are in the earths, the earth must be composed of things foreign to it in kind which grow out of these earths...if flame and smoke and ash are latent in woods, woods must necessarily be composed of things foreign to them in kind....Anaxagoras chooses to suppose that all things though latent are mixed up in things, and that is alone visible of which there are the largest number of bodies in the mixture and these more ready to hand and stationed in the first rank." (Lucretius.)

"The statements of Anaxagoras clearly differ from those of Empedocles ; for the latter says that fire, water, air, and earth are four elements, and that they are more simple than flesh and bone and other such of the homoeomeriae, whereas the former says that these are simple and are elements, while earth and fire and water are composite, being a mixture of all the seeds of the homoeomeriae." (Aristotle, *Gen. and Cor.*)

The homoeomeriae are assumed to be eternal, unchangeable, infinitely divisible, and of continuous extension :

The Homoeo. meriae are eternal, unchangeable, and continuous in space.

"He does not allow that void exists anywhere in things, or that there is a limit to the division of things." (Lucretius.)

The elements are moveable in space, but the power of motion is not inherent in them ; this is the prerogative of a special kind of matter conceived to be the lightest and most mobile of elements, and which is always present along with any other kind of matter, moving not only itself but also the matter associated with it.

Motion is due to the universal presence of a special element.

The cosmogony of Anaxagoras consists in the assumption of a mixture ($\mu \hat{\iota} \gamma \mu a$) in which the division and the confusion of the homoeomeriae being carried to an infinite degree, no special quality asserts itself, and from which by separation arise the different substances, this evolution being guided by an organising intelligence, not inherent but detached and independently existent. This "$\nu o \hat{v} s$" it is that makes the system of Anaxagoras metaphysically and ethically so interesting; from the physical point of view it constitutes a sharp antithesis to the hylozoism of the Ionian school as well as to the materialism of the Atomistic school.

The "$\nu o \hat{v} s$" of Anaxagoras.

"...were it never so true that it is out of some single element or even of several that all life and death proceed, yet why should such changes occur at all and what is the cause of these changes? for of course the subject itself cannot of itself make any change in itself. No Unitarian Materialist had the fortune to discern any such cause,...nor again did it seem satisfactory to entrust such an important matter to haphazard or spontaneous development. And so when someone said that Intelligence was present in Nature just as it is in Animals and was the cause of this universe and its wonderfully perfect order, it was the appearance of a rational man....Anaxagoras is the first we know of who clearly and unmistakeably embraced this theory." (Aristotle, *Metaphysics*.)

Three names are associated with the promulgation amongst the Greeks of an atomistic view of the ultimate constitution of matter: Leucippus, Democritus, and Epicurus.

The Atomistic Philosophers, Leucippus, Democritus, and Epicurus. Of Leucippus we know but little, not even the names of his writings. Tradition has closely allied him with Democritus, whose fellow-citizen and teacher he is supposed to have been, the philosophy of these two thinkers being generally treated as one and the same and as belonging to both.

Democritus (*circ.* 460—360) was born in Abdera, a town in Thrace, whither he returned after many and long journeys. By education and travel he equipped himself with the scientific knowledge of his time, and could say of himself:

" Among all my contemporaries I have travelled over the largest portion of the earth in search of things the most remote, and have seen the most climates and countries, heard the largest number of thinkers, and no one has excelled me in geometric construction and demonstration—not even the geometers of the Egyptians, with whom I spent in all five years as a guest."

He is supposed to have written a very great deal, his chief work being the "Great Universe" (μέγας διάκοσμος); but the merest fragments only have come down to us, and our knowledge of his philosophy is derived chiefly from Aristotle and commentaries.

His system was most unpopular at the time of its promulgation and for long after, and when more than 200 years later it was revived, it was in the form of the Epicurean philosophy into which it had been absorbed.

Epicurus (341—270), born at Samos, an island in the Aegean Sea near Asia Minor, at a very early age studied the works of Democritus. He taught in various places, and finally in Athens, which he had already visited as a youth and where he founded the school which is named after its garden[1]. His Natural

---

[1] An ancient and a modern estimate of Epicurus may find a place here; " When human life to view lay foully prostrate upon earth crushed down under the weight of religion, who showed her head from the quarters of Heaven with hideous aspect lowering upon mortals, a man of Greece ventured first to lift up his mortal eyes to her face and first to withstand her to her face. Him neither story of gods nor thunderbolts nor Heaven with threatening roar could quell, but only stirred up the

Philosophy met with exposition in the poem *De Rerum Natura* of the Roman, Titus Lucretius Carus (98—54), a man about whose life we know practically nothing and whose claim to any independent philosophy has been justly repudiated, but who has handed down to us another's philosophy in a form resplendent with imagery, such as never before and never since has fallen to the lot of any other naturalistic scheme of the universe.

Lucretius' exposition of the Atomistic Philosophy.

"Epicurus declares that the right study of nature must not arbitrarily propose new laws, but must everywhere base itself upon actually observed facts. So soon as we abandon the way of observation, we have lost the traces of nature, and are straying into the region of idle fantasies. In other respects Epicurus' theory of nature is almost entirely that of Democritus." (Lange, *History of Materialism.*)

In the account of the Greek atomistic doctrine about to be given, no attempt will be made to strictly separate or differentiate the Epicurean from the Democritean. All the parts will be omitted which deal with life, the soul, and the relations between men and gods; that is the very parts which for Epicurus are the main object, and to which the physical conception of matter—that which the chemist is mainly concerned with—is but subsidiary.

The characteristic tenets of the atomistic view of the constitution of matter and of the production of the substances such as we apprehend them, may all be found included in postulates concerning:

Firstly: The ultimate constituents of matter termed " atoms," which are eternal and unchangeable; which are characterised by size, form, and situation; and which by their combination form the substances of the world.

Secondly: The motion inherent in the atoms.

more the eager courage of his soul, filling him with desire to be the first to burst the fast bars of nature's portals. Therefore the living force of his soul gained the day : on he passed far beyond the flaming walls of the world and traversed throughout in mind and spirit the immeasurable universe; whence he returns a conqueror to tell us what can, what cannot come into being; in short on what principle each thing has its powers defined, its deepset boundary mark." (Lucretius.)

"In Athens he bought a garden, where he dwelt with his disciples. It is said to have borne as an inscription, ' Stranger, here will it be well with thee: here pleasure is the highest good.' Here lived Epicurus with his followers, temperately and simply, in harmonious effort, in heartfelt friendship, as in a united family. By his will he bequeathed the garden to his school, which for a long time still had its centre there. The whole of antiquity furnishes no brighter and purer example of fellowship than that of Epicurus and his school." (Lange, *History of Materialism.*)

Thirdly: The existence of void, in virtue of which the displacements and combinations of the atoms become possible.

What the scope and nature of the conceptions under each of these divisions was, and how these were brought into accordance with the facts observed, can be best gathered by hearing directly the report of Aristotle and Lucretius. This method is necessarily a long one but otherwise its advantages are obvious.

I. *The existence and the properties of atoms.*

1. All substances are primarily formed of atoms.

"Democritus and Leucippus say that all things are composed of indivisible bodies, and that these are infinite both in number and in their forms, and that the differences between things are due to the elements of which they are composed and to the position and arrangement of these elements." (Aristotle, *Gen. and Cor.*)

2. These atoms are so small that they are incapable of being sensually realised, but this is no argument against their reality, as there are plenty of instances of substances which we know to exist though they are present as such small particles as not to be perceived:

"After the revolution of many years a ring is thinned on the underside by wearing, the dripping from the eaves hollows a stone, the bent ploughshare of iron imperceptibly decreases in the fields, and we behold the stone-paved streets worn down by the feet of the multitude; the brass statues too at the gates show their right hands to be wasted by the touch of the numerous passers by who greet them. These things then we see are lessened, after they are thus worn down; but what bodies depart at any given time nature has jealously shut out the means of seeing. Lastly the bodies which time and nature add to things by little and little, constraining them to grow in due measure, no exertion of the eye-sight can behold; and so too wherever things grow old by age and decay, and when rocks hanging over the sea are eaten away by the fine salt spray, you cannot see what they lose at any given moment. Nature therefore works by unseen bodies." (Lucretius.)

3. Matter itself is eternal and indestructible, and so are the atoms.

"...the law of nature whose first principle we shall begin by thus stating, nothing is ever gotten out of nothing by divine power." (*Ibid.*)

F.

16

This statement is supported by the argument of permanency in nature.

"...if things came from nothing, any kind might be born of anything, nothing would require seed. Men for instance might rise out of the sea, the scaly race out of the earth, and birds might burst out of the sky...any tree might bear any fruit...nor would time be required for the growth of things... little babies would at once grow into men, and trees in a moment would rise and start out of the ground. But none of these events it is plain ever comes to pass, since all things grow step by step,...so that you may be sure that all things increase in size and are fed out of their own matter." "Again unless matter had been eternal, all things before this would have utterly returned to nothing and whatever things we see would have been born anew from nothing...for we see that anything is more quickly destroyed than again renewed." "...those which are first-beginnings of things no force can quench: they are sure to have the better by their solid body—these can neither be broken in pieces by the stroke of blows from without nor have their texture undone by aught piercing to their core nor give way before any other kind of assault.... First-beginnings therefore are of solid singleness, massed together and cohering closely by means of least parts, not compounded out of a union of those parts, but, rather, strong in everlasting singleness. From them nature allows nothing to be torn, nothing further to be worn away, reserving them as seeds for things." (*Ibid.*)

4. The atoms are unchangeable, which involves the property of indivisibility. The quotations just given bear as much if not more on the indivisibility of the atoms as on their indestructibility, and the following may be added:

4. The atoms are unchangeable and indivisible.

"Democritus says that no one of the first elements can arise out of any other, but that nevertheless the common primitive matter, differing in the size and form of its parts, is the principle of them all." (Aristotle, *Physics*.)

"...Since by the laws of nature it stands decreed what they (these things) can each do and what they cannot do, and since nothing is changed, but all things are so constant...they must sure enough have a body of unchangeable matter also. For if the first beginnings of things could in any way be vanquished and changed, it would be then uncertain too what could and what could not rise into being, in short on what principle each thing has its powers defined, its deepset boundary mark." (Lucretius.)

5. The atoms differ from each other in shape, size and arrangement.

5. The atoms differ from each other in shape, size and arrangement.

The quotation from Aristotle given above shows how the atomistic philosophy assumed but one kind of matter, which according to differences in the size and shape of the individual particles, the "atoms" of Democritus, the "first-beginnings" of Lucretius, formed the ultimate constituents of the different elements. Further testimony is afforded by the passages :

"Now mark and next in order apprehend of what kind and how widely differing in their forms are the beginnings of all things, how varied by manifold diversities of shape.... And quickly as we see wines flow through a strainer, sluggish oil on the other hand is slow to do so, because sure enough it consists of elements either larger in size or more hooked and tangled in one another.... The things which are able to affect the senses pleasantly, consist of smooth and round elements ; while all those on the other hand which are found to be bitter and harsh, are held in connexion by particles that are more hooked and for this reason are wont to tear open passages into our senses and in entering in to break through the body.... Again things which look hard and dense must consist of particles more hooked together, and be held in union because compacted throughout with branch-like elements.... Those things which are liquid and of fluid body ought to consist more of smooth and round elements." (*Ibid.*)

"...these philosophers say that the varieties in their elements are the causes of all other things. These varieties they say are but three in number, varieties of form, varieties of arrangement and varieties of position,...for instance *A* differs from *N* in shape, *AN* from *NA* in arrangement, and *Z* from *N* in position merely." (Aristotle, *Metaphysics.*)

6. The qualities of all substances depend on the kind of the constituent atoms and on their arrangement, though the properties of matter such as we apprehend by our senses are not due to the properties of the atoms themselves but to the manner of their combination.

6. The properties and the arrangement of constituent atoms determine the properties of all substances.

"Bodies again are partly first-beginnings of things, partly those which are formed of a union of first-beginnings.... And it often makes a great difference with what things and in what position the same first-beginnings are held in union and what motions they mutually impart and receive ; for the same make up heaven, sea, lands, rivers, sun ; the same make up corn, trees, living things ; but they are mixed up with different things and in different ways as they move. Nay you see throughout even in these verses of ours many elements common to many words, though you must needs admit that the lines and words differ one from the other both in meaning and in the sound wherewith they sound. So much can elements effect by a mere change of order, but those elements which are the first-beginnings of things can bring with them more combinations out of which different things can severally be produced." (Lucretius.)

"Democritus and Leucippus after conceiving the forms, produce change and generation from them ; deriving birth and destruction from their separation and concretion, and change from their arrangement and position." (Aristotle, *Gen. and Cor.*)

16—2

II. *The atoms are constantly in motion and this*

motion is a property inherent in them.

"Some have recourse to an energy that is always in action, as Leucippus and Plato; for they maintain that motion is always in existence: but why, and in what way, they do not state, nor how this is the case, nor do they assign the cause for this perpetuity of motion." (Aristotle, *Metaphysics.*)

"Solid bodies of matter fly about for ever unvanquished through all time.... The first-beginnings of things move first of themselves.... No rest is given the first bodies throughout the unfathomable void, but driven on rather in ceaseless and varied motion they partly, after they have pressed together, rebound leaving great spaces between, while in part they are so dashed away after the stroke as to leave but small spaces between." (Lucretius.)

The motion of these particles is illustrated by the familiar spectacle of the motion of the motes in a sunbeam:

"Of this truth, even as I relate it, we have a representation and picture always going on before our eyes and present to us; observe whenever the rays are let in and pour the sunlight through the dark chambers of houses you will see many minute bodies in many ways through the apparent void mingle in the midst of the light of the rays, and as in never-ending conflict skirmish and give battle combating in troops and never halting, driven about in frequent meetings and partings." (*Ibid.*)

Combination is supposed to be due to the coalescence between the moving particles. To the original Democritean conception of purely accidental collisions between the atoms moving in straight lines Epicurus adds the assumption of a slight declination from rectilinear motion in order to better account for such coalescence:

"When bodies are borne downwards sheer through void, at quite uncertain times and uncertain points of space they swerve a little from their equal poise: you just and only just can call it a change of inclination. If they were not used to swerve, they all would fall down, like drops of rain, through the deep void, and no clashing would have been begotten, nor blow produced among the first-beginnings: thus nature never would have produced aught." (*Ibid.*)

That a state of rest of a material substance, as we perceive it by our senses, is not incompatible with the continuous motion of its ultimate constituent particles, has been demonstrated by Lucretius by means of splendid imagery still used for the same purpose:

"And herein you need not wonder at this, that though the first-beginnings of things are all in motion, yet the sum is seen to rest in supreme repose, unless where a thing exhibits motions with its individual body. For all the nature of first things lies far away from our senses beneath their ken ; and therefore since they are themselves beyond what you can see, they must withdraw from sight their motions also ; and the more so that the things which we can see, do yet often conceal their motions when a great distance off. For often the woolly flocks as they crop the glad pastures on a hill, creep on whither the grass jewelled with fresh dew summons and invites each, and the lambs fed to the full gambol and playfully butt ; all which objects appear to us from a distance to be blended together and to rest like a white spot on a green hill. Again when mighty legions fill with their movements all parts of the plains waging the mimicry of war, the glitter then lifts itself up to the sky and the whole earth round gleams with brass and beneath a noise is raised by the mighty trampling of men and the mountains stricken by the shouting re-echo the voices to the stars of heaven, and horsemen fly about and suddenly wheeling scour across the middle of the plains, shaking them with the vehemence of their charge. And yet there is some spot on the high hills, seen from which they appear to stand still and to rest on the plains as a bright spot." (*Ibid.*)

III. *The atoms are separated from each other by void.*

"But Leucippus and his disciple Democritus say that the full and the empty exist as the primary elements of the world.... And...they say that the empty has an existence equally real as material body (each being a primary element of the universe) and these two elements they say are the causes of things." (Aristotle, *Metaphysics.*)

<small>III. The existence of void is postulated.</small>

Aristotle gives the proofs from reason and from experience that were adduced by the atomistic philosophers in support of the existence of void:

"Those who affirm the existence of void argue more legitimately. One thing they say is that there could be no movement in space if there were no void ; for it is impossible for the full to admit anything into itself. If it were to admit anything and there were to be two things in the same place, it would be possible for there to be any number whatever of bodies in the same place. But if this is possible, then the smallest thing may contain the largest. This is one way in which they argue the existence of void, but there is another.... They say that all increase of bulk depends on void ; for food is a body, and two bodies cannot occupy the same space. And they adduce as evidence the case of ashes[1], which can take in as much water as the empty vessel." (Aristotle, *Physics.*)

<small>(1) Motion requires the existence of void.</small>

<small>(2) The existence of void is required in the explanation of expansion.</small>

"All nature as it exists by itself has been founded on two things : there are bodies and there is void in which these bodies are placed and through

---

[1] A famous instance of what Aristotle quotes as fact and as the result of observation.

which they move about.... All things are not on all sides jammed together
and kept in by body : there is also void in things.... If there were not void,
things could not move at all ; for that which is the property of body, to let
and hinder, would be present to all things at all times ; nothing therefore
could go on, since no other thing would be the first to give way.... In rocks
and caverns the moisture of water oozes through, and all things weep with
abundant drops.... Food distributes itself through the whole body of living
things.... Now if there are no void parts, by what way can the bodies severally

(3) The exis-
tence of void
is required to
account for
differences in
density.

pass ? you would see it to be quite impossible. Once more,
why do we see one thing surpass another in weight though
not larger in size ? For if there is just as much body in
a ball of wool as there is in a lump of lead, it is natural it
should weigh the same, since the property of body is to weigh
all things downwards, while on the contrary the nature of
void is ever without weight. Therefore when a thing is just as large, yet is
found to be lighter, it proves sure enough that it has more of void in it ;
while on the other hand that which is heavier shows that there is in it more
of body and that it contains within it much less of void. Therefore that
which we are seeking with keen-sighted reason exists sure enough, mixed up
in things ; and we call it void." (Lucretius.)

This then is the case for the atomistic philosophy. That
which differentiated and absolutely separated it from the other
views held contemporarily is the assumption of the finite divisi.
bility of matter and of the existence of void. On this point
Lucretius, summarising the tenets of the anti-atomistic school,
says :

Essential
points of the
atomistic phi-
losophy.

"They have banished void from things and yet assign to them motions
and allow things soft and rare, air, sun, fire, earth, living things, corn, and
yet mix not up void in their body ; next they suppose that
there is no limit to the division of bodies and no stop set
to their breaking and that there exists no least at all in
things."

The atomistic school aimed at explaining not only the physics
of the universe, but also all the phenomena of life and death, by
the properties inherent in matter. In absolute contrast to this
materialistic scheme is the idealistic one of Plato.

Plato.

Plato (427—344), by birth an Athenian, a pupil of Socrates
and well acquainted also with the philosophical tenets of the
Heraclitean, Eleatic and Pythagorean schools, devoted
himself after years spent in travel to philosophical
teaching in his native town; the groves of Academus
where t•met gave its name to that celebrated school .which for

centuries has been known as that of the "Academy." Plato's contribution to the philosophy of the natural sciences is contained mainly in the dialogue, the *Timaeus*.

To Plato the individual things and occurrences were but transitory and unreal, the abstract idea of them was the eternal, the real. Hence he conceived the aim of philosophy

Plato's idealistic philosophy— the primal elements of the ideas were the primal elements of all things soever.

to be the study of these ideas, the discovery of the abiding principles of which the phenomena and things of this world were but the images; the study of these phenomena and of the things themselves and their classification could not be dispensed with, but was to serve not as an end, only as a means for learning something concerning that abiding and perfect principle of which they are the temporary and imperfect indications:

"Plato from youth upwards had been familiar...with the opinions of Heraclitus and his school, how the things of sense one and all together are in a perpetual flux, and that no precise knowledge about them is possible....He held that...common definition...was an impossibility in the class of any sensible objects, inasmuch as they are changing at every moment. Such existences (*i.e.* such things as *could* be so defined) Plato accordingly called Ideas: the things of sense he held to be apart and distinct from these, and were only called what they were because of them: for that, speaking generally, those things which were called by the same name as the Ideas existed only by virtue of 'participation.'... And since the Ideas are to all other things the causes of their existence, he held that the primal elements of the Ideas were the primal elements of all things soever." (Aristotle, *Metaphysics*.)

He investigates the laws of matter in the hope that he may more clearly ascertain the laws of spirit[1].

His method of investigation consists in the propounding

His method of studying na- ture is deduc- tive, starting from certain à priori mathe- matical con- ceptions and leading to the following re- sults:

of certain mathematical doctrines concerning the parts and elements of the universe: "not so much as assertions concerning physical facts of which the truth or falsehood is to be determined by reference to nature herself," but as *à priori* truths. The results of this method lead to the following con- ception of the structure of matter:

1.   The elements are four in number.

---

[1] The *Timaeus*, edited by R. D. Archer-Hind. *Introduction*, p. 47.

" Now that which came into being must be material and such as can be
seen and touched. Apart from fire nothing could ever become visible, nor
without something solid could it be tangible, and solid cannot
1. The ele-
ments are four exist without earth ; therefore did God when he set about to
in number. frame the body of the universe form it of fire and of earth...
God...set...air and water betwixt fire and earth, and making
them as far as possible exactly proportional, so that fire is to air as air to
water, and as air is to water, water is to earth, thus he compacted and
constructed a universe visible and tangible. For these reasons and out of
elements of this kind, four in number, the body of the universe was created,
being brought into concord through proportion[1]."

2. Matter is   2. Within the universe matter can neither be
eternal. created nor destroyed.

The whole of the matter available for creation and consisting
of the four elements was by God fashioned into the shape of
a perfect sphere :

"nothing went forth of it nor entered in from anywhere ; for there was
nothing. For by design was it created to supply its own sustenance by its
own wasting."

3. The regular solids are assigned to the forms of the elements[2].

" Let us assign the figures that have come into being in our theory to fire
and earth and water and air. To earth let us give the cubical form, for
earth is least mobile of the four and most plastic of
3. The regu- bodies : and that substance must possess this nature in the
lar solids are highest degree which has its bases most stable. Now of the
assigned to the
forms of the triangles which we assumed as our starting-point that with
elements. equal sides is more stable than that with unequal ; and of the
surfaces composed of the two triangles, the equilateral quad-
rangle necessarily is more stable than the equilateral triangle.... Therefore
in assigning this to earth we preserve the probability of our account ; and
also in giving to water the least mobile and to fire the most mobile of those
which remain ; while to air we give that which is intermediate. Again we
shall assign the smallest figure to fire, and the largest to water and the
intermediate to air : and the keenest to fire, the next to air, and the third to
water. Now among all these that which has the fewest bases must naturally
in all respects be the most cutting and keen of all, and also the most nimble,
seeing it is composed of the smallest number of similar parts ; and the
second must have these same qualities in the second degree, and the third
in the third degree. Let it be determined then, according to the right account
and the probable, that the solid body which has taken the form of the
pyramid is the element and seed of fire ; and the second in order of generation
let us say to be that of air, and the third that of water."

[1] The translation followed in this as in all subsequent quotations from the
Timaeus is that given in Mr Archer-Hind's edition.
[2] For diagrams see footnote p. 251.

4. The elements are made up of infinitely small particles, which are not further divisible.

"Now all those bodies we must conceive as being so small that each single body in the several kinds cannot for its smallness be seen by us at all; but when many are heaped together their united mass is seen."

4. The divisibility of matter is finite.

"There is this difference only between what Plato and Leucippus say on this point: Leucippus speaks of the indivisible particles as solids, and Plato as surfaces, and while Leucippus says that each of the indivisible solids is determined by forms which are infinite in number; according to Plato their number is finite. (This is the only difference), for both speak of indivisible particles and say they are determined by forms." (Aristotle, *Gen. and Cor.*)

5. The differences between the various kinds of the same elements are due to differences in the size of the bounding planes.

"...for the fact that within the several classes different kinds exist we must assign as its cause the structure of the elementary triangles; it does not originally produce in each kind of triangle one and the

5. The different kinds of the same elementary substances are due to differences in size of the constituent particles.

same size only, but some greater and some less; and there are just so many sizes as there are kinds in the classes: and when these are mixed up with themselves or with one another, an endless diversity arises, which must be examined by those who would put forward a probable theory concerning nature.... Of fire there are many kinds; for instance flame and that effluence from flame, which burns not but gives light to the eyes, and that which remains in the embers when the flame is out....Of water there are two primary divisions, the liquid and the fusible kind. The liquid sort owes its nature to possessing the smaller kinds of watery atoms, unequal in size; and so it can readily either move of itself or be moved by something else, owing to its lack of uniformity and the peculiar shape of its atoms. But that which consists of larger and uniform particles is more stable than the former and heavy, being stiffened by its uniformity."

6. The elements are capable of being transformed one into another, which is explained by the bounding surfaces going to form other solids.

"What we...have named water, by condensation, as we suppose, we see turning to stones and earth; and by rarefying and expanding this same element becomes wind and air; and air when inflamed

6. The elements may be transmuted one into the other.

becomes fire: and conversely fire contracted and quenched returns again to the form of air; also air concentrating and condensing becomes cloud and mist; and from these yet further compressed comes flowing water; and from water earth and stones once more: and so, it appears, they hand on one to another the cycle of generation. Thus then since these several bodies never assume one constant form, which of them can we positively affirm to be really *this* and not another without being shamed in our own eyes?... From all that we have already said in the matter of these four kinds, the facts would seem to be as follows : When earth meets with fire and is dissolved by the keenness of it, it would drift about, whether it were dissolved in fire itself, or in some mass of air or water, until the parts of it meeting and again being united become earth once more ; for it never could pass into any other kind. But when water is divided by fire or by air, it may be formed again and become one particle of fire and two of air : and the divisions of air may become for every particle broken up two particles of fire[1]. And again when fire is caught in air or in waters or in earth, a little in a great bulk, moving amid a rushing body, and contending with it is vanquished and broken up, two particles of fire combine into one figure of air ; and when air is vanquished and broken small, from two whole and one half particles one whole figure of water will be composed."

Fire by the sharpness of its angles and sides is enabled to break up all the shapes constituting other elements into tetrahedra, *i.e.* change them into fire, but on the other hand fire can change into air and air into water by coalescence into the forms peculiar to these.

7. The existence of void is denied.

7. The existence of void is denied.

Plato's natural philosophy has been practically without influence on the development of physical science ; it has been very different with that of his great pupil Aristotle (384—322).

From Stagira the town in Northern Greece where he was born, Aristotle takes the name of the Stagirite. Much of his life was spent at Athens. In his youth a pupil and friend of Plato, he returned to Athens after a long absence, part of which was occupied by a stay at the court of Philip of Macedon, where he acted as tutor to the young Alexander. At the Lyceum, a gymnasium in the vicinity of Athens, he founded the school called the Peripatetic. This name has been accounted for either by Aristotle's habit of walking to and fro during the delivery of his lectures, or by the name of the lecture place (ὁ περίπατος). Of Aristotle it has been said aptly that " he is felt to be so mighty, and is known to be so wrong."

To fully support this verdict cannot be attempted here, mere indication must suffice.

Aristotle.

---

[1] See footnote on next page.

¹ Here evidently only the number and kind of the bounding surfaces are considered; the eight triangles of the octahedron (air) go to form two tetrahedra (fire) each bounded by four triangles; the twenty triangles bounding the icosahedron (water) go to form two octahedra (air) requiring sixteen, and one tetrahedron (fire) requiring the remaining four triangles. The following diagrams and further quotations from the *Timaeus* may serve to explain paragraphs 4, 5, 6.

THE FIVE REGULAR SOLIDS.

TETRAHEDRON.
Plato's Fire.

"If four equilateral triangles are combined, so that three plane angles meet in a point, they make at each point one solid angle, that which comes immediately next to the most obtuse of plane angles; and when four such angles are produced there is formed the first solid figure, dividing its whole surface into four equal and similar parts."

OCTAHEDRON.
Plato's Air.

"The second is formed of the same triangles in sets of eight equilateral triangles, bounding every single solid angle by four planes; and with the formation of six such solid angles the second figure is also complete."

ICOSAHEDRON.
Plato's Water.

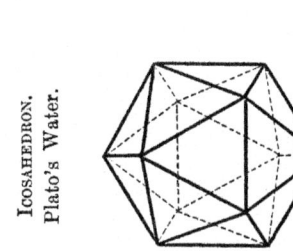

"The third is composed...of twelve solid angles, each contained by five plane equilateral triangles; and it has twenty equilateral surfaces."

CUBE.
Plato's Earth.

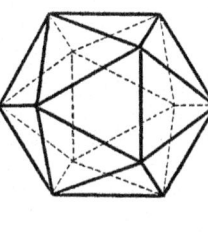

"The isosceles triangle...combined in sets of four, with the right angles meeting at the centre, thus form a single square. Six of these squares joined together formed eight solid angles, each produced by three plane right angles: and the shape of the body thus formed was cubical, having six square planes for its surfaces."

DODECAHEDRON.
Plato's Symbol of the Universe.

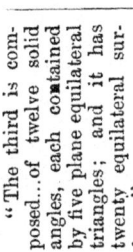

"And whereas a fifth figure yet alone remained, God used it for the universe in embellishing it with signs."

The scientist of to-day feels him to have been mighty because it was he who after a period of night, again recognised the importance of physical inquiries; because he undertook

Aristotle's claims to greatness.

and carried out the gigantic task of collecting and classifying the knowledge of his time; and above all because he, first of all men, in a manner wonderfully clear and luminous announced the principles of the inductive method:

"In direct opposition to Plato, who denying the validity of the senses, made intuition the ground of all true knowledge, Aristotle sought his basis in sensuous perception. His reliance was on Experience and Induction, the one furnishing the particular facts, from which the other found a pathway to general facts or laws." (G. H. Lewes, *Aristotle*.)

"Science commences when, from a great number of Experiences, one general conception is formed which will embrace all similar cases.... Thus if you know that a certain remedy has cured Callias of a certain disease, and that the same remedy has produced the same effect on Socrates, and on several other persons, that is *Experience*; but to know that a certain remedy will cure all persons attacked with disease is *Science*: for Experience is the Knowledge of individual things, Science is that of Universals." (Aristotle, *Metaphysics*.)

"Those who have devoted themselves more to the study of nature are better able to establish principles more widely applicable, though they who start from theories without examining the facts and take into account only a small number of phenomena, no doubt arrive more easily at their conclusions." (Aristotle, *Gen. and Cor.*)

And we know Aristotle to have been wrong because his induction stopped short at a critical point and lacked the control of verification; because when he had reached clearness

The weaknesses and fallacies of the Aristotelian method.

and symmetry in his definitions and verbal distinctions of the phenomena of nature he accepted these without heed to their correspondence with the facts; and because he considered the task of science completed and himself in the possession of all the facts required for solving its problems, thereby imposing a finality on his system which in the hands of his disciples for centuries acted as a check to the development of science.

"Looked at in a general way, the Aristotelian Method seems to be the method of positive science; but on closer meditation we shall detect their germinal difference to be the omission in Aristotle of the principle of rigorous verification of each inductive step.... When Aristotle therefore lays down as a canon the necessity of ascertaining generals from an examination of particulars, his canon, admirable indeed, needs to be accompanied by a distinct recognition of the equal necessity of verification." (G. H. Lewes, *History of Philosophy*.)

"He imposed many arbitrary rules upon nature; being everywhere more careful how one may give a ready answer, and make a positive assertion, than how he may apprehend the variety of Nature...the Physics of Aristotle in general sound only of Logical Terms.... He corrupted Natural Philosophy by his Logic, and made the world out of his Categories.... The philosophy of Aristotle, after having by hostile confutations destroyed all the rest,...has laid down the law on all points; which done, he proceeds himself to raise new questions of his own suggestion, and dispose of them likewise; so that nothing may remain that is not certain and decided; a practice which holds and is in use among his successors." (Bacon, *Novum Organum*.)

"Averroes (1126—1198) of Cordova maintained that Aristotle carried the sciences to the highest possible degree, measured their whole extent, and fixed their ultimate and permanent boundaries." (Whewell, *History of the Inductive Sciences*.)

What then were the tenets concerning the ultimate constitution of matter that Aristotle left with that impress of final and complete truth that made them for so long a fetter to posterity? His treatise on *Generation and Corruption* gives us most information on this point, but much is also contained in the *Physics* and the *Meteorology*. Shortly summarised these may be taken to be:

> *Aristotelian tenets concerning the ultimate constitution of matter.*

1. One kind of unqualified matter exists which is the carrier of different properties and is never disconnected from these.

> *1. Matter of one kind is associated with qualities, the different conjunctions of which give rise to the different substances.*

"We too assert that there is one primitive matter constituting perceptible bodies, but that it has no existence apart from these bodies, but is always found in connection with the one or the other of pairs of opposite qualities; and from it arise the so-called elements...and we hold that matter, the carrier of opposing qualities and inseparable from these, is the principle and first beginning (of things). Thus heat does not serve as matter for cold, nor cold for heat, but the underlying (principle) is the material for both." (*Gen. and Cor.*)

2. For some reason not very apparent the essential properties of which this fundamental matter acts as a carrier are hot and its contrary cold, wet and its contrary dry; and the possible combinations in sets of two of these qualities give rise to the four elements, earth and fire, air and water.

> *2. The four Aristotelian elements.*

"Since the possible combinations of four are six, and since it is not in the nature of contraries to be coupled (for hot and cold, and again dry and moist cannot exist), it is clear that there will be four elemental combinations, namely hot and dry, and hot and moist, and again cold and mois‚, and cold and dry. This is in complete agreement with the existence of the apparently simple bodies, fire, air, water, and earth ; for fire is hot and dry, and air is moist and hot (for air resembles vapour), and water is cold and moist, and earth cold and dry." (*Gen. and Cor.*)

3. The different kinds of the same element are due to different degrees of the same properties.

3. There are Varieties of the four elements which are due to the preponderance of one or other of the essential characteristic properties.

"Fire, air, water, and earth (such as we perceive them) are not simple (bodies) but mixed. The simple bodies are indeed similar to, but not identical with fire etc. such as we perceive them.... Still in a general way, out of the four elements each one belongs to one of the qualities (dry, hot etc.), earth belonging to dry rather than cold, water to cold rather than moist, air to moist rather than hot, and fire to hot rather than dry." (*Gen. and Cor.*)

4. The elements can change one into another.

4. The elements may be transmuted one into the other.

"So then it has been shown that any one of the elements can arise out of any other." (*Gen. and Cor.*)

This happens owing to the abstraction of certain qualities and the substitution of others, hence an element can more readily change into one with which it has one quality in common (cold water to cold earth, hot fire to hot air etc.), than into one completely its opposite (hot and dry fire to cold and wet water etc.).

5. The elements are separated in space.

5. The elements are separated in space.

"Now the simple bodies being four, each pair belongs to one of the two regions severally (of the universe) ; fire and air to the region which extends towards the outer limit (of the universe), and earth and water to the region which extends towards the centre. Fire and earth are the extremes and are the purest, water and air are intermediate and are less pure." (*Gen. and Cor.*)

6. Matter is capable of infinite division.

6. The divisibility of matter is infinite.

"Generally speaking, if a thing is conceivably infinitely divisible, nothing impossible will have happened, if it is actually so divided, for there is no impossibility in its being divided into ten thousand times ten thousand, although perhaps no one could perform the division." (*Gen. and Cor.*)

7. The existence of void is denied on grounds taken both from reason and from experience.

"Further, objects when thrown continue to move without the continued action of the originator of the motion; on one theory this is due to the counterpressure of the air which is displaced, on another to the fact that the air it displaces itself then pushes it forward with a motion more rapid than the original innate tendency of the object pushed, which would naturally carry it to its appropriate region (that is, which would make it fall towards the centre)." (*Physics.*)

7. The existence of void is denied.

The survey just given of the tenets of the different schools of Greek philosophy on the subject of the ultimate constitution of matter contains about all the essential elements of the various conceptions and hypotheses on this subject that the human mind is capable of framing. All that has been done since has been essentially a repetition of the old, at most a different combination of the same old constituent elements; but fortunately for the development of science, aim and method in the framing and in the application of such speculations have changed completely. Before tracing the history of these hypotheses through the middle ages and up to the beginning of the nineteenth century, it seems fit to close their history in antiquity by some evaluation of the achievements of Greek philosophy, which is practically the only one to be considered in this department. This cannot be done better than in the words of two great historians of the subject: Whewell and Lange:

"The early philosophers of Greece entered upon the work of physical speculation in a manner which showed the vigour and confidence of the questioning spirit, as yet untamed by labour and reverses. It was for later ages to learn that man must acquire slowly and patiently, letter by letter the alphabet in which nature writes her answers to such inquiries; the first students wished to divine at a single glance the whole import of her book. They endeavoured to discover the origin and principle of the universe.... The rude attempts at explanation which the first exercise of the speculative faculty produced might have been gradually concentrated and defined so as to fall in both with the requisitions of the reason and the testimony of the senses. But this was not the direction which the Greek speculators took. On the contrary, as soon as they had introduced into their philosophy any abstract and general conceptions, they proceeded to scrutinise these by the internal light of the mind alone, without any longer looking abroad into the world of sense. They ought to have reformed and fixed their usual conceptions by observation; they only analysed and expanded them by reflexion; they ought to have sought by trial among the Notions which passed through their mind, some one which admitted of an exact application to Facts; they selected arbitrarily and consequently erroneously the Notions

Whewell's disparagement of the natural philosophy of the Greeks.

according to which Facts should be assembled and arranged; they ought to have collected clear Fundamental Ideas from the world of things by *inductive* acts of thought, they only derived results by *deduction* from one or other of their familiar conceptions.... The whole mass of the Greek philosophy therefore shrinks into an almost imperceptible compass when viewed with reference to the progress of physical knowledge." (Whewell, *History of the Inductive Sciences.*)

"With the freedom and boldness of the Hellenic mind was united an innate ability to draw inferences, to enunciate clearly and sharply general propositions, to hold firmly and surely to the premisses of an

Lange's appreciation of the natural philosophy of the Greeks.

inquiry, and to arrange the results clearly and luminously; in a word, the gift of scientific deduction....A great work had to be done before the uncritical accumulations of observations and traditions could be transformed into our fruitful method of experiment. A school of vigorous thinking must first arise, in which men were content to dispense with premisses for the attainment of their immediate end. This school was founded by the Greeks, and it was they who gave us, at length, the most essential basis of deductive processes, the elements of mathematics and the principles of formal logic. The apparent inversion of the natural order, in the fact that mankind learnt to *deduce* correctly, before they learnt to find correct *starting-points* from which to reason, can be seen to be really natural only from a psychological survey of the whole history of thought." (Lange, *History of Materialism.*)

These verdicts apparently so different may yet both be accepted as true and just.

The transmission of Greek philosophy to the nations of the West has been the work of the Arabs, who after their conquest of

The transmission of the Aristotelian philosophy to the West.

Spain (711 A.D.), "aspired also to treasures intellectual, to arts and sciences." Arab scholars acted as translators and commentators of Aristotle, and set the example of absolute submission to the Peripatetic Philosophy. The physician and writer, Averroes, translated Aristotle from the Syriac, and also published a most important commentary on his works. His attitude, destined to remain for many centuries the absolutely dominant one, has by himself been described when he said:

"Aristotle initiated and perfected all the sciences, no writer before him being worthy of mention, no writer after him having in the course of fifteen centuries added anything of importance, or detected any serious errors." (Quoted in Lewes' *History of Philosophy.*)

But it would seem in the light of modern research, that, whilst Aristotelianism reigned supreme, the other systems of Greek philosophy, such as Platonism, were never completely neglected, and that even the Atomistic tenets of Democritus and Epicurus formed the basis of the philosophy professed by an important sect, that of the Motekallemîn.

For the chemist it is of most importance to realise that it was the Arabs who not only contributed enormously to the increase of chemical knowledge, but who also made the first systematic effort to explain the observed diversities in matter by the nature of the constituents of which it is composed, and who modified the Aristotelian doctrine of the four elements so as to make it available for this purpose.

For many centuries after, the development of chemistry was but an incident in the history of alchemy; Paracelsus made it the servant of medicine; and it was only towards the end of the seventeenth century that its claims to independent existence became recognised. The alchemical endeavours to transmute base metals into gold are quite compatible with the fundamental idea of Aristotelianism, that the same primordial matter carrying different qualities which could be added to it and subtracted from it constitutes the different substances we know.

Geber[1], the great eighth century Arabian chemist, who added much to the world's knowledge of chemical facts, and who was a prolific writer (*Summa Perfectionis Magisterii*; *De Investigatione Perfectionis Metallorum*; *De Investigatione Veritatis*, etc.), developed the peripatetic conception of the ultimate constitution of matter, so as to account for the observed differences between the various metals and to supply a theoretical basis for the possibility of their transmutation. The four elements of Aristotle are retained as the ultimate constituents, but the substances termed *mercury* and *sulphur* respectively are assumed as the more proximate ones; mercury being the vehicle of the qualities of ductility, fusibility and lustre, and sulphur the bearer of the quality of combustibility. The conception is, that the metal exhibits the sum of the properties of its constituents, and that the differences between individual metals are due to the relative quantities of these constituents and to the degree of purity exhibited by them. This *mercury* and *sulphur* were not supposed to be identical with the substances bearing these names. According to

The contribution of the Arabs to the development of chemistry.

Geber introduces the "principles" of mercury and sulphur to account for the observed differences between various metals.

---

[1] Kopp, *Geschichte der Chemie in der neueren Zeit* (p. 10).

this view the change of one metal into another should be possible and should consist in the addition or withdrawal of one of the two constituents, or in its purification.   It is interesting to note how enormous an advance Geber's theorising is on that of Aristotle. His hypothesis allows of deductive application in that it explains satisfactorily a whole number of phenomena exhibited by metals, *e.g.* their different combustibility, by a difference in the relative amount of sulphur contained in them.   Geber's doctrine was for many centuries retained in its original form and afterwards extended, first by making sulphur and mercury the proximate constituents of all substances and not of metals only, and later by the

A third " principle " salt is added to Geber's mercury and sulphur.

introduction of a third constituent principle or element, *salt*, representative of that which is permanent and unaltered by the action of heat.   Basil Valentine (second half of the fifteenth century) was the

Paracelsus the exponent of the "three principles" theory of matter.

first to definitely teach it, and the doctrine of the *Three Principles* became for about two centuries dominant in the science.   It was accepted in its entirety by Paracelsus (1493–1541) and promulgated

by him so vigorously as to become associated with his name.   Whether these three principles were accepted by chemists instead of the Aristotelian elements or alongside with these, is a question capable of being answered in any way one pleases.   Elaborate ambiguity, direct contradictions, and studied vagueness are the characteristics of the alchemical style, and hence it would be hopeless to attempt to "prove" anything by quoting from these writings, but the following is an exposition of one view of the three principles and of their relation to the Aristotelian elements:

"Know that all the seven Metalls are brought forth after this manner, out of a threefold matter, viz. Mercury, Sulphur, and Salt, yet in distinct and

An alchemical exposition of the relation between "the four elements" and the "three principles."

peculiar colours....Now this is not to bee understood so that of every Mercury, every Sulphur, or of every Salt, the 7 Metalls may be generated....concerning the generation of mineralls, and halfe metalls, nothing else need bee known than what was at first said concerning metals, viz. that they are in like manner produced of the three Principles, viz. Mercury, Sulphur, and Salt...and yet with their distinct colours. The

Generation of Gemmes is from the subtilty of the Earth of transparent and crystalline Mercury, Sulphur and Salt, even according to their distinct colours.   But the generation of common stones is of the subtilty of Water, of mucilagenous Mercury, Sulphur, and Salt." (Paracelsus, *Of the Nature of Things*, 1537.)

"...very few have hitherto showed whence the Principles arise, and it is a hard thing to judge of any of the principles, or anything else, whose originall and generation is unknowne.... Now the Principles of things, especially of Metalls, according to the ancient Philosophers are two, Sulphur and Mercury; but according to the latter Philosophers, three: Salt, Sulphur and Mercury. Now the originall of these Principles are the foure elements....There are foure Elements and every one of these foure hath in its center another element by which it is elementated: and these are the four statues of the world, separated from the chaos in the creation of the world by divine wisdom; and these uphold the fabric of the world by their contrary acting in equality, and proportion, and also by the inclination of celestiall virtues, bring forth all things, that are within, and upon the earth....We will now descend unto the Principles of things....But how they are produced of the foure elements, take it thus....The Fire began to act upon the Aire, and produced Sulphur, the Aire also began to act upon the Water, and brought forth Mercury, the Water also began to act upon the Earth and brought forth Salt. But the Earth since it had nothing to work upon, brought forth nothing, but that which was brought forth continued and abided in it. Wherefore there became only three Principles, and the Earth was made the Nurse and Mother of the rest. These three things are in all things, and without them there is nothing in the world, or ever shall bee naturally....These three Principles are altogether necessary because they are the near matter....The remote are the four elements out of which God alone is able to create things. Leave therefore the Elements, because out of them thou shalt do nothing, neither canst thou out of them produce anything but these three Principles, seeing Nature herselfe can produce nothing else out of them. If therefore thou canst out of the Elements produce nothing but these three Principles, wherefore then is that vaine labour of thine to seeke after, or to endeavour to make that which Nature hath already made to thy hands? Is it not better to goe three mile than four? Let it suffice thee then to have three Principles, out of which Nature doth produce all things in the earth, and upon the earth; which three we find to be entirely in everything. By the due separation, and conjunction of these Nature produceth as well Metalls, as Stones in the Minerall kingdome; but in the Vegetable Kingdome Trees, Herbs, and all such things; also in the Animal Kingdome the Body, Spirit, and Soule, which especially doth resemble the work of the Philosophers. The Body is Earth, the Spirit is Water, the Soule is Fire, or the Sulphur of Gold." (Michael Svandivogius, *A New Light of Alchymia*, 1607.)

Instances of rebellion against the Aristotelian system became common towards the end of the sixteenth century. Van Helmont (1577–1644) opposed it from the point of view of the chemist:

The beginning of the rebellion against the Aristotelian and the Paracelsian systems.

Van Helmont.

"...Van Helmont declared emphatically against the view that the three principles designated as Sulphur, Mercury, and Salt, are the fundamental constituents of all substances.

17—2

He pointed out that the action of heat, which according to the Paracelsian doctrine should make these principles evident, by no means always separates substances into their simple constituents, but often gives rise to new combinations.... He drew attention to the fact that according to this doctrine a great diversity of substances were comprised under the same name and looked upon as representing the same Principle; and that this diversity and variability of each of the three Principles was incompatible with the conception of 'fundamental constituent.' But he also declared with equal decision against the doctrine of Aristotle in the form in which at that time it found many supporters, namely that in the different substances there are contained four simplest constituents designated as fire, water, air, and earth. He attacked the view that fire is material and that it could in its own nature enter into the composition of substances; he denied that the substance designated as earth could be looked upon as an element." (Kopp, *Geschichte der Chemie in der neueren Zeit.*)

But the attack was being pressed by stronger hands: Bacon, Descartes, Gassendi, Boyle, each of them in his own way had a share in the work.

Francis Bacon (1561–1626) stands out as the champion, some would say the founder of the Inductive Method; as a most bitter

Bacon.

and uncompromising opponent of the Aristotelian Method, such at any rate as he found it practised in his day. The following may be added to the illustrations already given of Bacon's criticism of the principles and the practice of the Peripatetics:

"... It is...the peculiar manner and discipline of Aristotle and his school to teach men what to say, not what to think."

Criticism of
the Peripatetic
Philosophy:
(i) as followed
by Aristotle,
(ii) as applied
by Bacon's
contempora-
ries.

"In the Physics of Aristotle, you hear hardly anything but the words of logic....Nor let any weight be given to the fact that in his books on animals and his problems, and other of his treatises, there is frequent dealing with experiments. For he had come to his conclusion before; he did not consult experience, as he should have done, in order to the framing of his decisions and axioms; but having first determined the question according to his will, he then resorts to experience, and bending her into conformity with his placets leads her about like a captive in a procession; so that even on this count he is more guilty than his modern followers, the schoolmen, who have abandoned experience altogether." (*Novum Organum.*)

And the results of the supreme reign of this system led to a condition of Natural Philosophy described by Bacon thus:

"...The information of the sense itself, sometimes failing, sometimes false; observation careless, irregular, and led by chance; tradition, vain and fed on rumour; practice, slavishly bent upon its work; experiment, blind, stupid, vague, and prematurely broken off; lastly, natural history trivial and poor;—all these have contributed to supply the understanding with very bad materials for philosophy and the sciences. Then an attempt is made to mend the matter by a preposterous subtlety and winnowing of argument. But this comes too late, the case being already past remedy." (*The Great Instauration.*)

What then is it that Bacon attempts to put in the place of the system so justly and so vehemently attacked by him ? Strictly speaking, we are here concerned only with his views on the ultimate constitution of matter and with the difference between his tenets on this subject and those current up till then. But his share in the overthrow of the *four element* theory lay in his attack, not so much on the tenets as on the methods of the Peripatetics, and in the passionate advocacy of a different method propounded by him in language most beautiful and stately. Beside this the production of results by the application of his own method is of as little importance as is his partial acceptance of the Democritean atomistic doctrine, by him admitted because emanating from a sect to which Aristotle was opposed. Hence an account of the salient features of his method, as short as it is possible to make it, must precede the exposition of his views concerning the nature of matter. Bacon deplores the want of just appreciation of natural philosophy and indicates what to him seems the ultimate aim of the sciences :

"Natural philosophy, even among those who have attended to it, has scarcely ever possessed, especially in these later times, a disengaged and whole man,...but...has been made merely a passage and bridge to something else. And so this great mother of the sciences has with strange indignity been degraded to the offices of a servant; having to attend on the business of medicine or mathematics, and likewise to wash and imbue youthful and unripe wits with a sort of first dye, in order that they may be the fitter to receive another afterwards."

The Baconian Method: (1) a plea for the just appreciation of Natural Philosophy.

"Now the true and lawful goal of the sciences is none other than this: that human life be endowed with new discoveries and powers." (*Novum Organum.*)

And the method which should be employed to attain this object he describes in the *Novum Organum* thus :

"Now my method, though hard to practise, is easy to explain ; and it is this. I propose to establish progressive stages of certainty. The evidence of the sense, helped and guarded by a certain process of correction, I retain. But the mental operation which follows the act of sense I for the most part reject; and instead of it I open and lay out a new and certain path for the mind to proceed in, starting directly from the simple sensuous perception."

(2) Proposes to establish successive steps of certainty.

"There are and can be only two ways of searching into and discovering truth. The one flies from the senses and particulars to the most general axioms, and from these principles, the truth of which it takes for settled and immoveable, proceeds to judgment and to the discovery of middle axioms. And this way is now in fashion. The other derives axioms from the senses and particulars, rising by a gradual and unbroken ascent, so that it arrives at the most general axioms last of all. This is the true way, but as yet untried. ...Both ways set out from the senses and particulars, and rest in the highest generalities; but the difference between them is infinite. For the one just glances at experiment and particulars in passing, the other dwells duly and orderly among them. The one, again, begins at once by establishing certain abstract and useless generalities, the other rises by gradual steps to that which is prior and better known in the order of nature."

The evidence of the senses forms the great starting point:

"The information of the sense itself I sift and examine in many ways. For certain it is that the senses deceive; but then at the same time they supply the means of discovering their own errors....I have sought on all sides diligently and faithfully to provide helps for the sense-substitutes to supply its failures, rectifications to correct its errors; and this I endeavour to accomplish not so much by instruments as by experiments...such experiments, I mean, as are skilfully and artificially devised for the express purpose of determining the point in question. To the immediate and proper perception of the sense therefore I do not give much weight; but I contrive that the office of the sense shall be only to judge of the experiment, and that the experiment itself shall judge of the thing. And thus I conceive that I perform the office of a true priest of the sense...and a not unskilful interpreter of its oracles; and that while others only profess to uphold and cultivate the sense, I do so in fact....Moreover, whenever I come to a new experiment of any subtlety...I subjoin a clear account of the manner in which I made it; that men knowing exactly how each point was made out, may see whether there be any error connected with it, and may arouse themselves to devise proofs more trustworthy and exquisite, if such can be found."

(i) The evidence of the senses is the starting point but the information thereby gained must be checked by means of crucial experiments.

The next step is the framing of axioms:

"In establishing axioms, another form of induction must be devised than has hitherto been employed; and it must be used for proving and discovering

not first principles...only, but also the lesser axioms, and the middle, and
indeed all....The induction which is to be available for the

(ii) The framing of axioms follows next. discovery and demonstration of sciences and arts, must analyse nature by proper rejections and exclusions; and then, after a sufficient number of negatives, come to a conclusion on the affirmative instances."

The necessity of verification is recognised, as also the nature
and value of deduction to which a clear and definite position and
function is assigned:

(iii) Verification and deduction. "But in establishing axioms by this kind of induction, we must also examine and try whether the axiom so established be framed to the measure of those particulars only
from which it is derived, or whether it be larger and wider. And if it be
larger and wider, we must observe whether by indicating to us new particulars it confirm that wideness and largeness as by a collateral security."

"The true method of experience...[commences] with experience duly
ordered and digested, not bungling or erratic, and from it educing axioms,
and from established axioms again new experiments;...and...leads by an unbroken route through the woods of experience to the open ground of axioms."

"Now my directions for the interpretation of nature embrace two generic
divisions; the one how to educe and form axioms from experience; the other
how to deduce and derive new experiments from axioms. The former again
is divided into three ministrations: a ministration to the sense, a ministration to the memory, and a ministration to the mind or reason."

Such then was the tool offered by Bacon, nay, pressed by him
on his generation. In his own hands it produced results, in which
it is difficult to find matter for praise, very easy to find opportunity for much adverse criticism. But however arrived at, however maintained, his views of the fundamental properties of matter
present points of distinct interest. He was naturally attracted by
the atomistic doctrine of Democritus, of whom he speaks as being
of all the Greek philosophers the one who had the deepest insight
into nature, and whose doctrine he describes as a glimpse of truth
such as can be obtained by the intellect left to its own natural
impulses and not ascending by successive and connected steps:

"The doctrine of atoms from its going a step beyond the period at which
it was advanced was ridiculed by the vulgar, and severely handled in the
disputations of the learned, notwithstanding the profound

Bacon's appreciation of the Democritean Philosophy. acquaintance with physical science by which its author was allowed to be distinguished and from which he acquired the character of a magician....However, neither the hostility of Aristotle, with all his skill and vigour in disputation, nor
the majestic and lofty authority of Plato, could effect the subversion of

the doctrine of Democritus. And while the opinions of Plato and Aristotle were rehearsed with loud declamation and professional pomp in the schools, this of Democritus was always held in high honour by those of a deeper wisdom, who followed in silence a severer path of contemplation. In the days of Roman speculation it keeps its ground and its favour....The destruction of this philosophy was not effected by Aristotle and Plato, but by Genseric and Attila and their barbarians. For then when human knowledge had suffered shipwreck, these fragments of the Aristotelian and Platonic philosophy floated on the surface like things of some lighter and emptier sort and so were preserved; whilst more solid matters went to the bottom and were almost lost in oblivion. But to me the philosophy of Democritus seems worthy to be rescued from neglect." (*On Principles and Origins.*)

But whilst thus favourably disposed towards it, Bacon does not accept the atomistic doctrine in its entirety. On many points he differed from it fundamentally. His views may be summarised as follows :

1. The principle of the conservation of mass is· emphatically asserted.

Bacon's own views concerning the ultimate constitution of matter.

(1) The principle of the conservation of mass is asserted.

"There is nothing more true in nature than the twin propositions that *nothing is produced from nothing*, and *nothing is reduced to nothing*, but that the absolute quantum or sum total of matter remains unchanged, without increase or diminution." (*Novum Organum.*)

"In no transmutation of bodies is there any reduction either from nothing or to nothing."

"One axiom has been rightly received, namely that nothing is taken from or added to the sum of the Universe." (*History of Dense and Rare.*)

This axiom he then proceeds to apply deductively in conjunction with his experimental work on the densities of various substances, and thereby arrives at the highly interesting result that the four element theory must be regarded as improbable :

"The opinion that all sublunary bodies are composed of the four elements is ill borne out. For the cube of gold weighed 20 pennyweights; the common earth only a little more than 2; water 1 pennyweight 3 grains; air and fire are far more rarefied...the question is, therefore, how is it possible from a body of 2 pennyweights together with others far more rarefied, to educe by form a body which in an equal dimension weighs 20 pennyweights?"

And after considering and rejecting possible modes of explanation he says :

"It would be better therefore that they [the Peripatetics] should give up trifling, and that the dictatorship should cease." (*Ibid.*)

The transmutation of metals into gold, which is the densest metal, would involve a change of a certain weight of matter into an equal weight of gold occupying a much smaller volume; because of the inevitable great contraction resulting he considers the possibility of such a change as very doubtful.

"The manufacture of gold, or the transmutation of metals into gold, is to be much doubted of. For of all bodies gold is the heaviest and densest, and therefore to turn anything else into gold there must needs be condensation.... But the conversion of quicksilver or lead into silver (which is rarer than either of them) is a thing to be hoped for." (*Ibid.*)

The above may be considered two distinctly interesting instances of the application of a physical method to the solution of a chemical problem.

2. Matter is supposed not to be infinitely divisible, but made up of discrete ultimate particles.

(2) A corpuscular theory, different from the Democritean, is propounded.

Whilst inclining to the Democritean conception of atoms, his approval of this philosophy is not constant:

"The doctrine of Democritus concerning atoms is either true or useful for demonstration. For it is not easy either to grasp in thought or to express in words the genuine subtlety of nature, such as it is found in things, without supposing an atom. ...Atom...is taken for the last term or smallest portion of the division or fraction of bodies, or else for a body without vacuity." (*Thoughts on the Nature of Things.*)

"Men cease not from abstracting nature till they come to potential and uninformed matter, nor on the other hand from dissecting nature till they reach the atom; things which, even if true, can do but little for the welfare of mankind." (*Novum Organum.*)

But without being able to quite see what precise meaning he himself attached to the conception, we have Bacon's own statement for it that his theory of the ultimate constitution of bodies does not relate to atoms properly so called, but only to actually existing ultimate particles:

"Nor shall we be led to the doctrine of atoms, which implies the hypothesis of a vacuum and that of the unchangeableness of matter (both false assumptions); we shall be led only to real particles, such as really exist." (*Ibid.*)

3. The notion of a vacuum is rejected.

"There is no vacuum in nature, either collected or inter-
spersed. Within the bounds of dense and rare there is a
fold of matter, by which it folds and unfolds itself without
creating a vacuum." (*History of Dense and Rare.*)

(3) The denial of the existence of a vacuum.

4. Only one kind of primitive matter is assumed which can
change into the different kinds that we perceive:

(4) Assumption of one kind of primitive matter capable of transmutation.

"There are two opinions, nor can there be more, with
respect to atoms or the seeds of things; the one that of
Democritus, which attributed to atoms inequality and configuration, and by configuration position; the other perhaps
that of Pythagoras, which asserted that they were altogether
equal and similar. For he who assigns equality to atoms
necessarily places all things in numbers; but he who allows other attributes
has the benefit of the primitive natures of separate atoms, besides the
numbers or proportions of their conjunctions." (*Thoughts on the Nature of
Things.*)

He does not associate himself with this view of Democritus,
and his own is only one of the instances of the inherent desire for
simplicity, which leads to the ever-recurring attempt to represent
all the different elements as the condensation product of one kind
of fundamental matter:

"There is no doubt that the seeds of things, though equal, as soon as
they have thrown themselves into certain groups and knots, completely
assume the nature of dissimilar bodies, till those groups or knots are dissolved." (*Ibid.*)

No purpose would be served by giving here instances to show
how Bacon violated his own principles, how he underrated the
truly scientific work of his contemporary Gilbert, how he appro-
priated the work of others[1]. Admitting weaknesses and errors,
the fact remains that Bacon affirmed—or reaffirmed—in terms
not to be surpassed for clearness, for vigour and for beauty the
principles of the method by which all knowledge should be collected and worked up, a method which except in cases isolated
and not appreciated, was not the one followed then; and that
thereby he stands foremost in the ranks of those with whose
names we must associate the beginning of a new era.

René Descartes (Latinised Cartesius, 1596–1650), the founder
of the Cartesian Philosophy, a Frenchman by birth, educated by
the Jesuits; who in his youth for some time followed
the profession of a soldier, later on settled in Holland,

Descartes.

[1] *J. von Liebig, Über Francis Bacon und die Methode der Naturforschung.*

and finally attracted by Queen Christina went to Sweden where he died: is another of the great pioneers in the renaissance of philosophy and science. Like Bacon he is an opponent of the method then current[1]:

"Their [the present followers of Aristotle's] fashion of philosophising, however, is well suited to persons whose abilities fall below mediocrity ; for the obscurity of the distinctions and principles of which they make use enables them to speak of all things with as much confidence as if they really knew them, and to defend all that they say on any subject against the most subtle and skilful, without its being possible...to convict them of error....In philosophy, when we have true principles, we cannot fail by following them to meet sometimes with other truths, and we could not better prove the falsity of those of Aristotle, than by saying that men made no progress in knowledge by their means during the many ages they prosecuted them."

Descartes also propounded a new method, but one fundamentally different from that of Bacon. The great English philosopher inveighs against reference to first principles, against assigning too much time to mathematics:

The Cartesian and the Baconian methods compared.

"Nor...is it a less evil that in...philosophies...and contemplation...labour is spent in investigating and handling the first principles of things and the highest generalities of nature ; whereas utility and the means of working result entirely from things intermediate....Mathematics...ought only to give definiteness to natural philosophy, not to generate or give it birth." (Bacon, *Novum Organum.*)

Deduction versus induction; different value assigned to mathematics.

The Cartesian method on the other hand consists in seeking a few clear first principles, whose validity is beyond a doubt, and in applying these deductively to the explanation of everything occurring and observed. Experiments are used not as a beginning but as " crucial experiments " for deciding between several possible explanations of the same phenomenon, all of these explanations having been deduced from first principles. And the possibility of mathematical demonstration is the ultimate criterion of the validity of any inference:

"We must commence with the investigations of those first causes which are called *Principles.* Now these principles must possess *two conditions*: in the first place, they must be so clear and evident that the human mind, when it attentively considers them, cannot doubt of their truth ; in the second place, the knowledge of other things must be so dependent on them as that though the principles themselves may indeed be known apart from what

1 The quotations about to be given are from Descartes' works : *Discourse on Method* (Veitch's translation); *The Principles of Philosophy.*

depends on them, the latter cannot nevertheless be known apart from the former. It will accordingly be necessary thereafter to endeavour so to deduce from those principles the knowledge of the things that depend on them, as that there may be nothing in the whole series of deductions which is not perfectly manifest."

"I have adopted the following order: first, I have essayed to find in general the principles, or first causes of all that is or can be in the world... without educing them from any other source than from certain germs of truth naturally existing in our minds. In the second place, I examined what were the first and most ordinary effects that could be deduced from these causes; and it appears to me that in this way I have found heavens, stars, an earth, and even on the earth, water, air, fire, minerals."

And concerning the place of experiments and the function of crucial experiments:

"I remarked, moreover, with respect to experiments, that they become always more necessary the more one is advanced in knowledge; for, at the commencement, it is better to make use only of what is spontaneously presented to our senses....Turning over in my mind all the objects that had ever been presented to my senses, I freely venture to state that I have never observed any which I could not satisfactorily explain by the principles I had discovered. But it is necessary also to confess that the power of nature is so ample and vast, and these principles so simple and general, that I have hardly observed a single particular effect which I cannot at once recognise as capable of being deduced in many different modes from the principles, and that my greatest difficulty usually is to discover in which of these modes the effect is dependent upon them; for out of this difficulty I cannot otherwise extricate myself than by again seeking certain experiments, which may be such that their result is not the same, if it is in the one of these modes that we must explain it, as it would be if it were to be explained in the other."

The value assigned to mathematics in the Cartesian method is a high one:

"I will accept nothing as true but what is deduced (from the first principle that matter can be divided, figured and moved in all sorts of ways) by direct evidence which can take rank of a mathematical demonstration."

The application of his method led Descartes to results concerning the ultimate constitution of matter and the evolution of the material universe which are of historical interest only; but the system bearing his name is possessed of such completeness and originality that it should not be passed over, in spite of the fact that detaching a few passages dealing with what is of primary interest to the chemist cannot result in a satisfactory, harmonious representation. The Natural

Descartes' views concerning the ultimate constitution of matter. The one essential attribute of matter is extension.

Philosophy of Descartes is expounded in its main outlines in his work entitled "The Principles of Philosophy" (1644). *The one essential attribute of matter is assumed to be extension :*

"The nature of matter or body, considered in general, does not consist in its being hard, or ponderous, or coloured, or that which affects our senses in any other way, but simply in its being a substance extended in length, breadth, and depth.... The whole of corporeal substance is extended without limit,... the earth and heavens are made of the same matter ; and...although there were an infinity of worlds, they would all be composed of this matter ;... because...we cannot find in ourselves the idea of any other matter (than that, whose nature consists only in its being an extended substance)."

From this first principle, that extension is the only essential characteristic of matter, are deduced the further attributes and qualities in virtue of which the different substances exhibit the properties whereby we recognise and distinguish them. The most important points investigated and the chief results arrived at by Descartes may be summarised as follows :

1. The existence of void is denied.

1. The existence of void is denied.

"A vacuum or space in which there is absolutely no body is repugnant to reason...For since from this alone, that a body has extension in length, breadth, and depth, we have reason to conclude that it is a substance, it being absolutely contradictory that nothing should possess extension, we ought to form a similar inference regarding the space which is supposed void, *viz.*, that since there is extension in it there is necessarily also substance."

And pointing out the danger there is in assuming that space is empty, because it contains no matter that we can perceive ;

"If...we...suppose that in the space we called a vacuum, there is not only no sensible object, but no object at all, we will fall into the same error as if, because a pitcher in which there is nothing but air, is, in common speech, said to be empty, we were therefore to judge that the air contained in it is not a substance."

2. There is no limit to the divisibility of matter.

2. Matter is infinitely divisible.

"There cannot exist any atoms or parts of matter that are of their own nature indivisible. For however small we suppose these parts to be, yet, because they are necessarily extended, we are always able in thought to divide any one of them into two or more smaller parts, and may accordingly admit their divisibility. For there is nothing we can divide in thought which we do not thereby recognise to be divisible."

Moreover not only the possibility but the actual occurrence of division without limit follows of necessity in a system in which the existence of motion has to be reconciled with the absence of vacuum :

"After what has been demonstrated above, namely, that all places in space are full of matter, whose every part is so proportioned to the size of the space it occupies that it is impossible for it to fill a larger or to be contained in a smaller, or that another body should simultaneously find room in it, we are obliged to conclude that there must always be a circle of matter, or a ring of substance moving together at the same time, such that when a substance leaves its place to one which drives it away it takes that of another, and the other takes that of another again, and so on till the last, which at the same instant occupies the place vacated by the first."

Such motion may occur along perfect circles, when its conception is said to present no difficulties; or along any irregular path, provided this returns into itself. The essential requisite of all such actual motion is unlimited division of the matter, or at any rate of certain portions of it, so as to adapt its shape to any of the demands made upon it.

"It follows that matter divides itself into an indefinite and infinite number of parts, and we dare not doubt the actual occurrence of such division, though we cannot grasp it."

3.   Motion is considered the cause of the varieties exhibited by matter.

3. The differences observed in various kinds of matter are due to differences in motion.

"There is but one kind of matter in the universe, and this we know only by its being extended. All the properties we distinctly perceive to belong to it may be reduced to the capacity of being divided and moved according to its parts ; and thus it is capable of all those affections which we perceive can arise from the motion of its parts...all diversity of form depends on motion."

4.   The nature and cause of motion are investigated ; its total quantity recognised as constant ; and its laws formulated.

4. The Laws of Motion.

Descartes defines motion in the ordinary sense of the term to be the action by which a body passes from one place to another. It has been shown already how, in the absence of void, motion must always involve a simultaneous circular displacement and the adaptation through infinite divisibility of the moving matter to all possible shapes. The origin of motion is put down to a creative act of God, who also maintains its total quantity constant and equal to that originally created. Three laws of motion essentially the same as those of Newton are deduced and then applied to the explanation of the existence of substances hard and liquid respectively. It is beyond the scope of this chapter to show how these laws were arrived at and how applied.

Turning to Descartes' celebrated conception of vortex motion, whereby he explained the formation of the heavenly bodies as well as of everything on the earth, we find this to be intimately connected with his view of the shape of the constituent parts of matter and with that of the variety in the kinds of such ultimate particles.

5. Vortex motion is described and is made the basis of the Cartesian explanation of the phenomena of the universe.

5. Vortex motion.

"Let us then assume that the matter of the universe in which the planets are placed rotates incessantly in the manner of a whirl or vortex, in the centre of which is the sun, and that the parts nearer to the sun move faster than those further from it, and moreover that all the planets, amongst which we will count the earth, always remain 'suspended betwixt the same portions of the material of the skies, because thereby and without using any other means we shall easily explain all the things met with in them. For as in the bends of rivers where the water turning back on itself and eddying produces circles, if some straws or other light bodies are floating amidst this water, it may be seen that the stream draws them in and makes them circle round with it, and even among such straws it may be noticed that there are often some which also rotate about their own centre; and that those nearer to the centre of the whirl which contains them accomplish their revolution sooner than those further from it; and finally, though these water-whirls always incline to a circular motion, they seldom describe a perfect circle but a path extending sometimes more in length, sometimes more in breadth, such that all its points are not equally distant from the centre. One may easily imagine all these same things to happen to the planets; and nothing further would be needed for explaining all their phenomena."

6. The visible universe is assumed to be made up of three chief kinds of matter.

6. The assumption of three kinds of matter.

A vacuum being an impossibility, the whole universe must be filled with matter, which according to the simplest and hence most probable hypothesis, originally was divided into parts all the same and endowed with rotatory motion. In consequence of this motion the parts grinding against each other rubbed off their corners until they became spherical. The finer matter thus rubbed off serves to fill the interstices between these spheres and constitutes a second kind of matter, whilst the third is formed of parts more coarse and less fitted for motion than even the first. Luminous bodies as the sun and fixed stars, then the transparent sky, and finally the opaque earth and planets: these three kinds of matter give rise to three types of cosmic constituents.

" If we assume that all the matter of which the world is made up, had at the beginning been divided into many equal parts, then these could not at first have been all round, because spheres joined cannot constitute an entirely solid and continuous body such as this universe, in which, as I have shown above, void cannot exist. But whatever the original shape of these parts may have been, they must in time have become round because they are endowed with rotatory motions of different kinds, and so by-and-by as they collided their corners were ground away."

" But inasmuch as empty space can exist nowhere in the universe, and as these round particles of matter cannot join so intimately as not to leave several small interstices, it follows that these must be filled up by some other portion of this matter, which must be extremely finely divided, so as to change its configuration at all moments and adapt itself to that of the spaces into which it passes. Hence we must assume that what comes off from the corners of the particles of matter, as these gradually get rounded off by rubbing against each other, is so subtle, and acquires so great a velocity that the violence of its motion can shatter it into an infinite number of parts, which, being of no fixed size or shape, readily fill up all the small interstices into which the other portions of matter cannot penetrate."

" And we shall meet with a third kind in certain portions of matter whose particles, owing to their size and their form cannot be moved as readily as the preceding ones ; and I shall try to show that all the bodies in this visible universe are composed of these three kinds of matter as of three different elements, namely : the sun and the fixed stars have the form of the first of these elements ; the skies that of the second ; and the earth, the planets, and the comets that of the third."

The Cartesian system, whilst agreeing in some points with the atomistic doctrine, fundamentally differs from it in others :

The relations between the Cartesian and the atomistic theories. " I admit the existence of particles so small as to be perceived by none of our senses.... But it may be said that Democritus also assumed the existence of small particles of different figures, sizes, and motions, from the varied combinations of which all sensible bodies arose ; and that nevertheless his philosophy is commonly rejected. To which I reply that the philosophy of Democritus was never rejected by anyone because he admitted the existence of bodies smaller than those we can perceive and attributed to them diverse sizes, diverse figures, and diverse motions,...but because, in the first place, he supposed that the corpuscles were indivisible, on which ground I also reject it; because, in the second place, he imagined that there was a vacuum about them, which I show to be impossible; because, thirdly, he attributed gravity to these bodies, of which I deny the existence in any body, in so far as a body is considered by itself, because it is a quality that depends on the relations of situation and motion which several bodies bear to each other."

Descartes' own summary of his views concerning the ultimate constitution of matter seems to form a fit conclusion to the exposition here given of its main features:

Descartes' summary of his philosophy of matter.   "Though I have endeavoured to give an explanation of the whole nature of material things...I have merely considered the figure, motion, and size of bodies, and examined what must follow from their mutual concourse on the principles of mechanics."

Pierre Gassendi (1592—1655), born at Digne in Provence, in later life Provost of that town, for some time Professor of Mathematics at the Collège Royal in Paris, is generally **Gassendi expounds Epicurean Natural Philosophy, and recognises the importance of experiment.** accredited with the merit of having made known to his contemporaries the Atomistic doctrine of the ancients, till then more or less lost in oblivion. A violent opponent of Aristotelianism, and also differing from his great contemporary Descartes, he found himself most in harmony with the philosophy of Epicurus. He urged the importance and necessity of experimental research. His dictum preserved by Descartes, *there is nothing in the intellect which has not been in the senses*, embodies his attitude in this respect; and it would appear that, unlike other philosophers of his time, he put his precept into practice. We are told by himself how he helped to overthrow certain arguments used by those who denied the motion of the earth. The favourite among such arguments was that:

"...if the earth revolved, it would be impossible for a cannon-ball fired straight up into the air to fall back upon the cannon. Gassendi...had an experiment made : On a ship travelling with great speed a stone was thrown straight up into the air. It fell back, following the motion of the ship, upon the same part of the deck from which it had been thrown. A stone was dropped from the top of the mast, and it fell exactly at its foot." (Lange, *History of Materialism.*)

Gassendi added nothing really original to the stock of human

F.   18

knowledge; his doctrine of matter is purely that of Epicurus, and is, according to Lange's summary, as follows: The elements consist of atoms; the atoms are therefore the first principles, and constitute fundamental matter. Matter is the durable substratum, but the various forms it may assume arise and pass away. The atoms can neither be created nor destroyed, and are in substance identical but vary in figure. They are indivisible. They have by God been endowed with motion, and whilst among visible bodies one is always put into motion by another, the atoms are endowed with motion self-inherent. They are the seed of all things, and becoming and passing away is but a combination and a separation of atoms.

Robert Boyle (1627—1691) received a careful education, and after travel on the continent settled in Oxford, where he devoted

Robert Boyle.

himself to scientific research. He was one of the founders of the Royal Society of London (1663), and for some time filled the office of president. His published works are many, and cover nearly all provinces of physics and chemistry; there is nothing in the literature of these sciences to surpass, or even to equal them. Whilst in the grasp of the subject dealt with and in the method employed, we find all the lucidity, the directness, and the scientific penetration of the best work of our own times, the exposition is marked by that mixture of courtesy, grace, and quaintness which is met with in all that is best in the productions of the end of the 17th century. Amongst the writings of Boyle, those most important for the purposes of this chapter are: "The Sceptical Chymist" (1661), "The Usefulness of Natural Philosophy" (1663), "The Usefulness of Experimental Knowledge" (1671). It is in these works that we meet with (I) his refutation, based on appeal to experiment, of both the Peripatetic and the Paracelsian theory of "Element," and the substitution of that admirably clear and definite and entirely empirical conception which we still hold; and (II) the exposition, given with a certain hesitation, of theoretical views concerning the ultimate constitution of these elements and the fundamental processes resulting in chemical change.

His denunciation of the method of Aristotelianism is in sentiment and form strangely like that of Bacon and Descartes:

"I am not a little pleased to find that you are resolved, on this occasion, to insist rather on experiments than syllogisms...for...those dialectical subtleties that the schoolmen too often employ...are wont much more to declare the wit of him, that uses them, than increase the knowledge or remove the doubts of sober lovers of truth. And such captious subtleties do indeed often puzzle, and sometimes silence men, but rarely satisfy them; being like the tricks of jugglers, whereby men doubt not but they are cheated, though oftentimes they cannot declare by what slights they are imposed on." (*The Sceptical Chymist.*)

(I) Criticism of the Peripatetic and Paracelsian philosophy, whose tenets are incompatible with experiment.

And what could be more condemnatory than the sentiments which Boyle puts into the mouth of the supporter of that philosophy:

"It is much more high and philosophical to discover things *à priori* than *à posteriori*. And therefore the Peripatetics have not been very solicitous to gather experiments to prove their doctrines, contenting themselves with a few only, to satisfy those that are not capable of a nobler conviction. And indeed they employ experiments rather to illustrate than to demonstrate their doctrines." (*Ibid.*)

And he is no less severe on the studied vagueness and intentional obscurity of the writings of the Paracelsian school:

"I have, in the reading of Paracelsus, and other chymical authors, been troubled to find that such hard words and equivocal expressions, as you justly complain of, do, even when they treat of principles, seem to be studiously affected by those writers; whether to make themselves to be admired by their readers, and their art appear more venerable and mysterious, or (as they would have us think) to conceal from them a knowledge themselves judge inestimable." (*Ibid.*)

His attack on the fundamental tenets of these schools is all based on experimental evidence. He refuses to admit either the *tria prima*, salt, sulphur, and mercury of Paracelsus; or the *four elements* of Aristotle as the universal constituents of all matter. He does so because it can be proved experimentally that the bodies having these names cannot always be produced from all substances, and because in some cases a greater number can be obtained; moreover he denies the claim of these principles to the name of element, since it can be shown experimentally that they may themselves be resolved into simpler things still.

"Since, in the first place, it may justly be doubted, whether or no the fire be, as chymists suppose it, the genuine and universal resolver of mixt bodies; since we may doubt, in the next place, whether or no all the distinct substances, that may be obtained from a mixt body by the fire, were pre-existent there, in the forms, in which they were separated from it; since also, though

18—2

we should grant the substances separable from mixt bodies by the fire, to have been their component ingredients, yet the number of such substances does not appear the same in all mixt bodies; some of them being resoluble into more differing substances than three, and others not being resoluble into so many as three; and since, lastly, those very substances, that are thus separated, are not, for the most part, pure and elementary bodies, but new kinds of mixts; since, I say, these things are so, I hope you will allow me to infer, that the vulgar experiments (I might perchance have added, the arguments too), wont to be alledged by chymists to prove that their three hypostatical principles do adequately compose all mixt bodies, are not so demonstrative, as to induce a wary person to acquiesce in their doctrine, which, till they explain and prove it better, will, by its perplexing darkness, be more apt to puzzle than satisfy considering men, and will to them appear encumbered with no small difficulties."

"In the next place then I consider, that as there are some bodies, which yield not so many as the three principles; so there are many others, that in their resolution exhibit more principles than three; and that therefore the ternary number is not that of the universal and adequate principles of bodies." (*Ibid.*)

It was possible for Boyle to expose the shortcomings and fallacies of the then prevalent idea of *Element* or *Principle*, because he himself had formulated a conception of element such that now, two hundred years later, nothing has been added, nothing taken from it; and its basis being purely empirical, it will no doubt adapt itself to the requirements of the further growth of the science.

"I...mean by elements, as those chymists, that speak plainest, do by their principles, certain primitive and simple, or perfectly unmingled bodies; which not being made of any other bodies, or of one another, are the ingredients of which all those called perfectly mixt bodies are immediately compounded, and into which they are ulti- mately resolved.... I need not be so absurd, as to deny, that there are such bodies as earth and water, and quicksilver and sulphur : but I look upon earth and water, as component parts of the universe, or rather of the terres- trial globe, not of all mixt bodies." (*Ibid.*)

Definition of "element."

But what is the ultimate constitution of these elements them- selves? Is it corpuscular or continuous? Are the differences exhibited by them due to ultimate differences in the constituent matter, or to modifications of one and the same primitive matter? To find answers to these pressing questions, the domain of speculation must be entered. Boyle does so somewhat reluct- antly, and. a different tone is noticeable in his writings when he

(II) Views concerning the ultimate con- stitution of matter.

deals with experimental certainties and theoretical speculations respectively.

1. One kind of primitive matter is assumed:

"I consider, that if it be as true, as it is probable, that compounded bodies differ from one another but in the various textures resulting from the bigness, shape, motion, and contrivance of their small parts, it will not be irrational to conceive, that one and the same parcel of the universal matter may, by various alterations and contextures, be brought to deserve the name, sometimes of a sulphureous, and sometimes of a terrene, or aqueous body." (*Ibid.*)

1. One kind of primitive matter is assumed.

"...if it be granted rational to suppose,...that the elements consisted at first of certain small and primary coalitions of the minute particles of matter into corpuscles very numerous, and very like each other, it will not be absurd to conceive, that such primary clusters may be of far more sorts than three or five; and consequently, that we need not suppose, that in each of the compound bodies we are treating of, there should be found just three sorts of such primitive coalitions, as we are speaking of." (*Ibid.*)

2.. The view above expressed that the properties of the different elements are due to differences in the shape, size, and motion of the constituent particles is met with again and again.

2. Differences in shape, size and motion of the constituent particles.

"There are divers effects in nature, of which though the immediate cause may be plausibly assigned, yet if we further inquire into the causes of those causes, and desist not from ascending in the scale of causes, till we are arrived at the top of it, we shall perhaps find the more catholick and primary causes of things to be either certain, primitive, general, and fixt laws of nature (or rules of action and passion among the parcels of the universal matter), or else the shape, size, motion, and other primary affections of the smallest parts of matter, and of their first coalitions or clusters." (*The Usefulness of Natural Philosophy.*)

"Motion [is] the grand and primary instrument, whereby nature produces all the changes and other qualities, that are to be met with in the world.... The principles of particular bodies might be commodiously enough reduced to two, namely matter, and...the result, or aggregate, or complex of those accidents, which are the motion or rest,...the bigness, figure, texture, and the thence resulting qualities of the small parts." (*The Sceptical Chymist.*)

It is evident therefore that Boyle is an adherent of a corpuscular and dynamical theory of matter, of that in fact on which the Cartesian and Atomistic schools, then dividing the allegiance of the scientific world, agreed.

Boyle is an adherent of a corpuscular theory of matter.

"I considered, that the Atomical and Cartesian hypotheses, though they differed in some material points from one another, yet in opposition to the Peripatetick and other vulgar doctrines they might be looked upon as one philosophy: for they agree with one another, and differ from the schools in this grand and fundamental point, that not only they take care to explicate things intelligibly; but that whereas those other philosophers give only a general and superficial account of the phaenomena of nature...both the Cartesians and the Atomists explicate the same phaenomena by little bodies variously figured and moved. I know, that these two sects of modern Naturalists disagree about the notion of body in general, and consequently about the possibility of a true vacuum ; as also about the origin of motion, the indefinite divisibleness of matter, and some other points of less importance than these.... Both parties agree in deducing all the phaenomena of nature from matter and local motion.... I esteemed that, notwithstanding these things, wherein the Atomists and the Cartesians differed, they might be thought to agree in the main, and their hypotheses might by a person of a reconciling disposition be looked on as, upon the matter, one philosophy. Which because it explicates things by corpuscles, or minute bodies, may (not very unfitly) be called corpuscular." (*Some Specimens of an attempt to make Chymical Experiments useful to illustrate the notions of the Corpuscular Philosophy.*)

Boyle does not definitely range himself with either the Cartesian or the Atomistic school, but seems to have distinctly leaned towards the latter, which he knew well, and of whose fundamental tenets as contained in Lucretius, he gives, in the "Usefulness of Natural Philosophy," an admirably clear and concise synopsis. But whilst evidently quite willing to use the conception of atoms for the explanation of chemical phenomena, he insists on its hypothetical nature and hence on the possibility of explaining the same phenomena by other agencies also. He recognises the limitations of an hypothesis, which whilst in general terms referring the properties of substances to differences in the shape, size and motion of the ultimate particles, makes no attempt to connect definite properties of matter with specific fundamental differences of the constituent atoms. And lastly, he shrinks from committing himself to that pure materialism of the Epicurean doctrine which conceives the fundamental properties of the different atoms to be self-inherent

The Atomistic doctrine is preferred to the Cartesian.

" And here let us further consider, that as confidently as many Atomists, and other Naturalists, presume to know the true and genuine causes of the things they attempt to explicate ; yet very often the utmost they can attain to, in their explications, is, that the explicated phaenomena may be produced after such a manner, as they deliver, but not that they really are so. For as an artificer can set all the wheels of a clock a going, as well with springs as with weights, and may with violence discharge a bullet out of a barrel of a gun, not only by means of gunpowder, but of compressed air, and even of a spring: so the same effects may be produced by divers causes different from one another; and it will oftentimes be very difficult, if not impossible, for our dim reasons to discern surely, which of these several ways, whereby it is possible for nature to produce the same phaenomena, she has really made use of to exhibit them.... And as confident as those we speak of use to be, of knowing the true and adequate causes of things,...some modern philosophers, that much favour the doctrine [of Epicurus], do likewise imitate his example, in pretending to assign not precisely the true, but possible causes of the phaenomenon they endeavour to explain."

"It is one thing to be able to show it possible, for such and such effects to proceed from the various magnitudes, shapes, motions and concretions of atoms ; and another thing to be able to declare what precise, and determinate figures, sizes, and motions of atoms, will suffice to make out the proposed phenomena, without incongruity to any others to be met with in nature."

"Indeed, that the various coalitions of atoms, or at least small particles of matter, might have constituted the world, had not been perhaps a very absurd opinion for a philosopher, if he had, as reason requires, supposed that the great mass of lazy matter was created by God at the beginning, and by him put into a swift and various motion, whereby it was actually divided into small parts of several sizes and figures, whose motion and crossings of each other were so guided by God as to constitute, by their occursions and coalitions, the great inanimate parts of the universe and the principles of animated concretions." (*The Usefulness of Natural Philosophy.*)

Boyle's own conception of the things purely corporeal is more idealistic, and contains what must strike us as an echo of the principle of the guiding intelligence, the νοῦς of Anaxagoras.

"To acquaint you with divers of the conjectures (for I must yet call them no more) I have had concerning the principles of things purely corporeal: for though, because I seem not satisfied with the vulgar doctrines, either of the Peripatetick or Paracelsian schools, many of those, that know me,...have thought me wedded to the Epicurean Hypothesis (as others have mistaken me for a Helmontian).... I should tell you, that I have sometimes thought it not unfit, that to the principles, which may be assigned to things, as the world is now constituted, we should, if we consider the great mass of matter, as it was whilst the universe was in making, add another, which may conveniently enough be called an Architectonick principle or power; by which I mean those various determinations, and that skilful guidance of the motions of the small parts of the universal matter by the most wise Author of things, which were necessary at the beginning to turn that confused chaos into this orderly and beautiful world.... For I confess I cannot well conceive, how from matter, barely put into motion, and then left to itself, there could emerge such curious fabricks, as the bodies of men and perfect animals, and such yet more admirably contrived parcels of matter, as the seeds of living creatures." (*The Sceptical Chymist.*)

It seems clear from Boyle's way of writing on the subject that the atomistic hypothesis was well known to the scientific men of

the day; and that amongst chemists it held the field, side by side
with the Cartesian philosophy. But amongst mathematicians and
physicists adherence to the Cartesian tenets was almost universal.
The atomistic hypothesis scored its greatest victory when it was
accepted by the man who dealt the death-blow to the methods
and the results of the cosmogony based upon vortex motion. That
this acceptance was a somewhat cold and forced one was perhaps
all the more effective.

Isaac Newton (1642—1727), student and Fellow of Trinity
College, Cambridge; the holder at one time of the three offices of

Newton.

Lucasian Professor of Mathematics in the University
of Cambridge, representative of that University in
Parliament, and President of the Royal Society; for
some years Master of the Mint; not only appreciated and honoured
by his University and all his contemporaries, but placed by the
unanimous verdict of posterity as foremost of all Natural Philosophers, gave the atomic hypothesis firm foothold in physical
science.

What was it that led the man who uttered the famous saying
" *Hypotheses non fingo* " to deal with the highly speculative atoms ?
The necessity for the existence of a vacuum in interstellar space,
would seem to have been the starting point[1]. Newton substitutes
for the Cartesian cosmology one arrived at by a very different
method and leading to very different results.

"The hypothesis of vortices assumed *à priori* a matter devoid of all
quality other than that of extension, and from this logically proceeded
further; universal gravitation, established *à posteriori* from observation,
induction and calculation, requires a vacuum for giving an account of
planetary motion." (Pillon, *L'Évolution Historique de l'Atomisme.*)

Against the Cartesian inferences that matter is infinitely
divisible, unlimited in extension and continuous (vacuum cannot

*Newton an
adherent of
the atomistic
hypothesis
from the
necessity of
assuming
interstellar
space to be
a vacuum.*

exist) stand the conclusions of Newton. His calculation required that the earth and other planets
should in their motion round the sun experience no
resistance.

"This he thought was because they encounter no matter
in their path, because the interstellar space is an absolute
vacuum. Hence a vacuum is not only a possibility but is
a fact which may be affirmed as the result of observation

---

[1] Pillon, *L'Évolution Historique de l'Atomisme* (*L'Année philosophique*, 1891) has
been closely followed in this account of Newton's atomistic views.

and induction. But if so, then matter is neither unlimited in extent nor continuous." (*Ibid.*)

If a vacuum exists in the interstellar space, why not also among things terrestrial? The one leads inevitably to the other, and Newton accepts the conception of discontinuity in the substances met with in the earth, and thereby explains certain of the phenomena observed. In Book III of the *Principia*, amongst the corollaries to the theorem "*that the weights of bodies towards any the same planet, at equal distances from the centre of the planet, are proportional to the quantities of matter which they severally contain*," we find:

Newton extends the conception of the existence of a vacuum in interstellar space to terrestrial matter.

"All spaces are not equally full; for if all spaces were equally full then the specific gravity of the fluid which fills the region of the air, on account of the extreme density of the matter, would fall nothing short of the specific gravity of quicksilver or gold or any other the most dense body.... If all the solid particles of all bodies are of the same density nor can be rarefied without pores, a void space or vacuum must be granted. By bodies of the same density, I mean those whose *vires inertiae* are in the proportion of their bulks."

But given the existence of a vacuum, that of atoms necessarily follows. Newton accepts the consequence and utilises the conception of atoms which attract each other according to some fixed law—not that of the inverse square, but probably some higher power—for the explanation of a variety of phenomena. A Query from the "Opticks" dealing with this subject must be of special interest to the chemist.

Query 31. "Have not the small particles of bodies certain powers, virtues or forces, by which they act at a distance, not only upon the rays of light for reflecting, refracting and inflecting them, but also upon one another for producing a great part of the phaenomena of nature? For it's well known that bodies act one upon another by the attractions of gravity, magnetism and electricity; and these instances show the tenor and course of nature, and make it not improbable but that there may be more attractive powers than these. For nature is very consonant and conformable to herself. How these attractions may be perform'd, I do not here consider. What I call attraction may be perform'd by impulse, or by some other means unknown to me. I use that word here to signify only in general any force by which bodies tend towards one another, whatsoever be the cause. For we must learn from the phaenomena of nature what bodies attract one another, and what are the laws and properties of the attraction, before we enquire the cause by which the attraction is perform'd. The attractions of gravity, magnetism and electricity, reach to very sensible distances, and so have been observed by vulgar eyes, and there may be others which reach to so small distances as hitherto escape observation ; and perhaps electrical attraction may reach to such small distances, even without being excited by friction.

For when the salt of tartar runs *per deliquium*, is not this done by an attraction between the particles of the salt of tartar, and the particles of the water which float in the air in the form of vapours ?"

Other examples of a similar nature are given in great number and discussed fully.

"When spirit of vitriol poured upon common salt or saltpetre makes an ebullition with the salt and unites with it, and in distillation the spirit of the common salt or saltpetre comes over much easier than it would do before, and the acid part of the spirit of vitriol stays behind ; does not this argue that the fix'd alcaly of the salt attracts the acid spirit of the vitriol more strongly than its own spirit, and not being able to hold them both, lets go its own ?"

His views on the atomistic constitution of matter are summed up in the famous passage at the end of the " Opticks " :

"It seems probable to me, that God in the beginning form'd matter in solid, massy, hard, impenetrable, moveable particles, of such sizes and figures, and with such other properties, and in such proportion to space, as most conduced to the end for which he form'd them ; and that these primitive particles being solids, are in. comparably harder than any porous bodies compounded of them, even so very hard, as never to wear or break in pieces ; no ordinary power being able to divide what God himself made one in the first creation. While the particles continue entire, they may compose bodies of one and the same nature and texture in all ages : but should they wear away, or break in pieces, the nature of things depending on them, would be changed. Water and earth composed of old worn particles and fragments of particles, would not be of the same nature and texture now, with water and earth composed of entire particles, in the beginning. And therefore that nature may be lasting, the changes of corporeal things are to be placed only in the various separations and new associations and motions of these permanent particles ; compound bodies being apt to break, not in the midst of solid particles, but where those particles are laid together, and only touch in a few points.... These principles I consider not as occult qualities, supposed to result from the specific forms of things, but as general laws of nature, by which the things themselves are form'd : their truth appearing to us by phaenomena, though their causes be not yet discover'd. For these are manifest qualities, and their causes only are occult."

Newton's views concerning the atomistic constitution of matter.

The effect of Newton's acceptance of the atomistic hypothesis had a swift and deep influence on its fortunes. About half a century after his death it could be said:

Atomistic conception of matter becomes generally accepted at end of eighteenth century.

"The Plenum is to-day considered a chimera,...the Void is recognised; bodies the most hard are looked upon as full of holes like sieves, and in fact this is what they are. Atoms are accepted, indivisible and unchangeable, principles to which is due the permanence of the different elements and of the different kinds of beings; which make it that water is always water, fire is always fire, earth always earth, and the imperceptible germs which form man can by no means form a bird." (Voltaire, *Dictionnaire Philosophe.*)

CHAPTER X.

DALTON AND THE ATOMIC HYPOTHESIS.

" One-story intellects, two-story intellects, three-story intellects with skylights. All fact-collectors, who have no aim beyond their facts, are one-story men. Two-story men compare, reason, generalise, using the labours of the fact-collectors as well as their own. Three-story men idealise, imagine, predict; their best illumination comes from above, through the skylight."

O. W. HOLMES.

DALTON is commonly called the Founder of the Atomic Theory. The preceding chapter should have shown that he did not devise the hypothesis of the atomic constitution of matter, nor even revive it. Neither of these merits has ever been really claimed for him, but it must be a cause of regret that the account of this matter as often given, is apt to produce the impression that in order to explain the laws of chemical combination, Dalton revived the old Greek Atomic Hypothesis. This of course is not so. Dalton dealt from the outset with the atom as a conception generally known. Newton had given the atom a firm standing in the science of physics, but chemists also had used it freely for the explanation of the phenomena they had to deal with. All previous advance in the establishment of clearer conceptions concerning chemical combination and chemical change must be considered as due to the use of a corpuscular theory of matter :

Dalton is called the Founder of the Atomic Theory.

"I was invited to try, whether...I could, by the help of the corpuscular philosophy,...explicate some particular subjects more intelligibly, than they are wont to be accounted for, either by the schools or the chymists." (Boyle, "*Some specimens of an Attempt to make Chymical Experiments useful to illustrate the notions of the Corpuscular Philosophy,*" 1661.)

Boyle explains chemical changes by atomic structure.

And in discussing the qualitative differences between the properties of nitre and the properties of the constituents into which nitre is resolved by distillation and from which it may be reproduced by combination, the same author says:

"It may not be useless to take notice of the difference, that there may be betwixt those active parts of a body, which are of differing natures, when they are as it were sheathed up, or wedged in amongst others in the texture of a concrete; and the same particles (when extricated from these impediments), they are set at liberty to flock together, and by the exercise of their nimble motions display their proper, but formerly clogged activity." (*A Physico-Chymical Essay*, 1661.)

There is no doubt but that the atomic hypothesis had in the 18th century already rendered valuable service in the explanation of chemical phenomena; but at a time when chemistry was as yet mainly a qualitative science, these phenomena could be but of a qualitative nature only. Dalton took the atomic hypothesis into the domain of the quantitative, and therein lies his merit, one so great as to fully justify his being called the Founder of the Atomic Theory.

The difficulty met with when tracing the history of Dalton's discovery of the law of multiple ratios was how to disentangle the sequence in time of the evolution of his theoretical views on the nature of chemical combination, and the establishment of experimental facts concerning such combinations. The same difficulty is naturally encountered at this point. The whole historical aspect of this subject has been dealt with exhaustively and conclusively by Roscoe and Harden in a book entitled "A new view of the Origin of Dalton's Atomic Theory" (1896). The point at issue is stated to be:

Sequence of Dalton's two great achievements.

"Was the atomic theory founded on an experimental knowledge of the law of combination in multiple proportions, or did Dalton arrive at this law as a necessary consequence of the atomic structure of matter?"

The evidence of Dalton's contemporaries supports the first of these alternatives. Thomson in his "History of Chemistry," referring to the visit which he paid to Dalton in Manchester in 1804 "and from which he carried away that clear and accurate idea of the new theory" which he three years later brought to the notice of the larger public, says:

According to Dalton's contemporaries the discovery of multiple ratios preceded the formulation of the atomic hypothesis.

"Mr Dalton informed me that the atomic theory first occurred to him during his investigations of olefiant gas and carburetted hydrogen gas, at that time imperfectly understood, and the constitution of which was first fully developed by Mr Dalton himself. It was obvious from the experiments which he made upon them that the constituents of both were carbon and hydrogen and nothing else; he found, further, that if we reckon the carbon in each the same, then carburetted hydrogen contains exactly twice as much hydrogen as olefiant gas does. This determined him to state the ratio of these constituents in numbers, and to consider the olefiant gas a compound of one atom of carbon and one atom of hydrogen; and carburetted hydrogen of one atom of carbon and two atoms of hydrogen. The idea thus conceived was applied to carbonic oxide, water, ammonia, etc. and numbers representing the atomic weights of oxygen, azote, etc. deduced from the best analytical experiments which chemistry then possessed."

Henry, in his life of Dalton, says:

" My own belief is that during the three years (1802—1804) in which the main foundations of the atomic theory were laid, Dalton had patiently and maturely reflected on all the phenomena of chemical combination known to him, from his own researches and all those of others; and had grasped in his comprehensive survey as significant to him of a deeper meaning than to his predecessors, their empirical laws of constant and reciprocal proportions, no less than his own law of multiple proportions, and his own researches in the chemistry of aeriform bodies."

Against this is to be set Dalton's own testimony from which it would appear that:

"...It was the application of the principle of the Newtonian atom to the constitution of the gases contained in the atmosphere that led Dalton to his Atomic Theory." (Roscoe, *John Dalton.*)

" As the ensuing lectures on the subject of the *Chemical Elements* and their combinations will perhaps be thought by many to possess a good deal of novelty, as well as importance, it may be proper to give a brief historical sketch of the train of thought and experience which led me to the conclusions about to be detailed. Having been long accustomed to make meteorological observations, and to speculate upon the nature and constitution of the atmosphere, it often struck me with wonder how a *compound* atmosphere, or a mixture of two or more elastic fluids, should constitute apparently a homogeneous mass, or one in all mechanical relations agreeing with a simple atmosphere. Newton had demonstrated clearly in the 23rd Prop. of Book II. of the *Principia* that an elastic fluid is constituted of small particles or atoms of matter which repel each other by a force increasing in proportion as their distance diminishes." (*Dalton's Manuscript Notes*, Royal Institution Lecture 17, Jan. 27th, 1810.)

According to Dalton himself, theoretical speculations came first.

Dalton then proceeds to reconcile the phenomena exhibited in the atmosphere with Newton's theory. He says:

"I set to work to combine my atoms upon paper. I took an atom of water, another of oxygen, and another of azote, brought them together, and threw around them an atmosphere of heat...."

It is not necessary for the present purpose to further follow Dalton's manipulation "on paper" of the atoms constituting the gases of which the atmosphere is composed, and of the difficulties therein encountered by him. The point of importance is that:

"In 1801 I hit upon an hypothesis which completely obviated these difficulties."

And in 1810, in the "New System of Chemical Philosophy," the description of carburetted hydrogen is prefaced by the remark:

"No correct notion of the gas about to be described, seems to have been formed till the atomic theory was introduced and applied in the investigation."

Dalton's laboratory note-books preserved by the Literary and Philosophical Society of Manchester, have been carefully studied and the results elaborately set out by Roscoe and Harden. The evidence is all in favour of the theory having preceded the empirical law. The final inference is summed up in the following passage :

"The balance of evidence is...strongly in favour of the statement made in London by Dalton himself in 1810, that he was led to the atomic theory of chemistry in the first instance by purely physical considerations, in opposition to the view, hitherto held by chemists, that the discovery by Dalton of the fact of combination in multiple proportions led him to devise the atomic theory as an explanation.

Roscoe and Harden conclude that the hypothesis preceded the empirical law.

It therefore becomes necessary for us to modify our views as to the foundation of the atomic theory. There seems to be no doubt that the idea of atomic structure arose in Dalton's mind as a purely physical conception, forced upon him by his study of the physical properties of the atmosphere and other gases. Confronted, in the course of this study, with the problem of ascertaining the relative diameters of the particles, of which, he was firmly convinced, all gases were made up, he had recourse to the results of chemical analysis. Assisted by the assumption that combination always takes place in the simplest possible way, he thus arrived at the idea that chemical combination takes place between particles of different weights, and this it was which differentiated his theory from the historic speculations of the Greeks. The extension of this idea to substances in general necessarily led him to the law of combination in multiple proportions, and the comparison with experiment brilliantly confirmed the truth of his deduction. Once discovered, the principle of atomic union was found to be of universal application."

Turning from the difficult though interesting question of how Dalton arrived at his hypothesis, to the really important and fortunately not speculative one of what it consisted in, the main points of the hypothesis may be summarised as follows :

The main points of Dalton's atomic hypothesis :

1. Rejection of the idea of one kind of primitive matter.

1. The idea of one kind of primitive matter is rejected, and various kinds of fundamentally different elementary principles are assumed.

"It has been imagined by some philosophers that all matter, however unlike, is probably the same thing, and that the great variety of its appearances arise from certain powers communicated to it, and from the variety of combinations and arrangements of which it is susceptible.... This does not appear to have been his [Newton's] idea. Neither is it mine. I should apprehend there are a considerable number of what may properly be called *elementary* principles, which can never be metamorphosed one into another by any power we can control. We ought, however, to avail ourselves of every means to reduce the number of bodies or principles of this appearance as much as possible ; and, after all, we may not know what elements are absolutely undecomposable and what are refractory, because we do not apply the proper means for their reduction." (*Dalton's Manuscript Notes*, Royal Institution Lecture 18, Jan. 30, 1810.)

2. The divisibility of matter is finite.

"Matter, though divisible in an extreme degree, is nevertheless not infinitely divisible...there must be some point beyond which we cannot go in the division of matter. The existence of these ultimate particles of matter can scarcely be doubted, though they are probably much too small ever to be exhibited by microscopic improvements." (*Ibid.*)

2. Finite divisibility of matter.

3. The name of atom is given alike to the ultimate particles of elements and compounds.

"I have chosen the word *atom* to signify these ultimate particles in preference to *particle, molecule*, or any other diminutive term, because I conceive it is much more expressive ; it includes in itself the notion of *indivisible*, which the other terms do not. It may, perhaps, be said that I extend the application of it too far when I speak of *compound atoms* ; for instance, I call an ultimate particle of *carbonic acid* a *compound atom*. Now, though this atom may be divided, yet it ceases to become carbonic acid, being resolved by such division into charcoal and oxygen. Hence I conceive there is no inconsistency in speaking of compound atoms and that my meaning cannot be misunderstood." (*Ibid.*)

3. Name "atom" given to ultimate particles.

4.   The indestructibility of atoms is affirmed.

4. Indestructibility of the atoms. "No new creation or destruction of matter is within the reach of chemical agency. We might as well attempt to introduce a new planet into the solar system, or to annihilate one already in existence, as to create or destroy a particle of hydrogen." (Dalton, *New System of Chemical Philosophy*, 1808.)

5.   The atoms constituting one homogeneous substance are said to be all alike, but different from those constituting any other substance; hence it is recognised as an essential property of each kind of atom, that it has a weight, characteristic of and peculiar to itself.

5. Characteristic weight of the atoms.

"Whether the ultimate particles of a body, such as water, are all alike, that is, of the same figure, weight etc., is a question of some importance. From what is known, we have no reason to apprehend a diversity in these particulars : if it does exist in water, it must equally exist in the elements constituting water, namely, hydrogen and oxygen. Now it is scarcely possible to conceive how the aggregates of dissimilar particles should be so uniformly the same. If some of the particles of water were heavier than others, if a parcel of the liquid on any occasion were constituted principally of these heavier particles, it must be supposed to affect the specific gravity of the mass, a circumstance not known. Similar observations may be made on other substances. Therefore we may conclude that *the ultimate particles of all homogeneous bodies are perfectly alike in weight, figure etc.* In other words, every particle of water is like every other particle of water ; every particle of hydrogen is like every other particle of hydrogen etc." (*Ibid.*)

6.   Chemical combination occurs between *simple* numbers of elementary atoms of fixed characteristic weight.

6. Combination of atoms in simple numbers. "The hypothesis upon which the whole of Mr Dalton's notions respecting chemical elements is founded, is this : 'When two elements unite to form a third substance, it is to be presumed that *one* atom of one joins to *one* atom of the other unless when some reason can be assigned for supposing the contrary.'..."

"Whenever more than one compound is formed by the combination of two elements, then the next simple combination must...arise from the union of *one* atom of the one with *two* atoms of the other." (Thomson, *System of Chemistry*, 1807.)

7.   Whilst recognising the impossibility of ascertaining the absolute weight of atoms, it is asserted to be one of the fundamental problems of chemistry to determine the relative weights of the different kinds of atoms, as well as the number of each kind of elementary atoms entering into the composition of one compound atom ; but the available data are insufficient.

F.

19

" In all chemical investigations, it has justly been considered an important
object to ascertain the relative weights of the simples which constitute a
compound. But unfortunately the enquiry has terminated
here ; whereas from the relative weights in the mass, the
relative weights of the ultimate particles or atoms of the
bodies might have been inferred, from which their number
and weight in various other compounds would appear, in
order to assist and to guide future investigations, and to
correct their results. Now it is one great object of this work,
to show the importance and advantage of ascertaining the
relative weights of the ultimate particles, both of simple and compound
bodies, the number of simple elementary particles which constitute one
compound particle, and the number of less compound particles which enter
into the formation of one more compound particle." (Dalton, *New System
of Chemical Philosophy*, 1808.)

7. Determina-
tion of (i) $a:b$,
the relative
weights of ele-
mentary atoms
and (ii) $m + n$,
the complexity
of compound
atoms.

Here we find a clear statement of the problems which Dalton
attempted to solve by the application of his hypothesis :—

(i)   To determine the relative weights of the atoms.

(ii) To determine the number of constituent atoms in one
compound atom.

How is this to be done ? Suppose he was dealing with a
binary compound of the elements A and B, whose atoms weigh
a and b respectively, and let *one* atom of AB be formed by the
combination of m atoms of A with n atoms of B; then the weights
combining with each other in *one* atom will be ma of A and nb of
B. The quantities to be determined are the ratios $a:b$ and $m:n$.
What data are there at our disposal for so doing ? The ratios of
the quantities which combine to form one atom will also be the
ratio of the quantities combining to form any
number of atoms, and will be equal to $p:q$ the
experimentally ascertained ratio in which A and B
are present in the compound AB. We have then
$ma:nb = p:q$, and it is of course evident that since
both $m:n$ and $a:b$ are unknown, we have not sufficient data for
solving the equation. Hence in itself Dalton's original hypothesis is insufficient for the solution of the above two problems, and
therefore it must either be modified or burdened with subsidiary
hypotheses. Dalton took the latter course.

Insufficiency
of the equation
$ma : nb = p : q$
for determin-
ing $a : b$ and
$m : n$.

8. Rules are given concerning the number of elementary
atoms combining with each other. These rules are
based on the principle of " greatest simplicity," but
are otherwise quite arbitrary and hypothetical.

8. Rules of
chemical syn-
thesis.

"If there are two bodies, A and B, which are disposed to combine, the following is the order in which the combinations may take place, beginning with the most simple: namely,

1 atom of A + 1 atom of B = 1 atom of C, binary.

1 atom of A + 2 atoms of B = 1 atom of D, ternary.

2 atoms of A + 1 atom of B = 1 atom of E, ternary.

1 atom of A + 3 atoms of B = 1 atom of F, quaternary.

3 atoms of A + 1 atom of B = 1 atom of G, quaternary, etc.

The following general rules may be adopted as guides in all our investigations respecting chemical synthesis.

1st. When only one combination of two bodies can be obtained, it must be presumed to be a binary one, unless some cause appear to the contrary.

2nd. When two combinations are observed, they must be presumed to be a binary and a ternary.

3rd. When three combinations are obtained, we may expect one to be a binary and the other two ternary.

4th. When four combinations are observed, we should expect one binary, two ternary, and one quaternary, etc.

5th. A binary compound should always be specifically heavier than the mere mixture of its two ingredients.

6th. A ternary compound should be specifically heavier than the mixture of a binary and a simple, which would, if combined, constitute it, etc.

7th. The above rules and observations equally apply when two bodies such as C and D, D and E, etc. are combined." (Dalton, *New System of Chemical Philosophy*, 1808.)

The arbitrariness of these rules is self-evident. Why, if we know one compound only, should this be the binary one? Another may be discovered any day, and why should nature be so complacent, in the quite accidental sequence of discovery, as to always put us into the way of the binary compound first? Why any of these rules? No attempt is made to place them in connection with observed facts, and criteria are lacking for testing the validity of any one of them. Moreover they are not only arbitrary, but also insufficient and vague. What for instance is to constitute "a cause to the contrary"?

9. A symbolic notation[1] is devised for the purposes of representing the qualitative as well as the quantitative composition of compounds in terms of signs standing for the atomic weights of the different elements.

9. Symbolic notation which is qualitative and quantitative.

"It is deemed expedient to give plates, exhibiting the mode of combination in some of the more simple cases.... The elements or atoms of such bodies as are conceived at present to be simple,

[1] *Ante*, p. 193.

are denoted by a small circle, with some distinctive mark ; and the combinations consist in the juxtaposition of two or more of these...." (*Ibid.*).

This plate contains the arbitrary marks or signs chosen to represent the several chemical elements or ultimate particles :

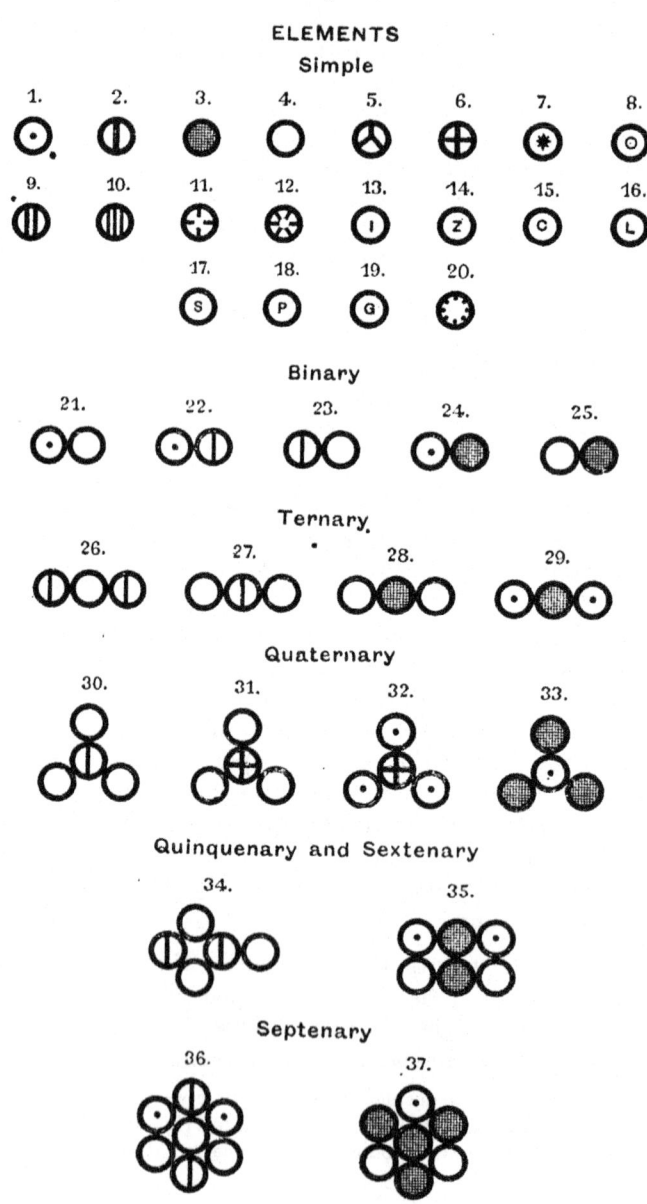

ELEMENTS
Simple

Binary

Ternary

Quaternary

Quinquenary and Sextenary

Septenary

Fig.				Fig.				
1. Hydrogen,	its rel. weight		1	11. Strontites,	its rel. weight			46
2. Azote,	„ „	„	5	12. Barytes,	„ „	„		68
3. Carbone or Charcoal,	„ „	„	5	13. Iron,	„ „	„		38
4. Oxygen,	„ „	„	7	14. Zinc,	„ „	„		56
5. Phosphorus,	„ „	„	9	15. Copper,	„ „	„		56
6. Sulphur,	„ „	„	13	16. Lead,	„ „	„		95
7. Magnesia,	„ „	„	20	17. Silver,	„ „	„		100
8. Lime,	„ „	„	23	18. Platina,	„ „	„		100
9. Soda,	„ „	„	28	19. Gold,	„ „	„		140
10. Potash,	„ „	„	42	20. Mercury,	„ „	„		167

21. An atom of water or steam, composed of 1 of oxygen and 1 of hydrogen, retained in physical contact by a strong affinity, and supposed to be surrounded by a common atmosphere of heat   its rel. weight   8

22. An atom of ammonia, composed of 1 of azote and 1 of hydrogen   „ „ „   6

23. An atom of nitrous gas, composed of 1 of azote and 1 of oxygen   „ „ „   12

24. An atom of olefiant gas, composed of one of carbone and 1 of hydrogen   „ „ „   6

25. An atom of carbonic oxide composed of 1 of carbone and 1 of oxygen   „ „ „   12

26. An atom of nitrous oxide, 2 azote + 1 oxygen   „ „ „   17

27. An atom of nitric acid, 1 azote + 2 oxygen   „ „ „   19

28. An atom of carbonic acid, 1 carbone + 2 oxygen   „ „ „   19

29. An atom of carburetted hydrogen, 1 carbone + 2 hydrogen   „ „ „   7

30. An atom of oxynitric acid, 1 azote + 3 oxygen   „ „ „   26

31. An atom of sulphuric acid, 1 sulphur + 3 oxygen   „ „ „   34

32. An atom of sulphuretted hydrogen, 1 sulphur + 3 hydrogen[1]   „ „ „   16

33. An atom of alcohol, 3 carbone + 1 hydrogen   „ „ „   16

34. An atom of nitrous acid, 1 nitric acid + 1 nitrous gas   „ „ „   31

35. An atom of acetous acid, 2 carbone + 2 water   „ „ „   26

36. An atom of nitrate of ammonia, 1 nitric acid + 1 ammonia + 1 water   „ „ „   33

37. An atom of sugar, 1 alcohol + 1 carbonic acid   „ „ „   35

[1] "The figure for sulphuretted hydrogen is incorrect: it ought to be 1 atom of hydrogen instead of 3, united to 1 of sulphur." (*New System of Chemical Philosophy*, 2, 1810, p. 450.)

How Dalton applied the system, the main features of which have been given in the above, and what were the results of his relative atomic weight determinations, will now

Dalton's application of his system to compounds of nitrogen.

be illustrated. In a passage already quoted he says :

"Whereas from the relative weights in the mass, the relative weights of the ultimate particles or atoms of the bodies might have been inferred, from which their number and weight in various other compounds would appear, in order to assist and to guide future investigations, and to correct their results."

And this is how in his "System of Chemical Philosophy" he deals under these aspects with different compounds of nitrogen :

"The compounds of oxygen with azote, hitherto discovered, are five ; they may be distinguished by the following names ; nitrous gas, nitric acid, nitrous oxide, nitrous acid, and oxynitric acid. In treating of these,

The complexity assigned to different oxides of nitrogen.

it has been usual to begin with that which contains the least oxygen (nitrous oxide), and to take the others in order as they contain more oxygen. Our plan requires a different principle of arrangement ; namely, to begin with that which is most simple, or which consists of the smallest number of elementary particles, which is commonly a binary compound, and then to proceed to the ternary and other higher compounds. According to this principle, it becomes necessary to ascertain, if possible, whether any of the above, and which of them, is a binary compound. As far as the specific gravities of the two simple gases are indicative of the weights of their atoms[1], we should conclude that an atom of azote is to one of oxygen as 6 to 7 nearly.... But the best criterion is derived from a comparison of the specific gravities of the compound gases themselves. Nitrous gas has the least specific gravity of any of them ; this indicates it to be a binary compound ; nitrous oxide and nitrous acid are both much heavier ; this indicates them to be ternary compounds ; and the latter being heavier than the former, indicates that oxygen is heavier than azote, as oxygen is known to abound most in the latter. Let us now see how far the facts already known will corroborate these observations."

Then follows a table embodying the constitution of these gases as given by Cavendish and Davy :

[1] Whatever his attitude towards this question may have been at different times, here certainly Dalton assumes that equal volumes of elementary gases contain equal numbers of atoms.

	Specific Gravities	Constitution by weight			Ratios
Nitrous gas	1·102	46·6 azote	+ 53·4	oxygen	6·1 : 7
		44·2 „	+ 55·8	„	5·5 : 7
		42·3 „	+ 57·7	„	5·1 : 7
Nitrous oxide	1·614	63·5 „	+ 36·5	„	2 × 6·1 : 7
		62·0 „	+ 38·0	„	2 × 5·7 : 7
		61·0 „	+ 39·0	„	2 × 5·4 : 7
Nitric acid	2·444	29·5 „	+ 70·5	„	5·8 : 7 × 2
		29·6 „	+ 70·4	„	5·9 : 7 × 2
		28·0 „	+ 72·0	„	5·4 : 7 × 2
		25·3 „	+ 74·6	„	4·7 : 7 × 2

"In the third column are given the ratios of the weights of azote and oxygen in each compound, derived from the preceding column, and reduced to the determined weight of an atom of oxygen, 7. This table corroborates the theoretic views above stated most remarkably. The weight of an atom of azote appears to be between 5·4 and 6·1 : and it is worthy of notice, that the theory does not differ more from the experiments than they differ from one another ; or, in other words, the mean weight of an atom of azote derived from the above experiments would equally accommodate the theory and the experiments. The mean is 5·6, to which all the others might be reduced. We should then have an atom of nitrous gas to weigh 12·6, consisting of 1 atom of azote and 1 of oxygen...nor has the weight of an atom of oxygen any influence on the theory of these compounds ; for it is obvious that if oxygen were taken 3, or 10, or any other number, still the ratios of azote to oxygen in the compounds would continue the same ; the only difference would be, that the weight of an atom of azote would rise or fall in proportion as that of oxygen. I have been solicitous to exhibit this view of the compounds of azote and oxygen, as derived from the experience of others, rather than from my own ; because, not having had any views at all similar to mine, the authors could not have favoured them by deducing the above results, if they had not been conformable to actual observation."

The atomic weight of nitrogen in terms of oxygen = 7.

Dalton then proceeds to give experimental reasons why the atomic weight of nitrogen derived from the composition of the oxides should be 5·1 rather than 5·6, the above mean number. It is evident that in so doing he is "favouring his own views." The three compounds water, ammonia, and nitrous gas being assumed to have the simplest possible, that is, binary composition, this hypothesis demands that the quantities

Comparison between the theoretical and the experimental composition of ammonia.

of hydrogen and nitrogen combining with 7 of oxygen in water and nitrous gas respectively, should be the same as the quantities of these elements combining with each other in ammonia. Such agreement is not found to exist, but the discrepancy is much less if 5·1 is accepted in place of 5·6 for the atomic weight of nitrogen derived from the oxides.

"By comparing the weight of azote with that of hydrogen...we find them as 4·7 : 1 nearly. This evidently marks the constitution of ammonia to be that of 1 atom of each of the elements combined. But we have before determined the element of azote to weigh 5·1, when treating of the compounds of azote and oxygen. This difference is probably to be ascribed to [experimental errors]."

This is, however, not so. The discrepancy is one which when expressed by the correct numbers now at our disposal, becomes very much greater, and is due to the fact also made apparent by the later data, that Dalton's own law of multiple ratios here modifies the simplest possible relations. The table on the next page shows the connection between the composition of nitrous gas, water, and ammonia, in terms of the data of Dalton and of those now accepted.

Dalton calculated the relative atomic weights of a fair number of elements, referring them all to the weight of the hydrogen atom as unity, and embodied them in successive tables. The data given for the non-metallic elements are set out in the table on page 298, which is taken from Roscoe and Harden's "Origin of Dalton's Atomic Theory."

Dalton's atomic weight tables.

The examination of this table is of historical interest, two points calling for special attention:

(i)   What is the value assigned to m and n (*ante*, p. 290), the number of elementary atoms entering into the composition of one compound atom of the substance whose quantitative composition furnishes the ratio $p : q$?

(ii)   What is the degree of accuracy with which the ratio $p : q$ has been determined?

Concerning (i) we find that in nearly all cases the formulae assigned are simpler than those now accepted; water, ammonia,

The Composition of Nitrous Gas, Water and Ammonia
according to:

I. DALTON.

	Nitrous Gas		Water	
	Oxygen	*Nitrogen*	*Hydrogen*	*Oxygen*
Assumed complexity of composition	Binary 1 atom : 1 atom		Binary 1 atom : 1 atom	
Composition by weight	7	: 5·6 (or 5·1)	1	: 7
∴ Atomic weights	7	: 5·6 (or 5·1)	1	: 7

	Ammonia	
	Nitrogen	*Hydrogen*
Assumed complexity of composition	Binary 1 atom : 1 atom	
∴ Composition calculated	5·6 (or 5·1) :	1
Composition found	4·7 :	1

II. THE DATA NOW ACCEPTED.

	Nitrous Gas		Water	
	Oxygen	*Nitrogen*	*Hydrogen*	*Oxygen*
Composition by weight	7·94	: 7·0	1	: 7·94
	= 7·94	: 4·66 × 1·5		
	= 7·94 × 2 :	4·66 × 3		

	Ammonia	
	Nitrogen	*Hydrogen*
Composition by weight	4·66 :	1
	= 4·66 × 3 :	3 × 1
	= 7·0 × 2 :	3 × 1

Hence all these compounds cannot be formulated as binary. With water assumed to be binary and represented by HO, where H = 1 and O = 7·94, the choice would be left between:

$$N = 4·66 \atop \left. \begin{matrix} \text{Ammonia} = NH \\ \text{Nitrous Gas} = N_3O_2 \end{matrix} \right\} \text{ or } \left\{ \begin{matrix} N = 7·0 \\ \text{Ammonia} = N_2H_3 \\ \text{Nitrous Gas} = NO \end{matrix} \right.$$

and neither set is compatible with Dalton's rules of combination (p. 291).

No.	Source	Date	Hydrogen	Oxygen	Azote	Carbon	Sulphur	Phosphorus
(1)	Notebook Sept. 6, 1803	1803	1	5·66	4	4·5	17	
(2)	Notebook Sept. 19, 1803	1803	1	5·66	4	4·4	14·4	7·2
(3)	Notebook Sept. 1803	1803	1	5·5	4			
(4)	Manchester Memoirs 1805	1805	1	5·5	4·2	4·3	14·4	7·2
(5)	Thomson's List, System of Chemistry, 1807, probably communicated, 1804	1807	1	6	5			
(6)	Notebook Aug. 23, „ 14, 1806	1806	1	7	5	5	22	9
(7)	Notebook Sept. 16, Oct. 22, 1806	1806	1	7	5	5	12	9·3
(8)	New System of Chemical Philosophy	1808	1	7	5	5	13	9
(9)	New System of Chemical Philosophy	1810	1	7	5	5·4	13	9
(10)	Numbers of (9) calculated to O = 16 and modern formulae		1·14	16	{11·4[1] / 15	12·3	29·7	{25·7[1] / 27
(11)	Modern numbers		1·008	16·000	14·0	12·0	32·0	31·0

[1] Two numbers are given for nitrogen and phosphorus, one calculated from Dalton's formula of the hydride, the other from that of the oxide.

olefiant gas are formulated as binary, nitric acid as ternary, etc.
(pp. 293, 295); but the formulae first decided upon
are consistently retained.

The formulae used are simpler than those now accepted, and the accuracy is low.

As regards the accuracy of the experimental work
involved, that is, as might be expected, of a very low
degree; but what is somewhat surprising is the
startlingly quick succession in the changes made; values evidently
are lightly accepted and as lightly again given up.

The system concerning the ultimate nature of chemical combination which Dalton had devised between the years 1801 and
1804 was retained by him unchanged to the end of his life.

The object of any hypothesis concerning the phenomena of
chemical combination, is to assign some cause more or less proximate for the phenomena observed and the laws formulated.
And a critical examination of the Daltonian hypothesis shows
us that it has achieved all that which is recognised as essential
for an hypothesis developing into a theory. The conception of
indestructible elementary atoms endowed with a
weight characteristic of each element, and which
unite in simple numbers, explains perfectly and
simply the conservation of mass and combination
according to fixed, multiple, and permanent ratios.

The atomic hypothesis accounts for all the laws of chemical combination.

All these laws follow as a natural consequence of such a constitution of matter. The connection between the hypothesis
and these empirical laws is so direct and so obvious, that to
specially elaborate it might tend to obscure rather than to
elucidate.

How the Daltonian Atomic Hypothesis adapted itself to the
explanation of phenomena discovered after its promulgation, and
how it lent itself to deductive application, will be dealt with in the
succeeding chapters.

The comparative insignificance of the modifications required to
enable it to perform its task towards an ever-growing number of
phenomena, in a science extending its scope in the marvellous
manner that chemistry has done since the time of the introduction
into it of this hypothesis, bears witness to the genius of its founder.
And now more than a hundred years later scientists of all countries,
in all possible variations of form, on all possible occasions reiterate
what Dalton himself and his great contemporary Berzelius have
said concerning this hypothesis:

" The doctrine of definite proportions appears to me *mysterious* unless we adopt the atomic hypothesis. It appears like the *mystical ratios* of Kepler, which Newton so happily elucidated. The prosecution of the investigation can terminate, I conceive, in nothing but the system which I adopt of particle applied to particle...." (Dalton to Berzelius, Manchester, 20th of September, 1812.)

"You are right in that the theory of multiple proportion is a mystery but for the Atomic Hypothesis, and as far as I have been able to judge all the results so far obtained have contributed to justify this hypothesis." (Berzelius to Dalton, London, 13th of October, 1812.)

In 1811 Berzelius had written:

" But such a doctrine of the composition of compounds would so illuminate the province of affinity, that supposing Dalton's hypothesis be found correct, we should have to look upon it as the greatest advance that chemistry has ever yet made in its development into a science."

And all the reviews of the progress made during the 19th century, not only in the domain of chemistry, but in the wider one of physical science, agree in assigning such pre-eminence to the Daltonian hypothesis.

CHAPTER XI.

GAY-LUSSAC AND THE LAW OF THE COMBINING
VOLUMES OF GASES.

" We believe in atoms, because Nature seems to use them, and we break them up continuously because we do not know where to stop. There are various methods of spanning the distance from nothing to something."

ANGUS SMITH.

IN 1802 Dalton, when discussing the various processes then known for determining the composition by volume of atmospheric air, and the results arrived at by their application, writes concerning the method of explosion with hydrogen thus :

"Volta's eudiometer is very accurate as well as elegant and expeditious : according to Monge, 100 oxygen require 196 measures of hydrogen ; according to Davy 192 ; but from the most attentive observations of my own, 185 are sufficient."

Earliest determinations of the volume ratio hydrogen : oxygen in water.

This placid manner of considering differences in results amounting to 5 per cent. strikes one as a strange characteristic of that stage in the development of quantitative methods. . Lavoisier had directed the attention of chemists to quantitative relations; Dalton himself bases all his theoretical work on such relations, but is obviously content with approximations very far from close. Cavendish, in his determination of this ratio, came, as he invariably did, very near to the correct value. In 1781 he found (but did not publish his results until 1784) that

"423 measures of inflammable air are nearly sufficient to completely phlogisticate [remove all the oxygen from] 1000 of common air."

That is, 423 volumes of hydrogen combine with 210 volumes of oxygen, which makes the ratio 201·5 : 100.

In 1805 Gay-Lussac[1] and Humboldt[2] published a memoir entitled "Experiments on the Ratio of the Constituents of the Atmosphere[3]." One of the methods employed for ascertaining this ratio consisted in the removal of the oxygen by explosion with hydrogen, and involved, therefore, a knowledge of the volume ratio in which these two gases combine. In the absence of concordant and reliable data, a new determination of this ratio had to be undertaken :

Gay-Lussac and Humboldt find the volume ratio hydrogen : oxygen = 200 : 100.

"In what ratio do oxygen and hydrogen unite to form water? To give an accurate answer to this important question, we have carried out the two following series of experiments. In the first series we inflamed in Volta's eudiometer 100 parts of oxygen and 300 parts of hydrogen, and in twelve experiments obtained the residues given under A. In the second series we inflamed a mixture of 200 parts of oxygen and 300 parts of hydrogen, the residues being those given under B :

A		B	
100·8	102·0	101·5	101·1
101·4	101·5	101·3	101·0
100·5	102·0	102·2	101·5
101·0	102·0	102·0	102·3
101·0	101·0	102·0	102·0
101·7	101·5	102·0	102·0
Mean = 101·3		Mean = 101·7	

and the absorption is.. 298·7 .... 298·3

Had our oxygen been quite pure, then, according to the first series, 100 of oxygen would in the mean have absorbed 198·7 parts of hydrogen ; but since potassium sulphide left ·004 of our oxygen unabsorbed, it follows that 99·6 of oxygen had united with 199·1 [*i.e.* 298·7 − 99·6 = 199·1] parts of hydrogen, and hence that 100 of oxygen had required for complete saturation 199·89 parts of hydrogen, for which 200 parts may be put without error.

[1] J. F. Gay-Lussac (1778—1850), who occupied different professorial chairs in Paris, has enriched the science of chemistry by important contributions to almost every one of its departments : by physical investigations into the behaviour of gases ; by the recognition of iodine as an element, and of cyanogen as a *compound radicle,* which latter conception he introduced into the science ; by laying the foundations of volumetric analysis, and much other work.

[2] Alexander von Humboldt (1769—1859), scientist and explorer, was born in Berlin. Between 1799 and 1804 he travelled in Central and South America, after which he lived till 1826 in Paris, engaged in collecting and publishing the results of his discoveries. In the *Kosmos,* which appeared between 1845 and 1858, he attempted the gigantic task of presenting to his contemporaries all that was then known concerning the physical universe.

[3] *J. Phys.,* Paris, 60, 1805 (p. 129).

If our hydrogen had been quite pure, then according to the 2nd series of experiments, it would have absorbed in the mean 98·3 parts of oxygen....The two results would agree absolutely if ·006 parts of nitrogen had been mixed with the hydrogen, and we are able to prove that nitrogen was in fact present."

Struck by the simplicity of the relations thus found, Gay-Lussac extended his investigations to the volume relations of other gaseous substances which are compounds of gaseous constituents, and by the end of 1808 he was able to publish results which clearly demonstrated the existence of a simple and general law[1].

Gay-Lussac investigates: (1) The volume relations be-tween the gaseous con-stituents of compounds.

"...It appears that it is only when the attraction [of the molecules for each other] is entirely destroyed, as in gases, that bodies under similar con-ditions obey simple and regular laws. At least, it is my intention to make known some new properties in gases, the effects of which are regular, by showing that these substances combine amongst themselves in very simple proportions, and that the contraction of volume which they experience on combination also follows a regular law."

From the experimental evidence on which he bases this generalisation, the following may be quoted:

"Suspecting, from the exact ratio of 100 of oxygen to 200 of hydrogen, which M. Humboldt and I had determined for the proportions of water, that other gases might also combine in simple ratios, I have made the following experiments. I prepared fluoboric (M. Thénard and I have given the name of fluoboric gas to that particular gas which we obtained by distilling pure fluoride of lime with vitreous boric acid), muriatic, and carbonic gases, and made them combine successively with ammonia gas. 100 parts of muriatic gas saturate precisely 100 parts of ammonia gas, and the salt which is formed from them is perfectly neutral, whether one or other of the gases is in excess. Fluoboric gas, on the contrary, unites in two proportions with ammonia gas. When the acid gas is first put into the graduated tube, and the other gas is then passed in, it is found that equal volumes of the two condense, and that the salt formed is neutral. But if we begin by first putting the ammonia gas into the tube, and then admitting the fluoboric gas in single bubbles, the first gas will then be in excess with regard to the second, and there will result a salt with excess of base, composed of 100 of fluoboric gas and 200 of ammonia gas[2]. If carbonic gas is brought into contact with ammonia gas, by passing it sometimes first, sometimes second

Experimental numbers for the volume ratios in which gases combine. (i) Direct deter-minations.

---

[1] "*Mémoire sur la combinaison des substances gazeuses,*" *Mémoires de la Société d'Arcueil*, II. 1809 (pp. 207—234); Alembic Club Reprints, No. 4.

[2] Interesting, as an early experimental confirmation of Dalton's law of multiple ratios.

into the tube, there is always formed a sub-carbonate composed of 100 parts of carbonic gas and 200 of ammonia gas. It may, however, be proved that neutral carbonate of ammonia would be composed of equal volumes of each of these components....Thus we may conclude that muriatic, fluoboric, and carbonic acids take exactly[1] their own volume of ammonia gas to form neutral salts, and that the last two take twice as much to form sub-salts....According to the experiments of M. Amédée Berthollet, ammonia is composed of

> 100 of nitrogen,
> 300 of hydrogen, by volume.

I have found that sulphuric acid is composed of

> 100 of sulphurous gas,
> 50 of oxygen gas.

When a mixture of 50 parts of oxygen and 100 of carbonic oxide (formed by the distillation of oxide of zinc with strongly calcined charcoal) is inflamed, these two gases are destroyed and their place taken by 100 parts of carbonic acid gas. Consequently carbonic acid may be considered as being composed of

> 100 of carbonic oxide gas,
> 50 of oxygen gas.

(ii) Indirect determinations.   Davy, from the analysis of various compounds of nitrogen with oxygen, has found the following proportions by weight :

	Nitrogen	Oxygen
Nitrous oxide	63·30	36·70
Nitrous gas	44·05	55·95
Nitric acid	29·50	70·50

Reducing these proportions to volumes we find :

	Nitrogen	Oxygen
Nitrous oxide	100	49·5
Nitrous gas	100	108·9
Nitric acid	100	204·7

The first and the last of these proportions differ only slightly from 100 to 50, and 100 to 200 ; it is only the second which diverges somewhat from 100 to 100. The difference, however, is not very great, and is such as we might

---

[1] The actual experimental numbers are of course only approximations to the simple whole numbers given by Gay-Lussac. A note to the German translation in *Gilbert's Annalen*, 36, 1810 (pp. 6—36), gives interesting details concerning the method used, and the actual results obtained in the determination of the volumetric composition of ammonia: " The first decomposition of ammonia was that communicated in 1785 to the Paris Academy by M. Berthollet. He passed electric sparks through ammonia gas until there was no further expansion ; thereby the volume increased in the ratio 1 : 1·94117 ; he then examined the mixed gases in Volta's eudiometer, and found that they were composed of ·725 of hydrogen, and of ·275 of nitrogen by volume. But Berthollet *junior* found as the mean of 6 such experiments made by him with absolutely pure ammonia, in graduated glass vessels, that after absorption of the undecomposed gas by muriatic acid, and after making the necessary corrections for temperature and pressure (because even with a good electrical machine the experiment takes 6 to 8 hours at least), the volume had increased in the ratio of 1 : 2·04643, and that the gaseous mixture consisted of 0·755 of hydrogen and 0·245 of nitrogen [but 0·755 : 0·245 = 3·08163 : 1]."

expect in experiments of this sort; and I have assured myself that it is actually nil. On burning the new combustible substance from potash[1] in 100 parts by volume of nitrous gas, there remained over exactly 50 parts of nitrogen, the weight of which, deducted from that of the nitrous gas, yields as result that this gas is composed of equal parts by volume of nitrogen and oxygen....Thus it appears evident to me that gases always combine in the simplest proportions when they act on one another; and we have seen in reality in all the preceding examples that the ratio of combination is 1 to 1, 1 to 2, or 1 to 3. It is very important to observe that in considering weights there is no simple and finite relation between the elements of any one compound; it is only when there is a second compound between the same elements that the new proportion of the element that has been added is a multiple of the first quantity. Gases, on the contrary, in whatever proportions they may combine, always give rise to compounds whose elements by volume are multiples of each other. Not only, however, do gases combine in very simple proportions, as we have just seen, but the apparent contraction of volume which they experience on combination has also a simple relation to the volume of the gases, or at least to that of one of them.

(2) The volume relations between the gaseous constituents and the gaseous compounds.

I have said, following M. Berthollet, that 100 parts of carbonic oxide gas...produce 100 parts of carbonic gas on combining with 50 of oxygen. It follows from this that the apparent contraction of the two gases is precisely equal to the volume of oxygen gas added....Ammonia gas is composed of three parts by volume of hydrogen and one of nitrogen, and its density compared to air is 0·596. But if we suppose the apparent contraction to be half of the whole volume, we find 0·594 for the density."

Gay-Lussac sums up his results thus:

" I have shown in this Memoir that the compounds of gaseous substances with each other are always formed in very simple ratios, so that representing one of the terms by unity, the other is one, or two, or at most three. These ratios by volume are not observed with solid or liquid substances, nor when we consider weights, and they form a new proof that it is only in the gaseous state that substances are in the same circumstances and obey regular laws....The apparent contraction of volume suffered by gases on combination is also very simply related to the volume of one of them, and this property likewise is peculiar to gaseous substances."

Summary of results of investigations on combining volumes of gases.

Parts of the tables with which Gay-Lussac concludes his paper are reproduced, chiefly because of the interest there is in comparing the experimental values for the densities of compound gases with those calculated from (i) the densities of their constituents, and (ii) the composition by volume according to the simple relations of the generalisation made.

[1] Davy had the year before isolated the metal potassium by the electrolysis of potash. The decomposition of nitrous gas (nitric oxide) or nitrous oxide by metallic potassium, is an experiment described in all text-books.

F.   20

I. *Gay-Lussac's Table showing composition of various compounds whose elements are gaseous.*

Substances	Composition by volume	Composition by weight
ihm chloride	100 ammonia : 100 muriatic acid	38·35 ammonia : 61·65 muriatic acid
Neutral ammonium carbonate	,, : 100 carbonic acid	28·19 ,, : 71·81 carbonic acid
Sub-carbonate of ammonia	,, : 50 ,,	43·98 ,, : 56·02 ,,
Fluoborate of ammonia	,, : 100 fluoboric acid	
Sifluoborate of ammonia	,, : 50 ,,	
Water	100 hydrogen : 50 oxygen	86·733 oxygen : 13·267 hydrogen
Nitrous oxide	100 nitrogen : 50 ,,	63·72 nitrogen : 36·28 oxygen
Nitrous gas	100 ,, : 100 ,,	46·757 ,, : 53·243 ,,
Nitric acid	100 ,, : 200 ,,	30·512 ,, : 69·488 ,,
Nitric acid	200 nitrous gas : 100 ,,	,, : ,,
Nitrous acid gas		34·507 ,, : 65·493 ,,
Ammonia	100 nitrogen : 300 hydrogen	81·525 ,, : 18·475 hydrogen
Sulphuric acid	100 sulphurous gas : 50 oxygen	42·016 sulphur : 57·984 oxygen
Sulphurous gas		52·083 ,, : 47·917 ,,
100 vols. of carbonic gas	100 carbonic oxide : 50 ,,	27·376 carbon : 72·624 oxygen
,, ,,	: 100 ,,	,, : ,,
100 vols. of carbonic oxide	: 50 ,,	42·99 ,, : 57·01 ,,

II. Gay-Lussac's Table giving densities of some Simple and Compound Gaseous Substances.

Substances	Densities determined by experiment		Densities calculated from the simple volume relations [given in the preceding table]	and on the supposition that the contraction of the gaseous constituents is equal to :
Atmospheric air	1·00000			
Oxygen	1·10359			
Nitrogen	0·96913	Biot and Arago		
Hydrogen	0·07321			
Carbonic gas	1·5196			
Ammonia gas	0·59669		0·59438	$\frac{1}{2}$ of the total vol.
Nitrous oxide	1·61414	Davy	1·52092	the vol. of the oxygen
	1·36293	Berthollet		
Nitrous gas	1·0388	Bérard, at Arcueil	1·03636	$\frac{1}{2}$ of the total vol.
Sulphurous gas	2·265	Kirwan		
Carbonic oxide	0·9569	Cruikshanks	0·96782	100 vols. carbonic oxide = 100 vols. carbonic acid − 50 vols. oxygen
Water vapour	0·6896	Tralès	0·625	the vol. of the oxygen

The following are examples of the wording[1], perhaps unnecessarily long and ponderous, used for Gay-Lussac's law :

" If gaseous substances enter into chemical combination, their volumes are in simple rational proportions, and if a gaseous substance is formed by their union, its volume also is rationally related to the volumes of the original gases." (Ostwald, *Outlines of General Chemistry*.)

Formulation of the law.

" If two gases take part in a chemical reaction, the volumes of the combining or mutually decomposing gases are either equal, or they bear a simple relation to each other ; and similarly the volume of the product of combination or decomposition, if it is capable of existing in the gaseous state, always exhibits a simple relation to the volume occupied by its constituents in a similar state before the act of decomposition or combination." (Lothar Meyer, *Modern Theories of Chemistry*.)

Similar conditions of temperature and pressure are of course assumed throughout.

Gay-Lussac knew Dalton's atomic hypothesis and recognised at once that his own law of combining volumes had a definite bearing on it. The inference from the connection between the hypothesis of the combination between simple numbers of elementary atoms, and the empirical law of the combination between volumes of gases which are in a simple ratio to each other,

Connection between Gay-Lussac's Law and Dalton's Atomic Hypothesis.

was not actually pointed out by Gay-Lussac himself; it was grasped and expounded by Berzelius, but did not meet with recognition by Dalton.

This inference is that the number of atoms contained in equal volumes of gases—elementary or compound—bear a simple numerical relation to each other. Obvious as the conclusion is, it

[1] Attention should here be drawn to the misuse of the term "two volumes" in connection with the enunciation of the law of combining volumes. So for instance : " Thus it appears that very simple relations exist, not only between the volumes of gases entering into combination, but also between these volumes and the volume occupied by the gas or vapour of the compound body. It should be remarked moreover that as far as we know at present, the volumes of the combining gases are always reduced to *two volumes* after combination." (Wurtz, *The Atomic Theory*.) Without any further qualification or explanation, it is clearly impossible to assign any meaning to " are always reduced to two volumes." What are two volumes, and what is the measure of the gases that are reduced ? But even when expanding the statement, and saying what it is presumably intended to mean, namely, "if the experimentally found ratios between the volumes of the constituents, and the volume of the compound gas, are expressed by the nearest simple whole numbers, the value for the compound will always be the number two "; this is not absolutely general and hence of no value. The following would constitute exceptions :

1 volume ethylene + 1 volume chlorine = 1 volume ethylene chloride.
1 volume carbonic oxide + 1 volume chlorine = 1 volume phosgene.
1 volume phosphorus + 6 volumes hydrogen = 4 volumes phosphine.

may yet be useful to give all the steps of the argument whereby it is arrived at :

Take any two gaseous constituents A and B, which unite to form the gaseous compound C; then by Dalton's atomic hypothesis :

p atoms of A combine with q atoms of B, to form 1 atom of C, and by Gay-Lussac's law of combining volumes :

r vols. of A combine with s vols. of B, to form t vols. of C, where p, q, r, s, t are all simple whole numbers.

Let the number of atoms in 1 vol. of A be X,

then „   „   „ r „ „ will be Xr,

and $\therefore$ to 1 atom of A there are $\dfrac{q}{p}$ atoms of B and $\dfrac{1}{p}$ atoms of C,

$\therefore$ „ Xr „ „ „ $\dfrac{Xrq}{p}$ „ „ $\dfrac{Xr}{p}$ „ „

but $\dfrac{Xrq}{p}$ being the number of atoms contained in s vols. of B

and $\dfrac{Xr}{p}$ „ „ „ „ „ „ „ t „ „ C,

$\therefore$ the number of atoms contained in

1 vol. of A, B, C is X, $\dfrac{Xrq}{ps}$, $\dfrac{Xr}{pt}$ respectively,

and the ratio of the number of atoms contained in equal vols. of A, B, C is

$$1 : \frac{rq}{ps} : \frac{r}{pt},$$

and since p, q, r, s, t are all simple whole numbers such as $1, 2, 3$, etc., it follows that the number of atoms contained in equal volumes of gases, elementary or compound, must be in the ratio of simple whole numbers.

There is nothing in the above to settle what that simple whole number should be, but the simplest, and at the same time the most legitimate assumption is that it should be unity, making the number of atoms[1] in equal volumes the same, and hence leading to direct proportionality between atomic weight and gaseous density.

Equal volumes of gases, elementary or compound, contain equal numbers of atoms.

Whilst it would require further explanation, why the number of atoms in a certain volume of one gas

[1] See Dalton's definition of "atom," p. 288, 3.

should be exactly twice the number contained in an equal volume of another gas, and three times that in a third, etc., the identity of this number may be taken as a characteristic of the gaseous state, furnishing at the same time a satisfactory explanation of the fact that all gases exhibit the same volume changes under changes in pressure (Boyle's, Mariotte's law), and changes in temperature (Charles', Gay-Lussac's law).

What was Dalton's attitude towards this contribution to the knowledge of the combining ratios of chemical substances ? It will be best to let him speak for himself:

"At the time I formed the theory of mixed gases I had a confused idea, as many have, I suppose, at this time, that the particles of elastic fluids are all of the same size; that a given volume of oxygenous gas contains just as many particles as the same volume of hydrogenous...But from a train of reasoning similar to that exhibited at page 71, I became convinced that different gases have *not* their particles of the same size : and that the following may be adopted as a maxim, till some reason appears to the contrary : namely,—

"That every species of pure elastic fluid has its particles globular and all of a size ; but that no two species agree in the size of their particles, the pressure and temperature being the same." (Dalton, *A New System of Chemical Philosophy*, 1808.)

What is the train of reasoning referred to ?

Dalton denies equality of number of atoms in equal volumes of gases, because incompatible with indivisibility of elementary atoms.   "It is evident the number of ultimate particles or molecules in a given weight or volume of one gas is not the same as in another ; for, if equal measures of azotic and oxygenous gases were mixed, and could be instantly united chemically, they would form nearly two measures of nitrous gas, having the same weight as the two original measures ; but the number of ultimate particles could at most be one half of that before the union." (*Ibid.*)

Here Dalton, before the publication of Gay-Lussac's work, at a time when the assumption of equal numbers of constituent particles in equal volumes of gases was still merely a conjecture of his own, perceives a real difficulty, a fundamental contradiction between this view and his atomic hypothesis. Whenever the volume of the compound gas is greater than that of any one of its constituents, the number of compound atoms formed, of which each one must contain at least one atom of every constituent, would be greater than the available number of atoms of the constituent of lesser volume.

This clearly cannot happen, unless the elementary atom in the process of combination splits into parts, which according to Dalton's fundamental hypothesis is impossible. Thus :

1 vol.   nitrogen + 1 vol.   oxygen = 2 vols. nitrous gas (nitrous oxide).
X atoms   ,,   + X atoms   ,,   = $2X$ atoms nitrous gas, each containing $\frac{1}{2}$ atom oxygen and $\frac{1}{2}$ atom of nitrogen.

1 vol.   oxygen + 2 vols.   hydrogen = 2 vols. water.
X atoms   ,,   + $2X$ atoms   ,,   = $2X$ atoms water, each containing $\frac{1}{2}$ atom oxygen and 1 atom hydrogen.

1 vol.   nitrogen + 3 vols.   hydrogen = 2 vols. ammonia.
X atoms   ,,   + $3X$ atoms   ,,   = $2X$ atoms ammonia, each containing $\frac{1}{2}$ atom nitrogen and $1\frac{1}{2}$ atoms hydrogen.

These considerations which would seem to have induced Dalton to give up his own hypothesis concerning the equality of the number of atoms in equal volumes of the gases, account also for his uncompromising rejection of Gay-Lussac's work later on. Of the logical validity of the inference drawn from Gay-Lussac's work there could be no doubt; and Dalton raises no objections on this ground. The correctly drawn inference when applied deductively leads to a conflict with the fundamental hypothesis, hence a source of error must lie in the premises ; either the Daltonian hypothesis requires modification, or the experimental law is not correct. Dalton preferred to believe the latter, and after criticising the numbers determined or accepted by Gay-Lussac, and comparing them with those obtained by other investigators,—and there was no difficulty whatever about finding plenty which differed greatly from Gay-Lussac's—he proceeds to say :

Dalton believes experimental foundation of Gay-Lussac's law to be incorrect.

"The truth is, I believe, that gases do not unite in equal or exact measures in any one instance ; when they appear to do so, it is owing to the inaccuracy of our experiments. In no case, perhaps, is there a nearer approach to mathematical exactness, than in that of one measure of oxygen to two of hydrogen; but here the most exact experiments I have ever made gave 1·97 hydrogen to 1 oxygen." (Dalton, *A New System of Chemical Philosophy*, 1810.)

And to this attitude he adhered, in spite of the fact that his contemporaries looked upon Gay-Lussac's work as a powerful corroboration of the atomic hypothesis, a view which finds expression in a letter of Berzelius to Dalton, quoted in Roscoe's "John Dalton and the Rise of Modern Chemistry."

"I believe however that in the theory such as the science owes it to you, there are parts which require some modification. That part, for instance, which compels you to declare inaccurate the experiments of Gay-Lussac on the combining volumes of gases. I should have thought rather, that these experiments are the most beautiful proof of the probability of the atomistic theory ; and besides I must confess, that I should not so easily believe Gay-Lussac at fault, especially in a matter where it is only a case of measuring well or badly."

And Gay-Lussac's work has stood the test of further investigations. His data have been corroborated and extended, and the generalisation known as "Gay-Lussac's law of the combining volumes of gases" ranks as one of the fundamental laws of the science of chemistry.

Gay-Lussac's data corroborated.

Many ingenious experiments have been devised by Hofmann for ascertaining these ratios, and for demonstrating them by lecture experiments[1]. His methods and apparatus are still used for this purpose, and good approximations to whole numbers have always been obtained in such work. But extreme accuracy was not aimed at by Hofmann, and until comparatively recently, the classing of the law of combining volumes as an exact law rested on no good evidence. And yet this is a matter of great practical importance to the chemist because of its bearing on the determination of combining weights. The combining weights of all gaseous elements were supposed to be to each other in the *exact* ratio of their respective gaseous densities, a supposition only warranted if the ratio of their combining volumes is *exactly* that of *simple whole numbers*.

Is the law "exact" or "approximate"?

The determination of the *exact* ratio by volume in which hydrogen and oxygen combine to form water, was taken in hand almost simultaneously by various investigators, and their final results, obtained by very different methods, show excellent agreement.

Accurate determination of the composition of water by volume.

The method used by Dr Scott[2] was a direct one, consisting in the explosion of known volumes of the

[1] A. W. Hofmann, *Introduction to Modern Chemistry*, 1865.
[2] "On the Composition of Water by Volume," London, *Phil. Trans. R. Soc.*, 184, 1893 (p. 543).

two gases which had been prepared with all possible care to ensure purity, and in the measurement of the residual gas in which the impurities present were determined. The table on page 314 gives the results of a series of twelve experiments, in which the source of the hydrogen was palladium hydride, and that of the oxygen silver oxide prepared with special precautions against the formation of carbonate.

When tested by the general conclusions concerning the characteristics of exact and approximate laws, arrived at in chap. III (p. 100), these experiments must be interpreted as proving a real discrepancy between actually existing relations and the generalisation of Gay-Lussac.

The mean values obtained for the ratio by volume in which hydrogen and oxygen combine to form water is:

Scott   2·00245 : 1.

Leduc   2·0037 : 1.

        2·0024 : 1 (corrected by Morley for the deviations from Boyle's law, not taken into account by Leduc).

Morley 2·00268 : 1.

A similar deviation from whole numbers has been observed by Dr Scott in the combining ratio of carbonic oxide and oxygen;

Gay-Lussac's Law found "approximate." and there is little doubt that the same would be found to hold in the case of all such ratios, were they but amenable to very exact direct measurements.

The following might be taken as indirect evidence for such an occurrence in the case of the combining volumes of oxygen and nitrogen. The combining weight of nitrogen (oxygen = 16) has been determined by a variety of chemical methods yielding concordant results:

Discrepancy between chemical and physical value for combining weight of nitrogen, assuming accuracy of Gay-Lussac's law at atmospheric pressure.

Stas (silver : silver nitrate)	
,, (potassium chloride : potassium nitrate)	
,, (sodium chloride : sodium nitrate)	... 14·041
,, (lithium chloride : lithium nitrate)	
Dean .... (silver : silver cyanide)	. 14·031
,, (silver cyanide : potassium bromide)	... 14·055

The value obtained from the relative density of oxygen and nitrogen differs appreciably from the above:

Scott's Determination of the Composition by Volume of Water.

Hydrogen volume as measured and corrected to 0°C. and 760 mm.	Oxygen volume as measured and corrected to 0°C. and 760 mm.	Hydrogen in residue	Oxygen in residue	Impurities determined by potash and pyrogallol	Ratio of combining volume of hydrogen to that of oxygen when the impurity was assumed to be	
					equally distributed in both gases	all in the hydrogen
6863·8	3443·8		15·4	0.3	2·0020	2·0019
6870·0	3432·9		2·1	·0		2·0024
6870·1	3439·7		9·2	·0		2·0026
6848·7	3422·1		2·9	·0		2·0030
6792·5	3386·6	13·5		·0		2·0022
6809·2	3399·5	1·5		·0		2·0025
6793·9	3399·6		7·7	·0		2·0029
6789·6	3389·5	2·9		·0		2·0023
6808·5	3396·4	6·0		·0		2·0028
6793·1	3395·8		2·1	·0		2·0017
6786·5	3395·0		5·4	·0		2·0022
6814·8	3411·9		9·3	·0		2·0028

Mean = 2·00245

Specific Gravity Nitrogen at atmospheric pressure.
Oxygen = 16.

Rayleigh[1]	14·003
Leduc	14·005

The discrepancy between the chemical and the physical values is about ·2 per cent., and on the supposition of the correctness of the chemical value for the combining weight of nitrogen, the value for the ratio of the combining volumes nitrogen : oxygen can be calculated with the following result :

$$\text{Nitrogen} : \text{oxygen} = 1\cdot00257 : 1,$$

a deviation from the simple whole number of the same order of magnitude as that found directly in the case of hydrogen and oxygen, but the existence of which is not supported by the evidence got from the measurement of the compressibilities of the two gases.

It must be obvious that since the volume changes accompanying changes in temperature and pressure, are only "nearly" and not "exactly" the same for all gases, Gay-Lussac's law, even if "exact" for one set of physical conditions, would not be so for others. Hence the inference from it, that equal volumes of different gases at the same temperature and pressure contain equal numbers of constituent particles, also cannot be perfectly generally true, and would hold only at very low pressures at which the conditions approximate to those of an ideal gas. And it is only the density ratios under these conditions which lend themselves to the purpose of the comparison of the weights of these ultimate particles. Berthelot has calculated[2] these values which he terms "limiting densities" by multiplying Leduc's "normal densities," that is the values found experimentally at 0°C.

Is Gay-Lussac's law exact at conditions approximating to those of ideal gas?

---

[1] The agreement between the results of these two investigators is not quite so good when the standard is air.

	Specific Gravity (air=1)		Specific Gravity (oxygen=1)	
	Rayleigh	Leduc	Rayleigh	Leduc
Oxygen	1·10535	1·10523		
Nitrogen	0·96737	0·96717	0·87507	0·87508

which is due no doubt, as has been pointed out by Leduc, to differences in the standard ; Lord Rayleigh's sample of air being slightly less dense, owing to a slightly smaller percentage of oxygen contained in it.

[2] Paris, *C.-R. Acad. Sci.*, 126, 1898 (p. 1030).

and atmospheric pressure by a factor $(1 - e)$, where e represents
the deviation, between 0 atm. and 1 atm., of the
compressibility of the gas from that of a perfect gas.
A similar correction has been applied by Lord
Rayleigh to his own experimental numbers[1]. The
ratios for these densities give the following combining weight values :

Combining
weights   of
some gaseous
elements from
the values of
the "limiting
densities."

Combining Weight		Oxygen standard	Hydrogen	Nitrogen	Chlorine
(i) From ratio of limiting densities	Berthelot	16·000	1·0074	14·005	35·479
	Rayleigh		1·0088	14·009	
(ii) By chemical methods		16·000	1·0076	14·04	35·45

An inspection of these numbers shows that whilst for hydrogen
the agreement between the " physical " and the " chemical " value
is very good, this is not the case for nitrogen and for chlorine.
Considering only nitrogen, for which the perfect agreement of
Lord Rayleigh's and Leduc's results excludes the possibility of
an explanation by a sufficiently large experimental error in the
density, it follows either that the calculation of the " limiting
densities" leaves out of account some factor of influence, or that
the "chemical value" is in need of further revision still. As
Lord Rayleigh has pointed out quite recently, this is a question
deserving the attention of chemists. If Gay-Lussac's law be
strictly true, it seems impossible that the atomic weight of nitrogen
can be 14·04. This conclusion is supported by the results of a
redetermination of this value, just published.

MM. Guye and St Bogdan in a preliminary notice (*C.-R. Acad.
Sci.* 138, 1904, p. 1494) give some account of their analyses of
nitrous oxide. The ratio measured was $N_2O : O$, and it is pointed
out that this method of determining the atomic weight of nitrogen
has the advantage over all others used before, that no error is
introduced due to incorrectness in the values of antecedent data.
The numbers obtained for the atomic weight of nitrogen varied
between 13·992 and 14·023.

"The mean 14·007 must not be looked upon as final. Nevertheless,
considering that it agrees within $\frac{2}{10000}$ with the result of physico-chemical
methods, we conclude that the value now accepted for the atomic weight of
nitrogen must certainly be revised, so as to bring it very near to 14·01."

[1] London, *Proc. R. Soc.*, 73, 1904 (p. 153).

CHAPTER XII.

AVOGADRO AND THE MOLECULAR HYPOTHESIS.

" To search thro' all...
And reach the law within the law."

TENNYSON.

THE Daltonian conception of the indivisible elementary atoms of fixed characteristic weight, which by their combination in simple numbers give rise to the compound atoms, owed its rapid and signal success to the ease and simplicity with which it accounted for the then known fundamental laws of chemical combination. In its original form it proved itself, however, inadequate to deal with Gay-Lussac's law of the combining volumes of gases, an experimental relation discovered soon after the promulgation of the Daltonian hypothesis. The difficulty lay in the fact that the inferences from the newly discovered relation of combining volumes were incompatible with the chemical indivisibility of the elementary atoms as postulated by Dalton. Clearly the thing required was a modification of this part of the original hypothesis, *i.e.* the assumption of two orders of finite particles, (i) those which are the result of mechanical division carried to its utmost conceivable limits, but which still may have parts, and (ii) these constituent parts themselves, which though incapable of independent existence, are the units of chemical interaction.

It has been shown how Dalton, though urged thereto by

[side note: Indivisibility of elementary atom incompatible with inference from law of combining volumes.]

Berzelius, refused to introduce any modification into his original hypothesis. This step was taken by Amadeo Avogadro[1] in 1811, and by Ampère[2] in 1814. Avogadro's paper on the subject is the earlier and more important. His line of argument and his final results are contained in the following portions of his Memoir[3]:

Distinction between two orders of finite particles. Avogadro's 1811 paper.

M. Gay-Lussac has shown...that gases always unite in a very simple proportion by volume, and that when the result of the union is a gas, its volume also is very simply related to those of its components. But the quantitative proportions of substances in compounds seem only to depend on the relative number of molecules[4] which combine, and on the number of composite molecules which result. It must then be admitted that very simple relations also exist between the volumes of gaseous substances and the numbers of simple or compound molecules which form them. The first hypothesis to present itself in this connection, and apparently even the only admissible one, is the supposition that the number of integral molecules in any gases is always the same for equal volumes, or always proportional to the volumes. Indeed, if we were to suppose that the number of molecules contained in a given volume were different for different gases, it would scarcely be possible to conceive that the law regulating the distance of molecules could give in all cases relations so simple as those which the facts just detailed compel us to acknowledge between the volume and the number of molecules. ...Setting out from this hypothesis, it is apparent that we have the means of determining very easily the relative masses of the molecules of substances obtainable in the gaseous state, and the relative number of these molecules in compounds ; for the ratios of the masses of the molecules are then the same as those of the densities of the different gases at equal temperature and pressure, and the relative number of molecules in a compound is given at once by the ratio of the volumes of the gases that form it. For example, since the numbers 1·10359 and 0·07321 express the densities of the two gases oxygen and hydrogen compared to that of atmospheric air as unity, and the ratio of the two numbers consequently represents the ratio between the masses of equal volumes of these two gases, it will also represent on our hypothesis the ratio of the masses of their molecules. Thus the mass of the molecule of oxygen will be about fifteen times that of the molecule of hydrogen, or, more exactly, as 15·074 to 1....On the other hand, since we

Avogadro's hypothesis: Equal volumes of gases contain equal numbers of molecules.

Determination of relative molecular weights from gaseous densities.

[1] Amadeo Avogadro (1776–1856) Professor of Physics at Turin.
[2] Ampère (1775–1836) Professor of Mathematics first at Lyons and then at the Paris École Polytechnique.
[3] The translation followed is that of The Alembic Club Reprints, No. 4.
[4] Avogadro's terminology of the different orders of ultimate particles in terms of the one now employed is: *Molecule* = molecule or atom ; *constituent molecule* = molecule of an element ; *integral molecule* = molecule of a compound ; *elementary molecule* = atom of an element.

know that the ratio of the volumes of hydrogen and oxygen in the formation of water is 2 to 1, it follows that water results from the union of each molecule of oxygen with two molecules of hydrogen.

...There is, [however], a consideration which appears at first sight to be opposed to the admission of our hypothesis with respect to compound substances...the volume of water in the gaseous state is, as M. Gay-Lussac has shown, twice as great as the volume of oxygen which enters into it....But a

Divisibility of elementary molecules.

means of explaining facts of this type in conformity with our hypothesis presents itself naturally enough : we suppose, namely, that the constituent molecules of any simple gas whatever...are not formed of a solitary elementary molecule [atom of an element], but are made up of a certain number of these molecules [elementary atoms] united by attraction to form a single one [elementary molecule]; and further, that when molecules of another substance unite with the former to form a compound molecule, the integral molecule [molecule of a compound] which should result splits up into two or more parts composed of half, quarter, etc. the number of elementary molecules [atoms],...so that the number of integral molecules of the compound becomes double, quadruple, etc. what it would have been if there had been no splitting up, and exactly what is necessary to satisfy the volume of the resulting gas. On reviewing the [volumes of the] various compound gases most generally known, I only find examples of duplication of the volume relatively to the volume of that one of the constituents which combines with one or more volumes of the other. We have already seen this for water. In the same way, we know that the volume of ammonia gas is twice that of the nitrogen which enters into it. M. Gay-Lussac has also shown that the volume of nitrous oxide is equal to that of the nitrogen which forms part of it, and consequently is twice that of the oxygen. Finally, nitrous gas, which contains equal volumes of nitrogen and oxygen, has a volume equal to the sum of the two constituent gases, that is to say, double of each of them.

Thus in all these cases there must be a division of the [elementary] molecule into two; but it is possible that in other cases the division might be into 4, 8, etc....M. Gay-Lussac clearly saw that according to the facts, the diminution of volume on the combination of gases cannot represent the approximation of their elementary molecules. The division of [elementary] molecules on combination explains to us how these things may be made independent of each other....Dalton, on arbitrary suppositions as to the most likely relative number of molecules in compounds, has endeavoured to fix ratios between the masses of the molecules of simple substances. Our hypothesis, supposing it well founded, puts us in a position to confirm or rectify his results from precise data, and, above all, to assign the magnitude of compound molecules according to the volumes of the gaseous compounds, which depend partly on the division of molecules entirely unsuspected by this physicist. Thus Dalton supposes that water is formed by the union of hydrogen and oxygen, molecule to molecule. From this, and from the ratio by weight of the two components, it would follow that the mass of the molecule of oxygen would be to that of hydrogen as $7\frac{1}{2}$ to 1 nearly, or, according to Dalton's evaluation, as 6 to 1. This ratio on our hypothesis is, as we saw, twice as great, namely, as 15 to 1."

Avogadro takes the molecular weight of hydrogen as unity, whereas we now regard it as 2, and so he gets for water a different weight from that now accepted:

"As for the molecule of water, its mass ought to be roughly expressed by $15 + 2 = 17$, if there were no division of the molecule into two; but on account of this division it is reduced to half, $8\frac{1}{2}$, or more exactly 8·537, as may also be found directly by dividing the density of aqueous vapour 0·625 (Gay-Lussac) by the density of hydrogen 0·0732....In the case of ammonia, Dalton's supposition as to the relative number of molecules in its composition is on our hypothesis entirely at fault. He supposes nitrogen and hydrogen to be united in it molecule to molecule, whereas we have seen that one molecule of nitrogen unites with three molecules of hydrogen. According to him the molecule of ammonia would be $5 + 1 = 6$: according to us it should be $\dfrac{13 + 3}{2} = 8$, or more exactly 8·119, as may also be deduced directly from the density of ammonia gas. The division of the molecule, which does not enter into Dalton's calculations, partly corrects in this case also the error which would result from his other suppositions."

After many other examples in which the hypothesis is applied legitimately, and a great many others where this is not the case, Avogadro in his concluding paragraph says:

"It will have been in general remarked on reading this Memoir that there are many points of agreement between our special results and those of Dalton....This agreement is an argument in favour of our hypothesis, which is at bottom merely Dalton's system furnished with a new means of precision from the connection we have found between it and the general fact established by M. Gay-Lussac."

It may be well to enumerate and state separately the points of which Avogadro says, "it will have been in general remarked"...and to show in what way he modified the Daltonian hypothesis and what he added to it:

Avogadro's extensions and modifications of the Daltonian hypothesis.

(1) He does away with Dalton's artificial distinction between the ultimate particles of a compound and those of an element, according to which the atom of a compound was assumed to have parts, whilst that of the element had not. According to Avogadro both have parts which in the case of a compound are different, in the case of an element the same. Hence there are

(1) Two orders of ultimate particles, now termed molecules and atoms.

two orders of ultimate particles, now termed molecules and atoms respectively.

(2) The molecule of an element on interacting with another molecule splits into parts, the constituent atoms. The number of atoms in one molecule of an element is generally 2, but may be 4, 8, etc. It should here be noted that Avogadro expressly guards against the assumption that the number of constituent atoms in an elementary molecule must always be 2 ; whether it is by accident or by design that he mentions only powers of 2, we cannot tell.

(2) In chemical change, elementary molecules break up.

(3) Equal volumes of gases, elementary or compound, at the same temperature and pressure contain equal numbers of molecules. This assumption, of such fundamental importance to chemistry that in spite of its being a hypothesis only it is commonly known under the name of "Avogadro's law," is also developed in Ampère's paper in a form so clear and so attractive as to justify quotation :

(3) Equal volumes of gases contain equal numbers of molecules.

"When bodies pass into the gaseous state, their several particles are separated by the expansive force of heat to much greater distances from each other than when the forces of cohesion or attraction exercise an appreciable influence; so that these distances depend entirely upon the temperature and pressure to which the gas is subjected, and under equal conditions of pressure and temperature the particles of all gases, whether simple or compound, are equidistant from each other. The number of particles is on this supposition proportional to the volumes of the gases. Whatever be the theoretical reasons which to me seem to support it, the above conclusion cannot be considered anything but a hypothesis. But if on comparing the inferences which follow from it as a necessary consequence with the phenomena or the properties such as we observe them, the hypothesis should agree with all the known results of experience; if the inferences drawn from it be confirmed by subsequent experiment, it will acquire a degree of probability approximating to what in physics is called certainty."

(4) The relative weights of gaseous molecules can be determined by measurement of the relative weights of equal volumes of the gaseous substances, that is by comparison of gaseous densities.

(4) and (5) Relative weights of gaseous molecules, and relative number interacting, can be determined.

(5) The number of gaseous volumes interacting indicates the relative number of molecules interacting, and similarly the volume of the compound gas formed when compared with that of the con-

21

stituents gives the number, whole or fractional, of elementary molecules entering into the composition of one compound molecule. Avogadro's own statement that the above relations are to take the place of Dalton's arbitrary rules concerning the number of atoms combining, calls for the criticism that in the two cases different orders of ultimate particles are being dealt with, molecules and atoms respectively. Avogadro tells us that two molecules of hydrogen combine with one molecule of oxygen to form two molecules of water, and that each of these molecules of water formed contains one molecule of hydrogen and one half molecule of oxygen. No arbitrary assumptions are made in the course of the argument leading to this result; but the result such as it is tells us nothing about the number of hydrogen atoms in the hydrogen and water molecules, and whilst indicating that the oxygen molecule must contain at least two atoms, the possibility of its containing four or six, etc., *i.e.* any number divisible by 2, is not excluded. Dalton deals throughout with the atoms directly.

All the points just enumerated are readily seen to reduce themselves to two new ideas introduced by Avogadro into the science, namely: (i) The equality of the number of elementary or compound molecules contained in equal gaseous volumes, (ii) the complexity of the molecules of elementary substances.

Avogadro's two conjectures now generally accepted.

These speculations of Avogadro produced practically no effect at the time of their promulgation, and half a century elapsed before they met with the recognition which they deserved. How this came to be so will be dealt with later, together with the exposition of the manner in which Avogadro's hypothesis—or as it is also called, Avogadro's law—is applied, and how the molecular hypothesis has grown into the molecular theory of to-day. It may, however, be useful to forestall this by a short summary of the main reasons for the present universal acceptance of the two Avogadrian ideas above named.

(1) The empirical laws connecting volume, temperature and pressure (Boyle's and Charles' laws), which are the same for all gases, find an easy explanation in the assumption of a common constitution of these bodies. When we remember that the volume of the liquid water formed is so small as to be negligible when compared with that of its gaseous constituents, hydrogen and

oxygen, we must recognise that the actual volume occupied by the particles constituting a gas may be conceived as negligible relatively to that occupied by the gas as a whole. Hence if, according to Avogadro's hypothesis, the number of these particles in equal volumes is the same whatever the nature of the gas, the distances between the particles must also be the same; and the fact that these initially equal distances are by similar changes in the conditions maintained equal, seems but natural.

The hypothesis accounts for laws of gaseous compressibility, expansion, and combining volumes.

And the assumption of equal numbers of molecules in equal volumes of gases, elementary or compound, together with that of the complexity and consequent divisibility of the elementary molecules, obviously gives a satisfactory explanation of Gay-Lussac's law of combining volumes.

(2) The relation between a gaseous volume and the number of molecules contained in it, known as Avogadro's hypothesis (or law), follows as a deduction from the kinetic theory of gases, and as such possesses all the probability attaching to this theory as a whole. The assumptions of the kinetic hypothesis (*ante*, pp. 29, 97) lead by calculation[1] to a result identical with Avogadro's law, the outline of the argument being as follows: According to the kinetic theory, which conceives gaseous pressure to be due to the impacts of the moving molecules on the sides of the containing vessel, this pressure must be proportional to the number of such blows in unit time, and to the average magnitude of each blow. Of these two factors, the second varies as mv, where m is the mass of each molecule and v its velocity, whilst the number of impacts occurring in unit time varies as n the number of molecules present in unit volume, and as v the average velocity with which these move. Hence p, the pressure exerted by a gas, is directly proportional to mnv^2, and for equal volumes of gases which exert the pressure p_1 and p_2 respectively we have

The hypothesis follows as deduction from the kinetic theory.

$$p_1 : p_2 = m_1 n_1 v_1^2 : m_2 n_2 v_2^2.$$

But, according to the kinetic hypothesis, temperature is proportional to the energy of motion of the constituent particles, and the quantity $\frac{1}{2}mv^2$, termed the kinetic energy, is the mechanical

[1] Ostwald, *Outlines of General Chemistry*, 1890 (pp. 60—64). Walker, *Introduction to Physical Chemistry*, 1903 (pp. 90, 91).

measure of temperature. Hence if the two gases above considered are at the same temperature and at the same pressure, $m_1v_1^2 = m_2v_2^2$, and $p_1 = p_2$, and therefore $n_1 = n_2$, that is, when two gases are at the same pressure and temperature, the number of molecules in unit volume is the same for both gases.

The train of reasoning necessary for arriving at this inference has been indicated without resorting to any symbols or equations whatever:

"If I have two vessels containing gas at the same pressure and the same temperature (suppose that hydrogen is in one and oxygen in the other), then I know that the temperature of the hydrogen is the same as the temperature of the oxygen, and that the pressure of the hydrogen is the same as the pressure of the oxygen. I also know (because the temperatures are equal) that the average energy of a particle of the hydrogen is the same as that of a particle of the oxygen. Now the pressure is made up by multiplying the energy by the number of particles in both gases; and as the pressure in both cases is the same, therefore the number of particles is the same." (W. K. Clifford, *Atoms, Lectures and Essays*, 1879.)

(3) The complexity of elementary molecules, first recognised by Avogadro, is supported by evidence derived from the department of chemical dynamics. The *atomicity* of elements, that is, the number of atoms in the gaseous molecule of an element (*post*, chap. XVII) can be found from a study of the volume relations between the gaseous element and its gaseous compounds, and knowledge of the atomicity leads to inferences concerning the minimum number of atoms participating in, and the minimum number of molecules resulting from, chemical changes in which these elements play a part. So in the case of the formation of hydriodic acid from its constituents, at temperatures at which iodine is gaseous, the fact that

Equilibrium between HI and its components is such as is deduced from supposed complexity of hydrogen and iodine molecules.

1 vol. hydrogen + 1 vol. iodine = 2 vols. hydriodic acid

leads to the recognition that the hydrogen and the iodine molecule must each consist of (at least) 2 atoms, and that the formation of hydriodic acid must therefore be represented by the equation

$$H_2 + I_2 = 2HI$$

and not by $\quad\quad H + I = HI.$

The reaction is never complete, and it is reversible; hydrogen and iodine, when heated, combine partially to form hydriodic acid;

gaseous hydriodic acid, when heated, decomposes partially into its component gases; the amount of combination and of decomposition depends on the temperature, and at any definite temperature the equilibrium condition, that is the distribution into hydrogen, iodine, and hydriodic acid, is independent of the original distribution. It is the same whether we started from the elementary substances, or from the compound, or from a mixture of the two. The view at present held is that such a state of equilibrium is the result of *balanced actions*, of two reactions the opposite of one another, which occur simultaneously and at the same rate. In every time unit, the number of hydriodic acid molecules formed is equal to the number of such molecules decomposed. Hence equilibrium results when the velocity of the reaction

$$\text{hydrogen} + \text{iodine} = \text{hydriodic acid}$$

is the same as that of the reverse reaction

$$\text{hydriodic acid} = \text{hydrogen} + \text{iodine}.$$

In accordance with our present views on affinity, on the basis of Berthollet's, and Guldberg and Waage's law of mass action, the amount of each of these reactions occurring in any time interval will depend on the actually present number of molecules of the substances required for the reaction and on a coefficient, the value of which is conditioned by the nature of the reaction and the physical conditions, and which therefore for the same physical conditions is a constant.

If u, v, w are the active masses of hydrogen, iodine, and hydriodic acid respectively, and c, c' the velocity coefficients for the formation and decomposition of hydriodic acid, it is evident that the equilibrium equation will be different according to the molecular formulation of the reaction; that for $H_2 + I_2 = 2HI$, it will be

$$cuv = c'w^2,$$

$$\frac{uv}{w^2} = \text{constant} \quad \dots\dots\dots\dots\dots\dots(i),$$

whilst for $H + I = HI$, it will be

$$cuv = c'w,$$

$$\frac{uv}{w} = \text{constant} \quad \dots\dots\dots\dots\dots\dots(ii).$$

Experimental measurements made in various investigations on the equilibrium conditions of the system hydrogen, iodine, and hydriodic acid[1], are in good agreement with the requirements of equation (i), and therefore indirectly support the molecular complexity of hydrogen and iodine, which was expressed in the formula $H_2 + I_2 = 2HI$.

(4) The application of Avogadro's hypothesis, according to which the ratio of gaseous densities is also the ratio of the molecular weights, leads to values in perfect agreement with those obtained from the molecular formulae chosen on chemical grounds (*ante*, pp. 196 *et seq.*). The evidence here is based on many thousands of data accumulated mainly though not solely in the province of carbon compounds. Where examples are so abundant, a very few of the simplest will serve best for the purpose of illustration. When determined according to Avogadro's law, the molecular weight of benzene is 78 ($H_2 = 2$), and hence if the atomic weights of carbon and hydrogen are 12 and 1 respectively, the molecular formula becomes C_6H_6, which is in perfect agreement with the fact that the hydrogen of benzene can be replaced by other elements or groups of elements in 6, and not more than 6 stages. The molecular weights of hydrochloric acid and of water lead to the formulae HCl and H_2O, which represent the facts that HCl is a monobasic acid, and that the hydrogen of water can be replaced in two stages.

Avogadro's hypothesis leads to molecular formulae identical with those chosen on chemical grounds.

"The formula H_2O has the advantage [over HO, then commonly used] of representing a fact general in organic chemistry, namely that every monatomic radicle [*i.e.* radicle equivalent to 1 atom of hydrogen] has two oxides, one representing a molecule of water in which 1 volume or atom has been replaced by the equivalent of the radicle, the other representing a molecule of water in which both volumes or atoms of hydrogen have been replaced.

1 molecule of
water = 2 vols.   Oxides of the radicle *ethyl*

H	C₂H₅		C₂H₅	
O	O	$= \begin{matrix} 2 \text{ vols.} \\ \text{alcohol} \end{matrix}$	O	$= \begin{matrix} 2 \text{ vols.} \\ \text{ether.} \end{matrix}$
H	H		C₂H₅	

---

[1] Walker, *Introduction to Physical Chemistry*, 1903 (p. 253). Ostwald, *Outlines of General Chemistry*, 1890 (p. 308); *Lehrbuch*, vol. ii, part ii, 1896–1902 (p. 494).

Oxides of the radicle *acetyl*

H		C_2H_3O		C_2H_3O
O		O $= \begin{matrix} \text{2 vols.} \\ \text{acetic acid} \end{matrix}$		O $= \begin{matrix} \text{2 vols.} \\ \text{acetic anhydride} \end{matrix}$
H		H		C_2H_3O

Oxides of the radicle *potassium*

H		K		K
O		O $= \begin{matrix} \text{potassium} \\ \text{hydrate} \end{matrix}$		O $= \begin{matrix} \text{potassium} \\ \text{oxide.”} \end{matrix}$
H		H		K

(Gerhardt, *Traité de Chimie Organique*, IV, p. 582.)

(5) The recognition of the complexity of elementary molecules helps us to account simply for a number of chemical phenomena.

Hydrogen and oxygen when mixed remain uncombined; the temperature must be raised before the reaction resulting in the formation of water occurs. If the combination consisted simply in the coalescence of the elementary particles, it would not be very obvious why it should occur at one temperature and not at another. But if the reaction is supposed to consist of two parts, decomposition of the elementary molecules, and combination of the parts of these molecules, the effect of the rise of temperature might be to facilitate the first of these reactions. The argument is not invalidated by a refusal to believe that a decomposition actually precedes the combination; it is enough to grant that the attraction between the hydrogen and oxygen atoms is reduced to a degree such that the attraction between the component parts of the different molecules is sufficient to overcome it.

Complexity of elementary molecules accounts for certain chemical reactions. (i) Combination of H_2 and O_2.

The same line of argument may be followed for the purpose of explaining *nascent action*. This is a name given to all those phenomena in which a substance at the moment of its liberation from compounds performs reactions it is incapable of in its ordinary condition. Thus to cite simple and well-known cases: Hydrogen bubbled through has no action on silver chloride suspended in a liquid; hydrogen evolved within the liquid, in contact with the silver chloride, produces metallic silver, hydrochloric acid being formed at the same time.

(ii) Nascent action.

Silver chloride + molecular hydrogen…no change.

{Zinc + sulphuric acid = zinc sulphate + hydrogen.
{Silver chloride + hydrogen = silver + hydrochloric acid.

Oxygen bubbled through a liquid containing lead hydrate in suspension produces no effect; oxygen evolved in contact with it (by the action of chlorine on water) changes the lead hydrate to lead peroxide.

Lead hydrate + molecular oxygen…no change.

{Chlorine + water = hydrochloric acid + oxygen.
{Lead hydrate + oxygen = lead peroxide + hydrochloric acid.

Of such nascent hydrogen and oxygen it has been assumed that they are liberated in the atomic state, and that the actions described are performed before the atoms have coalesced to form molecules. That the elements do not produce the same effect when in the ordinary, the presumed molecular state, is supposed to be due to the fact that here also a decomposition of the complex molecule[1] must accompany, though it need not actually precede, the action of its constituent parts. But the elements at the moment of liberation from compounds may for a time, however short, be assumed to be in the atomic and not in the molecular condition, and if so, the first stage of the reaction required in the case of elements in the molecular state becomes unnecessary.

"A binary association of atoms might allow us also to account to a certain extent for the affinity possessed by substances in the nascent state. If two free molecules of bromine and of hydrogen, BB' and HH', are brought together, the affinity of B for B' and of H for H', may suffice to prevent the combination of B and B' with H and H'; but if the substances present are H and B, these two which have no affinity to overcome will be able to combine readily. This is what will occur when the hydrogen is in the nascent state; that is, whenever it is evolved from hydrochloric acid by the action of a metal, we shall have the equation

$$HCl + M = ClM + H,$$

and there will be a tendency to a reconstruction of a binary molecule either by combination with bromine or with another atom of hydrogen." (Laurent, *Ann. Chim. Phys.*, 1846.)

---

[1] The assumption of such molecular decomposition is however not essential to the explanation of *nascent action*, which can be accounted for simply by a consideration of thermal effects.

Brodie has brought together and investigated a number of cases to prove that "the particles of an element have a chemical affinity for each other[1]." Thus in cases of sub-

(iii) Affinity between par- ticles of an element.

stitution by chlorine, for every hydrogen particle substituted two chlorine particles must enter into the reaction.

"When one particle of chlorine combines with the hydrogen of an organic body, another particle of chlorine is thrown into an opposite chemical condition, and therefore rendered capable of combining with the remaining elements of the same....

[Solution of bichromate of potash is stable, so is an acid solution of peroxide of barium,] but together they are both decomposed,

$$3ClH + 2CrO_3 + 3HO_2 = Cr_2Cl_3 + 6HO + 6O_2.$$

I regard the oxygen itself as the true reducing agent, and I believe that one particle of oxygen removes the other from combination, in the same sense and for the same reason as would a particle of hydrogen, were zinc thrown into the acid solution, the group HOO replacing the group ZnOH in the de- composition[2]."

[1] "On the Condition of certain Elements at the Moment of Chemical Change," London, *J. Chem. Soc.*, 4, 1852 (p. 194).
[2] In Brodie's formulae, $O = 8$.

CHAPTER XIII.

CANNIZZARO AND THE APPLICATION OF AVOGADRO'S HYPOTHESIS TO THE DETERMINATION OF MOLECULAR AND ATOMIC WEIGHTS.

" Da in der Chemie nur zu haüfig, und fast gewohnheitsmässig Hypothesen für Thatsachen angesehen, oder wenigstens wie solche gehandhabt werden, ist es vor allem nöthig, sich darüber klar zu werden, wo in der Chemie das Gebiet der Thatsachen aufhört und das der Betrachtungen und Hypothesen anfängt."

KEKULÉ.

AVOGADRO'S work met with no appreciation at the time of its promulgation. Nearly half a century had to elapse before its value was understood and before it began to be used legitimately and systematically. How was it that with the material at hand for emerging to clearness and unity, chemists should for so long have put up with vagueness and uncertainty? And what was it that led to the eventual triumph of the Avogadrian hypothesis? To give a clear and complete answer to these questions would involve a detailed exposition of the history of the science between 1811 and 1860. No more can be attempted here than a short account of the nature of the chief obstacles Avogadro's hypothesis had to encounter, and of how these were overcome. To begin with, what was the position in 1811 after Avogadro's paper had

Position of theory of chemical composition in 1814.

been published, or let us say in 1814, when Ampère had added his contribution to the solution of the problem? We have Dalton's hypothesis concerning the combination of simple numbers of elementary atoms to form compound atoms, together with his arbitrary rules for ascertaining these numbers. We have Gay-Lussac's law of the simple combining volumes of gases, which leads of necessity to the assumption of equal numbers of atoms in equal volumes of elementary and compound gases, and hence to the inference that whenever the volume of a compound gas is greater than the volume of one of its constituents, there must occur a division into parts of the elementary atoms which enter into the composition of a greater number of compound atoms; a result which is incompatible with Dalton's fundamental assumption of the indivisibility of the elementary atom. We have Avogadro proposing the necessary modification which the Daltonian hypothesis evidently required, namely that of the complex nature of such elementary particles; two kinds of ultimate particles are assumed: the molecule, beyond which physical division cannot proceed, and the atom, which with other atoms—of the same kind in elements, of different kinds in compounds—forms the molecule, and which is not further divisible either physically or chemically. This modification is sufficient to remove all difficulties from the inference which follows from the combination of the Daltonian hypothesis of atoms with the experimental law of Gay-Lussac, and according to which equal volumes of gases, elementary or compound, contain an equal number of molecules—not of atoms—and which affords a means of directly determining the relative weight of molecules—not of atoms.

Turning to the reasons for the neglect of Avogadro's hypothesis by his own and the next generation of chemists, it is safe to name as the chief of these its then very limited applicability. The number of gaseous and gasifiable substances then known was comparatively small; organic chemistry, which supplies the greatest number of these, was only beginning to be worked at to any degree. The problem then most pressing was that of the determination, not of relative molecular weights but of relative atomic weights, *i.e.* the collection of reliable data concerning the fundamental magnitudes forming all the different compounds, of which but few were volatile.

Reasons for neglect of Avogadro's work. Limited applicability. Paramount importance of atomic weight determinations.

The name most prominently associated with the accomplishment of this gigantic piece of work is that of Berzelius. We have seen before how eagerly he welcomed Dalton's discovery of the law of multiple ratios, how he proceeded to accumulate experimental data in proof of it, how he at once appreciated the vast bearings of the Daltonian hypothesis and saw its application to the law of

Berzelius' atomic weight determinations.

equivalent ratios. From this time onwards through nearly thirty years he worked at the determination of the relative atomic weights, that is of the quantities which the Daltonian hypothesis had made into the fundamental units of the science of chemistry. And in 1845, when himself reviewing his work in this department, he tells us:

"I resolved to make the analysis of a number of salts whereby that of others might become superfluous....I soon convinced myself by new experiments that Dalton's numbers were wanting in that accuracy which was requisite for the practical application of his theory....I recognised that if the newly-arisen light was to spread, it would be necessary to ascertain with the utmost accuracy the atomic weights of all elementary substances, and particularly those of the more common ones. Without such work, no day would follow the dawn. This was therefore the most important object of chemical investigation at the time, and I devoted myself to it with unresting labour....After work extending over ten years...I was able in 1818 to publish a table which contained the atomic weights, as calculated from my experiments, of about 2000 simple and compound substances." (*Lehrbuch.*)

Berzelius, who, following up the idea of Dalton, devised the system of symbolic notation (*ante*, p. 194) still in use, was able to represent the composition of the ultimate particles by a formula giving the names of the constituent elements, the number of atoms of each present, and the weight of each of these atoms

Berzelius' formulae. Criteria for their selection.

relatively to a standard atom. How did he arrive at these formulae? The problem may here again be stated in the form in which it has been put once before. If two elements A and B combine to form a compound C, if m and n are the numbers of atoms of A and B entering into the composition of 1 atom of C, a and b the weights of these atoms, and p and q the experimentally found ratio of the quantities of A and B which combine, we wish from the relation $ma : nb = p : q$ to determine the values of m and n and of $a : b$.

Berzelius recognised the insufficiency and arbitrariness of Dalton's rules for the fixing of the values of m and n, and raised special objection to the assumption that if one compound only is known to exist, that must be the one in which $m = n = 1$. Yet he used these rules, supplemented by some of his own, whereby he of

m and n fixed by rules similar to Dalton's, and according to combining volumes.

course altered nothing in the principle of the method. But while allowing himself to be helped by such rules he would not be bound by them, but checked the results they yielded at every step, comparing them with those arrived at by other methods. Such another method was the application of Gay-Lussac's law of combining volumes, towards which for reasons to be dealt with presently he took a somewhat strange attitude, admitting as he did the equality of the number of constituent particles in elementary gases, whilst denying the same for compound gases. But using this law to the extent that he did, the result was that in the case of combination between gaseous substances the ratio $m : n$, that of the number of atoms uniting, was equal to that of the combining volumes; hence from the fact that two volumes of hydrogen combine with one volume of oxygen, and three volumes of hydrogen with one volume of nitrogen, the inference follows that two atoms of hydrogen unite with one atom of oxygen, three atoms of hydrogen with one atom of nitrogen. The relations between atomic weight and heat-capacity, and between crystalline form and chemical composition which bear the names of Dulong and Petit's law of atomic heat, and Mitscherlich's law of isomorphism, and which will be dealt with in the two subsequent chapters, supplied him after 1819 with valuable aid for settling atomic weights and formulae. But all these methods he used subject to the general principle that *the formula given to a compound should indicate its chemical behaviour*, and especially that *substances similar in their properties should be represented by similar formulae*.

Heat capacity and crystalline forms used in settling atomic weight and formulae, but chemical considerations paramount.

The study of the oxides supplied him with the largest number of his data. In the table of 1818, having ascertained experimentally that the quantity of oxygen combined with the same amount of iron in ferrous and ferric compounds respectively is $2 : 3$, and applying the principle of "greatest simplicity," he assigned to these oxides the formulae FeO_2, FeO_3. The chemical similarity of ferric compounds with those of chromium led to the formula CrO_3 for the basic oxide of chromium; while strong bases, such as the oxides of zinc, manganese, etc., were formulated on the type of ferrous oxide (ZnO_2, MnO_2, etc.), and hence all these metals were given atomic weights twice as great as those now accepted. But Berzelius himself in a table published in 1826 halved the values, and his argument in justification of the change made is so important an exposition of his method as to justify quotation.

"It is known that the oxide of chromium contains three atoms of oxygen. Chromic acid for the same number of chromium atoms contains twice as much oxygen, which would be six atoms; but in its neutral salts chromic acid neutralises an amount of a base containing one-third as much oxygen as it contains itself, a relation found to hold in the case of all acids with three atoms of oxygen[1] (*e.g.* sulphuric acid and sulphates). In order to harmonise the multiple relation between the amount of oxygen in the oxide and in the acid, it is most probable that the acid contains three atoms of oxygen to one atom of chromium, and the oxide three atoms of oxygen to two of chromium. Isomorphous with the oxide of chromium are those of manganese, iron and aluminium; these also we know to contain three atoms of oxygen, and consequently must represent them as containing two atoms of the radicle. But if the ferric oxide consists of $2Fe+3O$, the ferrous oxide is $Fe+O$, and the whole series of oxides isomorphous with it contains one atom of the radicle and one atom of oxygen."

Berzelius' reason for halving in 1826 most of the atomic weights given to the metals in 1818.

He shows that with the exception of cobalt and silver the law of specific heat justifies the change made[2].

"Acting on this decision, I have assigned atomic weights to a large number of elements, such that the stronger basic oxide is composed of one atom of the radicle and one of oxygen....To light upon what is true is a matter of luck, the full value of which is however only realised when we can prove that what we have found *is* true. Unfortunately, in these matters the certainty of our knowledge is as yet at so low a level that all we can do is to follow along the lines of greatest probability." (*Jahresbericht*, 1828.)

It will be seen that Berzelius used the term "atomic weight" for elements and compounds alike, and that the compounds considered are non-volatile oxides to which Avogadro's hypothesis is not applicable.

But Berzelius does not and cannot admit the possibility of the division of elementary atoms, and hence his strange attitude towards the inference from Gay-Lussac's law of volumes. The cause for this is to be found in his inability to reconcile the splitting up of the ultimate particles of elementary substances when they form a binary compound, with his theoretical views concerning the nature of chemical action. The theory known in the history of the science by the

Berzelius on theoretical grounds does not admit divisibility of particles of elements in chemical change.

[1] For the manner in which this relation had been experimentally determined in the case of lead oxide and lead sulphate, see *ante*, p. 182.

[2] The value used for the specific heat of cobalt was erroneous owing to the impurity of the specimen, and the atomic weight of silver was subsequently again halved, making its oxide Ag_2O.

name of the "dualistic system" maintains that every compound substance is made up of two parts oppositely electrically charged, and that all chemical combination consists in the coalescence of two such parts, which may themselves be elementary or compound, but which must always be distinguished by the nature of this charge, thus:

$$\overset{+}{Fe} + \overset{-}{O} = FeO \text{ which still retains a positive charge.}$$

$$\overset{+}{S} + \overset{-}{O_3} = SO_3 \text{ which still retains a negative charge.}$$

$$\overset{+}{FeO} + \overset{-}{SO_3} = FeSO_4, \text{ etc.}$$

$$(\overset{+}{BaO})(\overset{-}{N_2O_5}) + (\overset{+}{K_2O})(\overset{-}{SO_3}) = (\overset{+}{BaO})(\overset{-}{SO_3}) + (\overset{+}{K_2O})(\overset{-}{N_2O_5}).$$

The strictly dualistic view is incompatible with a reaction such as

$$2H_2 + O_2 = H_2O + H_2O,$$

because, according to theory, the compound nature of the oxygen could only be due to a different electrical charge of the component parts of the OO, which is not in agreement with its elementary nature and with the identity of the two water particles formed from it. Hence Berzelius retains Dalton's original view of the absolute indivisibility of the ultimate particles of elements. There can be no manner of doubt that in accepting a view of the simple constitution of elementary gases which he denied to the compounds, a distinction for which there was no ground, considering the identity of the gaseous laws for elements and compounds, he was gravely inconsistent.

Equality of number of constituent particles in equal volumes of gases restricted first to elements, then to "permanent" gases.

But even amongst elementary gases he was led before long to make another arbitrary distinction, admitting the equality of the number of constituent atoms in equal volumes for the so-called permanent gases, oxygen, nitrogen and hydrogen, and denying it for all others. The reason for this is to be found in the results of Dumas' work on vapour densities.

J. B. Dumas[1] (1800–1884) in a paper entitled "Memoir on

---

[1] The whole of his scientific work was carried out in Paris, where he made his mark in all departments of chemistry, pure and applied. The discovery of the relations in organic chemistry included under the name of *substitution* is the most prominent among a number of important results.

some points of the atomistic theory[1]," had set himself a startlingly ambitious task:

"The object of these researches is to replace by definite conceptions the arbitrary data on which nearly the whole of the atomic theory is based."

His appreciation of the situation given a few years later is correct and much to the point.

"Though it is quite easy to establish the ratio in which elements combine, it is very difficult to estimate the actual number of atoms which enter into each of these combinations. Berzelius in his treatise on chemical proportions, which marks so important an epoch in the history of the science...was the first to attack this difficult problem in its full scope. Without any rules to guide, he fixed by intuition the atomic weight of each substance, and usually allowed himself to be influenced by analogies which subsequent experience has only tended to confirm. But chemists have always wished that this arbitrary method, so successfully used by Berzelius, might be supplanted by something more fixed, more accessible to all kinds of intellect, and less subject to the capricious modifications of each writer." (*Ann. chim. phys.*, 50, 1832, p. 170.)

Dumas on uncertainty in atomic weight determinations.

This certainty, the absence of which Berzelius himself so fully and clearly realised, Dumas sees in the acceptance of Avogadro's and Ampère's work. He wishes to make the equality of the number of the constituent particles in equal volumes of gases the basis of atomic weight determinations. He seems to recognise fully the difficulty arising from our ignorance of the number of atoms constituting elementary molecules, and hence the necessity for supplementing the knowledge derived from the determination of the gaseous densities of the elements by the investigation of the amount of splitting up that these must be supposed to undergo on entering into the composition of the compound.

"We are obliged to look upon the molecules of elementary gases as capable of further division, a separation which occurs at the moment of combination and which varies with the nature of the compound formed.... When thus considered, it is evident that in the present state of the science the determination of the true atomic weights of gases or vapours offers insurmountable obstacles. Indeed if the molecules of elements on passing into the gaseous state still remain clustered in certain numbers, we may compare these substances under conditions such that they contain the same number of such clusters, but it is impossible in the present state of our knowledge

Difficulty about finding complexity of elementary gaseous molecule.

---

[1] *Ann. chim. phys.*, Paris, 33, 1826 (p. 337).

to ascertain the number of the elementary molecules contained in each of these."

He thinks that the determination of the density of vapours and gases, elementary as well as compound, is required to throw light on the composition of the elementary molecule. But his practice did not agree with his theory; he as well as all his contemporaries argued from the premise that the vapour densities of elements are proportional not only to their molecular weights but also to their atomic weights, which of course involved the unwarranted assumption that all elementary gaseous molecules are composed of the same number of atoms, *i.e.* of two. Hence it would follow:

Dumas erroneously supposes atomic weights of gaseous elements always proportional to densities.

$$\frac{\text{molecular weight of } A}{\text{molecular weight of } B} = \frac{\text{gaseous (or vapour) density of } A}{\text{gaseous (or vapour) density of } B};$$

and *if* the molecular weight is always $= 2$. atomic weight

$$\frac{\text{atomic weight of } A}{\text{atomic weight of } B} = \frac{\text{density } A}{\text{density } B};$$

$$\therefore \frac{\text{atomic weight of } B}{\text{atomic weight of hydrogen}} = \text{specific gravity of gas (or vapour) } B_{(H=1)}.$$

Dumas by the application of a method devised by himself for the determination of vapour densities of substances volatile at high temperatures only, obtained data which were partly corroborated, partly extended by Mitscherlich, and which, together with the atomic weights deduced from them on the erroneous inference of direct proportionality between vapour density and atomic weight, are given in the table on page 338. D indicates the values due to Dumas, M those afterwards determined by Mitscherlich. The two last columns, which have been added for purposes of comparison, contain the corresponding "chemical" results of Berzelius. Obviously Dumas' values here lead to atomic weights quite incompatible with the chemical relations existing, and with the results arrived at by other physical methods, *e.g.* with the value for mercury obtained by the specific heat method (*post*, chap. XIV). Berzelius' refusal to accept the atomic weights of Dumas

Discrepancy between Dumas' and Berzelius' atomic weights.

F.

22

Atomic Weights deduced from Vapour Densities of the Elements.

Substance and Authority		Temperature of Determination	Specific Gravity, air=1	Atomic Weight = spec. grav. (H=1) = 14·4. spec. grav.(air=1)	Formula of Compound	Berzelius' Atomic Weight	Berzelius' Formula of Compound
Sulphur	D	506° C.	6·512	94·4	H_6S	32·24	H_2S analogous to H_2O
	D	493°	6·495				
	D	524°	6·617				
	D	524°	6·581				
	M		6·90				
Iodine	D	185°	8·716	125·5	HI	126·54	HI
Bromine	M		5·54	79·8	HBr	78·40	HBr
Mercury	D		6·976	100·8	$Hg_2O + Hg_4O$	202·68	HgO and Hg_2O analogous to CuO and Cu_2O, Ag_2O
	M		7·03				
Phosphorus	D	313°	4·420	68·51	H_6P	31·44	H_3P analogous to H_3N
	D	500°	4·355				
	M		4·58				
Arsenic	M		10·6	152·6	H_6As	75·34	H_3As analogous to H_3N

was legitimate, and has been justified by all that has happened since[1].

These results, together with Dumas' loose way of using the terms atom and half-atom, provoked, not unduly, the severe criticism of Berzelius that:

"Till now it has been usual to discard a hypothesis as soon as it leads to absurdities, but to some modern investigators this course seems too inconvenient."

Hence a gallant attempt to rescue Avogadro's hypothesis was unsuccessful, and indeed led to contradictions calculated to bring discredit on the cause advocated; and it is no wonder that considering the complex and involved nature of Berzelius' arguments in choosing the values for atomic weights, and the failure of Dumas' attempt "to replace the arbitrary data by definite conceptions,"

Desire to discard the hypothetical atomic weights. Wollaston's "equivalents."

the desire to do without any such hypothetical quantities should in many quarters have been strong. As far back as 1814 Wollaston[2], already dissatisfied with Dalton's arbitrary rules, had proposed to discard the hypothetical conception of atomic weights, and to substitute for it the empirically determined combining proportions, by him termed *equivalents*.

"According to Mr Dalton's theory, by which these facts" [*i.e.* Richter's law of permanent proportions and Dalton's law of multiple proportions] "are best explained, chemical union in the state of neutralisation takes place between single atoms of the substances combined; and in cases where there is a redundance of either ingredient, then two or more atoms of this kind are united to only one of the other. According to this view, when we estimate the relative weights of equivalents, Mr Dalton conceives that we are estimating the aggregate weights of a given number of atoms, and consequently the proportion which the ultimate single atoms bear to each other. But since it is impossible in several instances, where only two combinations of the same ingredients are known, to discover which of the two compounds is to be regarded as consisting of a pair of single atoms, and since the decision of these questions is purely theoretical, and by no means necessary to the formation of a table adapted to most practical purposes, I have not been desirous of warping my numbers according to an atomic theory, but have endeavoured to make practical convenience my sole guide."

---

[1] Gaudin in a paper published in 1833 had recognised that Dumas' vapour density determinations lead to the inference that whilst the molecules of bromine and iodine, like those of hydrogen, oxygen, nitrogen, and chlorine, are diatomic, the molecule of mercury is monatomic, and that of sulphur hexatomic.

[2] "A Synoptic Scale of Chemical Equivalents," London, *Phil. Trans. R. Soc.*, 104, 1814 (p. 1).

The fact that more than one equivalent number exists for elements uniting in various ratios must have been very clear to the man who had himself been one of the discoverers of the law of multiple proportions, but beyond saying:

"I have considered the doctrine of simple multiples, on which that of atoms is founded, merely as a valuable assistant in determining, by simple division, the amount of those quantities that are liable to such definite deviations from the original law of Richter,"

Wollaston pays no attention to this matter and ignores the difficulty which arises in deciding between these possible multiples of the equivalents.

L. Gmelin, the author of the most popular text-book of the time[1], made himself in Germany the exponent of similar views.

"All speculations upon relative atomic values were to be banished, and only the most sober possible formulation of chemical compounds attempted.

Gmelin's equivalents. The immediate result of this reaction was the halving of a large number of the atomic weights which Berzelius had introduced into the science. In place of the values assumed by him for carbon, oxygen, sulphur, and most of the metals, other values only half as great were taken; these *equivalents* were: $C=6$, $O=8$, $S=16$, $Ca=20$, $Mg=12$, and so on." (E. v. Meyer, *History of Chemistry*.)

What was it that brought about the return to atomic weights? Organic chemistry had accumulated the material adequate to serve as an empirical basis for Avogadro's hypothesis. It was found that in the majority of cases the formulae chosen on purely chemical grounds as best representing the nature of a substance, stand for quantities which in the gaseous state occupy equal volumes. Avogadro's law could therefore be looked upon as arrived at inductively, and its deductive application seemed legitimate.

Return to atomic weights. Laurent and Gerhardt.

Formulae when altered in accordance with the requirements of this law, that is so as to represent quantities which in the gaseous state occupy a standard volume, when judged by the supreme test of chemical suitability were not found wanting. Two names are most prominently associated with the work of establishing the molecular theory, those of Laurent (1807—1853)[2] and Gerhardt (1816—1856)[3]. Laurent clearly distinguished between atom, molecule, and equivalent,

Distinction between atom, molecule, and equivalent.

[1] *Handbuch der Chemie*, 1817; 4th edition, 1843; English translation of 4th ed. 1848—72.   [2] *Ante*, p. 194.
[3] Pupil of Liebig, for some time Professor at Montpellier, and at the end of his life Professor at Strassburg.

and furnished those admirably concise definitions of these quantities which are still current, and which until lately it was usual to meet —where most certainly they were quite out of place—on the first page of elementary text-books of chemistry.

"The *atom* of M. Gerhardt represents the smallest quantity of a simple body which can exist in a combination; my *molecule* represents the smallest quantity of a simple body which must be employed to perform a chemical reaction[1]." (Laurent, 1846.)

"According to our views, the conception of *equivalent* implies that of similarity of function; it is known that the same element can play the part of one or other of several very different elements, and it may happen that to each of these different functions corresponds a different weight of the first element." (Gerhardt, 1849.)

As far back as 1842, at the time when he had not yet arrived at clearness in the distinction between "equivalent" and "atomic weight," Gerhardt[2] had doubled the equivalents of carbonic acid and of water, and in 1843 he followed this up by doubling the values of the symbol weights used by French chemists ($H = 0.5$, $O = 8$, $C = 6$, etc.), making them thereby identical with Berzelius' atomic weights. The reasons for this change were derived from the study of chemical reactions. He found that the amounts of water, carbonic acid, ammonia, and sulphurous acid evolved in the interactions between the quantities of organic compounds represented by the formulae assigned to them on chemical grounds, were always *two* (or multiples of two) equivalent weights of

Gerhardt returns to Berzelius' atomic weights for O, C, and S.

$$H_2O, \ CO_2, \ SO_2, \ \text{when } H = 0.5, \ O = 8, \ C = 6, \ S = 16.$$

"In the decomposition of organic substances we never meet with cases in which CO_2, C_3O_6, C_5H_{10} separate.... Whenever water is split off or added in a reaction, the quantity is always H_4O_2 or a multiple of this."

The following few examples are taken from the very large number of reactions which Gerhardt quotes in support of the above statements:

Benzoic acid = carbonic acid + benzene.
$$C_{14}H_{12}O_4 \qquad C_2O_4 \qquad C_{12}H_{12}$$

Salicylic acid = carbonic acid + phenol.
$$C_{14}H_{12}O_6 \qquad C_2O_4 \qquad C_{12}H_{12}O_2$$

---

[1] This definition of molecule has been severely criticised by Lothar Meyer, who calls it inaccurate and insufficient. (*Modern Theories of Chemistry,* 1888, p. 12.)

[2] *J. Prakt. Chem.*, Leipzig, 27, 1842 (p. 439); 30, 1843 (p. 1).

Grape sugar = carbonic acid + alcohol.

$$C_{24}H_{48}O_{24} \qquad 4C_2O_4 \qquad 4C_4H_{12}O_2$$

Acid oxalate of ammonia = oxamic acid + water.

$$C_4H_4O_8 . N_2H_6 \qquad C_4H_4O_6 . N_2H_2 \quad H_4O_2$$

Glycerine + water = formic acid + acetic acid + hydrogen.

$$C_6H_{16}O_6 \quad H_4O_2 \quad C_2H_4O_4 \qquad C_4H_8O_4 \qquad H_8$$

"M. Gerhardt has observed that in almost all the reactions of organic chemistry, H_4O_2, C_2O_2, C_2O_4, H_4S_2, S_2O_4, S_2O_6, or multiples of these are used; whence he concludes that these formulae should represent the equivalents (it would be better to say the molecules) of oxygen, water, etc." (Laurent, 1846.)

Obviously the formulae thereby became unnecessarily complex, but if the atomic weights of the constituent elements were doubled, H_2O, CO, CO_2, H_2S, SO_2, SO_3 are the quantities which in conformity with the arguments given above, represent the molecules. This change in atomic weight leads to a change in the formula for water from HO (then commonly used in accordance with Gmelin's equivalents, $H = 1$, $O = 8$) to H_2O, and also to the halving of a large number of complicated organic formulae, the remnant of a clumsy notation forced on the chemical world in subjection to Berzelius' dualistic theory (*ante*, p. 335). Gerhardt justifies these innovations, and draws attention to the fact that the new formulae represent quantities which in the gaseous state occupy the same volume.

"I represent the molecule of water by OH_2 making $H = 1$ and $O = 16$. There are two points to consider in this notation OH_2: the first relates to the number of atoms of hydrogen thereby assigned to one molecule of water ; the second relates to the molecular weight of the compounds derived from water by the substitution of another radicle for the radicle hydrogen." (*Traité de Chimie Organique*, 4, 1856, pp. 582 *et seq.*)

Changes in molecular formulae due to doubling of symbol weights of O, C, and S.

Concerning the first point, Gerhardt advocates the acceptance of the formula OH_2 on the ground that the hydrogen of water can be replaced in two successive stages, giving rise to the hydroxide and oxide respectively. The argument in the original form has been quoted already (*ante*, p. 326).

Justification of Gerhardt's molecular formulae.

"The second point in which my notation differs fundamentally from any used before, is that with OH_2 for the unit molecule, I admit that in the case of a number of substances the molecule, *i.e.* the smallest possible amount of the substance which can take part in a reaction, weighs only half the amount usually assigned to it. In my opinion the molecule of alcohol is C_2H_6O, the molecule of acetic acid is $C_2H_4O_2$ and not $C_4H_8O_4$; if the molecule of water is OH_2, the formulae for a great number of substances must be halved in order that they may be correct.

I shall be asked to prove why, if the molecule of water is represented by OH_2, it should be necessary to halve so many formulae, particularly those of the alcohols, aldehydes, hydrocarbons, and of a great number of acids and their salts. The proofs rest on the chemical and physical properties of these substances. The following may be taken as examples: When we study and compare the composition of *equal* gaseous volumes of volatile substances derived from organic and mineral acids, especially their neutral ethers and their chlorides, the quantities under consideration are exactly the same as those by which I represent the molecules of these acids. Thus:

$$\text{A molecule of sulphuric acid} = SH_2O_4 = O_2 \begin{cases} SO_2 \\ H_2 \end{cases}$$

$$\text{A molecule of acetic acid} = C_2H_4O_2 = O \begin{cases} C_2H_3O \\ H \end{cases}$$

$$\text{Equal volumes, } i.e.\ 2\text{ vols.}[1]\text{ of methyl sulphate} = O_2 \begin{cases} SO_2 \\ (CH_3)_2 \end{cases}$$

$$\text{,, \quad ,, \quad ,, \quad ,, \quad ,, methyl acetate} = O \begin{cases} C_2H_3O \\ CH_3 \end{cases}$$

,,   ,,   ,,   ,,   ,, sulphuryl chloride $= Cl_2 . SO_2$

,,   ,,   ,,   ,,   ,, acetyl chloride $= Cl . C_2H_3O.$

The whole question of polybasic acids is involved in the necessity of halving the formula of acetic acid." (*Ibid.*)

Gerhardt also points out how a comparative study of other physical properties (boiling point, specific volume) supports his choice of molecular formulae.

Gerhardt and Laurent when justifying their notation, of which they say that "it represents like substances in like manner"

Guiding principle in choice of atomic weights and molecular formulae. enunciate a general principle to guide chemists in the framing of formulae and the choice of atomic weights.

"Since it is our object to discover the weight of particles, we shall consider that we have attained it if for each substance, simple or compound, we find a proportional number concordant with the form, the volume, and the specific heat of the substance; and if this number allows us to represent in a very simple manner the reactions and the formulae.... Finally we wish to appeal to the judgment of chemists, who must decide in each case whether our method is correct or not." (Laurent, *Ann. Chim. Phys.*, Paris, (3), 18, 1846, p. 266.)

[1] The common volume is called *2 volumes* because it is equal to that occupied by H_2, 2 parts by weight of hydrogen.

There can be no doubt as to the soundness of these views, which recommend the choice of such atomic weights as are in agreement with the requirements of the law of isomorphism (*post*, chap. XV), the law of Avogadro, often referred to as the law of volumes, and the law of heat capacity (*post*, chap. XIV), and which above everything are chemically adequate. But apparently the chemical public was not yet ready for the change. In his great text-book on organic chemistry, the publication of which was begun in 1853, Gerhardt retained Gmelin's equivalent-weight notation. He is reported to have said in private conversation that unless he had done so no one would have bought the work; in the introduction to the book the matter is put more formally:

Gerhardt's notation not followed. Confusion and diversity prevail.

> "I have even sacrificed my notation, retaining the old formulae the better to show by example how irrational they are, and leaving to time the consummation of a reform which chemists have not yet adopted."

Confusion continued to reign for some time longer. The terms equivalent, atomic weight, molecular weight were used and abused in every conceivable sense; sometimes even employed as synonymous. Lothar Meyer in his note to a German reprint of Cannizzaro's paper to which reference will presently be made, tells how about the middle of the century HO might have stood for water or hydrogen peroxide, C_2H_4 for marsh gas or ethylene, and how in certain text-books[1] a whole page was covered by the different formulae assigned to and used for acetic acid.

In 1860, in response to the wish for agreement on the subject of the formulae used and the values assigned to the different symbols, a congress met at Karlsruhe, and was attended by all the great chemists of the day. Lothar Meyer, the author of "Modern Theories of Chemistry," that most brilliant exposition of the atomic and molecular theory, then a young *Privat-docent*, was present, and in the note above-mentioned he tells us that at the congress nothing definite was achieved, but that at its close copies were distributed of a paper by Cannizzaro[2] published

Congress held in 1860 to promote agreement in symbols and formulae.

---

[1] Kekulé, in his *Lehrbuch der Organischen Chemie*, 1, 1861, p. 58, gives 19 different formulae.

[2] Professor of chemistry in Rome; in 1896 chemists of all nationalities joined in celebrating his 70th birthday "with an impressiveness worthy of the high scientific value of the man who was honoured."

in 1858 and entitled "Sketch of a Course of Theoretical Chemistry held at the University of Genoa[1]," and how on reading this paper all that had seemed to him confused and contradictory became clear and harmonious; and as with him, so it must have been

Cannizzaro's sketch of a course of theoretical chemistry.

with many others.   It is certain that to Cannizzaro belongs the merit of having been the first to clearly expound the atomic and molecular hypotheses in their relation to each other, of having lucidly stated the misconceptions which had delayed the general acceptance of these theories, and of having cast all the arguments dealing with the nature and method of molecular and atomic weight determinations into that form in which they are now familiar to us.   Cannizzaro's exposition begins by assigning to Avogadro's hypothesis the place of paramount importance in the science.

"It seems to me that the progress of chemistry within the last year has served to confirm the hypothesis of Avogadro, Ampère and Dumas concerning

Avogadro's hypothesis considered of paramount importance.

the similar constitution of gaseous substances; namely the assumption that equal volumes of gases, whether elementary or compound, contain an equal number of molecules.   But they by no means contain an equal number of atoms, the reason for this being that the molecules of different substances, or even of the same substance in the different states which it can assume...may consist of a different number of atoms of the same or of different kinds.   In order to bring my pupils to this same conviction, I have let them follow the same path that had led me to it, namely that of the historical examination of chemical theories.

"I begin by showing how from a consideration of the physical properties of gases, together with Gay-Lussac's law concerning the relation between the volume of a compound and that of its constituents, there has arisen as it were of itself that hypothesis which was first enunciated by Avogadro and shortly afterwards by Ampère.   Whilst expounding in detail the line of argument followed by these two physicists I proceed to prove that it is not in contradiction to a single known fact, provided only that we do as they did: (i) Distinguish between the molecules and the atoms; (ii) avoid confounding the criteria for comparing the weights and numbers of molecules with those

Current misconceptions.

employed for ascertaining the weights of atoms; (iii) abandon the erroneous view that whilst the molecules of a compound may consist of any number of atoms, those of the different elements must consist of one atom only, or at any rate of an identical number of atoms."

After reviewing historically the causes for the non-acceptance

---

[1] *Sunto di un corso di filosofia chimica fatto nella Reale Università di Genova,* 1858, German translation in Ostwald's *Klassiker der Exacten Wissenschaften,* No. 30.

of Avogadro's hypothesis, and dealing in particular with the attitude taken up by Berzelius, who whilst at first accepting the hypothesis for elements but repudiating it for compounds, ended by admitting its validity only in the case of the so-called permanent elementary gases, Cannizzaro says:

"I then prove that in order to reconcile all the experimental facts known to Berzelius, nothing was required but to differentiate the atoms from the molecules; there was no need for the assumption [made by Berzelius] of a fundamental difference between the permanent and the easily condensed, the simple and the complex gases, an assumption which is in direct contradiction to the physical properties common to all elastic fluids."

Then turning to the positive side he proceeds :

"From the historical examination of chemical theories, as well as from the results of physical investigations, I draw the inference that in order to bring all the branches of chemistry into agreement it becomes imperative that in the determination of the weight and the number of molecules the theory of Avogadro and Ampère should be used in its full scope. I then set myself the task of showing that the results thereby obtained are in complete accordance with all the physical and chemical laws so far discovered."

The determination of molecular and atomic weights on the basis of Avogadro's hypothesis, according to the principles indicated by Cannizzaro and still followed, must now be dealt with. By Laurent's definition the molecule is "the smallest quantity of a substance which must be employed to perform a chemical reaction"; the molecule being a definite quantity of matter we are at once led to the idea of molecular weight. In the present state of the science, chemistry offers no means for the actual determination of the weight of the molecule, and physical methods have only led to the assigning of approximate limiting values for this quantity[1]. With regard to the value of the molecular weight of one substance relatively to another the position is different; chemical and physical methods become available. For gaseous and gasifiable substances Avogadro's hypothesis supplies a direct means for the comparison of molecular weights. Cannizzaro thus deals with this matter:

Determination of molecular weights by Avogadro's law.

[1] "There are about 640 trillions of hydrogen molecules in 1 milligram; the unit of the usual atomic weights is thus equal to about a 1,300-trillionth of a milligram, or as we may more shortly express it, a quadrillion of hydrogen atoms weigh about ¾ gram." (O. E. *Meyer, The Kinetic Theory of Gases*, 1899, p. 335.)

"In my fifth lecture I begin to use the hypothesis of Avogadro and Ampère for the determination of molecular weights, irrespective of whether the composition of the molecules is known or not. According to the above-mentioned hypothesis the molecular weights are proportional to the vapour-densities of the respective substances. In order that the vapour-densities may express the molecular weights, it is more to the purpose to compare them all with the weight of a simple gas chosen as standard than with the weight of a mixture such as air. And since hydrogen is the lightest of all gases, it might be taken as unit in vapour-density determinations, and the values obtained would then represent molecular weights, the value of the hydrogen molecule being taken as 1. But I prefer to take for the common unit of the weights of whole molecules and of their parts, not the weight of the hydrogen molecule itself, but that of half such a molecule; hence I refer the vapour densities of the various substances to that of hydrogen taken as 2. It is sufficient then to multiply the vapour-densities referred to air by 14·438 in order to transform them into the values referred to hydrogen taken as unity, or to multiply them by 28·87 in order to obtain them referred to hydrogen taken as 2.... I arrange the two sets of numbers which represent these weights in the following manner:

Cannizzaro chooses for unit half the weight of the hydrogen molecule.

Names of Substances	Densities or weights of one volume, that of hydrogen being put = 1, or Molecular Weights referred to the weight of a *whole* hydrogen molecule taken as unit	Densities referred to that of hydrogen = 2, or Molecular Weights referred to the weight of a *half* hydrogen molecule taken as unit
Hydrogen	1	2
Oxygen	16	32
Electrolysed Oxygen (= Ozone)	64	128
Sulphur below 1000° C.	96	192
Sulphur above 1000° C.	32	64
Chlorine	35·5	71
Bromine	80	160
Arsenic	150	300
Mercury	100	200
Water	9	18
Hydrochloric acid	18·25	36·5
Acetic acid	30	60

If we wish to refer the vapour-densities to hydrogen as unit, and the molecular weights to the half-hydrogen molecule, we may say that all molecular weights are expressed by the weight of 2 volumes. Some of the examples contained in the above table show that the same substance in its different allotropic forms may have different molecular weights; but I must not pass over in silence the fact that the experimental numbers on which this assumption is based require further confirmation. I represent this matter as if the study of the various substances had been begun by the determination of their molecular weights, irrespective of whether the substances are elementary or compound. The numbers quoted in this table are strictly comparable, because they are all referred to the same unit. In order to impress this on the memory of my students I use the following simple device; I say to them: Imagine it could be proved that half a molecule of hydrogen weighed one-millionth of a milligram, then all the numbers in the preceding table become concrete quantities, and express the absolute weight of the molecules in millionths of a milligram. The same would be the case if the common unit had any other concrete value. It is thus I lead them to the clear conception that these numbers are all comparable, whatever may be the concrete value of the common unit. But as soon as this device has served its purpose I hasten to do away with it, and I add immediately that as a matter of fact it is impossible to determine the absolute value of this unit."

In the above, except for a discussion of the accuracy of the values obtained, the question of molecular weight determination on the basis of Avogadro's hypothesis is really dealt with exhaustively; but the importance of the subject makes it desirable to enumerate the various points involved, and to discuss them separately.

Points involved in such molecular weight determinations.
1. The standard.

1. As the ordinary units of mass are not available for the determination of the weights of individual molecules, and as molecular weight in the sense in which we use it denotes the ratio between the weight of the molecule considered and that of an arbitrarily chosen standard molecule, that standard has to be selected. Cannizzaro's argument in favour of making the weight of half the hydrogen molecule the standard has met with general acceptance; but the simple relation between molecular, atomic and combining weights makes everything that has been said under the head of equivalent and combining weights as to the advantage of an oxygen over a hydrogen standard equally applicable here, and the standard which some chemists hope may become universal is the weight of half an oxygen molecule $= 16$. Hence the following definition of molecular weight: The molecular weight of a substance, elementary or compound, is the weight of its molecule referred to the weight of the hydrogen molecule taken as 2 (oxygen $= 31.76$), or to that of the oxygen molecule taken as 32 (hydrogen $= 2.0152$).

2. The physical part of the definition, which tells us how the quantity may be measured, asserts that "in the case of gaseous and gasifiable substances the molecular weights may be determined by finding the weights of these substances which as gases at the same temperature and pressure occupy the same volume as 1 molecular weight of the standard, *i.e.* as 2 of hydrogen or 32 of oxygen"; this volume being that often referred to as *2 volumes*. The definition in terms of actual experimental values can be cast into various forms whose interdependence may be seen from the following simple equations:

2. Measurement.

$$\frac{\text{Molecular wt. of } A}{\text{molecular wt. of standard}} = \frac{\text{wt. of } X \text{ molecules of } A}{\text{wt. of } X \text{ molecules of standard}},$$

but by Avogadro's hypothesis, if A and the standard substance are gaseous:

$$\frac{\text{wt. of } X \text{ molecules of } A}{\text{wt. of } X \text{ molecules of standard}} = \frac{\text{gaseous density of } A}{\text{gaseous density of standard}},$$

and if the molecular weight of the standard is that of hydrogen $= 2$, we have

$$\frac{\text{molecular wt. } A}{2} = \frac{\text{gaseous density } A}{\text{gaseous density hydrogen}} = \text{specific gravity } A_{(\text{hydrogen}=1)};$$

$\therefore$ molecular wt. of $A = M = 2 \times$ spec. grav. of $A_{(\text{hydrogen}=1)}$..............(i),

but spec. grav. $A_{(\text{hydrogen}=1)} =$ spec. grav. $A_{(\text{air}=1)} \times$ spec. grav. $\text{air}_{(\text{hydrogen}=1)}$

$= $ spec. grav. $A_{(\text{air}=1)} \times 14\cdot39$;

$\therefore M = 2$ spec. grav. $A_{(\text{air}=1)} \times 14\cdot39$

$= 28\cdot78^1 \times$ spec. grav. $A_{(\text{air}=1)}$...... ....(ii).

Calling the weight of 1 litre of a gas at 0° and 760 mm. its *normal density*, we have the relation

$$M = 2\frac{\text{normal density } A}{\text{normal density hydrogen}} = \frac{2 \times \text{normal density } A}{\cdot08987}$$

$= 22\cdot25^1 \times$ normal density of A...................................(iii).

Hence the molecular weight of a gaseous substance is:

(i) Twice the spec. grav. of the gas referred to hydrogen as unity, or

---

[1] The values used until recently were 28·87 and 22·32 respectively; the discrepancy is due to the change in the standard value for the density of hydrogen, which according to Morley's determinations is ·08987, whilst the value used before was ·08958·

(ii)   28·8 times the spec. grav. of the gas referred to air as unity, or

(iii)   The weight of 22·25 litres of the gas at $0°$ and 760 mm.

It must seem doubtful whether anything is gained by this variety of definition.   (i) and (iii) do not even involve a difference in calculation; all that is done is that in (iii) the constants of the numerator are reduced to one by performing the division once for all.   In (ii) it is assumed that the specific gravity is known in terms of air and not of hydrogen; but since by far the greater number of specific gravity determinations of gases are màde by measuring the density itself, and referring this to the density of the standard supposed to be known, it does not seem clear why we should prefer the arithmetic involved in

Various forms of molecular weight equation.

$$\frac{\text{density of } A}{\text{density of air}} \cdot \frac{\text{density of air}}{\text{density of hydrogen}},$$

to that involved in the simpler expression

$$\frac{\text{density of } A}{\text{density of hydrogen}}.$$

Practically this amounts to saying that tables of the weights of 1 litre of the various gases should be given instead of the more usual ones of the specific gravities of gases relatively to air, generally called *vapour densities* (air=1).

3.   The only experimental datum to be determined is the gaseous density; no knowledge of the composition of the molecule is required.   But the value so obtained cannot, except in the case of permanent gases at low pressures, attain to any great accuracy.   The determination of gaseous densities at higher temperatures is attended with great difficulties, such as the exact measure of the temperature, the uncertainty as to whether the gas dealt with is contaminated with air, etc. etc.; besides which, in the application to molecular weight determinations, an error arises from supposing that Avogadro's law holds at all temperatures, which would only be true if the coefficients of expansion and of compressibility for the gas investigated were identical with those of the gas of known molecular weight with which the comparison is made.

3. Experimental work required.

4. In tables of molecular weights it is not unusual to find two columns headed "Found" and "Calculated" respectively, the two sets of values generally differing from each *4. Accuracy of the result.* other by a few per cent., the calculated value being regarded as the exact one. This latter represents the results of correcting the vapour density value from a knowledge of the composition of the substance, and of the combining weights of the constituent elements. The fundamental conception of a molecule being that of the aggregate of a small number of atoms, its weight must be the sum of the weights of the constituent atoms. But from the relation $ma : nb = p : q$ (*ante*, p. 290) where $p : q$ is the ratio of the combining weights, a and b that of the atomic weights, and m and n are simple whole numbers, it follows that if the standard is the same for combining weights as for atomic weights, the molecular weight must be equal to the sum of simple integral numbers of combining weights *The molecular weight of phosphorus fluoride.* of the constituent elements. The following example is given to illustrate how such a corrected value is obtained from the above enumerated data:

Prof. Thorpe in his investigation of a fluoride of phosphorus obtained the following results:

Specific gravity (H=1) at 16° C. found 62·98
   „   „   „ 16°   „   63·33 } Mean = 63·23 ;
   „   „   „ 19·4°   „   63·39

∴ **Molecular Weight from vapour density** = 63·23 × 2 = **126·46**.

Composition: a certain quantity of the substance yielded

(1) 0·4605 grs.   silicon fluoride   corresponding to 0·336 grs. fluorine,

(2) 0·3167 „ magnesium pyrophosphate   „   „   0·108 „ phosphorus.

∴ 126·46, the approximate molecular weight, contains

   95·69   fluorine and   30·77   phosphorus, or very nearly
94·5 (=5 × 18·9) fluorine and 30·8 (=5 × 6·16) phosphorus,

the numbers 6·16 and 18·9 being the exact combining weights of phosphorus and fluorine when H=1·00;

∴ **Calculated Molecular Weight** = 5 × 6·16 + 5 × 18·9 = **125·3**.

Other examples are given in the following table:

The Determination of Molecular Weights.

I	II	III	IV	V	VI	VII
Name of Substance	Specific Gravity (air=1)[1]	Molecular Weight = Spec. Grav. × 28·87 (hydrogen=2)[1]	Composition and Weight of substance containing 1·00 of hydrogen[2]	Molecular Composition calculated from columns III and IV	Molecular Composition in integrals of the accurate combining weights in IV	Molecular Weight calculated (hydrogen=2·00)[1]
Hydrogen	·0693	2·000				
Hydrochloric Acid	1·247	36·01	hydrogen chlorine 1·00 + 35·18 = 36·18	hydrogen chlorine ·995 + 35·01 = 36·01	hydrogen chlorine 1·00 + 35·18	36·18 = 1 × 36·18
Water	·623	17·99	oxygen 1·00 + 7·94 = 8·94	oxygen 2·02 + 15·97 = 17·99	oxygen 2·00 + 15·88	17·88 = 2 × 8·94
Ammonia	·597	17·24	nitrogen 1·00 + 4·64 = 5·64	nitrogen 3·06 + 14·18 = 17·24	nitrogen 3·00 + 13·93	16·93 = 3 × 5·64
Methane	·555	16·03	carbon 1·00 + 2·98 = 3·98	carbon 4·04 + 11·99 = 16·03	carbon 4·00 + 11·92	15·92 = 4 × 3·98
Ethane	1·075	31·05	carbon 1·00 + 3·97 = 4·97	carbon 6·26 + 24·79 = 31·05	carbon 6·00 + 23·84	29·84 = 6 × 4·97
Pentane	2·483	71·71	carbon 1·00 + 4·97 = 5·97	carbon 12·03 + 59·67 = 71·71	carbon 12·00 + 59·60	71·60 = 12 × 5·97
Ethylene	·971	28·04	carbon 1·00 + 5·96 = 6·96	carbon 4·02 + 24·02 = 28·04	carbon 4·00 + 23·84	27·84 = 4 × 6·96
Benzene	2·752	79·48	carbon 1·00 + 11·92 = 12·92	carbon 6·16 + 73·32 = 79·48	carbon 6·00 + 71·52	77·52 = 6 × 12·92
Ethyl Chloride	2·219	64·09	carbon 1·00 + 4·77, chlorine + 7·04 } = 12·81	carbon 5·04 + 23·85, chlorine + 35·20 } = 64·09	carbon chlorine 5·00 + 23·84 + 35·18	64·02 = 5 × 12·81

[1] H=2·000 has been used as standard instead of O=32·00, in conformity with the special constants used by Cannizzaro in the passages and tables reproduced from his paper. Hence to make the combining weight values given in the Appendix to chap. VIII available for the purposes of this table, the numbers must all be reduced in the ratio of 16 : 15·88.

[2] These ... thrs have bn ... by the analysis of the substances, corrected from our knowledge of the a ... ing weights of the ... onst ... el ... , ... :

(a) The ... on of ... aid may be ... lad from the following values:

... ity hydrogen = ·0896 grs.; ... ity chlorine = 3·012 grs.; ratio of combining volumes = 1 : 1;

∴ hydrogen : chlorine = 1·00 : 35·82,

but the accurate ... ing ... ight of ... he (H=1·00) is 35·18;

∴ the ... ... in is hydrogen : chlorine = 100 : 35·18.

(b) The analysis of ... ... by exploding the ... ... with ... ... of oxygen. The volume of oxygen ... and ... for the ... of all the carbon and hydrogen into carbonic ... aid and ... air is ... and the volume of carbonic acid formed is ... by ... ... with From the ... ada, ... with our ... of the ... ... of ... and carbonic aid, and of the densities of hydrogen and oxygen, we can calculate the composition by weight of the substances analysed. So Dalton found that 1 volume of methane required for its complete combustion 2 volumes of oxygen, and yielded 1 volume of carbonic acid (*ante*, p. 156). Since carbonic acid contains its own volume of oxygen, and since water is formed by the combination of 2 volumes of hydrogen with 1 volume of oxygen, therefore of the 2 volumes of oxygen used, 1 volume had gone to form 1 volume of carbonic acid, 1 volume had combined with 2 volumes of hydrogen; and the ratio by weight of hydrogen to carbon in methane is that of the weight of 2 volumes of hydrogen to the weight of carbon contained in 1 volume of carbonic acid;

but ... ight of 2 litres of hydrogen = $2 \times ·0896$ grs. = ·1792 grs.

,,	,, 1	,, ,, carbonic aid	= 1·977 ,,
,,	,, 1	,, ,, oxygen	= 1·429 ,,
,,	,, 1	,, ,, carbonic aid	= ·548 ,,

∴ wt. of carbon in 1 ,, ,, carbonic aid = ·548 ,,

∴ in methane:

hydrogen : carbon = ·1792 : ·548

= 1·00 : 3·056,

but the accurate combining weight of carbon is 5·96, which makes the accurate value for the composition of methane:

hydrogen : carbon = 1·00 : 2·98.

The number of molecular weights thus determined is enormous and continually increasing; carbon compounds supply by far the largest number of these, but that of inorganic compounds is not insignificant. Tables are given in the more important text-books[1].

Turning from the molecule to its constituent the atom, which, according to Laurent, is "the smallest quantity of a simple body which can exist in a compound," and remembering

Determination of atomic weights by Avogadro's law. the definition of the molecule as "the smallest quantity of a simple body which must be employed to perform a chemical reaction," it follows that the smallest quantity which can exist in a compound must necessarily be that which enters into the composition of a molecule, and hence the justification for the following variant definition, which asserts that "the atom is the smallest amount of an element which combines with other atoms to form a molecule." The obvious unit for expressing this quantity will be that of the standard molecular weight. Adopting the same method as that just followed in the exposition concerning molecular weights, we must begin with a quotation from Cannizzaro's paper.

"We next proceed to the investigation of the composition of the molecules. Whenever the substance cannot be decomposed, it must be assumed that the

Cannizzaro's exposition of the principle of the method. Quantities of elements contained in molecular weights of compounds. whole weight of its molecules is composed of one kind of matter only; but if the substance is a compound we analyse it and thereby determine the invariable ratio by weight of its constituents, and we then proceed to divide the molecular weight into parts proportional to those of the relative weights of the constituents and thus obtain the quantities of the elements contained in a molecular weight of a compound, all of them referred to the same unit in terms of which all molecular weights are expressed; according to this method I compile the table [given on the next page].

Compare the various quantities of one and the same element contained in the molecular weight of the element or in that of its compounds; it is at once apparent that the several quantities of the same element contained in these various molecular weights are all whole multiples of one and the same quantity, which since it always enters into compounds undivided can justly be termed the atom.

---

[1] Ostwald, *Outlines of General Chemistry*, 1890 (pp. 56, 57). Wurtz, *The Atomic Theory*, 1892 (pp. 104—109). Lothar Meyer, *Outlines of Theoretical Chemistry*, 1892 (pp. 39, 40).

Name of the substance	Weight of one volume, or molecular weight referred to the weight of a half hydrogen molecule = 1	Weights of the constituents of one volume or of one molecule, all referred to the weight of a half hydrogen molecule = 1		
Hydrogen	2	2	Hydrogen	
Oxygen	32	32	Oxygen	
Electrified Oxygen	128	128	,,	
Sulphur under 1000°	192	192	Sulphur	
Sulphur above 1000°	64	64	Sulphur	
Phosphorus	124	124	Phosphorus	
Chlorine	71	71	Chlorine	
Bromine	160	160	Bromine	
Iodine	254	254	Iodine	
Nitrogen	28	28	Nitrogen	
Arsenic	300	300	Arsenic	
Mercury	200	200	Mercury	
Hydrochloric acid	36·5	35·5 Chlorine	1 Hydrogen	
Hydrobromic acid	81	80 Bromine	1 Hydrogen	
Hydriodic acid	128	127 Iodine	1 Hydrogen	
Water	18	16 Oxygen	2 Hydrogen	
Ammonia	17	14 Nitrogen	3 Hydrogen	
Arsenine	78	75 Arsenic	3 Hydrogen	
Phosphine	35	32 Phosphorus	3 Hydrogen	
Mercurous chloride	235·5	35·5 Chlorine	200 Mercury	
Mercuric chloride	271	71 Chlorine	200 Mercury	
Arsenic chloride	181·5	106·5 Chlorine	75 Arsenic	
Phosphorous chloride	138·5	106·5 Chlorine	32 Phosphorous	
Ferric chloride	325	213 Chlorine	112 Iron	
Nitrous oxide	44	16 Oxygen	28 Nitrogen	
Nitric oxide	30	16 Oxygen	14 Nitrogen	
Carbonic oxide	28	16 Oxygen	12 Carbon	
Carbonic acid	44	32 Oxygen	12 Carbon	
Ethylene	28	4 Hydrogen	24 Carbon	
Propylene	42	6 Hydrogen	36 Carbon	
Acetic acid	60	4 Hydrogen 32 Oxygen 24 Carbon		
Acetic anhydride	102	6 ,, 48 ,, 48 ,,		
Alcohol	46	6 ,, 16 ,, 24 ,,		
Ether	74	10 ,, 16 ,, 48 ,,		

Hydrogen in molecule of hydrochloric acid chosen for unit. From this it follows that all the different quantities of hydrogen contained in the molecules of the different substances are whole multiples of the quantity contained in the hydrochloric acid molecule, which justifies the choice of this amount as the common unit for molecules and atoms. The hydrogen atom is contained twice in the molecule of uncombined hydrogen.

23—2

In fact we find that
one molecule of

free hydrogen	contains	2 parts by weight of hydrogen	=	2×1
hydrochloric acid	,,	1 ,, ,, ,, ,, ,,	=	1×1
hydrobromic acid	,,	1 ,, ,, ,, ,, ,,	=	1×1
hydriodic acid	,,	1 ,, ,, ,, ,, ,,	=	1×1
hydrocyanic acid	,,	1 ,, ,, ,, ,, ,,	=	1×1
water	,,	2 ,, ,, ,, ,, ,,	=	2×1
sulphuretted hydrogen	,,	2 ,, ,, ,, ,, ,,	=	2×1
formic acid	,,	2 ,, ,, ,, ,, ,,	=	2×1
ammonia	,,	3 ,, ,, ,, ,, ,,	=	3×1
phosphine	,,	3 ,, ,, ,, ,, ,,	=	3×1
acetic acid	,,	4 ,, ,, ,, ,, ,,	=	4×1
ethylene	,,	4 ,, ,, ,, ,, ,,	=	4×1
alcohol	,,	6 ,, ,, ,, ,, ,,	=	6×1
ether	,,	10 ,, ,, ,, ,, ,,	=	10×1

In like manner we prove that the quantities of chlorine in the different molecules are all whole multiples of the quantity contained in the molecule of hydrochloric acid, *i.e.* of 35·5, and that the various quantities of oxygen contained in the molecules are all multiples of that contained in water, *i.e.* of 16, a quantity which is half of that contained in the oxygen molecule....

One molecule of

free chlorine	contains	71	parts by weight of chlorine	=	$2 \times 35·5$
hydrochloric acid	,,	35·5	,, ,, ,, ,, ,,	=	$1 \times 35·5$
mercuric chloride	,,	71	,, ,, ,, ,, ,,	=	$2 \times 35·5$
arsenic chloride	,,	106·5	,, ,, ,, ,, ,,	=	$3 \times 35·5$
tin chloride	,,	142	,, ,, ,, ,, ,,	=	$4 \times 35·5$
free oxygen	,,	32	,, ,, ,, ,, oxygen	=	2×16
ozone	,,	128	,, ,, ,, ,, ,,	=	8×16
water	,,	16	,, ,, ,, ,, ,,	=	1×16
ether	,,	16	,, ,, ,, ,, ,,	=	1×16
acetic acid	,,	32	,, ,, ,, ,, ,,	=	2×16

In this manner we ascertain the smallest quantity of each element which always enters undivided into the molecules of the substance of which it forms a constituent, and which is justly named the atom. Hence in order to determine the atomic weight of each element, we must know the molecular weights and the composition of all or at least of most of its compounds....

And I make it my special aim to impress on my students the difference between molecule and atom; as a matter of fact we may know the atomic weight of an element without knowing its molecular weight, as is the case with carbon. Since a large number of the compounds of this element are volatile, we can compare their molecular weights and their composition, and we find that all the different quantities of

Atomic wt. of an element deduced from molecular wt. and composition of its compounds.

Atomic wt. of an element may be known when molecular wt. is not.

carbon contained in those molecules are whole multiples of 12, which quantity is therefore the atomic weight of carbon, designated by C....

It is easy to convince ourselves of this by putting down the values for the molecular weights derived from the vapour-densities, and those for the weights of the constituents contained in them.

Name of the carbon compound	Molecular weight referred to the hydrogen atom	Weights of the constituents referred to the weight of the hydrogen atom taken as unit	Formulae when $H=1$, $O=16$, $C=12$, $S=32$
Carbonic oxide	28	12 carbon, 16 oxygen	CO
Carbonic acid	44	12 „ 32 „	CO_2
Carbon bisulphide	76	12 „ 64 sulphur	CS_2
Marsh gas	16	12 „ 4 hydrogen	CH_4
Ethylene	28	24 „ 4 „	C_2H_4
Propylene	42	36 „ 6 „	C_3H_6
Ether	74	48 „ 10 hydrogen, 16 oxygen	$C_4H_{10}O$

If the weight of the carbon molecule were known, this might be included in the list of the molecules of the carbon compounds, but no greater advantage would be derived thereby than that resulting from the addition of any other compound; because it would only once more confirm the fact that the quantity of carbon in any molecule containing that element is 12 or $n \times 12 = C_n$, where n is a whole number.

But since we do not know the vapour-density of uncombined carbon, we have no means of ascertaining its molecular weight; and therefore we cannot tell how many atoms are contained in its molecule. Analogy serves no purpose here, since we observe that the molecules of the substances most closely resembling each other (such as sulphur and oxygen), and even those of one and the same substance in its different allotropic modifications may consist of a different number of atoms. Nor have we any data for predicting the vapour-density of carbon; all we can say is, that on the basis of my numbers it must be either 12 or a whole multiple of 12. The number quoted in different text-books of chemistry as the theoretical vapour-density of carbon is quite arbitrary and absolutely useless for the purposes of chemical calculations. It is of no value for the calculation or the verification of the molecular weights of the different carbon compounds, because all these may be ascertained without our knowing anything concerning that of carbon itself; it is equally useless for the determination of the atomic weight of carbon, because this can be derived from the comparison of the composition of a certain number of its compounds, and because a knowledge of the molecular weight of this element would afford merely one more datum beyond those already sufficient for the solution of the problem.

...It is true that a difficulty sometimes arises when we have to decide whether the quantity of one element entering into combination with another consists of 1, 2, 3... or *n* atoms. In order to settle this point,

we must compare the composition of all other molecules containing the same element and ascertain that weight of the element which always enters undivided into the composition of the molecules. If it should happen that we cannot ascertain the vapour-density of the other compounds of the element whose atomic weight is being determined, we must have recourse to other means for ascertaining the molecular weights and for deriving the atomic weight from them."

Here also Cannizzaro's treatment, except that he does not consider the accuracy of the data determined, is exhaustive, and resolves itself into a consideration of the following

points:

1. What is the atomic weight, and how can it be determined?

The atomic weight is the weight of one atom of the element referred to the weight of the hydrogen atom taken $= 1.00$ (or oxygen $= 16.00$), and is determined by finding the least amount of the element present in the molecular weight of any of its compounds, which if these are volatile is the least amount present in *2 volumes* (*ante*, p. 349) of any of these compounds in the gaseous state.

2. The value so obtained is the maximum possible atomic weight and may yet prove to be a multiple of the real value.

If very few volatile compounds are known, there must be great uncertainty as to whether the data found give the true value or not. If a single volatile compound exists, it might well be one which contains in the molecule two atoms of the element considered.

3. The data required are the molecular weights of as many volatile compounds of the element as possible, together with the composition of these compounds expressed as parts of the molecular weights.

It has been discussed before, how for the determination of the molecular weight, the vapour density value is all that is required. The other datum, the composition by weight, must be ascertained by chemical means; analysis or synthesis is used, and the special procedure employed depends on the nature of the substance. For example chlorine is usually estimated as silver chloride, hydrogen is estimated as water, carbon as carbonic acid, sulphur as barium sulphate, phosphorus as magnesium pyro-phosphate, all these being substances of accurately known composition.

4. The accuracy of the atomic weights thus obtained is conditioned by that of the data involved, and cannot be greater than that of the least accurate of these, which is generally the molecular weight.

5. The values for the atomic weights obtained from the data discussed under 3, may be corrected from their relation to the accurately known combining weights.

The relation between the atomic weight and the combining weight obtained from the formula $ma : nb = p : q$, when the standard is the same, is that of simple whole numbers (*ante*, p. 290). The manner of applying the corrections required, is shown in the table given on the next page.

Cannizzaro in his tables gives the corrected values of the atomic weights and not the directly obtained empirical numbers.

6. The molecular weights of the elements themselves need not be known, only those of their compounds.

In the case of the non-volatile element carbon, owing to the enormous number of its volatile compounds, the evidence that the atomic weight is 12 and not a submultiple of 12, is the strongest possible. Whilst the non-volatile elements, chromium, iron, copper, lead, form volatile compounds (chloride and oxy-chloride of chromium, chlorides of iron, chloride of copper, ethide and methide of lead), the volatile elements, sodium and potassium do not do so; and finally the alkaline earth metals, magnesium and silver, are neither volatile nor do they form volatile compounds.

Throughout the latter part of his paper Cannizzaro is occupied in showing how "all the inferences drawn from Avogadro's hypothesis are in complete agreement with all the physical and chemical laws so far discovered."

The Atomic Weight of Chlorine.

Name of Compound	Vapour Density = wt. of 1 litre at 0° C. and 760 mm.	Molecular Weight = V.D. × 22·39 (O = 32·00)	Composition — Method and Data	Ratio	Molecular Composition	Combining Weights of the constituent elements (O = 8·00)	Molecular Composition in terms of the comb. wts. of the constituent elements
Free Chlorine	3·21	71·8			chlorine 71·8 $2 \times 35·9$	chlorine 35·45	chlorine 71·90 $2 \times 35·45$
Hydrochloric Acid	1·62	36·2	Volume composition of hydrochloric acid = 1 hydrogen : 1 chlorine. Gaseous densities: H = ·0899, Cl = 3·21	Hydrogen : chlorine ·0899 : 3·21	Hydrogen 1 : chlorine : 35·2	Hydrogen chlorine 1·008 35·45	Hydrogen : chlorine 1·008 : 35·45
Mercuric Chloride	12·22	273·6	12·048 grs. of the chloride when distilled with lime yielded 8·889 grs. of mercury	Mercury : chlorine 8·889 : 3·159	Mercury : chlorine 201·8 : 71·8 $2 \times 35·9$	Mercury chlorine 100·00 35·45	Mercury : chlorine 200·00 : 70·90 $2 \times 35·45$
Arsenic Chloride	8·13	182·1	4·298 grs. of the chloride required for precipitation of the chlorine 7·673 grs. of silver, and therefore contain 2·521 grs. of chlorine	Arsenic : chlorine 1·777 : 2·521	Arsenic : chlorine 75·8 : 106·3 $3 \times 35·41$	Arsenic chlorine 25·00 35·45	Arsenic : chlorine 75·00 : 106·35 $3 \times 35·45$
Stannic Chloride	11·93	267·2	2·665 grs. of the chloride required for precipitation of all the chlorine contained in it 4·427 grs. of silver, and therefore contain 1·454 grs. of chlorine	Tin : chlorine 1·211 : 1·454	Tin : chlorine 121·3 : 145·9 $4 \times 35·5$	Tin chlorine 29·76 35·45	Tin : chlorine 119·04 : 141·80 $4 \times 35·45$

∴ Maximum Atomic Weight of Chlorine = 35·45

CHAPTER XIV.

PETIT AND DULONG AND THE LAW OF ATOMIC HEAT.

DALTON and Berzelius determined atomic weights from *à priori* considerations concerning the nature of these quantities. The table published by Berzelius in 1818 supplied

Relations be-
tween atomic
weights and
physical pro-
perties.

the material necessary for the recognition of the connection between the values of the atomic weights and those of other physical properties of the elements. The generalisations thus made gave empirical laws which could then be applied to the determination of new atomic weights and to the correction of old ones. The first of such relations was that recognised as existing between the atomic weight and the heat capacity[1] of solid elements.

---

[1] The unit used in the measurement of heat is the calorie. The calorie is the amount of heat required to raise 1 gram of liquid water through 1 degree centigrade, but the heat capacity of water itself varies;—the amount of heat required to raise 1 gram of water from $0°$ C. to $1°$ C. is not the same as the amount required to raise 1 gram from say $50°$ C. to $51°$ C. Hence the definition to become complete requires a statement of the temperature to which it refers. Three values are in use, referring to :—

(i) The heat capacity of water between $0°$ C. and $1°$ C.

(ii) The mean heat capacity between $0°$ C. and $100°$ C. *i.e.* 1/100 of the amount of heat required to raise 1 gram of water from $0°$ C. to $100°$ C.

(iii) The heat capacity between $18°$ C. and $19°$ C., which possesses the advantage of being the temperature at which most measurements are actually made, and of being practically identical with (ii).

Heat capacity is the amount of heat required to raise unit weight of a substance through $1°$ C., and the recognition that this value varies with the nature of the substance goes back to Kirwan in 1780. Specific heat is the ratio between the amount of heat required to raise any weight of the substance through any temperature range and the amount of heat required to produce the same effect in the same weight of an arbitrarily chosen standard:

$$\text{Specific heat of substance } A = \frac{\text{Heat capacity of substance } A}{\text{Heat capacity of standard substance}}.$$

The standard usually chosen is water, whose heat capacity forms the unit of measurement and is therefore equal to 1, with the result that using this particular system of units the numbers expressing specific heat and heat capacity become the

Petit[1] and Dulong[2] in 1819 published a paper entitled " Researches on some important Points in the Theory of Heat[3]," in the introductory part of which they say:

"We have attempted to investigate the connection between certain properties of matter and the best established results of the Atomic Theory, and have been led to discover simple relations between hitherto unconnected phenomena."

Petit and Dulong establish a relation between atomic weight and heat capacity of solid elements. Using the method of cooling, they determined the heat capacity of certain solid elements and embodied the results together with the relations established between these values and the corresponding atomic weights in the table:

Name of Element	Specific Heat	Atomic Weight $O=1$	Product of Atomic Weight and Specific Heat
Bismuth	·0288	13·30	·3830
Lead	·0293	12·95	·3794
Gold	·0298	12·43	·3704
Platinum	·0314	11·16	·3740
Tin	·0514	7·35	·3779
Silver	·0557	6·75	·3759
Zinc	·0927	4·03	·3736
Tellurium	·0912	4·03	·3675
Copper	·0949	3·957	·3755
Nickel	·1035	3·69	·3819
Iron	·1100	3·392	·3731
Cobalt	·1498	2·46	·3685
Sulphur	·1880	2·011	·3780

" The result required is obtained by multiplying each of the experimentally determined heat capacities by the corresponding atomic weights. These products are given in the last column of the table....The products which represent the heat capacities of different kinds of atoms approximate so nearly to equality that it is impossible to account for the very slight differences

---

same; but whilst specific heat is a pure number independent of any particular system of units employed, the heat capacity is a quantity of heat. The relation between heat capacity and specific heat is the same as that between density and specific gravity, and in both cases there is the same regrettable tendency to use the two sets of terms as synonymous, simply because of the choice of unit of measurement.

[1] A. T. Petit (1791—1820), Professor of Physics at the Paris " École Polytechnique."

[2] P. L. Dulong (1785—1838), Director of the above school.

[3] *Ann. Chim. Phys.*, Paris, 10, 1819 (p. 395).

otherwise than by the unavoidable errors in the measurement of either the heat capacity or the atomic weight....Hence the following

The atoms of all substances have exactly the same capacity for heat. law may be enunciated: 'The atoms of all substances have exactly the same capacity for heat.'...Remembering...the uncertainty still attaching to atomic weight determinations, it will be evident that the law just established would have to be differently enunciated if an hypothesis concerning the density of the ultimate particles, different from ours, were accepted; but the law would in any case express a simple relation between the weights and the specific heats of the elementary atoms. But having to choose between equally probable hypotheses, we have been obliged to decide in favour of that which established the simplest relation between the elements we were comparing."

Petit and Dulong, in order to maintain the constancy of the value for the atomic heat, changed certain of the Berzelian atomic weight values. What these changes were, and how far they were justified, is shown in the table on p. 364.

It will be seen that the values for the atomic weights of the metals relatively to that of sulphur had been halved, a change

Constancy of atomic heat only shown if certain atomic weights altered. made later on by Berzelius himself in the 1827 table for the reasons of chemical analogy quoted in a previous chapter (p. 334); but Petit and Dulong also altered some of these values relatively to each other. The number for cobalt was that obtained on further division by $1\frac{1}{2}$, a change which since it obscured the evident analogies between the compounds of cobalt and nickel could not be accepted; the error lay in the heat capacity and not in the atomic weight value. The halving of the atomic weight of silver has been justified by later research, though the double value continued to be used for some time longer. For tellurium the agreement of the atomic heat value with that of the other elements was due to compensating errors in the two factors.

The empirical basis of the law of constant atomic heat was much extended and strengthened by the work of Regnault, who

Regnault's measurements of heat capacities. in 1840 published the result of a large number of determinations made by the "method of mixture[1]." Berzelius in commenting on Regnault's work in his annual report of 1842 said that in repeating and extending Dulong and Petit's work most attention had been paid to the physical part of the work and comparatively little to the

---

[1] *Ann. Chim. Phys.*, Paris, 73, 1840 (p. 1).

The Atomic Heat of Solid Elements.

Name of Element	Heat Capacity		Atomic Weight			Atomic Heat			
	Petit and Dulong 1819	Standard Values 1904	Berzelius (O=100) 1818	Petit and Dulong (O=1) 1819	Standard Values (O=16) 1904	Petit and Dulong's and Berzelius' numbers	Petit and Dulong's numbers	Petit and Dulong's and present Standard Values	Standard Values 1904
	1	2	3	4	5	1 & 3	1 & 4	1 & 5	2 & 5
Bismuth	0·0288	0·0305	1773·8	13·30	208·5	51·07	0·3830	6·01	6·20
Lead	0·0293	0·0315	2589·0	12·93	206·9	75·86	0·3794	6·05	6·52
Gld	0·0298	0·03035	2486·0	12·43	197·2	74·07	0·3704	5·87	6·25
Platinum	0·0314	0·03147	1215·23	12·16	194·8	38·13	0·3740	6·11	6·29
Tin	0·0514	0·0559	1470·58	7·35	119·0	75·59	0·3779	6·11	6·65
Silver	0·0557	0·0559	2703·21	6·75	107·93	150·57	0·3759	6·01	6·03
Zinc	0·0927	0·0939	806·45	4·03	65·4	74·75	0·3736	6·06	6·11
Tellurium	0·0912	0·0475	806·45	4·03	127·6	73·54	0·3675	11·64	6·05
Copper	0·0949	0·09232	791·39	3·957	63·6	75·10	0·3755	6·02	5·88
Nickel	0·1035	0·10842	739·51	3·69	58·7	76·38	0·3819	5·99	6·40
Iron	0·1100	0·10983	678·43	3·392	55·9	74·62	0·3731	6·15	6·28
Cobalt	0·1498	0·10303	738·00	2·46	59·0	110·56	0·3685	8·83	6·29
Sulphur	0·1880	0·1712	201·16	2·011	32·06	37·83	0·3780	6·03	5·49

purity of the substances used. But the results afforded strong corroboration of the approximate constancy of the product, atomic weight × heat capacity, and the supposition seemed warranted that if both factors were known absolutely correctly, their product would also be absolutely constant. It was known that the specific heat varies greatly with the temperature, and Regnault having proved the influence of physical conditions on the value of this property (*e.g.* difference between amorphous carbon and diamond, etc.) Berzelius pointed out that while the atomic weight factor is

Berzelius' attitude to-wards atomic weight changes made by Regnault.

a true constant influenced only by the errors in the measurements made, the heat capacity factor is itself a variable quantity. Regnault had altered certain of the then current atomic weight values, in order to bring the elements they referred to within the scope of the law; Berzelius discusses these proposed changes. The further halving of the atomic weight of silver which made its sulphide Ag_2S, and thereby brought out existing analogies with cuprous sulphide Cu_2S[1], was accepted; but he emphatically refused to double the atomic weight of carbon, whereby the oxides of this element would have become CO_2 and CO_4 respectively, and great complication would have been introduced into the formulae of carbon compounds.

Regnault's results are given in the table on p. 366, which is taken from Berzelius' *Jahresbericht* (21, 1842, p. 6).

Kopp[2], by his extensive and classical researches on the specific heats of solids, introduced no fundamental change into the position

Kopp's specific heat measure-ments. C, B, Si exceptions to law of con-stant heat capacity.

of the subject beyond that of recognising carbon as an undoubted exception to the law of constant atomic heat, and classing boron and silicon in this respect with carbon. The influence of the state of aggregation and of temperature he considered as insignificant, the abnormal value of the atomic heat of certain elements being put down as essentially due to their chemical nature.

"Each element in the solid state and at sufficient distance from its

[1] The mineral argentite, Ag_2S, crystallises in the cubic system, and is iso-morphous with artificially prepared Cu_2S.
    The mineral daleminzite, Ag_2S, crystallises in the rhombic system, and is isomorphous with the native ore chalcocite or copper glance, Cu_2S.
[2] Hermann Kopp (1817—1892) the great historian of chemistry, for many years Professor at Giessen.

Regnault's Table of Atomic Heats.

Name of Substance	Specific Heat		Atomic Weight O = 100		Atomic Heat Atomic Weight (R) × Specific Heat (R)
	Regnault	Dulong and Petit	Berzelius	Altered by Regnault	
Sulphur	0·20259	0·1880	201·17		40·754
Phosphorus	0·1887	0·385	196·14		37·024
Carbon	0·2411	0·25	76·44	152·18	36·873
Iodine	0·05412	0·089	789·75		42·703
Selenium	0·0837		494·58		41·403
Arsenic	0·08140	0·081	470·04		38·261
Tungsten	0·03636		1183·00		43·002
Molybdenum	0·07218		598·52		43·163
Tellurium	0·05155	0·0912	801·76		41·549
Antimony	0·05077	0·0507	806·45		40·944
Gold	0·03244	0·0298	1243·01		40·328
Platinum	0·03243	0·0314	1233·50		39·993
Iridium	0·03683	.	1233·50		45·428
Palladium	0·05927		665·90		39·468
Silver	0·05701	0·0557	1351·61	675·80	38·527
Mercury	0·03332	0·0330	1265·82		42·129
Bismuth	0·03084	0·0288	886·92	1330·48	41·028
Uranium	0·06190		2711·36	677·84	41·960
Copper	0·09515	0·0949	395·70		37·849
Lead	0·03140	0·0293	1294·50		40·647
Tin	0·05623	0·0514	735·29		41·345
Zinc	0·09555	0·0927	403·23		38·526
Cadmium	0·05669		696·77		39·502
Nickel	0·10863	0·1035	369·68		40·160
Cobalt	0·10696	0·1498	368·99		39·468
Iron	0·11379	0·1100	339·21		38·597
Manganese	0·14411		345·89		49·848

melting-point, has *one* specific or atomic heat, which may indeed somewhat vary with physical conditions, different temperature or density for instance; but not so markedly as to be regarded when considering in what relations the specific heat stands to the atomic weight." ("On the Specific Heat of Solid Bodies," *Phil. Trans. R. Soc.*, 155, 1865, p. 71.)

By using a method of indirect determination Kopp found the value for the atomic heat of a number of elements not solid at ordinary temperatures (*e.g.* hydrogen, oxygen, chlorine, etc.), or not obtainable in a state of sufficient purity (calcium, barium, etc.).

Indirect determinations.

This method, which will be dealt with in a later part of this chapter, is based on the empirical law that the heat capacity of a molecule is the sum of the heat capacities of its constituent atoms, and hence that if we know the heat capacity of the compound and the heat capacities of all the constituent elements except one, we have sufficient data for calculating the value of this unknown quantity. A number of these elements were found to have an atomic heat considerably below the value of the constant, and comparable with that of carbon, boron and silicon.

The numerical results of Kopp's work concerning the atomic heat of solid elements may be summarised thus:

(i) Elements whose atomic heat is approximately 6·4 (standard atomic weight, H = 1), the actual values ranging from Al = 5·87 to Mo = 6·93 : Ag, Al, As, Au, Ba, Bi, Ca, Cd, Cl, Co, Cr, Cu, Fe, Hg, I, Ir, K, Li, Mg, Mn, Mo, N, Nb, Ni, Os, Pb, Pd, Pt, Rb, Rh, Sb, Se, Sr, Te, Ti, Tl, W, Zn, Zr.

(ii) Elements of lower atomic heat: C = 1·8, H* = 2·3, B = 2·7, Si = 3·8, O* = 4·0, F* = 5·0, P = 5·4, S = 5·4.

The elements marked with an asterisk are cases of indirect determination. Concerning the worth to be assigned to these numbers see *post*, p. 381.

The heat capacities of carbon, boron and silicon were investigated by Weber[1], who proved that contrary to Kopp's conclusions specific heat varies enormously with temperature, and that within a suitable temperature range values can be obtained for the atomic heat of the above elements in fair agreement with Dulong and Petit's law. Referring to the facts that were then known concerning the specific heat values of the elements he says :

Weber establishes for C, B, and Si influence of temperature on specific heat.

"The physical state plays as great a part as the chemical nature of the elements as regards the specific heats, and the law of Dulong and Petit cannot therefore be regarded as the universal expression of the law of specific heat."

Working at lower temperatures with Bunsen's ice-calorimeter and at higher temperatures by the method of mixture, and measuring the specific heat through an extremely large temperature range, he obtained results of which some typical cases are reproduced in the following table.

[1] "The Specific Heat of the Elements Carbon, Boron, and Silicon," *Phil. Mag.*, London, 4, 49 (pp. 161, 276).

Carbon.

DIAMOND				GRAPHITE		
Temperature	Specific Heat	Atomic Heat		Temperature	Specific Heat	Atomic Heat
− 50·5°	0·0635	0·7620		− 50·3°	0·1138	1·3656
− 10·6°	0·0955	1·1460		− 10·7°	0·1437	1·7244
10·7°	0·1128	1·3536		10·8°	0·1604	1·9248
33·4°	0·1318	1·5816		61·3°	0·1990	2·3880
58·3°	0·1532	1·8384		138·5°	0·2542	3·0504
85·5°	0·1765	2·1180		201·6°	0·2966	3·5592
140·0°	0·2218	2·6616		249·3°	0·3250	3·9000
206·1°	0·2733	3·2796		633·9°	0·4404	5·2848
247·0°	0·3026	3·6312		640·0°	0·4504	5·4048
598·3°	0·4378	5·2536		811·7°	0·4560	5·4720
615·2°	0·4438	5·3256		832·3°	0·4518	5·4216
804·6°	0·4444	5·3328		974·2°	0·4652	5·5824
808·4°	0·4535	5·4420		981·6°	0·4688	5·6256
983·1°	0·4557	5·4684				
986·8°	0·4622	5·5464				

"All opaque modifications of carbon (graphitic, dense and porous) have the same specific heats."

Crystallised Silicon.

Temperature	Specific Heat	Atomic Heat
− 39·8°	0·1360	3·8080
21·6°	0·1697	4·7516
57·1°	0·1833	5·1324
86·0°	0·1901	5·3228
128·7°	0·1964	5·4992
184·3°	0·2011	5·6308
232·4°	0·2029	5·6812

Crystallised Boron.

Temperature	Specific Heat	Atomic Heat
− 39·6°	0·1915	2·1065
26·6°	0·2382	2·6202
76·7°	0·2737	3·0107
125·8°	0·3069	3·3759
177·2°	0·3378	3·6258
233·2°	0·3663	4·3293

"Considering the rapid increase in the value of the specific heat of boron as the temperature increases until a certain point is reached, and the gradual diminution in the velocity of this increase after that point, I feel justified in concluding that it is very probable indeed that at higher temperatures the specific heat attains a constant value [of about] 0·50, [and that] the atomic heat of boron is about 5·5."

From these numbers he draws the inference:

"The idea that the temperature exercises but an insignificant influence on the magnitude of the specific heats of the elements, and that this influence may be overlooked without introducing any serious error into the deter-

mination of the specific heat, may now no longer be entertained....The greater
the interval of temperature for which the specific heat is
determined, the greater is the number representing the
specific heat....The values of the specific heats of the elements
carbon, boron and silicon change with the temperature; these
values gradually increase with an increase of temperature
until a point is reached at which they are constant. This
point is situated at about 600° for carbon and boron, and at
about 200° for silicon. The specific heat of carbon at 600° is seven times,
that of boron two and a half times, as great as at $-50°$."

Specific heat of C, B, Si, increases with temperature until limiting constant value is reached.

Beryllium, on the supposition that its oxide is BeO, and hence
its atomic weight 9·1, would be another element which at a low
temperature range had an atomic heat considerably less than 6·4.
But it has been shown that the specific heat increases rapidly
with the temperature, and using Humpidge's values (London,
Proc. R. Soc., 38, 1885, p. 188), the following table can be com-
piled:

Temperature	Specific Heat	Atomic Heat	
		Oxide = BeO, At. Wt. = 9·1	Ox'de = Be$_2$O$_3$, At. Wt. = 13·6
0°	0·3756	3·42	5·11
100°	0·4702	4·28	6·39
200°	0·5420	4·93	7·39
300°	0·5910	5·38	8·07
400°	0·6172	5·61	8·42
500°	0·6206	5·65	8·47

Obviously the atomic weight 9·1 leads to an atomic heat value
in better agreement with Dulong and Petit's law than does the
higher value 13·6.

Reviewing all the data at our disposal it is clear that some
simple relation does exist between "atomic weight" and "heat
capacity," but it is equally clear that the numbers as we determine
them only approach to a constant, and that the utmost the facts
justify is the statement that the atomic heats of all
solid elements when determined under *comparable*
conditions are the same. Unfortunately we have not
the data for absolutely deciding what does constitute
comparable conditions, though the temperature at

Atomic heat only constant under comparable conditions.

F. 24

which the variation in specific heat has reached a minimum, and at which the elements are still far from the melting-point, has so far been empirically found to come nearest to these requirements. There is great need for a theory which, whilst accounting for the observed approximate constancy of the atomic heat, would also supply a satisfactory explanation of the observed deviations from an absolutely constant value, and would repre-

Dulong and Petit's law lacks theoretical foundation. sent the magnitude of these deviations as a function of the chemical nature of the elements and of the physical conditions of the experiment. No such theory is as yet available, but some general considerations allow us to appreciate the points which it would have to take into account. The amount of heat that has to be added to a substance in order to raise its temperature is only partially expended in increasing the kinetic energy of the molecules, some going to increase that of the constituent atoms; and further, rise of temperature involves increase of volume, necessitating external work to overcome the atmospheric pressure, and internal work to overcome molecular attraction. In the case of solids, for which the coefficient of expansion is small, the external work is also small; whilst from the large difference between the atomic heats of the same substances in the gaseous and in the solid condition (atomic heat gaseous $I = 3\cdot3$, solid $I = 6\cdot9$; gaseous $Br = 4\cdot7$, solid $Br = 6\cdot7$), it must be inferred that the value for one or both of the other factors is great.

Whether for any set of elements the conditions can be found for which Petit and Dulong's law holds absolutely has been investigated by Professor Tilden[1]. Joly's compensating calorimeter[2] was used for the range between ordinary temperature and 100°, when results were obtained which in accuracy surpass

---

[1] "The Specific Heats of Metals, and the Relation of Specific Heat to Atomic Weight," London, *Phil. Trans. R. Soc.*, 194A, 1900 (p. 233).

[2] "The theory of the method is, briefly, as follows: A substance at the temperature t_1, of the air brought into an atmosphere of saturated steam will, in attaining the temperature t_2, of the latter, condense a certain weight of steam w, such that $w\lambda$, where λ is the latent heat of vapour of water, represents a quantity of heat equal to the calorific capacity of the substance between the limits of temperature. Hence, if S be the specific heat of the body, W its weight,

$$w\lambda = WS\,(t_2^\circ - t_1^\circ).$$

From this S is deduced by measuring w, t_2°, t_1°, W, and knowing the value of λ from recorded experiments" (J. Joly, "On the Steam Calorimeter," London, *Proc. R. Soc.*, 97, 1890, p. 218).

anything previously done in the province of specific heat measurements, the values found agreeing to the third place of
decimals, and differing by only a few units in the
fourth. The method of mixture was employed in
the case of low temperature measurements, for the
range between − 78·4° and 15°, − 182·5° and 15°.
Cobalt and nickel, prepared with the utmost regard
to purity, and brought by fusion and subsequent solidification to
the same degree of aggregation, were used.

Tilden investigates whether for Co and Ni, under special conditions, the law is exact.

"The results obtained with these two metals might be made the means
of further testing the validity of the law of Dulong and Petit, inasmuch as
temperatures at which the specific heats would be determined are not only
very remote, but about equally remote, from the melting-points of these two
metals...and since their atomic weights, though not known exactly, are
undoubtedly very near together, as are also the densities of the metals and
other of their physical properties."

The following table is given as an example of the concordance
of the results obtained in individual series of determinations:

	Weight of the metal *in vacuo* = W	Weight *in vacuo* of water condensed by metal = w	Temperature of calorimeter full of air = $t_1°$	Temperature of the steam in the calorimeter = $t_2°$	Specific Heat $= \dfrac{w\lambda}{W(t_2° - t_1°)}$
Cobalt	17·1332	·2617	20·84	99·98	·10315
	17·1332	·2582	21·81	99·98	·10306
	17·1319	·2601	20·98	99·78	·10303
	17·1317	·2590	21·15	99·74	·10289
Nickel	15·1549	·2318	24·83	100·08	·10866
	15·1535	·2379	22·52	100·10	·10818
	15·1564	·2359	23·56	100·05	·10878

The final results for the atomic heat values in terms of the
best available data for the atomic weights of cobalt and nickel
are:

24—2

	Temperature	Specific Heat	Atomic Weight (H = 1·000)	Atomic Heat
Cobalt	From 100° to 15° „ 15° „ − 78·4°[1] „ − 78·4° „ − 182·5°[2]	·10303 ·0939 ·0712	58·55	6·0324 5·4978 4·1687
Nickel	From 100° to 15° „ 15° „ − 78·4° „ − 78·4° „ − 182·5°	·10842 ·0975 ·0719	58·24	6·3143 5·6784 4·1874

Professor Tilden interprets these results in the following manner:

"The value of the specific heat for nickel declines more rapidly than for cobalt, and the consequence is that the specific heats of the two metals steadily approach each other. If the numbers given above for the specific heats are multiplied by the atomic weights, the products are very nearly identical."

And referring to the results obtained at the lowest temperatures he says:

"Hence the absolute atomic heat of the two metals, cobalt and nickel, is almost exactly 4.

It appears probable therefore, that if the experiments could be carried further, the specific heats would stand in exactly the inverse ratio of the atomic weights.

Probability that at very low temperatures the atomic heats of Co and Ni would be identical.

It remains to be seen whether the value of the atomic heat for other metals agrees with this. If such turns out to be the case, the original expression of the law of Dulong and Petit, which is only applicable at atmospheric and higher temperatures, would be completely justified."

This expectation has not been realised. Prof. Tilden's measurements of the specific heats of aluminium, nickel, cobalt, silver, and platinum, through a large temperature range, from − 182° to at least 550°, have given results which are summarised thus:

"One important result of the extension of the experiments to other metals is that the assumption of a constant atomic heat at the absolute zero, which seemed justified in the case of cobalt and nickel, is apparently untenable.

Plotting the specific heats against absolute temperatures...curves are obtained from which it is obvious that unless some remarkable change in the

---

[1] Temperature of solid carbon dioxide.
[2] Boiling point of oxygen.

specific heats of silver and platinum occurs below $-182°$ C., the curves representing atomic heats cannot meet at the absolute zero.

It appears therefore that the usual application of the law of Dulong and Petit to the rectification of atomic weights is a rough empirical rule which,

Dulong and Petit's law is a rough empirical rule.
setting aside boron, carbon, silicon, and beryllium, is only available when the specific heats have been determined at comparatively low temperatures, usually, and most conveniently, between 0° and 100°." (*Phil. Trans. R. Soc.*, 201A, 1903, p. 37.)

Having given the empirical foundation for Dulong and Petit's law, the next step must be its deductive application. Inspection of the equations, and attention to the conditions

The law applied to atomic weight determinations.
under which these are applicable, brings out the following facts:

1. That the only measurement to be made is that of a heat capacity or specific heat.

2. That the atomic weight value thus obtained is only an approximate one.

3. That the method is only applicable to elements whose heat capacity in the solid state at a suitable temperature can be determined.

Considering the first of these points, the experimental work consists in the measurement of a physical quantity, and the methods available for this purpose are described in any text-book of physics. The choice of method and of the modifi-

1. Experimental work required.
cations required to suit the special case dealt with must depend on the chemical nature of the element, whether acted on by water or not, the quantity of it available, the temperature range of the experiment, etc.

With regard to the second point, it is obvious that the atomic weight number as directly obtained may be far from correct; but the value of the method lies in indicating which

2. Accuracy of the result. Number obtained is the "approximate" atomic weight.
multiple of the accurately determined combining weight should be taken for the atomic weight. It enables us to fix the value $m : n$, in the equation, $ma : nb = p : q$ (*ante*, p. 290). Thus, taking the case of copper, which forms two compounds with chlorine:

Copper combined with 35·45 chlorine in cuprous chloride = 63·6

„   „   cupric chloride = 31·8.

Hence 63·6 and 31·8 are the values for $a\dfrac{m}{n}$, but from the heat capacity of solid copper we get a approximately, and hence can calculate, $m:n$, and the accurate value for a, the atomic weight of copper.

Cannizzaro (*loc. cit.*), in using the case of the chlorides of copper as an example of the application of Dulong and Petit's law, casts the same argument into the following form:

" The specific heat of copper in the free state...confirms the atomistic conception of its chlorides based on analogies with the corresponding chlorides of mercury. Their composition leads us to the inference that they have the formulae CuCl and $CuCl_2$, and that the atomic weight of copper is 63, a fact made apparent by the following relations :

	Ratio between the components expressed by numbers whose sum is =100	Ratio between the components expressed by the atomic weights
Cuprous chloride Cupric chloride	Chlorine Copper 36·04 : 63·96 52·98 : 47·02	35·5 : 63 = Cl : Cu 71 : 63 = Cl_2 : Cu

But the number 63 for the atomic weight, when multiplied by the specific heat of copper, gives a product nearly equal to that of the atomic weight of mercury or of iodine multiplied by their respective heats. We get:

$$63 \quad \times \quad 0\cdot09515 \quad = 6$$
At. wt. of copper Spec. heat of copper."

The determination of the atomic weight of germanium is an example of more recent date (*post*, chap. XVI).

Turning to the last point, that of the applicability of the method, the statement made above that it is applicable in the case of solid elements whose specific heat is known

3. Applica-
bility of the
method.

needs no amplification. But what we are really concerned with are the special cases in which the method is not only available, but of practical importance. ·Besides affording in all cases to which it is applicable·

an important check on the results obtained by other methods, it
assumes special importance in the case of elements

(i) Elements for which no volatile compounds known. which form no volatile compounds, and whose atomic
weights cannot therefore be determined on the basis
of Avogadro's law.

The following table sets out the data referring to the atomic
weight determination of some such elements by means of this method:

Element	Comb. Wt. i.e. quantity combining with 35·45 of chlorine $= a \cdot \dfrac{m}{n}$	Specific Heat of solid metal $=$ Sp. Ht.	Approximate At. Wt. from Atomic Heat $= \dfrac{6\cdot4}{\text{Sp. Ht.}} = a'$	Relation between At. Wt. and Comb. Wt. $= \dfrac{n}{m} =$ integer nearest to $\dfrac{a'}{\text{comb. wt.}}$	Accurate At. Wt. $= a$
Sodium	23·06	0·2934	21·8	1	23·06
Potassium	39·15	0·1655	38·8	1	39·1
Lithium	7·03	0·9408	6·8	1	7·03
Calcium	20·0	0·1686 to 0·1732	37·6	2	40·0
Magnesium	12·18	0·2499	25·6	2	24·36

Atomic Weights determined by Avogadro's Law and Dulong and Petit's Law.

	Spec. Grav. of Vapour air = 1	Molecular Weight of Compound = Spec. Grav. × 28·8	Molecular Composition in integral multiples of accurate comb. weights	Heat Capacity of Metal	Approximate Atomic Weight of Metal $\dfrac{64}{=\text{Heat Capacity}}$	Accurate Atomic Weight of Metal
Aluminium Chloride	9·34	269	Aluminium = 54·2 }266·9 Chlorine = 212·7			
Aluminium Bromide	18·62	536	Aluminium = 54·2 }533·9 Bromine = 479·7	0·225	28·4	27·1
Aluminium Iodide	27	778	Aluminium = 54·2 }815·3 Iodine = 761·1			
Chromium Oxychloride	5·55	159	Chromium = 52·1 Oxygen = 32·0 }155·0 Chlorine = 70·9	0·122	52·1	52·1

Dulong and Petit's law also renders important service in the case of elements of which we only know one or few volatile compounds, especially if these latter belong to a class of analogous compounds (as chlorides, bromides, iodides) in which the molecular constitution is likely to be the same. In such cases we have no right to assume that any one of the compounds investigated by us contains in the molecule one atom only of the element considered; the specific heat method must be called in to decide the point. The table on the preceding page gives illustrative examples of the application of this method.

(ii) Elements for which insufficient number of volatile compounds known.

The problem of the relation between the heat capacity of a compound and that of its constituent elements had already attracted the attention of Petit and Dulong, who say in their memoir of 1819:

Relation between heat capacity of compound and that of its constituents.

"The observations which we have made so far seem to establish the very remarkable law that there always exists a very simple relation between the heat capacity of atoms of compounds and that of elementary atoms."

From this they draw the important deduction:

"That the greater or lesser quantities of heat developed at the moment of the combination of bodies bear no relation to the capacity of the elements, and that in most cases this loss of heat is not followed by any diminution in the capacity of the resulting compounds."

Neumann[1], in 1831, enunciated the relation between the specific heat and the formula weight of a compound known as Neumann's law, but which, referring as it does only to chemically similar substances, is much more limited in scope than the older generalisation of Petit and Dulong. He says:

Neumann's law.

"I find that for compound substances a simple relation exists between the specific heats and the stoichiometric quantities; and I call stoichiometric quantities the amounts of substances which, as for instance in the case of the anhydrous carbonates, contain the same amount of oxygen, whilst in the case of sulphur compounds the amount of sulphur is the measure of the stoichiometric quantity. For chemically similar substances the specific heats are inversely proportional to the stoichiometric quantities; or what comes to the

---

[1] *Poggend. Ann.*, Leipzig, 23, 1831 (p. 1).

same thing, the stoichiometric quantities of chemically similar substances
have the same heat capacity.

The investigation of the carbonates first led me to the discovery of
this law.

Name of Substance	Stoichio-metric Quantity $(O=1)$	Specific Heat	Stoichiometric Quantity × Specific Heat
Calcite (Calcium Carbonate)	6·32	0·2044	1·292
Dolomite (Calcium-Magnesium Carbonate)	5·88	0·2161	1·271
Magnesite (Magnesium Carbonate)	5·75	0·2270	1·305
Chalybite (Ferrous Carbonate)	7·15	0·1819	1·300
Calamine (Zinc Carbonate)	7·79	0·1712	1·335
Witherite (Barium Carbonate)	12·31	0·108	1·329
Cerussite (Lead Carbonate)	16·68	0·081	1·35
			1·300 = Mean
Barytes (Barium Sulphate)	14·58	0·1068	1·557
Anhydrite (Calcium Sulphate)	8·57	0·1854	1·589
Celestite (Strontium Sulphate)	11·48	0·130	1·492
Anglesite (Lead Sulphate)	18·95	0·83	1·572
			1·546 = Mean

Regnault in extending his investigations to the specific heats
of compounds corroborated Neumann's result, finding that substances containing the same number of similar atoms united in a
similar way possess the same heat capacity. For him also similarity of composition was an essential requisite for the exhibition
of this simple relation. It was reserved for Kopp[1] to definitely
prove that the connection between the specific heat
and the composition of a compound holds in the
perfectly simple and general manner enunciated by
Petit and Dulong, and that heat capacity is of the
nature of an additive property. Each elementary
atom retains in the compound the same heat capacity it had
possessed in the uncombined condition, the number and kind of

*Kopp recog-
nises heat
capacity to be
an additive
property.*

[1] " Investigations of the Specific Heat of Solid Bodies," London, *Phil. Trans.
R. Soc.*, 155, 1865 (p. 71).

other atoms present and their form of combination exerting no influence on the value of this property[1].

Kopp's final result that "each element has essentially the same specific or atomic heat in compounds as it has in the free state," is based on experimental evidence which may be conveniently classified as follows:

Experimental evidence for Kopp's law.

(i) Molecular heat divided by number of atoms = 6·4.

(i) The quotient obtained by dividing the molecular heat[2] of a compound by the number of elementary atoms in one molecule is approximately equal to 6·4; equal, that is, to the atomic heat of an element according to Dulong and Petit's law. Thus the molecular heat of the chlorides RCl has been found $= 12.8$ on the average, and that of the chlorides $RCl_2 = 18.5$. Now $12.8 \div 2 = 6.4$, and $18.5 \div 3 = 6.2$. The same regularity is met with in metallic bromides, iodides and arsenides, and is even found in the case of compounds which contain as many as seven or nine elementary atoms. The molecular heat of ZnK_2Cl_4 is 43·4 and that of PtK_2Cl_6 is 55·2; now $43.4 \div 7 = 6.2$ and $55.2 \div 9 = 6.1$.

(ii) Quotient is less than 6·4 for compounds containing C, B, Si, etc.

(ii) The regularity just discussed was found to be far from general, but this very fact constitutes an extremely strong proof of the additive nature of heat capacity, in that elements which in the free state have atomic heat values less than 6·4 carry that smaller value into the compound; and obviously elements not amenable to direct determination may be of the same nature as carbon, boron and silicon, *i.e.* have the characteristic property of a low atomic heat and retain it in the combined condition.

"But this regularity, though met with in many compounds, is by no means quite universal. The oxygen compounds of the metals correspond to it in general the less, the greater the number of oxygen atoms they contain as compared with that of metal. The mean atomic heat of the oxides RO is 11·1, and the quotient $11.1 \div 2 = 5.6$. The quotient for the oxides R_2O_3 is

---

[1] The most striking case of additive properties, *i.e.* of properties whose value for the molecule is equal to the sum of the values for the constituent atoms, is that of mass; the mass of the molecule is absolutely equal to the sum of the masses of the constituent atoms. In the case of other properties classed as additive, such as refractive indices, specific volumes, heat capacities, the relation is not one of strict equality, but there is an essential difference between these and other properties such as solubility, for which there is *no* relation between the numerical values for the molecule and for the constituent atoms.

[2] Kopp applies the term "atomic heat" to elements and compounds alike; in the quotations given "molecular heat" is substituted in the case of compounds.

only $27\cdot2\div5=5\cdot4$; for the oxides RO_2, only $13\cdot7\div3=4\cdot6$; for the oxides RO_3, only $18\cdot8\div4=4\cdot7$. Still smaller is the quotient for compounds which contain boron in addition to oxygen: *e.g.* for the compounds RBO_2, it is only $16\cdot8\div4=4\cdot2$; for boracic acid, B_2O_3, it is only $16\cdot6\div5=3\cdot3$....This quotient is near $6\cdot4$ in those compounds which only contain elements whose atomic heats, corresponding to Dulong and Petit's law, are nearly $=6\cdot4$; it is smaller in compounds which contain elements...having a much smaller atomic heat than $6\cdot4$, and which are recognised as exceptions to this law, either directly, if their specific heat has been determined for the solid condition, or indirectly, if it be determined in the manner to be subsequently described."

(iii) The empirical results for the heat capacities of the quantities represented by the molecular weights of compounds show good agreement with the theoretical values calculated from the atomic heats of the constituents on the supposition of the maintenance of these values in the compound. The following table brings out these relations in the case of compounds containing only elements of directly determined atomic heat values.

(iii) Molecular heat found agrees with that calculated from atomic heats of constituents.

Name and Formula of Compound	Specific Heat	Molecular Heat = Molecular Wt. × Specific Heat	Atomic Heat of Constituents	Molecular Heat of Compound from the Atomic Heats of its Constituents
Ferrous Sulphide FeS	·1357	11·9	Fe = 6·27, S = 5·22	$6\cdot27+\ \ 5\cdot22\ \ =11\cdot49$
Mercuric ,, HgS	·0517	12·0	Hg = 6·38, ,,	$6\cdot38+\ \ 5\cdot22\ \ =11\cdot6$
Nickel ,, NiS	·1281	11·6	Ni = 6·42, ,,	$6\cdot42+\ \ 5\cdot22\ \ =11\cdot64$
Lead ,, PbS	·0490	11·7	Pb = 6·52, ,,	$6\cdot52+\ \ 5\cdot22\ \ =11\cdot74$
Bismuth ,, Bi_2S_3	·0600	31·0	Bi = 6·41, ,,	$6\cdot41\times2+5\cdot22\times3=28\cdot48$
Antimony ,, Sb_2S_3	·0907	30·8	Sb = 6·38, ,,	$6\cdot38\times2+5\cdot22\times3=28\cdot42$

Water is contained in solid compounds with the molecular heat of ice.

"The various determinations of the specific heat of ice give the molecular heat of H_2O as $8\cdot6$ for temperatures distant from $0°$, and as $9\cdot1$–$9\cdot8$ for temperatures nearer to $0°$. The molecular heat has been found:

For crystallised calcium chloride...$CaCl_2+6H_2O$ ...75·6
For the anhydrous chlorides ......RCl_2...............18·5
    Difference for......$6H_2O$ =   57·1
      ∴ H_2O =   9·5

For crystallised gypsum ....$CaSO_4+2H_2O$...45·8
For the anhydrous sulphates ......RSO_4 ....26·1
    Difference for......$2H_2O$ =   19·7
      ∴ H_2O =   9·9"

The results of a recent investigation by Prof. Tilden[1] show good agreement between the directly determined and the calculated molecular heat of compounds.

"The average specific heats of...the elements were determined over various intervals from the boiling-point of liquid oxygen to nearly 500° C....a range of about 600° C. From these mean specific heats the true specific heats at intervals of 100° C. absolute temperature were calculated, and from the specific heats the atomic heats were deduced. The mean specific heats of the compounds, formed by their union, were also determined, and from these data the molecular heats of the compounds calculated. On comparing the sum of the atomic heats of the elements present with the molecular heat of the compound at the successive temperatures, it was found that there is throughout a close concordance. The order of difference may be shown by one example:

Nickel Telluride, NiTe.

Temperature, absolute	Sum of atomic heat of Ni and Te	Molecular heat of NiTe
100°	9·20	8·38
200°	11·08	11·35
300°	12·22	12·41
400°	13·00	12·92
500°	13·49	13·15
600°	13·85	13·28
700°	14·11	13·35

The results of these experiments show that Neumann's law is approximately true, not only at temperatures from 0° to 100° C., but at all temperatures. They thus support the view that the specific heat of a solid is determined by the nature of the atoms composing the physical molecules, and is not a measure of the work done in thermal expansion."

Proceeding to the application of Neumann's and Kopp's law of molecular heat, we find that it is used for the purpose of the indirect determination of atomic heats and for that of ascertaining the relative complexity of molecular structure.

Neumann's and Kopp's law applied. 1. The indirect determination of atomic heats.

The indirect determination of atomic heat may be accomplished in two ways. If we have a compound of known composition and known molecular heat, containing not more than one element of unknown atomic heat, we may calculate the value for this by subtracting from the molecular heat the known atomic heats of all other elements.

[1] London, *Proc. R. Soc.*, 73, 1904 (p. 226).

Formula of Chloride	Molecular Weight	Specific Heat	Molecular Heat	At. Ht. of Metal (found)	Atomic Heat of Chlorine calculated
$HgCl$	235·75	·0521	12·3	6·38	$12·36 - 6·38 = 5·92$
$AgCl$	138·39	·0911	13·1	6·16	$13·1 - 6·16 = 6·9$
KCl	74·6	·171	12·8	6·47	$12·8 - 6·47 = 6·33$
$HgCl_2$	271·2	·0689	18·7	6·38	$\frac{1}{2}(18·7 - 6·38) = 6·16$
$MgCl_2$	59·8	·191	18·2	5·88	$\frac{1}{2}(18·2 - 5·88) = 6·16$

Or "we may ascertain the difference between the molecular heats of analogous compounds of an element of known and of an element of unknown atomic heat, in which case the difference is taken as being the difference between the atomic heats of these two elements."

Molecular Heat of compound containing element whose atomic heat is required	Molecular Heat of analogous compounds	Atomic Heat known		Atomic Heat calculated
Barium chloride, $BaCl_2 = 18·8$	$MgCl_2 = 18·2$	$Mg = 5·9$	$Ba =$	$5·9 + (18·8 - 18·2) = 6·8$
	$ZnCl_2 = 18·6$	$Zn = 6·1$	$=$	$6·1 + (18·8 - 18·6) = 6·3$
	$HgCl_2 = 18·7$	$Hg = 6·4$	$=$	$6·4 + (18·8 - 18·7) = 6·5$
Rubidium chloride, $RbCl = 13·5$	$KCl = 12·8$	$K = 6·47$	$Rb =$	$6·47 + (13·5 - 12·8) = 7·17$
Rubidium carbonate, $Rb_2CO_3 = 28·4$	$K_2CO_3 = 28·5$			$= \frac{1}{2}\{12·94 + (28·4 - 28·5)\} = 6·42$

The legitimacy and value of such calculations is, however, open to grave doubt.

"Notwithstanding the validity of Neumann's law, the attempts which have been made to deduce the atomic heats of elements, such as oxygen, which do not admit of experiment in the solid state, cannot, however, be regarded as satisfactory. It is obvious that in such calculations whatever change in the molecular heat of the compound is induced by slight alteration of density, or of structure in the solid, is concentrated upon one element in the compound of any two, when it is assumed that the other enters into combination with the atomic heat it possesses in the elemental state. Taking the figures for the compounds containing silver, for example, the value deduced for the atomic heat of silver is found to vary considerably according to the nature of the compound selected....From the mean molecular heat of the telluride...the result is 8·39....In the silver-aluminium Ag_3Al, [it] comes out as 6·19, and in the aluminium-silver $AgAl_{12}$, as 9·67." (Tilden, *Relation of Specific Heat and Atomic Weight*, 1904.)

In order to ascertain the relative complexity of a compound molecule, we have to proceed as follows:

If $A_m X_n$ is the formula for a compound between

2. The determination of relative molecular complexity. the elements A and X we may from a determination of its specific heat ascertain the ratio $m : n$ provided that the atomic weight of A is known, on the assumption that the constant atomic heat of each constituent atom is 6·4; multiplication of the weight of the compound containing one atom of A by the specific heat will give a value for the quantity $6\cdot4 + \left(6\cdot4 \cdot \dfrac{n}{m}\right)$, and hence for $m : n$. Obviously this method only lends itself to the purpose of ascertaining the *relative* number of atoms in the molecule, not their *absolute* number.

Taking barium chloride as example, is its formula BaCl, or $BaCl_2$, or Ba_2Cl, etc.? The atomic weight of chlorine is 35·45; in barium chloride 35·45 of chlorine are combined with 68·7 of barium; the specific heat of barium chloride is ·0902:

$$\therefore \text{ heat capacity } ClBa_{\underset{m}{n}} = (35\cdot45 + 68\cdot7) \times \cdot0902 = 9\cdot4,$$

$$\therefore \frac{n}{m} = \frac{9\cdot4 - 6\cdot4}{6\cdot4} = \frac{1}{2} \text{ nearly.}$$

Hence barium chloride contains half an atom of barium to every atom of chlorine, and its formula is $BaCl_2$, or possibly Ba_2Cl_4, or quite generally Ba_nCl_{2n}.

"The heat capacity of the atoms is not appreciably altered on entering into combination, and since this quantity is practically the same for all elements, it follows that, in order to raise their temperature 1°, the molecules will require quantities of heat proportional to the number of atoms contained in them. If Hg=200, that is, if the formulae for the two chlorides and iodides of mercury are HgCl, HgI, $HgCl_2$, HgI_2, it follows that the molecules of the two first require twice as much heat as a single atom, and those of the two last three times as much. That this is the case is shown in the table [on p. 383].

It is true that from the considerations of the heat capacity only, we cannot distinguish whether the [two first compounds are HgCl and HgI, or Hg_2Cl_2 and Hg_2I_2]...the only thing that can be said for certain is that the number of atoms of the [halogen]...is the same as that of [mercury].

Formulae of Mercury Compounds	Molecular Weights of these Compounds $= p$	Heat Capacity of Unit Weight $= c$	Heat Capacity of the Molecule $= p \times c$	Number of Atoms in the Molecule $= n$	Heat Capacity of each Atom $= \dfrac{p \times c}{n}$
HgCl	235·5	0·05205	12·257745	2	6·128872
HgI	327	0·03949	12·91323	2	6·45661
HgCl$_2$	271	0·06889	18·66919	3	6·22306
HgI$_2$	454	0·04197	19·05438	3	6·35146

(Cannizzaro, *loc. cit.*)

"In a great number of cases the molecular heat of compounds gives more or less accurately a measure for the degree of complexity of their composition. And this is the case also with such compounds as are comparable in their chemical deportment to undecomposed bodies. If cyanogen or ammonium had not been decomposed, or could not be so with the means at present offered by chemistry, the greater molecular heats of their compounds, compared with those of analogous chlorine or potassium compounds (compare mercuric chloride 18·7, mercuric cyanide 25·2; potassium chloride 12·8, ammonium chloride 20·0), and of cyanogen and ammonium as compared with potassium and chlorine, would indicate the more complex nature of those so-called compound radicles." (Kopp.)

Molecular heat of CN and NH₄ compounds indicates complexity of these compound radicles.

The remarkable facts of the approximate identity of the heat capacity of the substances by us termed elementary, and of the maintenance of this value in compounds, have led Kopp to speculations concerning the ultimate structure of "elements" which may fitly close this chapter.

The ultimate structure of elements in the light of their heat capacity.

"The magnitude of the molecular heat is exactly a measure of the complexity or of the degree of composition. If Dulong and Petit's law were valid, it could be concluded with great positiveness that the so-called elements, if they are compounds of unknown and simpler substances, are compounds of the same order. It would be a remarkable result that the act of chemical decomposition had everywhere found its limit at such bodies as these which, if compound at all, have with every difference of chemical deportment the same degree of composition. Imagine the simplest bodies, probably as yet unknown to us, the true chemical elements, forming a horizontal spreading layer, and piled above them the simpler and then the more complicated compounds; the universal validity of Dulong and Petit's law would include the proof that all elements at present assumed...lay in the same layer, and that chemistry in recognising hydrogen, oxygen, sulphur, chlorine, and the different metals as undecomposable bodies had penetrated to the same depth in that field of inquiry, and had found at the same depth the limit to its penetration[1]."

[1] Kopp, who looked upon all the elements of low atomic heat, such as carbon, boron and silicon, as undoubted exceptions to Dulong and Petit's law, infers that the elementary atoms are of different degrees of complexity. But now that these elements have been brought within the scope of Dulong and Petit's law, the parts of Kopp's speculations given above, which deal with the problem on the basis of the supposed universal validity of that law, are the most interesting.

CHAPTER XV.

MITSCHERLICH AND THE CONNECTION BETWEEN CRYSTALLINE FORM AND CHEMICAL COMPOSITION.

"All things whate'er they be
Have order among themselves, and this is form."

DANTE, *Paradiso.*

PETIT and Dulong's discovery of the relation between the atomic weights of solid elements and their heat capacity was quickly followed by the establishment of a connection between the atomic weight and another important physical property, that of crystalline form.

All matter may be classified as being either *homogeneous* (ὁμός = the same, and γένος = kind), when the physical properties measured at any point are the same as those measured in the same direction at any other point, or *heterogeneous* (ἕτερος = other), when this is not the case. Ice in contact with liquid water is a heterogeneous system; so is a piece of granite, in which the constituent particles of quartz, felspar and mica may be detected by the naked eye. Solutions, gaseous mixtures, chemical elements and compounds in the same state of aggregation and in the absence of isomeric modifications, are instances of homogeneous systems.

Homogeneous solids are further classified into *amorphous* (ἀ- = without, and μορφή = form), and *crystalline*. Substances in the amorphous condition are distinguished by the absence of definite external form, and by the fact that the *directional* properties are the same in all directions; characteristics which are explained by the supposition of a haphazard arrangement of the constituent particles. On the other hand, matter in

[margin note: Matter classified into "homogeneous" and "heterogeneous."]

[margin note: Homogeneous solids classified into "amorphous" and "crystalline."]

the crystalline state assumes a definite geometrical form, and generally exhibits differences in the values of the directional properties according to the direction in which these are measured[1]. This is considered to be due to a regular arrangement of the component particles. Thus a piece of ordinary

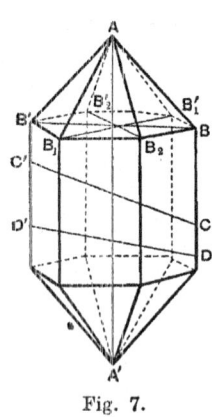

Fig. 7.

glass has no natural geometrical shape, and its coefficient of expansion, conductivity for heat, refractive index, absorptive power for light, etc. are the same in all directions. On the other hand, rock crystal (quartz = SiO_2) is found in nature in the shape of hexagonal prisms terminated by hexagonal pyramids (fig. 7), and the value of any one of the above properties measured along AA', the axis of the prism, differs from that found in any direction inclined to it such as BB', CC', DD', the difference being greatest in a direction at right angles to it, *i.e.* between AA' and BB', or between AA', and B_1B_1', etc.

An experiment due to H. de Sénarmont (fig. 8) illustrates this difference in a simple and telling manner.

Differences in physical properties according to direction in which measured.

A is a plate of quartz cut perpendicular to AA' the axis of the crystal.

B is a plate of quartz cut parallel to AA' the axis of the crystal.

The plates are coated with a layer of white wax. They are pierced at the centre, and a wire is inserted which is heated by an electric current. The wax melts around the place where the heat is applied, and on the plate A leaves a circle, and on the plate B an ellipse, showing that in the latter case the heat travels more readily along the axis than across it[2].

The mineral cordierite (fig. 9), $H_2(Mg Fe)_4 Al_8 Si_{10} O_{37}$, appears blue in transmitted light, when viewed parallel to C, yellowish-green

---

[1] The essential properties of all crystalline substances are of two sorts: (1) general properties, such as density, specific heat, melting point, chemical composition, which do not involve any particular direction, but represent the nature of the substance in the aggregate; (2) directional properties, such as cohesion, elasticity, refraction and absorption of light, expansion, conductivity of heat, crystalline form, etc., which are measured in some definite direction.

[2] Tyndall, *Heat a Mode of Motion*, 1880 (p. 244).

along *B*, and bluish-green along *A*, intermediate directions giving intermediate effects.

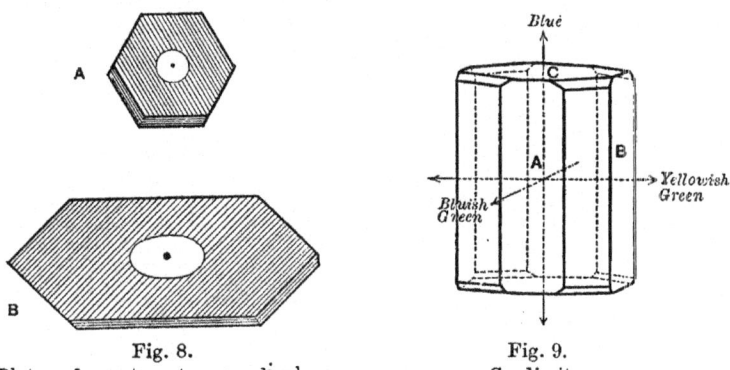

Fig. 8.
Plates of quartz cut perpendicular
and parallel to the crystal axis.

Fig. 9.
Cordierite.

Thus external form is not the only, nor even the most important characteristic of the crystalline state, but being the most conspicuous, it was the first to arrest attention. From this point of view, such knowledge as we now have of the laws of crystal structure and of the relations between external form and physical properties leads to our present conception of *crystal*, which is:

Definition of "crystal," "face" and "zone." "A solid body bounded by plane surfaces arranged according to definite laws, and possessed of definite physical properties; both external form and physical properties resulting from, and being the expression of, definite internal structure."

The bounding planes are termed *faces*. A set so arranged as to be all parallel to an imaginary straight line through the interior of the crystal is called a *zone*, and the straight line is termed the *zone axis*. The faces of a zone therefore, if they meet, intersect in parallel edges. However large may be the number of faces on any one crystal, it is found that they are generally arranged in a very few zones. In figs. 10 and 11 the planes aa' are in a zone, and so are the planes raz', $za'r'$.

The faces may vary in shape and size, but for the same substance they are always inclined to each other at angles characteristic of that substance. The angular relations

Constancy of angles. between the planes designated in the subjoined figures (10 and 11) by the same letters are the same.

25—2

The first figure is the representation of a crystal of quartz (SiO_2) such as is actually found in nature, in which, owing to uneven development of the different faces, considerable distortion has occurred; the second figure is the representation of the "ideal" equably developed geometrical figure corresponding to it.

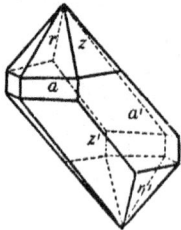 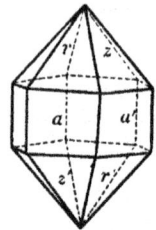

Fig. 10.   Fig. 11.

Distorted and ideal crystals of Quartz.

The same is shown in figs. 12 and 13, which represent misshapen octahedra of spinel ($MgAl_2O_4$) with the geometrically perfect form inscribed.   It will be seen that the faces of the spinel crystals are parallel to the faces of the octahedron, that they are inclined to each other at the same angles as the faces of the octahedron, and that the variations from the ideal figure are caused by differences in the size of the faces, an unequal development which has been found to be due to the conditions under which the crystals have been formed.

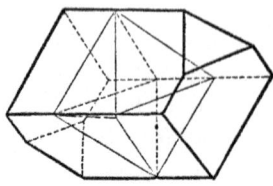 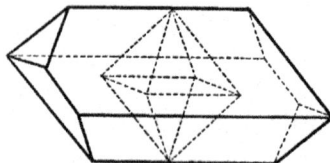

Fig. 12.   Fig. 13.

Distorted and ideal Octahedra.

The examination of the large variety of crystals found in nature or prepared in the laboratory shows that *similar* faces—recognised as similar by their lustre, striations, or other markings (fig. 14)—are almost always found regularly repeated at different parts

Recurrence of similar faces in definite order produces symmetry.

of the crystals, which consequently assume a *symmetrical* appearance.

"Similar" in this connection means "identical" in physical properties and mutual inclination, but not necessarily identical in shape and size.

Simple forms and combinations. A complete group of similar faces is called a *simple form.* Every crystal is either a simple form (figs. 15, 16, 18, 19), or a *combination* of two or more simple forms (figs. 17, 20).

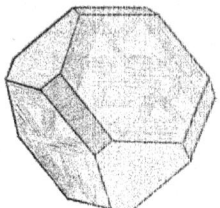

Fig. 14.

Crystal of the mineral Blende (ZnS), showing 3 forms, each with characteristic markings.

Simple Forms.

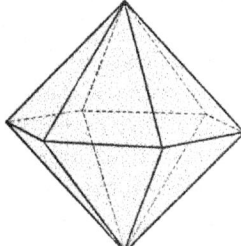

Fig. 15.
Hexagonal Bipyramid.

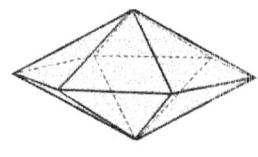

Fig. 16.
Hexagonal Bipyramid.

Combination.

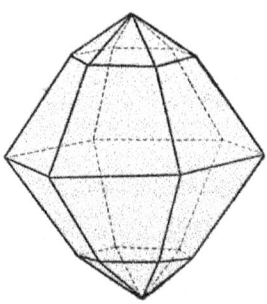

Fig. 17.
Two Hexagonal Bipyramids.

Simple Forms.   *Combination.*

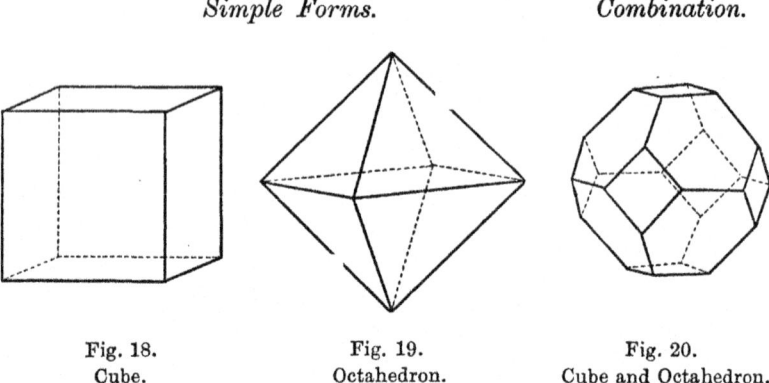

| Fig. 18. | Fig. 19. | Fig. 20. |
| Cube. | Octahedron. | Cube and Octahedron. |

In this repetition of similar faces, crystals exhibit different *grades of symmetry.* All the different grades of symmetry met with may be referred to *three elements of symmetry,* which are (1) symmetry about a central point, (2) symmetry about one or more axes, (3) symmetry about one or more planes.

Grades of symmetry referred to three elements of symmetry.

A *centre of symmetry* (giving rise to centro-symmetry) is present when to every face there corresponds a parallel face at the other side of the crystal.

Centro-symmetry.

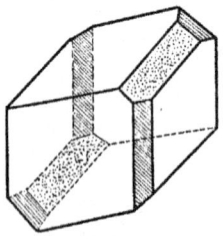

Fig. 21.
Showing centro-symmetry; to every plane there is a corresponding parallel plane.

An *axis of symmetry* is present when the crystal can be rotated about a line, through an angle of $\dfrac{360°}{n}$, so that faces and edges and corners are brought into the position of those similar to them, the aspect of the crystal being again the same as before. Thus fig. 22

Axes of symmetry.

represents the section of a crystal which is symmetrical about
an axis perpendicular to the paper through O, because a rotation

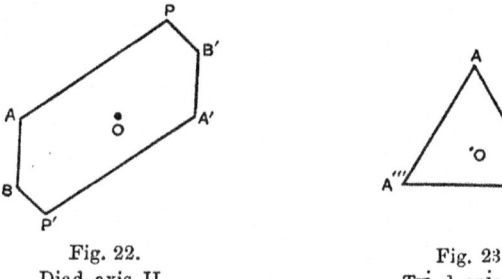

Fig. 22.
Diad axis II.

Fig. 23.
Triad axis III.

of $180°$ about this axis interchanges P and P', A and A', etc.   In
the fraction $\dfrac{360°}{n}$, n can be 2, 3, 4, or 6, corresponding to axes
termed *diad* (II), *triad* (III), *tetrad* (IV), and *hexad* (VI) respec‑
tively.   Such axes are represented in figs. 22—25 by lines
perpendicular to the paper through O.

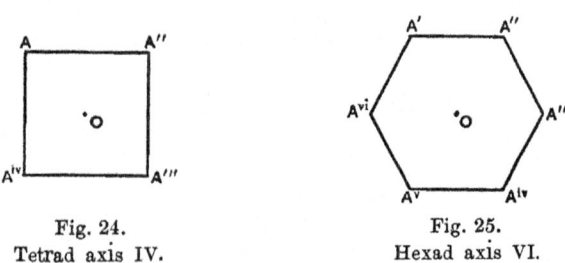

Fig. 24.
Tetrad axis IV.

Fig. 25.
Hexad axis VI.

A *plane of symmetry* is present when a crystal can be divided
in such a manner that the two portions bear to each other the
relation of an object and its image in a mirror
which is represented by the dividing plane. Fig. 26
is the section of a crystal such as is usually found,
with the faces unequally developed; fig. 27 is the section of the
corresponding geometrically perfect, the ideal form, obtained by
replacing each set of similar faces by planes parallel to them and
equidistant from any one point within the crystal chosen as the

Planes of sym-
metry.

centre. Then pp' represents a plane of symmetry because of the equality of the corresponding angles A and a, B and b, C and c.

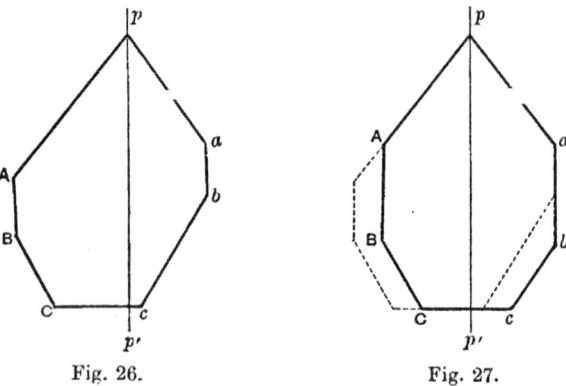

Fig. 26.   Fig. 27.

Sections showing plane of symmetry.

Crystals may have from 1 to 9 planes of symmetry (figs. 28—30).

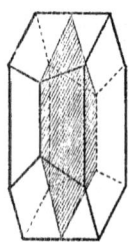 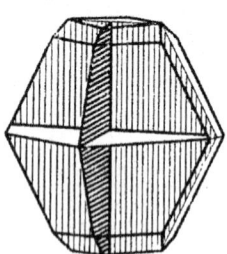 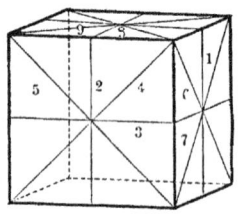

Fig. 28.   Fig. 29.   Fig. 30.

1 plane of   3 planes of symmetry at   9 planes of symmetry,
symmetry.   right angles to one   represented in section
                        another.   by the fine lines 1 to 9.

It has been stated before (p. 387) that in crystals of the *same* substance corresponding planes always intersect in the same angles, and hence that these angles are characteristic of the particular crystalline form assumed by each substance. In apparent contradiction to this is the fact that different specimens of crystals of the same substance do not always exhibit the same forms;

Occurrence of the same substance in a variety of crystalline forms.

thus calcite is found in a great number of different forms, of which figs. 31—33 are examples.

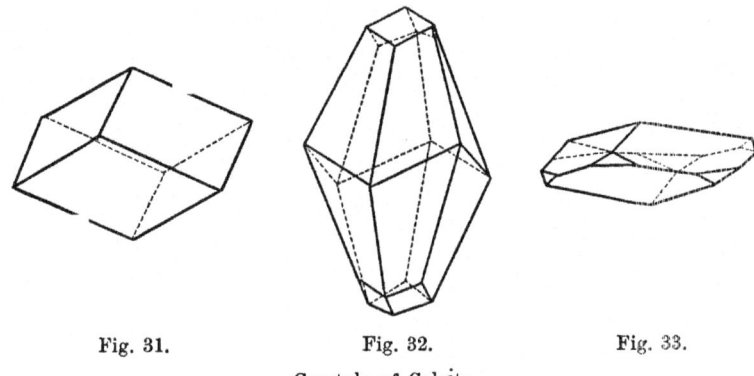

Fig. 31. Fig. 32. Fig. 33.

Crystals of Calcite.

But however great may be the variety of forms assumed in the same and in different crystals of any one substance, these all exhibit the same grade of symmetry, and are connected according to a simple law.

This law becomes evident when, following the method of geometry, we refer the faces of the crystal to a set of axes, the so-called *crystallographic axes* (generally 3, sometimes 4) which are imaginary lines passing through a point inside the crystal termed the origin, and drawn parallel to any 3 edges of the crystal that do not lie in the same plane. In choosing the edges parallel to which we draw the axes, we always select when possible edges parallel to axes of symmetry, especially those at right angles to one another.

Faces referred to crystallographic axes.—Parameters.

In fig. 34, X, Y, Z represent 3 such axes, and ABC is a plane which cuts all the 3 axes and is called the *parametral plane*, the lengths $OA = a$, $OB = b$, $OC = c$, which it cuts off from the axes X, Y, Z being called the *parameters*. Since the characteristic inclination of a crystal face is not changed by shifting it parallel to itself, it is obvious that it is not the absolute but the relative length of the parameters a, b, and c with which we are concerned, and it is usual to express this ratio in terms of b as unity.

Any other plane *CDE* actually found on a crystal whose parametral plane is *ABC* may, by shifting parallel to itself, be brought into a position such that the lengths cut off from the axes other than those to which it is parallel all lie within the extremities of the parameters, and these intercepts will be fractions of the parameters given by

Indices.—Law of Rational Indices.

$$OD = \frac{a}{h}, \quad OE = \frac{b}{k}, \quad OC = \frac{c}{l}.$$

It is found that h, k, l are always small whole numbers, 1, 2, 3, etc. or zero, in which last case the plane does not cut the particular

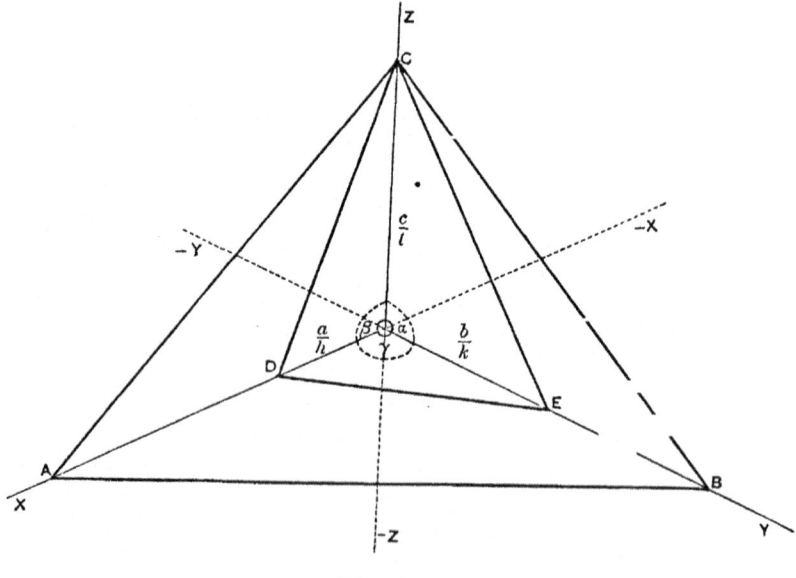

Fig. 34.

axis involved, but is parallel to it. In the above figure, for the plane *CDE*, $h = 3$, $k = 2$, $l = 1$. The quantities h, k, l are called *indices*, and the fact that they are rational whole numbers determines the fundamental law of crystallography, termed the *law of rational indices*.

The axial ratio $a : b : c$, and the angles between the axes

$YZ = \alpha$, $ZX = \beta$, $XY = \gamma$ are characteristic of the crystals of each

Crystallographic constants.— Symbols.

individual substance, and are its *crystallographic constants.* The position of any crystal face referred in the manner above described to three axes is therefore represented by

$$\frac{a}{h}, \frac{b}{k}, \frac{c}{l},$$

and since for any given substance a, b, and c are constants, the simple whole numbers h, k and l are sufficient identification, and written together thus, hkl or (hkl), they constitute the *symbol of the face.* Intercepts measured in the negative direction are indicated by corresponding indices having a negative sign placed over them, thus $(\bar{h}kl)$, $(11\bar{1})$, etc. For the parametral plane the symbol is (111); for the plane CDE in fig. 34, it is (321); the greater the index, the less is the corresponding intercept.

The symbol of a face when enclosed in a twisted bracket, thus $\{hkl\}$, becomes the *symbol of the form,* that is of the complete group of similar faces which have been developed together in virtue of the symmetry exhibited by the crystal.

Forms are classified according to the number of axes which each of their component faces intersects.

Classification of forms.

Pyramids $\{hkl\}$ are forms whose faces cut all three axes (figs. 36, 41, 45, 48, 51, 52, 58, 59).

Prisms $\{hk0\}$ or *Domes* $\{h0l\}$ or $\{0kl\}$ are forms whose faces cut two axes and are parallel to the third (figs. 43, 71, 72, 73).

Pinacoids ($\pi i \nu a \xi$, a board), $\{h00\}$ or $\{0k0\}$ or $\{00l\}$ are forms whose faces cut one axis and are parallel to the other two. The pair of parallel faces $(00l)$ and $(00\bar{l})$ constitute a "basal pinacoid," and each one of them is called a "basal plane" (figs. 9, 18, 73).

Crystallographic Classification is based primarily on relations of symmetry; all crystals which have the same degree of symmetry are placed in the same *class.* Thirty-

Crystallographic classification; 32 classes divided into 7 systems.

two classes are obtained by all the possible consistent combinations of the three elements of symmetry (*ante,* p. 390). Of these 32 possible classes, all save one or two are actually represented among known crystals. Certain of these classes have in common some physical properties, of which those that can be examined optically have been most studied; and they agree further in certain common crystallographic features, viz. the relative length and the inclination of the crystallographic axes to which they can be referred. Hence the 32 classes may be grouped under 7 *systems*.

That class of each system which exhibits the highest possible grade of symmetry is often termed holosymmetrical or *holohedral* (ὅλος = whole, ἕδρα = base or face). The less sym-

Holohedrism and hemi-hedrism.

metrical classes may in some cases be regarded as derived from the holohedral class by the suppression of one-half or three-quarters of its faces, and are therefore termed *hemihedral* (ἡμι- = half) or *tetartohedral* (τέταρτος = quarter) respectively.

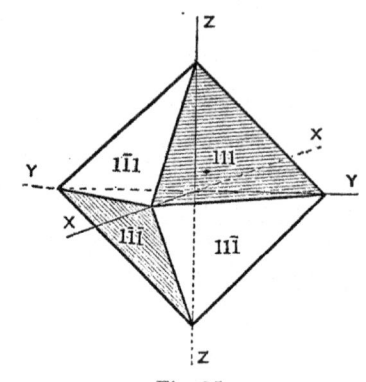

Fig. 35.

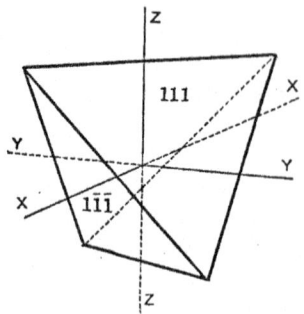

Fig. 36.

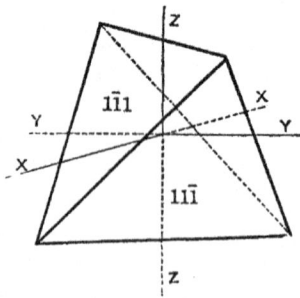

Fig. 37.

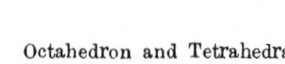

Octahedron and Tetrahedra.

Thus the tetrahedra (figs. 36, 37) are hemihedral forms constructed from alternate faces of the octahedron (fig. 35). They have 6 planes and 7 axes of symmetry, but no centre of symmetry; whilst the holohedral form from which they are derived possesses 9 planes of symmetry, centro-symmetry, and 13 axes of symmetry. But the axial relations are the same, *i.e.* both the octahedron and the tetrahedron can be referred to 3 equal crystallographic axes at right angles, tetrad axes in the case of the octahedron, diad axes in the case of the tetrahedron.

Fig. 38, which is a combination of the two tetrahedra, could geometrically not be distinguished from the octahedron, but in boracite, $Mg_7B_{16}Cl_2O_{30}$ (fig. 39), in which these forms occur in combination with the cube, their hemihedral nature is made evident by physical differences, the faces o being bright, and the faces ω dull.

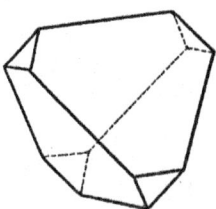

Fig. 38.
Combination of two Tetrahedra.

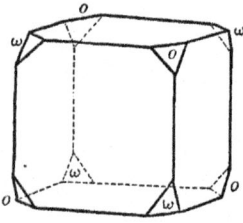

Fig. 39.
Boracite.

In the following presentation of the results of crystallographic classification, nothing will be attempted beyond a statement of the names and characteristics of the 7 crystallographic systems, together with examples of some typical forms of one important class belonging to each.

I. *Anorthic (Asymmetric or Triclinic) System.*

This includes 2 classes. . The crystals are referable to 3 un-equal oblique axes (fig. 40), and no symmetry higher than centro-symmetry is possible. Since there is no

Anorthic System.

plane of symmetry, the forms consist of single pairs of parallel faces, and each crystal must therefore be a combination of forms (fig. 41). The crystals are defined by the numerical values of: (i) the parameters $a : b : c$; (ii) the axial angles α, β, γ; (iii) the symbols of the component forms.

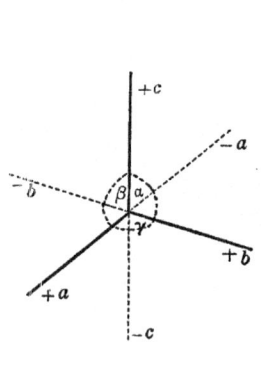

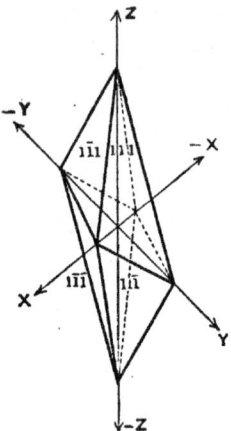

Fig. 40.
Anorthic Axes.

Fig. 41.
Anorthic Bipyramid.
Combination of $\{111\}$ $\{11\bar{1}\}$ $\{1\bar{1}1\}$ $\{1\bar{1}\bar{1}\}$.

Examples: Copper sulphate, $CuSO_4 . 5H_2O$ (fig. 42), and potassium bichromate, $K_2Cr_2O_7$ (fig. 43).

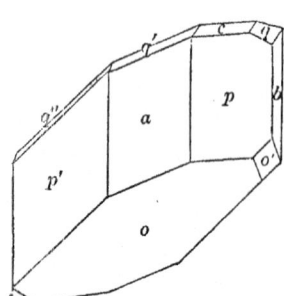

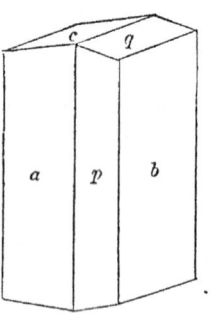

Fig. 42.
Copper Sulphate.
$a : b : c = 0\cdot5656 : 1 : 0\cdot5499$
$\alpha = 97° 39'$, $\beta = 106° 49'$, $\gamma = 77° 37'$.
Combination of $p\,\{110\}$, $p'\,\{1\bar{1}0\}$,
$a\,\{100\}$, $b\,\{010\}$, $c\,\{001\}$, $q\,\{011\}$,
$q'\,\{0\bar{1}1\}$, $q''\,\{0\bar{2}1\}$, $o\,\{11\bar{1}\}$,
$o'\,\{13\bar{1}\}$.

Fig. 43.
Potassium Bichromate.
$a : b : c = 0\cdot5575 : 1 : 0\cdot5511$
$\alpha = 82° 0'$, $\beta = 90° 51'$, $\gamma = 83° 47'$.
Combination of $a\,\{100\}$, $b\,\{010\}$,
$c\,\{001\}$, $q\,\{011\}$, $p\,\{110\}$.

II. *Oblique (Monoclinic or Monosymmetric) System.*

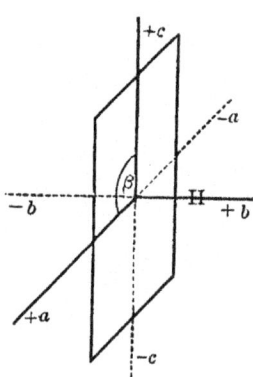

Fig. 44.
Oblique Axes
and Plane of Symmetry.

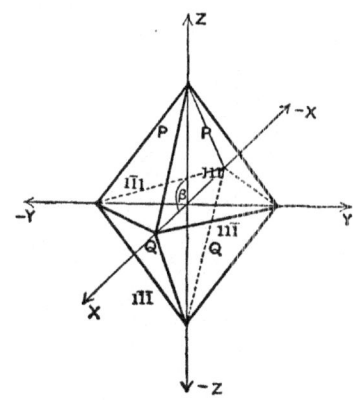

Fig. 45.
Oblique Bipyramid.
Combination of
P {111} and Q {$\bar{1}$11}.

This includes 3 classes. The crystals are referable to 3 unequal axes, two of which (*a* and *c*) intersect in an oblique angle β,
whilst the third (*b*) is perpendicular to the other

Oblique Sys-
tem.

two, $\therefore \alpha = \gamma = 90°$. The holohedral class has centro-
symmetry, 1 plane of symmetry (fig. 44) and 1 diad
axis (II, fig. 44 and fig. 22, p. 391),
which is the crystallographic axis *b*.
These elements of symmetry can produce forms consisting of at most two
pairs of parallel faces. Here again,
therefore, there are no closed simple
forms, and each crystal must be a
combination (fig. 45). The crystals are
defined by the numerical values of:
(i) the parameters $a : b : c$; (ii) the
axial angle β; (iii) the symbols of the
component forms.

Examples: Potassium (or caesium
or rubidium) -magnesium (or zinc or
copper) sulphate, $K_2Mg(SO_4)_2 . 6H_2O$
(table to p. 415); sulphur from fusion
(p. 440, fig. 88); gypsum (fig. 46).

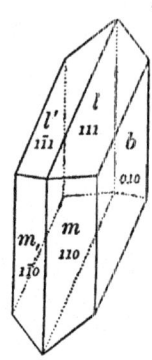

Fig. 46.
Gypsum.
$a : b : c = 0.690 : 1 : 0.412$
$\beta = 80° 42'$.

Combination of *l* {111},
m {110}, *b* {010}.

III. *Rhombic (Orthorhombic or Prismatic) System.*

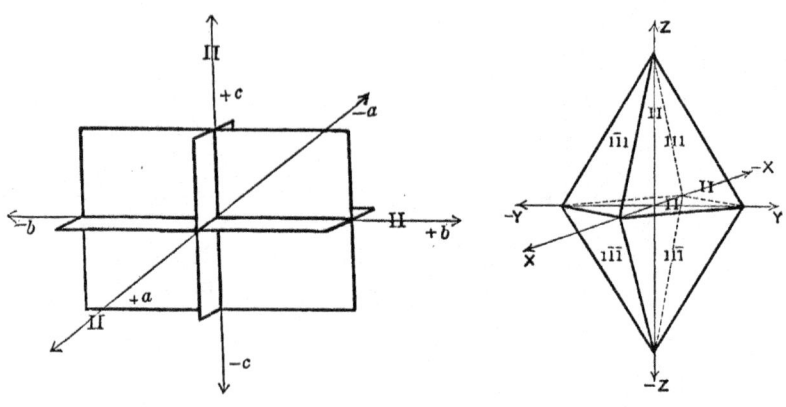

Fig. 47.
Rhombic Axes and Planes of Symmetry.

Fig. 48.
Rhombic Bipyramid {111}.

This includes 3 classes. The crystals are referable to 3 unequal axes (a, b, c) at right angles $(\alpha = \beta = \gamma = 90°)$. The holohedral class has centro-symmetry and 3 planes of symmetry intersecting at right angles, and 3 diad axes (II), which are also the crystallographic axes (fig. 47). The rhombic form {hkl} is a closed form, an 8-sided bipyramid (fig. 48). The crystals are defined by the numerical values of: (i) the parameters $a : b : c$; (ii) the symbols of the component forms.

Rhombic System.

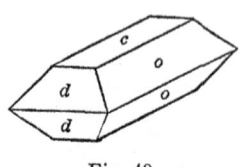

Fig. 49.
Barytes.

$a : b : c = 0·815 : 1 : 1·314$.

Combination of c {001}, o {011}, d {102}.

Examples: Acid sodium phosphate (or arsenate), $NaH_2PO_4 . H_2O$ (p. 412, fig. 72); aragonite, $CaCO_3$ (p. 438, fig. 86); anhydrous sodium (or silver) sulphate (or selenate), Na_2SO_4 (p. 419, fig. 76); sulphur from solution (p. 439, fig. 87); barytes, $BaSO_4$ (fig. 49).

IV. *Tetragonal (Pyramidal or Quadratic) System.*

This includes 7 classes. The crystals are referable to 3 axes at right angles, of which 2 are equal $(a = b$, fig. 50). The

holohedral class has centro-symmetry; 5 planes of symmetry;
4 diad (II) and one tetrad (IV) axis. In fig. 51

Tetragonal System. X, Y, Z, A_1, A_2 represent lines normal to the 5 planes
of symmetry. These lines are also the 5 axes of
symmetry, of which 3 (X, Y, Z) coincide with the crystallographic axes. The Z-axis, termed the *principal crystallographic*

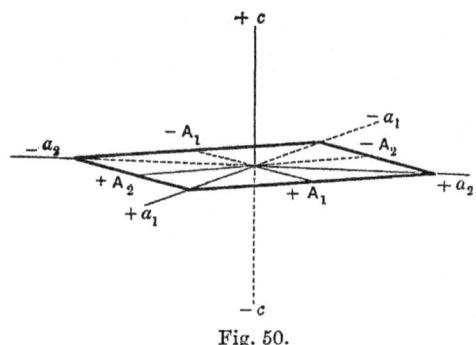

Fig. 50.
Tetragonal Axes.

axis, is the tetrad axis of symmetry. The form $\{hkl\}$ is either
an 8-sided bipyramid $\{111\}$ (fig. 51) or a 16-sided bipyramid,
such as $\{321\}$ (fig. 52). The crystals are defined by the numerical values of: (i) the parameters $a:c$; (ii) the symbols of the
component forms.

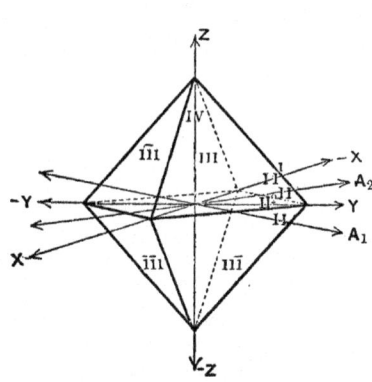

Fig. 51.
Tetragonal 8-sided Bipyramid $\{111\}$.

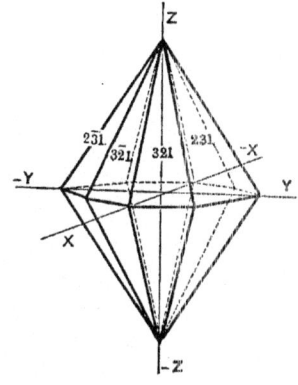

Fig. 52.
Tetragonal 16-sided Bipyramid $\{321\}$.

F.

Examples: Acid potassium phosphate (or arsenate), KH_2PO_4 (p. 411, fig. 71); rutile and anatase, TiO_2 (p. 440, figs. 89, 90); cassiterite, SnO_2 (fig. 53).

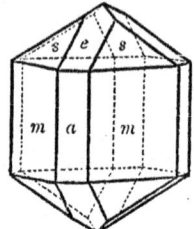

Fig. 53.
Cassiterite.
$a : c = 1 : 0.672.$
Combination of s {111}, m {110}, a {100}, e {101}.

V. *Rhombohedral System.*

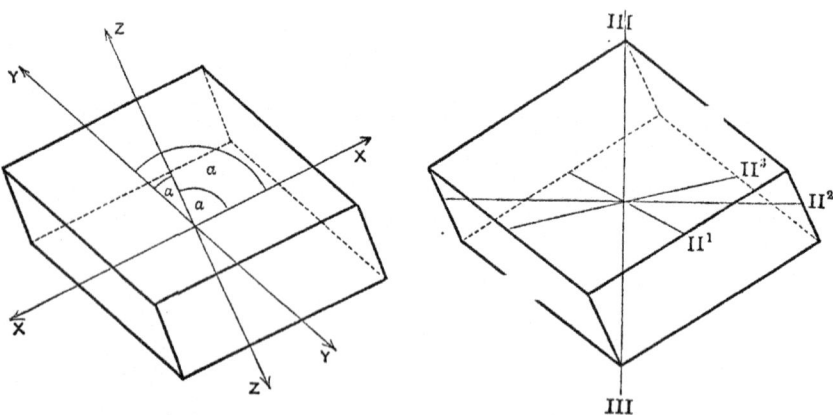

Fig. 54.

Crystallographic Axes.

Fig. 55.

Rhombohedra showing :

Axes of Symmetry.

This includes 7 classes. The crystals are referable to 3 equal axes equally inclined to one another at an angle α which is not 90° (fig. 54). These crystallographic axes, which are lines parallel to the edges of the rhombohedron, are *not* axes of symmetry. An important class has centro-symmetry; 1 triad axis perpendicular to 3 equally inclined diad axes (III, II¹, II², II³, figs. 55 and 22, 23, p. 391); 3 planes

Rhombohedral System.

of symmetry which intersect in the triad axis at angles of 60° (fig. 55). The simplest closed form is the rhombohedron {100}. The crystals are defined by the numerical values of: (i) the angle α; (ii) the symbols of the component forms[1].

Examples: Antimony, Sb; bismuth, Bi; calcite, $CaCO_3$ (p. 393, figs. 31—33; p. 413, fig. 74).

VI. *Hexagonal System.*

This includes 5 classes. The crystals are referable to 4 axes, of which 3 are of equal length ($a_1 = a_2 = a_3$), whilst the fourth (c) is

Hexagonal System.

different. The c axis, termed the *principal axis*, is perpendicular to the a axes, which latter intersect one another at angles of 120° (fig. 56).

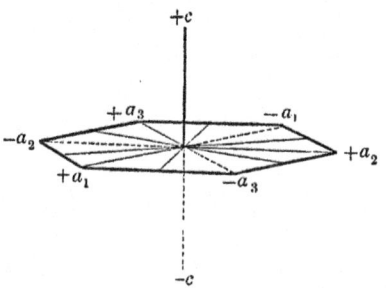

Fig. 56.
Hexagonal Axes.

The holohedral class has centro-symmetry; 7 planes of symmetry; 1 hexad (VI) and 6 diad (II) axes. In fig. 57 X, Y, U, Z are the crystallographic axes; X, Y, U, Z, A^1, A^2, and A^3 are lines perpendicular to the 7 planes of symmetry, and are themselves axes of symmetry. Z, the hexad axis, is also the principal crystallographic axis, whilst of the 6 diad axes 3 coincide with the

[1] The rhombohedral forms are not always treated as a separate crystal system. Some authorities deal with them as a special kind of hemihedral development of hexagonal forms.

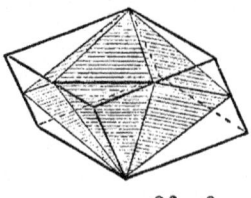

crystallographic axes X, Y, U (fig. 57). The form $\{hklm\}$ is either a 12-sided bipyramid such as $\{11\bar{2}1\}$ (fig. 58), or a 24-sided bi-

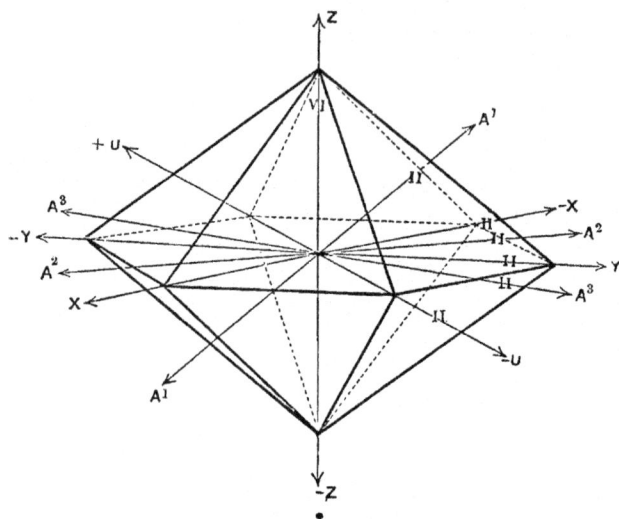

Fig. 57.
Axes of Symmetry of Hexagonal System.

pyramid such as $\{21\bar{3}1\}$ (fig. 59). The crystals are defined by the numerical values of: (i) the parameters $a:c$; (ii) the symbols of the component forms.

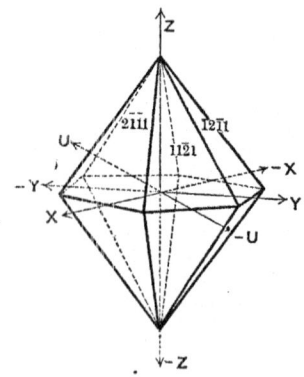

Fig. 58.
Hexagonal 12-sided Bipyramid $\{11\bar{2}1\}$.

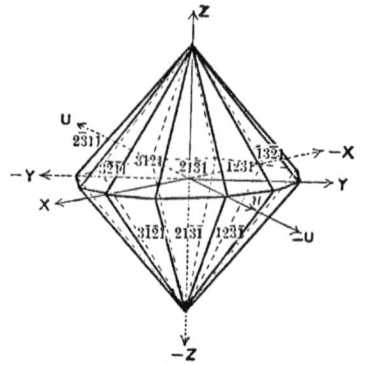

Fig. 59.
Hexagonal 24-sided Bipyramid $\{21\bar{3}1\}$.

Example: Beryl, $Be_3Al_2Si_6O_{18}$ (fig. 60).

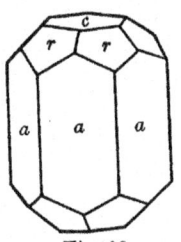

Fig. 60.
Beryl.
$a : c = 1 : 0.4989.$
Combination of $a\{10\bar{1}0\}$, $r\{11\bar{2}1\}$, $c\{0001\}$.

VII. *Cubic (Regular or Tesseral) System.*

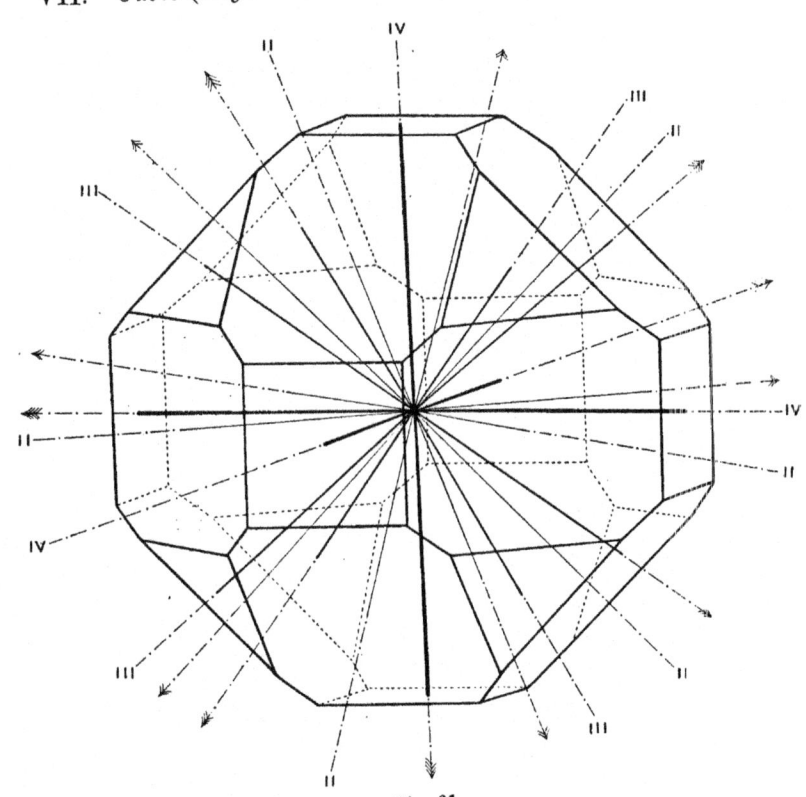

Fig. 61.
Crystallographic Axes and Axes of Symmetry of Cubic System.

.. This includes 5 classes. The crystals are referable to 3 equal rectangular axes ($a = b = c$; $\alpha = \beta = \gamma = 90°$). The holohedral class

Cubic System.

has centro-symmetry; 9 planes of symmetry (fig. 30, p. 392), of which the three parallel to the faces of the cube (1—3) intersect at right angles, whilst the 6 (4—9) which are parallel to the faces of the dodecahedron intersect at angles of 60°; 3 tetrad axes (IV) perpendicular to the faces of the cube, 4 triad axes (III) perpendicular to the faces of the octahedron, and 6 diad axes (II) perpendicular to the faces of the dodecahedron (fig. 61).

There are 7 simple forms which may occur singly or in combination.

The 7 simple forms of the cubic system.

(1) The planes are so inclined as to cut all 3 axes, producing forms of the general symbol $\{hkl\}$.

(i) The distances at which the axes are cut are all equal, $h = k = l$, when the symbol becomes $\{111\}$, the so-called parametral plane, which is the *Octahedron* (fig. 62).

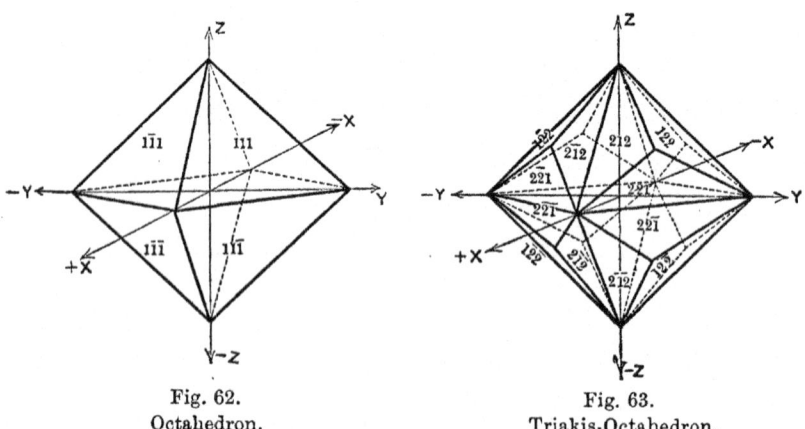

Fig. 62.
Octahedron.

Fig. 63.
Triakis-Octahedron.

(ii) The distances at which 2 of the axes are cut are the same, whilst that at which the 3rd is cut is different.

(a) The *Triakis-Octahedron* $\{hhk\}$. For the special case depicted by fig. 63, the indices are $\{221\}$, but other values are found, *e.g.* $\{441\}$.

(b) The *Icositetrahedron* $\{hkk\}$. The indices represented by fig. 64 are $\{211\}$, but $\{311\}$ and $\{322\}$ are also met with.

(iii)   The distances at which the 3 axes are cut are all unequal; the resulting form is the *Hexakis-Octahedron* {hkl}, that represented in fig. 65 having the indices {321}.

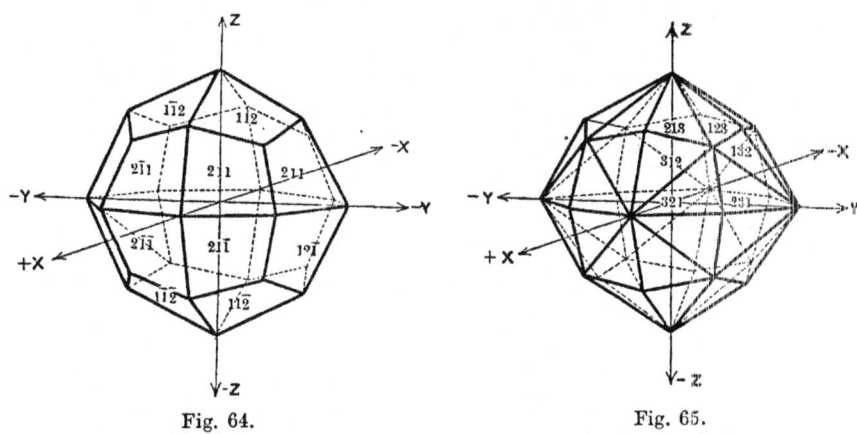

Fig. 64.
Icositetrahedron.

Fig. 65.
Hexakis-Octahedron.

(2)   The planes cut 2 of the axes and are parallel to the 3rd, producing forms of the general symbol {hk0}.

(i)   The *Rhombic Dodecahedron* {110}, the form in which the 2 axes are cut at equal distances (fig. 66).

(ii)   The *Tetrakis-Hexahedron* {hk0}, the form in which the 2 axes are cut at different distances.   Fig. 67 represents the most

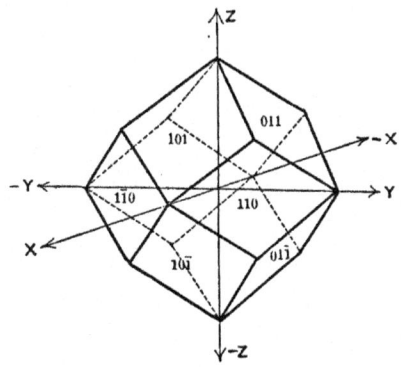

Fig. 66.
Rhombic Dodecahedron.

common and simple case, for which the indices are {210}, but {310} is also known.

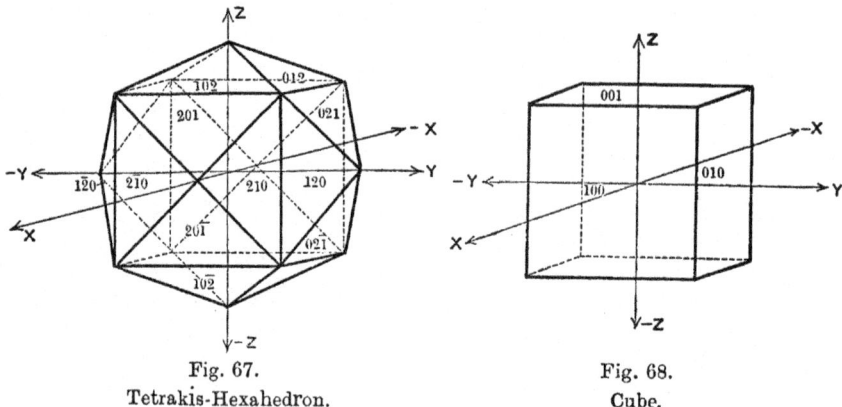

Fig. 67.
Tetrakis-Hexahedron.

Fig. 68.
Cube.

(3) The planes cut 1 axis and are parallel to the other 2, producing the *Hexahedron* or *Cube* (fig. 68).

The crystals of the cubic system are defined by the numerical values of the indices of the component forms.

Examples: Galena, PbS, found in cubes, or octahedra, or combinations of these two with or without the dodecahedron.—Alum, $R_2'SO_4 . R_2'''(SO_4)_3 . 24H_2O$, crystallising in octahedra.—Fluor spar, CaF_2, found as cubes, or octahedra, or in a variety of combinations, {210} being a common form.—Garnet occurs frequently as simple rhombic dodecahedra, fig. 70 representing a garnet crystal which is a combination of the dodecahedron, the icositetrahedron and hexakis-octahedron.

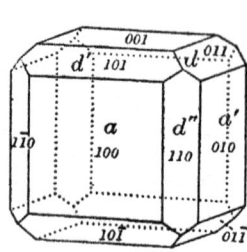

Fig. 69.
Cuprite.
Combination of a {100}, d {110}.

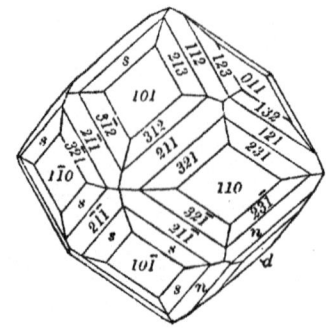

Fig. 70.
Garnet.
Combination of d {110}, s {321}, n {211}.

ISOMORPHISM.

A. Historical. The discovery and establishment of the law.

The name of Haüy (1743—1822), the founder of the science of crystallography, is associated with the view current at the beginning of the last century, that identity of crystalline form (except in the cubic system) implies identity of chemical composition, and conversely, that difference of crystalline form implies difference of chemical composition. This supposed law was supported by a mass of facts, but there were also on record well-defined and undoubted exceptions. As far back as 1772 Romé de l'Isle had observed that copper sulphate and ferrous sulphate crystallise from a mixed solution in the form of the latter[1], and in 1788 Klaproth established the chemical identity of rhombohedral calcite and rhombic aragonite.

E. Mitscherlich (1794—1863), a pupil of Berzelius, Professor of Chemistry in Berlin, eminent as a discoverer in the province of physical chemistry, at the very outset of his scientific career published what must be looked upon as the most important of his contributions to the development of the science, the paper entitled "On the Relation which exists between Crystalline Form and Chemical Proportions[2]."

He had undertaken to ascertain whether crystalline form is or is not independent of the chemical nature of the constituent elements. He chose for investigation the phosphates and arsenates, salts for which Berzelius had established a close parallelism between composition and properties.

Berzelius had shown that the ratio of the quantities of oxygen combined with the same amounts of arsenic or of phosphorus in the *-ous* and *-ic* compounds respectively was as $3:5$; and further, that the salts of phosphoric and arsenic acids showed remarkable analogies in composition as well as in properties. Each of these acids had been found to form three series of salts.

Haüy's views on the relation between crystalline form and chemical composition.

Mitscherlich investigates influence of chemical composition on crystalline form for phosphates and arsenates.

---

[1] Copper sulphate ($CuSO_4 . 5H_2O$) crystallises in the anorthic, ferrous sulphate ($FeSO_4 . 7H_2O$) in the oblique system.
[2] *Ann. Chim. Phys.*, Paris, 14, 1820 (p. 172); 19, 1821 (p. 350).

(i) The so-called neutral salts[1], in which the ratio between the oxygen of the base and that of the acid was $2:5$, which crystallise from solutions containing an excess of the carbonate of the base, and which precipitate barium chloride, leaving neutral solutions[2]. These salts contain the same amount of water of crystallisation, and all have the remarkable property of exhibiting both acid and alkaline reaction towards litmus.

(ii) The salts obtained from the above by the addition of as much acid as they already contain, and in which the ratio between the oxygen of the base and that of the acid is $1:5$.[3] They do not precipitate barium chloride[4], and their solutions are acid.

(iii) The salts of calcium, zinc, silver, mercury, etc. in which the ratio between the oxygen of the base and that of the acid is $3:5$. These are obtained when soluble salts of the above elements are precipitated by the so-called neutral phosphates or arsenates described under (i), the residual solution becoming acid[5]. This remarkable exhibition of similarity in properties attending similarity in composition was described by Mitscherlich in the words:

"Every arsenate has its corresponding phosphate, composed according to the same proportions, combined with the same amount of water of crystallisation, and endowed with the same physical properties; in fact the two series of salts differ in no respect, except that the radicle of the acid in the one series is phosphorus, whilst in the other it is arsenic."

[1] Represented in the Berzelian notation by $NaO_2 . PO_5$ and $NaO_2 . AsO_5$. The atomic weights then used were in the case of Na and K four times as great as the present ones, in the case of Zn, P and As twice as great.

[2] A reaction represented in our present notation by:

$$\left.\begin{matrix} Na_2HPO_4 \\ Na_2HAsO_4 \end{matrix}\right\} + BaCl_2 = \left.\begin{matrix} BaHPO_4 \\ BaHAsO_4 \end{matrix}\right\} + 2NaCl.$$
(neutral)   (insoluble)

[3] In the Berzelian notation: $NaO_2 . 2PO_5$ and $NaO_2 . 2AsO_5$.

[4] In modern notation:

$$\left.\begin{matrix} 2NaH_2PO_4 \\ 2NaH_2AsO_4 \end{matrix}\right\} + BaCl_2 = \left.\begin{matrix} BaH_4(PO_4)_2 \\ BaH_4(AsO_4)_2 \end{matrix}\right\} + 2NaCl.$$
(acid)   (soluble)

[5] In the Berzelian notation:

$$3(ZnO_2 . 2SO_3) + 3(NaO_2 . PO_5) = 3ZnO_2 . 2PO_5 + 3(NaO_2 . 2SO_3) + PO_5,$$

or in modern notation:

$$3ZnSO_4 + 3Na_2HPO_4 = Zn_3(PO_4)_2 + 3Na_2SO_4 + H_3PO_4.$$

As can throughout be substituted for P.

Mitscherlich found that the striking analogies between the composition and chemical properties of the two series of salts were accompanied by others equally striking between the crys-talline forms. He carried out the crystallographic investigation of the biphosphates and biarsenates of potassium, sodium, and ammonium, the so-called acid salts[1] described under (ii) (p. 410), which crystallise in the tetragonal and rhombic systems respec-tively. The results obtained in the case of the potassium salts he summarises thus:

"The primitive form of these salts is an octahedron[2] with a square base, but this is very rarely met with; the form in which they usually occur is

Crystallo-
graphic con-
stants of the
acid arsenate
and phosphate
of potassium
found the
same.

a prism with square base terminated by the faces of the octahedron. I have never observed other modifications. I have as-certained by my measurements that the angle between l' and l'' is 90°, and that the inclination between P' and each of the contiguous octahedral faces is the same, from which it follows that the primitive form is an octahedron with square base[2]. I found in 30 measurements, as many for the biphosphate as for the biarsenate, that the inclination of the plane P to that situated on the other side of the axis varies be-tween 93° 50' and 93° 30', in the majority of cases between 93° 40' and 93° 31'. The mean of all the measurements is 93° 36'.[3] The form of the crystals is not affected by the addition of acid or of base to the solution from which they crystallise, as is shown by the following numbers:

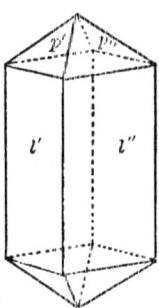

Fig. 71.
Biphosphate
or Biarsenate
of Potash.

----

[1] Represented in our present notation by: KH_2PO_4 and KH_2AsO_4, $NaH_2PO_4 \cdot H_2O$ and $NaH_2AsO_4 \cdot H_2O$, $(NH_4)H_2PO_4$ and $(NH_4)H_2AsO_4$.

[2] Mitscherlich's "octahedron" is an eight-sided bipyramid {hkl}, and not a regular octahedron according to modern nomenclature, which reserves this term for {111} of the cubic system (*ante*, p. 406).

[3] The angles given here and in all fur-ther quotations from Mitscherlich are the "Euclidean angles," such as $\beta\ (= ABC)$. The actual angles measured with the reflecting goniometer, and used in all modern works, are the supplements of the Euclidean angles, such as $a\ (= ABD)$, and are called "normal angles," because they are equal to a', the angle between the perpendiculars drawn from a point inside the crystal on to AB, CB, the faces which are inclined at the angle β.

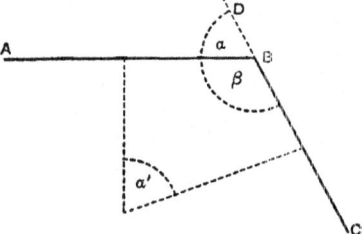

	Acid Arsenate of Potash $P' : P''$		Acid Phosphate of Potash $P' : P''$
1st series	93° 32'	....	93° 40'
	33'		39'
	33'		
	32'		
Mean	93° 32½'	....	93° 39½'
2nd series	93° 37'	....	93° 36'
	38'		37'
Mean	93° 37½'	....	93° 36½'
3rd series	93° 40'		
	38'		
	38'		
Mean	93° 38⅔' "		

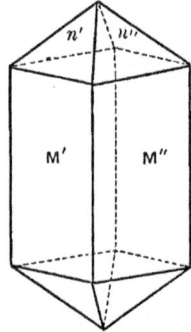

Fig. 72.
Biphosphate or Biarsenate of Soda.

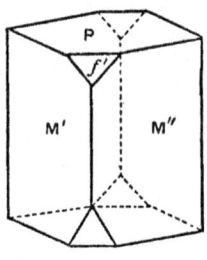

Fig. 73.
Biphosphate of Soda.

"The primitive form of these salts, a regular octahedron, is never found isolated; what usually occurs is the prism with the planes *n*."

$$M' : M'' = 78° 30'$$
$$n' : n'' = 122° 36'$$

"The primitive form of these not uncommon crystals is a right prism with rhombic bases."

$$M' : M'' = 93° 54'$$
$$M : P = 90°$$
$$f' : P = 134° 18'$$

In the case of the sodium salts, the phosphate gave two distinct kinds of crystals (figs. 72 and 73), both supposed to be rhombic[1], but of the different axial relations:

(i) $a:b:c = 0.817 : 1 : 0.500,$

(ii) $a:b:c = 0.9341 : 1 : 0.9572.$

The arsenate appears in one only of these forms (fig. 72).

Mitscherlich himself found similar analogies in the case of a variety of other substances, such as native sulphates and carbonates, and the name of *isomorphism* ($\acute{\iota}\sigma o\varsigma$ = equal, $\mu o\rho\phi\acute{\eta}$ = form) was introduced by him to designate these phenomena. The answer to the problem he had set himself to solve, namely, whether crystalline form depends on the *number*, and not on the *nature* of the constituent atoms, was given in the following words:

Name isomorphism occurrence of chemically different substances in same crystalline form.

"The same number of atoms combined in the same manner produce the same crystalline form; the crystalline form is independent of the chemical nature of the atoms, and is only determined solely by their number and mode of combination."

The validity and importance of Mitscherlich's work met with immediate and fairly general recognition, Berzelius being foremost amongst those who acclaimed the great discovery. Data in support of the law were rapidly accumulated, and it was applied deductively to atomic weight determinations.

But Mitscherlich himself had from the outset recognised the fact that the relations between crystalline form and composition discovered by him were not cases of absolute, but only of approximate isomorphism[2]. Concerning the isomorphous carbonates of calcium, magnesium, iron and zinc, which all crystallise in rhombohedra, he says:

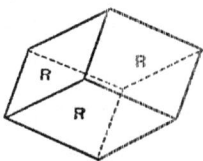

Fig. 74.
Rhombohedron of Calcite.

Deviations from identity of crystalline form.

"I now regard it as a constant feature that in some of these carbonates, the primitive form of which is a rhombohedron, the angles differ from those of the rhombohedron of calcium carbonate by as much as 2°, though

[1] The 2nd modification of $NaH_2PO_4 . H_2O$ is now considered to belong to the oblique system.

[2] The name *homoeomorphism* has been suggested as more suitable, but is not used.

iron carbonate does not differ from dolomite by more than a few minutes[1]. I have also observed a difference of 2° between barium sulphate and strontium sulphate[2]."

A similar difference was found between the neutral arsenate and phosphate of ammonia[3].

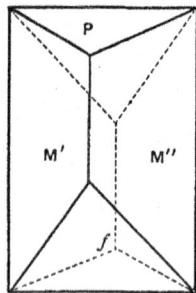

Fig. 75.
Arsenate or Phosphate
of Ammonia.

"The primitive form of the crystals of these two salts is an oblique prism with rhombic base. They are never found without secondary planes: that most commonly found is the plane f [fig. 75].

Inclination of the planes:

	$M':M''$	$P:f$
Arsenate of Ammonia	85° 54'	109° 8'
Phosphate of Ammonia	84° 30'	109° 44'."

Mitscherlich's memoir ends with the words:

"I must leave it for future investigations to show how this law is modified by the small differences which have been detected between the neutral phosphates and arsenates of ammonia, between calcite and magnesite, etc. etc."

Subsequent investigations have but tended to confirm the existence of such differences. Improvement of the experimental methods employed in the measurement of the crystallographic elements has resulted in reducing the experimental error, that is, in closer and closer approximation between the data obtained for different specimens of the crystals of each of the members of an isomorphous series, and has thereby emphasised the greater

Accuracy of the law of isomorphism. Tutton's investigations.

[1] The data are:

Calcite	$CaCO_3$	105° 5'
Dialogite	$MnCO_3$	106° 51'
Chalybite	$FeCO_3$	107° 0'
Magnesite	$MgCO_3$	107° 20'
Smithsonite	$ZnCO_3$	107° 40'
Dolomite	$(CaMg)CO_3$	106° 15'

Angle of Rhombic Prism

Barytes	$BaSO_4$	101° 40'
Celestine	$SrSO_4$	104° 40'

[3] The salts referred to under (i) on p. 410 and represented in our present notation by $(NH_4)_2HPO_4$ and $(NH_4)_2HAsO_4$.

constant discrepancy between the mean values for the different substances.

Mr A. E. Tutton, in his recent investigations on the sulphates and selenates of potassium, rubidium, and caesium, and on the double sulphates which these elements form with magnesium, zinc[1], etc., has studied the influence of the change in the atomic weight of the metals on the crystallographic properties of their isomorphous salts. His object was:

"To examine in great detail some well-defined...series of isomorphous salts...with the view of ascertaining whether the replacement of one metal by another of the same family group but of higher atomic weight was attended by a change in the values of the angles of sufficient magnitude to be far removed from the narrow limits of experimental error."

He chose a series of salts crystallising in one of the systems of lesser symmetry, since:

"In the case of substances crystallising in the higher systems of symmetry, the values of analogous angles upon the crystals of the various members of the series are so nearly identical that the differences frequently fall within the limits of the few minutes usually assigned to experimental... error."

A very few data from amongst the enormous number which Mr Tutton makes the basis of his generalisation are given in a special table, merely as an indication of the degree of accuracy attained, and of the magnitude of the differences established. These data refer to certain salts of the type $R_2SO_4 . MSO_4 . 6H_2O$, where $R = K$, Rb, Cs; and $M = Mg$, Zn, Fe, Co, Ni, Cu, Cd, Mn. The results obtained for all the other salts of these series are of the same kind, and fully justify the conclusions drawn, of which the most important are:

The change in atomic weight from potassium to rubidium and from rubidium to caesium has a regular influence on the axial angle β and on the other angles of the crystals. This is apparent even when, as in the case of the salts containing copper, the absolute values of the angles differ considerably from those of the salts containing other diad elements. The substitution of one of the second (diad) metals for another of the same group is practically without influence on the angles and

Influence of change of atomic weight of substituting elements in isomorphous compounds.

---

[1] London, *J. Chem. Soc.*, 63, 1893 (p. 337); 65, 1894 (p. 628); 71, 1897 (p. 846).

other properties of the crystals; the effect of introducing copper is however noticeable, a discrepancy which finds its explanation in the fact that copper belongs to quite a different group of chemical elements. The *habit* of the crystals, by which is meant the extent to which certain planes are usually developed in a given compound, also varies progressively with the atomic weight; a statement that may be verified by reference to the illustrations in the table.

The influence of change in the atomic weight of the alkali metals is also shown in all the physical properties investigated, *i.e.* in the volume relations (relative density, molecular volume, etc.), optical properties (refractive indices, optic axial angles, etc.), and thermal properties (coefficients of expansion, thermal conductivities, etc.). Mr Tutton's conclusion with regard to the salts of this series is that:

"The alkali metal R exerts a predominating influence in determining the character of the crystals of the isomorphous monoclinic series of double sulphates $R_2SO_4 . MSO_4 . 6H_2O$, and the whole of the crystallographical properties of the potassium, rubidium, and caesium salts containing the same second metal M are in the case of every such group throughout the series functions of the atomic weights of the alkali metal which they contain."

The investigation of the simple sulphates and of the selenates of the same three alkali metals yielded similar results, the substitution of sulphur for selenium producing a change of a few minutes only in the crystallographic angles. The following is the final conclusion derived from both investigations:

"The difference in the nature of the elements of the same family group which is manifested in their regular varying atomic weights, is also expressed in the similarly regular variation of the characters of the crystals of an isomorphous series of salts of which these elements are the interchangeable constituents."

B.   The application of the Law of Isomorphism.

Turning now to the deductive application of Mitscherlich's law, we find it used for purposes of: I. Classification, II. Atomic weight determinations.

I. Isomorphism made the basis of a classification of the elements.

The term "isomorphous elements" is commonly used to designate "equivalency of the elements in isomorphous compounds";

Classification of "isomorphous elements." it does not refer to the crystalline form of the elements themselves, only to that of the compounds into the composition of which they enter; but in some few cases the form of the elements themselves lends additional support to the conclusions arrived at. Thus the rhombic sulphides of antimony and of bismuth are isomorphous, and so are the elements themselves, which crystallise as rhombohedra.

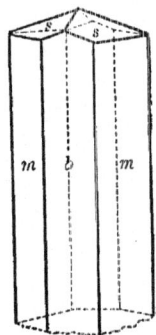

Fig. 76.
Rhombic crystal of Stibnite.

	Prism angle $m : m$	Axial ratios $a : b : c$
Stibnite (Sb_2S_3)	89° 34'	0·993 : 1 : 1·018
Bismuthite (Bi_2S_3)	88° 8'	0·968 : 1 : 0·985

	Rhombohedral angle
Antimony (Sb)	92° 53'
Bismuth (Bi)	92° 20'.

Arzruni[1] arranges the elements in 10 isomorphous series, which include them all except B, Sc, O, and C.

Arzruni's isomorphous series.

    I.   H, K, Rb, Cs, Am, Tl; Na, Li, Ag.

    II.   Be, Zn, Cd, Mg, Mn, Fe, Os, Ru, Ni, Pd, Co, Pt, Cu, Ca, Sr, Ba, Pb.

    III.   La, Ce, Di, Y, Er.

    IV.   Al, Fe, Cr, Co, Mn, Ir, Rh, Ga, In.

    V.   Cu, Hg, Pb, Ag, Au.

    VI.   Si, Ti, Ge, Zr, Sn, Pb, Th, Mo, Mn, U, Ru, Rh, Ir, Os, Pd, Pt, Te (?).

    VII.   N, P, V, As, Sb, Bi.

---

[1] *Physikalische Chemie der Krystalle*, 1893 (p. 100 *et seq.*).

27

F.

VIII. Nb, Ta.

IX. S, Se, Cr, Mn, W; Te (?), As, Sb.

X. Fl, Cr, Br, I, Mn; Cy.

Inspection of the above table shows that the same element may appear as a member of different isomorphous series.

Chromium is found: *Firstly*, in series IV with Al and Fe etc.,
because of the isomorphism, of (i) the rhombohe-
The same element found in different series. dral oxides Al_2O_3 (corundum), Fe_2O_3 (haematite)
and Cr_2O_3; (ii) the oblique cyanides K_3FeCy_6,
K_3CrCy_6 etc.; (iii) the cubic oxides $R''R'''O_4$ termed
spinels, of which we have the representatives $MgO.Al_2O_3$ (spinel),
$FeO.Fe_2O_3$ (magnetite), $FeO.Cr_2O_3$ (chromite) etc.; (iv) the
cubic double sulphates $R_2'SO_4.R_2'''(SO_4)_3.24H_2O$ termed alums.
Secondly, in series IX with S, Se, Mn etc. because of the isomorphism of (i) the rhombic potassium salts K_2SO_4, K_2SeO_4,
K_2CrO_4 and K_2MnO_4 (p. 419); (ii) the corresponding oblique
sodium salts which crystallise with $10H_2O$.

Manganese occurs in series II, IV, VI, IX, and X, belonging in
each of its states of oxidation to another isomorphous class. As
MnO in the rhombohedral carbonates (*ante*, p. 414), it is isomorphous with CaO, FeO etc.; as Mn_2O_3 in the spinels and alums
it is isomorphous with Al_2O_3, Fe_2O_3 etc.; as MnO_3 it is isomorphous
with SO_3, SeO_3 and CrO_3 (p. 419); as Mn_2O_7 it is isomorphous
with Cl_2O_7 (p. 419).

On the other hand the inference would be erroneous that
because manganese can replace chlorine, sulphur, and iron in
isomorphous compounds, these elements themselves are isomorphous.

The following examples may serve as instances of the utility
of the above classification.

Applications of classification of elements according to isomorphism.

1. Support to periodic law classification.

1. The periodic law which brings (i) silver into
the same group with sodium. (ii) chromium with
sulphur, and (iii) manganese with chlorine, finds
support in the analogy between these elements as
revealed by the isomorphism of (i) the sulphates
and selenates of silver and sodium, (ii) the chromates
and sulphates, (iii) the permanganates and per-
chlorates.

Rhombic crystal characteristic of Na_2SO_4, Ag_2SO_4, Na_2SeO_4, Ag_2SeO_4 :

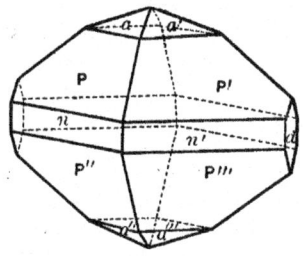

Fig. 77.

	Inclination of $P : P'$	$P : P''$	Axial ratios $a : b : c$
Na_2SO_4	...135° 41'	...123° 43'	0·4734 : 1 : 0·8005
Ag_2SO_4	...136° 20'	...125° 11'	0·4614 : 1 : 0·8077
Na_2SeO_4	...134° 22'	...123° 13'	0·4910 : 1 : 0·8155
Ag_2SeO_4	...135° 42'	...123° 30'	0·4734 : 1 : 0·7963

"In the case of all four compounds the number and development of the planes is identically the same[1]."

Rhombic crystal characteristic of K_2SO_4, K_2SeO_4, K_2CrO_4, K_2MnO_4 :

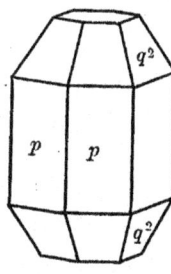

Fig. 78.

	Inclination of $q^2 : q^2$	$p : p$	Axial ratios $a : b : c$
K_2SO_4	...112° 22'	...120° 24'	0·5727 : 1 : 0·7464
K_2SeO_4	...111° 48'	...120° 25'	0·5724 : 1 : 0·7296
K_2CrO_4	...111° 10'	...120° 41'	0·5695 : 1 : 0·7297
K_2MnO_4	...113° 0'	...121° 10½'	0·5638 : 1 : 0·7571

"The crystals of potassium manganate have the same secondary planes, form the same combinations as the sulphate, selenate, and chromate of potassium, and exhibit down to the smallest details the same modifications in the size of the planes[2]."

Rhombic crystal characteristic of $KClO_4$ and $KMnO_4$:

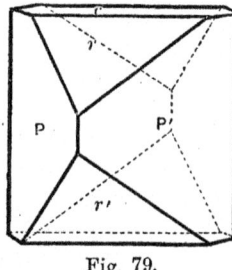

Fig. 79.

	Inclination of $P : P'$	$r : r'$	Axial ratios $a : b : c$
$KClO_4$	...103° 58'	...101° 19'	0·7817 : 1 : 0·6408
$KMnO_4$[3]	...103° 2'	...101° 40'	0·7949 : 1 : 0·6476

[1] Mitscherlich, *Poggend. Annal.*, 12, 1828 (p. 138).

[2] Mitscherlich, *ibid.*, 25, 1832 (p. 293).

[3] The two isomorphous salts should be consistently represented alike, either by the single or the double formulae ; by $KClO_4$ and $KMnO_4$, or by $K_2Cl_2O_8$ and $K_2Mn_2O_8$. But whilst the perchlorate is always written $KClO_4$, both permanganate formulae are met with. The formula $KMnO_4$ is in agreement with the results of molecular weight determinations by the lowering of the solidification point of a non-dissociating solvent.

2. Mitscherlich's recognition of the isomorphism with potassium sulphate of the green potassium salt of a manganese acid, and of the isomorphism with potassium perchlorate

2. Formula for manganic acid.

of the red potassium salt of another manganese acid, has revealed the true composition of these two acids.

"The green crystals are manganate of potassium, which is isomorphous with the sulphate of potassium. The red crystals have the same form as the crystals of oxidised potassium chlorate; accurate analysis has shown that oxidised chloric acid, as well as this higher stage of oxidation of manganese, contains 7 proportions of oxygen. It seems to me, therefore, suitable that the stage of oxidation of manganese which corresponds to sulphuric, selenic, and chromic acid should be called manganic acid, and the highest stage of oxidation of manganese, permanganic acid."

3. From the isomorphism of the salts of an acid of selenium with sulphates and chromates, Mitscherlich was led to the discovery of that acid[1].

3. Discovery of selenic acid.

"The crystals had the form of sulphate of potash and were analogous to that salt in relation to polarised light, and in that they yielded an insoluble precipitate with the salts of baryta. From these characters it results that the crystals were a compound of potash and a new acid of selenium, isomorphic with sulphuric acid. As the new acid contains more oxygen than that discovered by Berzelius, it must be called *selenic* acid, while to the latter the term *selenious* acid is appropriate."

The new acid was next prepared by the decomposition of the insoluble lead salt by sulphuretted hydrogen, its properties were investigated, and its composition was determined by application of the law of isomorphism.

"Since the neutral salts of selenic acid are isomorphous with the sulphates, the composition of selenic acid and selenates may be expected to obey the laws of isomorphism. Consequently selenic acid should contain one-half more oxygen than selenious acid for the same quantity of selenium; and the oxygen contained in the base of the selenates ought to be one-third of the oxygen of the acid. Experiment fully confirms this hypothesis."

II. Isomorphism made the basis of atomic weight determinations.

The principle of such determinations may be thus put: If we

[1] "On a New Acid of Selenium," *Edinb. J. Sci.*, 8, 1828 (p. 294).

know the percentage composition of two or more substances
recognised as isomorphous, and if we also know the

Isomorphism used for atomic weight determinations. atomic weight of at least one of the elements
replacing one another in these compounds, we can
calculate the atomic weight of the other element;
provided always that we may assume that these compounds really
have an analogous composition.

1. *Instances of such atomic weight determinations.*

(i)   Mitscherlich's study of the neutral salts of sulphuric and
of selenic acid led him to declare these substances as isomorphous,
and therefore to assume that their composition was

Mitscherlich's determination of the atomic weight of selenium. analogous.   Hence the quantities of sulphur and of
selenium which are "equivalent" in these com-
pounds are quantities representing the aggregate
weight of an equal number of atoms, and are therefore
in the ratio of their atomic weights.

The composition of these two substances he found to be :

Potassium Sulphate 100 parts contain:	Potassium Selenate 100 parts contain, or 127·01 parts contain	
Potassium.........44·83	Potassium.........35·29.........44·83	
Oxygen ....36·78	Oxygen ....28·96.........36·78	
Sulphur............18·39	Selenium ....35·75.........45·40	
100	100	127·01

The quantities of sulphur and of selenium equivalent as
regards combination with 44·83 potassium and 36·78 oxygen are
18·39 and 45·40,

∴ atomic weight sulphur : atomic weight selenium = 18·39 : 45·40 ;

but the atomic weight of sulphur having been found by the
application of other methods of atomic weight determinations = 32·0,

$$\text{atomic weight selenium} = \frac{32 \cdot 0 \times 45 \cdot 40}{18 \cdot 39} = 79.$$

(ii)   It has been shown before (p. 334) how the isomorphism
of sulphates and chromates led Berzelius to a modification of

the formula of basic chromic oxide and to a subsequent halving of the atomic weights of most of the

Effect on Berzelius' atomic weight tables of the recognition of the isomorphism of Cr and S.

Roscoe's work on vanadium.

metals.

(iii) An example of historical interest in the application of the law of isomorphism to atomic weight determinations is furnished by Sir Henry Roscoe's classical research on vanadium[1].

"The vanadium compounds have proved a remarkable and unexplained exception to the law of isomorphism. Vanadite[2] (lead vanadate and lead chloride) is isomorphous with apatite, pyromorphite and mimetisite[2], minerals consisting of calcium phosphatochloride, lead phosphatochloride, and lead arsenatochloride. These all crystallise together, their form being a hexagonal prism terminated by six-sided pyramids[3]. According to the work of Berzelius (1831) the formula of vanadic acid is VO_3[4]. If this is so, we have dissimilarly constituted substances acting as isomorphous bodies and crystallising together, or else Berzelius is mistaken, and the formula of vanadic acid is V_2O_5[4] corresponding to arsenic and phosphoric pentoxides."

Berzelius had found (1) that vanadic acid when reduced in hydrogen at a red heat suffers a constant loss of weight, leaving behind a lower oxide, and (2) that this lower oxide when heated in chlorine gives a volatile chlorine compound and a residue of vanadic acid exactly equal to one-third of the quantity originally taken for reduction. Hence if the volatile chlorine compound is a chloride, a compound of vanadium and chlorine *only*,—which is what Berzelius assumed it to be—the oxygen combined in the lower oxide with *all* the vanadium originally present is only

[1] "Researches on Vanadium," London, *Proc. R. Soc.*, 1867 (p. 220).
[2] The names now used are : Vanadinite and Mimetite.
[3]

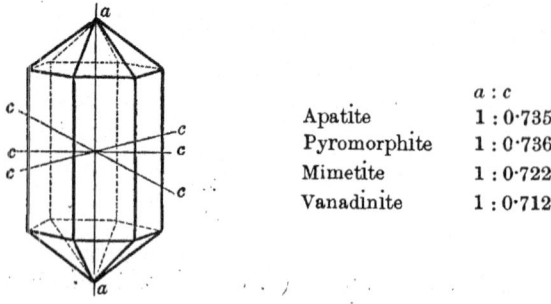

	$a : c$
Apatite	1 : 0·735
Pyromorphite	1 : 0·736
Mimetite	1 : 0·722
Vanadinite	1 : 0·712

[4] Berzelius' formula is referred to O = 8, Roscoe's to O = 16.

sufficient for combination with one-third that amount of vanadium in the acid. Berzelius concluded therefore that the number of atoms of oxygen in the oxide was to that in the acid in the ratio of $1:3$, and that the formula for the acid was VO_3, for the oxide VO, and for the chloride VCl_3, thus

$$3VO_3 + 6H = 3VO + 6HO$$
$$3VO + 6Cl = VO_3 + 2VCl_3.$$

120·927 of the acid having yielded on reduction 100 of the oxide, Berzelius got for the atomic weight of vanadium[1] calculated from the equation

$$(V + 3 \times 8):(V + 8) = 120\cdot927:100$$
$$V = \mathbf{68\cdot5}.$$

But Roscoe succeeded in proving that "Berzelius had been mistaken," that the substance[2] he had represented by the symbol V was really VO_2, and hence that the atomic weight assigned to it would give for the element vanadium $68\cdot5 - 16 = 52\cdot5$; and similarly that the presumed trichloride VCl_3 was an oxychloride VO_2Cl_3.

He found in four experiments that a sum of 24·5561 grams of vanadic acid yielded on reduction 20·2556 grams of the lower oxide, and assigning the formula V_2O_5 to the acid $(O = 16)$, the calculation of the atomic weight of vanadium became:

$$(V_2 + 5 \times 16):(V_2 + 3 \times 16) = 24\cdot5561:20\cdot2556$$
$$V = \mathbf{51\cdot36}.$$

And with $V = 51\cdot4$, $O = 16$, $VO = 67\cdot4$, the reactions became

$$3V_2O_5 + 6H_2 = 3V_2O_3 + 6H_2O,$$
$$3V_2O_3 + 6Cl_2 = V_2O_5 + 4VOCl_3.$$

The formulae for the isomorphous substances then showed the required agreement

Vanadinite.........$3\{Pb_3(VO_4)_2\} + PbCl_2 = Pb_5V_3O_{12}Cl.$

Apatite ....$3\{Ca_3(PO_4)_2\} + CaF_2 = Ca_5P_3O_{12}F.$

Pyromorphite......$3\{Pb_3(PO_4)_2\} + PbCl_2 = Pb_5P_3O_{12}Cl.$

Mimetite.........$3\{Pb_3(AsO_4)_2\} + PbCl_2 = Pb_5As_3O_{12}Cl.$

---

[1] The number so obtained is too high, owing probably to the fact that Berzelius' material was contaminated with phosphoric acid, whereby the loss of weight on reduction would not reach its correct value.

[2] The element had at that time not been actually isolated.

Here then a research which had for its starting point the recognition of the validity of Mitscherlich's law of isomorphism led to a fundamental change in the atomic weight of vanadium, to a different representation of its reactions, and to a different formulation of its compounds.

2. *Experimental work required.*

Besides the establishment of the antecedent data, which are the atomic weights of members of the isomorphous series,

Experimental work required in the application of isomorphism to atomic weight determinations. this consists in: (i) Recognition of the existence of isomorphism between two or more compounds of which one contains the element of unknown atomic weight. (ii) Analysis of the isomorphous compounds. Since comparatively little need be said, or can be said concerning (ii), it will be best to dispose of it first.

Analysis of the isomorphous compounds.

The methods used vary in the individual cases, and so does the degree of accuracy attained.

Put quite generally, let $X_m A_n B_p C_q \ldots$ and $Y_m A_n B_p C_q \ldots$ be the formulae of the isomorphous compounds, where $X, Y, A, B, C \ldots$

Calculation of atomic weight from results of analyses of isomorphous substances. are the atomic weights of the constituent elements, and m, n, p, $q \ldots$ simple whole numbers, and let $x, y, a, b, c \ldots$ be the weights of the corresponding elements found in a given weight of the compounds.

Then we may either find the ratios $x : a$ and $y : a$, and calculate these to the same amount of a, as was done (p. 421) in finding the quantities of sulphur and of selenium present in the neutral sulphate and selenate of potassium in combination with the *same* amount of potassium. Or if we know all the atomic weights except X, and make a complete analysis of the compound containing Y, we can find values for m, n, p, $q \ldots$, that is we can get a general formula for the two isomorphous compounds; if then in the compound $X_m A_n B_p C_q \ldots$ we can find the ratio between any two constituents or sums of constituents of which one may be the compound, we can calculate X the required atomic weight from any one of the relations:

$$x : a = mX : nA$$

$$(a + b) : (x + a + b + c + \ldots) = (nA + pB) : (mX + nA + pB + qC + \ldots)$$

$$(x + a) : (x + a + b + c + \ldots) = (mX + nA) : (mX + nA + pB + qC + \ldots)$$

Thus in the case of the isomorphous double sulphates which crystallise in regular octahedra and which are named *alums*, the analysis of salts which contain only elements of known atomic weight (*e.g.* sodium, potassium, etc., iron, chromium, etc.) has shown that the class may be represented by the general formula

$$X_2'SO_4 . X_2(SO_4)_3 . 24\,H_2O,$$

where X' is the atomic weight of elements of the type of K, Na, etc., and X the atomic weight of elements of the type of Fe, Cr, etc. Hence, if we find a double sulphate crystallographically recognised as an alum, and if we know X', we can find X by determining q the amount of sulphuric acid radicle present in a certain weight p of the pure salt, which can be done by weighing the insoluble barium sulphate obtained on addition of a soluble barium salt to the solution of the alum. Then $\dfrac{p}{q}.4(32+64)$ represents the formula weight of $X_2'SO_4 . X_2(SO_4)_3 . 24H_2O$ of the alum, *i.e.* the amount containing $4SO_4 = 4(32+64)$ of sulphuric acid radicle, and consequently:

$$\text{At. wt. required} = X = \tfrac{1}{2}\{X_2'X_2(SO_4)_4 . 24H_2O - X_2'(SO_4)_4 . 24H_2O\}$$

$$= \tfrac{1}{2}\left\{\frac{p}{q}.2(32+64) - 384 - 432 - 2X'\right\}.$$

Or, since the ignition of alums in which $X' =$ ammonium gives a residue of non-volatile sesquioxide of X, thus

$$(NH_4)_2SO_4 . X_2(SO_4)_3 . 24H_2O = X_2O_3 + \text{volatile products,}$$

we may determine q the weight of sesquioxide left by the ignition of a weight p of the pure alum, and calculate X from the equation:

$$p : q = (NH_4)_2SO_4 . X_2(SO_4)_3 . 24H_2O : X_2O_3$$

$$= X_2 + (2\times18) + (4\times96) + (24\times18) : X_2 + 3\times16.$$

Thus Lecoq de Boisbaudran, the discoverer of *gallium*, found that this element forms alums, and that 3·1044 grams of *gallium ammonium alum* when ignited left ·5885 grams of *gallium oxide*, numbers which when substituted for p and q in the above equation give

$$X = \text{atomic weight of gallium} = \mathbf{70\cdot1.}$$

Recognition of the existence of isomorphism.

It is here that the great practical difficulty comes in. What are the criteria that are applied, and how far is the

Recognition of the existence of isomorphism: (*a*) Similarity of crystalline form.

evidence obtained conclusive?

(*a*)   Great similarity of crystalline form.

The experimental data are those of the crystal angles from which the crystallographic constants are calculated. The necessary measurements are made by the instrument termed the "reflecting goniometer," the principle of which we owe to Wollaston[1].

The advisability, or rather necessity of speaking of "similarity," and not "identity," of form should be evident from what has been said on this point before (p. 413). Moreover Mitscherlich had already observed and pointed out that the magnitude of the deviations from identity varied. He had found that whilst the characteristic angles of the acid ammonium phosphate and arsenate were almost identical (89° 35' and 89° 34'), the differences in the neutral salts, *i.e.* in the oblique $(NH_4)_2 HPO_4$ and $(NH_4)_2 HAsO_4$, amounted to more than a whole degree (85° 54' and 84° 30'). Similarly in the class of rhombohedral carbonates (*ante*, p. 413) the difference of 2° 35' between the characteristic angles of calcite (105° 5') and smithsonite (107° 40') is of another order of magnitude than 20', the difference between chalybite (107° 0') and magnesite (107° 20').

All other classes of isomorphous substances exhibit similar variations in the differences between their crystallographic constants. The following data refer to the class of sulphates termed vitriols, which crystallise in the rhombic system (fig. 80) and have the general formula $X'' SO_4 . 7H_2O$.

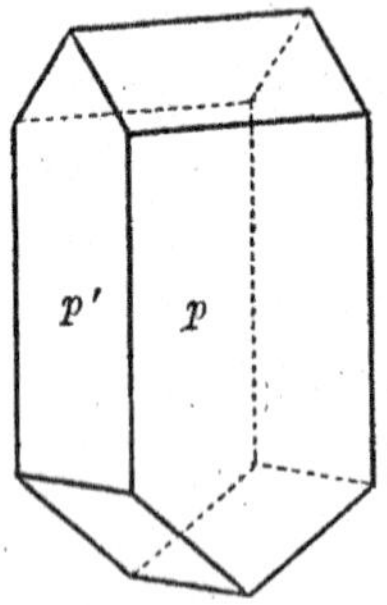

Fig. 80.

Inclination of the rhombic prism[2] $p : p'$:

MgSO$_4$.7H$_2$O........89° 26'

NiSO$_4$.7H$_2$O ....88° 56'

ZnSO$_4$.7H$_2$O ....88° 53'.

[1] Lewis, *Crystallography*, chap. xx., p. 589; Miers, *Mineralogy*, chap. iv., p. 99.

[2] These crystals belong to that sub-class of the rhombic system which possesses three diad axes at right angles to one another, but has no plane of symmetry.

Therefore the question arises, what is to constitute "similarity," what amount of difference is to be considered as still compatible with the existence of similarity? The arbitrary and fluctuating judgment of the individual is appealed to, with the result that similarity of form cannot by itself determine isomorphism.

Uncertainty concerning amount of crystallographic differences compatible with existence of isomorphism.

(*b*)   Analogous composition.

Mitscherlich had made this a requirement for the existence of isomorphism. His own formulation of the law of isomorphism restricts its scope to "an equal number of atoms united in the same manner," and he had selected the phosphates and arsenates for investigation because of the undoubted analogy of composition established for these salts by Berzelius. Chemical analogy, *i.e.* similarity of properties, is implicitly assumed. But here again the interpretation of "analogous" is arbitrary, and hence variable. Whilst some authorities demand of isomorphous substances that atoms should be replaced by an equal number of others of the same valency and of chemical similarity, such as is the case in

(b). Analogous composition.

Cerussite	$PbCO_3$		
Witherite	$BaCO_3$	rhombic	
Strontianite	$SrCO_3$		
Aragonite	$CaCO_3$		

Chloanthite	$NiAs_2$	cubic
Smaltite	$CoAs_2$	

others give a less strict interpretation, and will accept substances as isomorphous in which Pb is replaced by Ag_2, K by NH_4, such as is the case in

Galena	PbS	octahedra
Argentite	Ag_2S	and cubes

Sylvin	KCl	cubes and octahedra.
Salammoniac	NH_4Cl	
Potass. cyanide	KCN	

(*c*)   Complete miscibility.

(c) Complete miscibility.

Mitscherlich says: "Whilst substances of different crystalline form cannot combine otherwise than in fixed ratios, substances of the same crystalline form can crystallise together in all ratios," and he points out that this power of continuous replacement of one of the members of a class by another is possessed by the group of rhombohedral carbonates, which he therefore considered isomorphous, in spite of the differences in the rhombohedral angles.

Extensive recent research has however shown that the property of forming mixed crystals is possessed to a greater or lesser degree by almost all substances, whether they have the same crystalline form or not[1], but that this power increases with the similarity of the substances involved and reaches its highest degree, which is complete miscibility, when the atoms substituting one another are those of chemically similar elements and when the compounds are crystallographically very similar. Thus octahedral salammoniac forms perfectly homogeneous coloured mixed crystals with a number of chlorides such as $FeCl_2$, $FeCl_3$, $MnCl_2$, etc. with which it is *not* isomorphous, but the amount of these other chlorides taken up by the NH_4Cl is always small.

Property of forming mixed crystals not restricted to isomorphous substances.

Here, again, we deal with a property which is a matter of degree and which without the consideration of concomitant phenomena cannot be taken as a criterion. It will be necessary to return to this point and to consider further the phenomena of complete and partial miscibility of isomorphous crystals.

In crystalline minerals which are rarely pure specimens of one compound, the substitution of one element by an equivalent amount of another occurs in varying degrees and must be represented in the formula assigned.

Calculation of formula for minerals which are iso-morphous mixtures.

Thus $(Fe, Mn) CO_3$ is the formula most commonly used for the rhombohedral carbonate containing both iron and manganese, and in which the relative amounts of these two metals might vary from $Fe = 48 \cdot 2 \, \%$ and $Mn = 0$ to $Fe = 0$ and $Mn = 47 \cdot 8 \, \%$. Another, but less common mode of representation is $x \, FeCO_3 + y \, MnCO_3$. Both formulae are arrived at in the following way by calculation from analysis which in a special instance yielded the numbers given below:

---

[1] It is a case of the formation of *solid solutions*, i.e. of "solid homogeneous mixtures of several substances, the relative quantities of which can vary without loss of the homogeneity." Such solid solutions can be amorphous or crystalline.

$$\text{FeO} = 36\cdot81\ldots\ldots\frac{36\cdot81}{\text{FeO}} = \frac{36\cdot81}{72} = \cdot511$$

$$\left.\vphantom{\begin{array}{c}a\\b\end{array}}\right\} \cdot867$$

$$\text{MnO} = 25\cdot31\ldots\ldots\frac{25\cdot31}{\text{MnO}} = \frac{25\cdot31}{71} = \cdot356$$

$$\underline{\hphantom{\text{CO}_2} \text{CO}_2 = 38\cdot35\ldots\ldots\frac{38\cdot35}{\text{CO}_2} = \frac{38\cdot35}{44} = \cdot871}$$

$$(\text{Fe.Mn})\text{CO}_3 = 100\cdot47$$

$\therefore$ Equivalents basic oxide : equivalents acidic oxide $= \cdot867 : \cdot871$

$$= 1 : 1.$$

$\therefore$ Formula is $X''O \cdot CO_2 = X''CO_3$ where $X'' = $ Fe or Mn

and in the formula $x\text{FeCO}_3 + y\text{MnCO}_3$ the ratio $x : y$ is given by:

$$x : y = \cdot511 : \cdot356$$

$$= 59 \quad : 41 \,[1].$$

(*d*) Isomorphous overgrowth.

Kopp, in an address to the German Chemical Society given in
1879, on the subject of the application of isomorphism
to atomic weight determinations, puts at the outset
the momentous question: What compounds are
isomorphous?

(*d*) Isomorphous overgrowth.

On grounds such as those set forth in the preceding pages he
decides that the required answer cannot be obtained by making
equal (or approximately equal) crystalline form and analogous
composition the only requirements. He seeks for some property
which it is possible "to establish quite objectively," and believes
he has found it by "designating as isomorphous, substances which
have the same crystallising power, so that they can one in the
place of the other contribute in the same manner to the formation
of a crystal." This power is manifested by the formation of mixed
crystals and by *isomorphous overgrowth*. Kopp then proceeds to
discuss at length isomorphous overgrowth, which may be looked
upon as an addition made by him to Mitscherlich's original three
criteria for the existence of isomorphism. The well-known case of
the continuation in growth of a colourless crystal of aluminium
alum when placed in a solution of violet chromium alum is quoted,
and other cases are added in which a crystal maintaining its
original form continues to grow when placed in a solution of

[1] Roscoe and Schorlemmer, *Treatise on Chemistry*, i., 1894 (pp. 868 *et seq.*).

a different coloured isomorphous substance. Tetragonal colourless $ZnSO_4 . 6H_2O$ can be coated with the green $NiSO_4 . 6H_2O$; pale amethyst coloured triclinic $MnSO_4 . 5H_2O$ can be surrounded by a layer of blue $CuSO_4 . 5H_2O$, etc. To Kopp the value of the test seems to lie in the fact that:

Instances of isomorphous overgrowth; value of the test.

" If we consider as isomorphous, substances which are capable of forming mixed crystals and of overgrowing one another, we can detect the existence of isomorphism without knowing any particulars about the crystalline form, or about the composition of the substances considered."

But the fact that there are cases on record in which a substance whilst maintaining its original form continues to increase in size by the superposition of a layer of a substance undoubtedly not isomorphous with the nucleus,— *e.g.* rhombic pseudo-hexagonal K_2SO_4 can be coated with a layer of hexagonal $NaKSO_4$—deprives this method also of strict general validity.

Isomorphous overgrowth not restricted to isomorphous substances.

Analogies in molecular volume $\left(i.e. \dfrac{\text{molecular weight}}{\text{specific gravity}} \right)$, in cleavage, in etching figures, have also been used as indications of the existence of isomorphism, but the results have not been such as to much advance the subject.

It must therefore be admitted that none of the properties dealt with so far supply definite, conclusive criteria for the recognition of isomorphism, and that generally recourse must be had to summation and subsequent evaluation of evidence derived from as many different sources as possible.

But it would seem as if the " purely objective test" demanded by Kopp had at last been supplied theoretically for all, practically for a certain number of cases.

(e) Proportionality between the composition and the properties of mixed crystals.

Retgers, from general theoretical considerations and extensive experimental investigations, has supplied a firm basis for the recognition of isomorphism[1]. Of the criteria hitherto applied, the two first, *i.e.* similarity (or agreement) in crystalline form and chemical composition, he rejects as inadequate.

(e) Relation between composition and properties of isomorphous mixed crystals.

[1] "Das spezifische Gewicht isomorpher Mischungen," *Zs. Phys. Chem.*, Leipzig, 3, 1889 (p. 497); " Zur Kenntnis des Isomorphismus, i.—xii.," *ibid.*, 4—20, 1889—1896.

"The process hitherto exclusively used for the recognition of isomorphism, wherein the only thing considered is the similarity in form of the pure substances, cannot lead to correct and suggestive results."

The power of forming mixed crystals is what he considers as all-important.

" I believe that the essential and fundamental condition of isomorphism is to be found in the capacity which two substances may possess of forming mixed crystals in any ratio whatever. But I am of opinion that from the practical point of view we should have recourse to a more precise criterion than that of mixed crystallisation, viz. to that of the investigation of the physical properties of the mixed series obtained....In the investigation of the physical properties of isomorphous mixtures the following points must be considered: *Firstly*, we must find the law connecting the chemical composition and the physical properties of undoubtedly isomorphous mixed crystals, and *secondly* we must apply the law thus found to deciding in doubtful cases whether isomorphism exists or not."

Retgers on necessity of finding empirical law connecting composition and properties of isomorphous mixed crystals, and then applying it.

The law required, which is arrived at by induction from the results of his own investigations and from those of others, he formulates thus:

"If we plot the percentages of one constituent of the different mixtures as abscissae, and the corresponding magnitudes of the physical properties as ordinates, these different points lie on a continuous curve."

The expression "physical properties" is taken in its widest sense including geometrical properties (angles, and ratios of the crystallographic axes), optical properties (refractive indices, co-efficients of dispersion), specific gravities, thermal properties, moduli of elasticity and electrical conductivities.

" I hold that we can attain to some true insight into the often complicated cases of isomorphism, if we prepare the most complete possible series of mixed crystals and investigate it chemically and physically. Theoretically any physical property is available for the physical investigation...but the most suitable will be a property which appears in a purely additive form, such as the refractive index, specific gravity, optical rotatory power. The relation, which may often be a very complicated one, will then appear clearly in the graphical representation."

The following may serve as instances of the results of such investigations:

The data concerning *the relation between the composition of mixed crystals and their geometrical properties* are not numerous, and as far as they go they are either not reliable or

Composition and geometrical properties.

contradictory. The best investigated case, that of the crystallographic angle of mixed crystals of the rhombic sulphates of magnesium and zinc[1], shows the progressive change accompanying the corresponding change in composition; but the differences of angle are so small as to be scarcely beyond the limits of experimental error. On the other hand, for the mixed crystals of potassium perchlorate and potassium permanganate[2] the crystallographic constants are even outside the value of that of either of the pure substances.

Relation between Composition and Geometrical Properties of Mixed Crystals.

Percentage Composition of Mixed Crystals $x MgSO_4 . 7H_2O + (100 - x) ZnSO_4 . 7H_2O$		Prism angle $p : p'$ (fig. 80, p. 426)
x	$100 - x$	
100	0	89° 25′
78·88	21·12	89° 17′ 30″
74·44	25·56	89° 15′
62·70	37·30	89° 11′
57·59	42·41	89° 8′
42·80	57·20	89° 3′ 30″
35·64	64·36	89° 1′
18·11	81·89	88° 54′
0	100	88° 48′

Percentage Composition of Mixed Crystals $x KClO_4 + (100 - x) KMnO_4$		Prism angle $p : p'$ (fig. 79, p. 419)
x	$100 - x$	
100	0	**76° 2′**
99·69	0·31	75° 2′
90·4	9·6	75° 52′
82·8	17·2	76° 11′
0	100	**77° 8′**

Parallelism between the composition of mixed crystals and the values of the various optical properties has been established in a number of investigations.

[1] Dufet, *Bull. Min.*, 1, 1878 (pp. 58—61).
[2] Groth, *Poggend. Annal.*, 133, 1868 (p. 192).

Dufet, by measurement of the refractive indices of mixed
crystals of the sulphates of magnesium and zinc,

<small>Composition and optical properties.</small> and of magnesium and nickel respectively, found
that:

" The differences between the refractive indices of a mixed crystal of two isomorphous salts and those of the components are in the inverse ratio of the number of the equivalents of the two salts contained in the mixtures."

Hence the curve (fig. 81) plotted by taking the molecular weights as abscissae, and the refractive indices as ordinates, is a straight line.

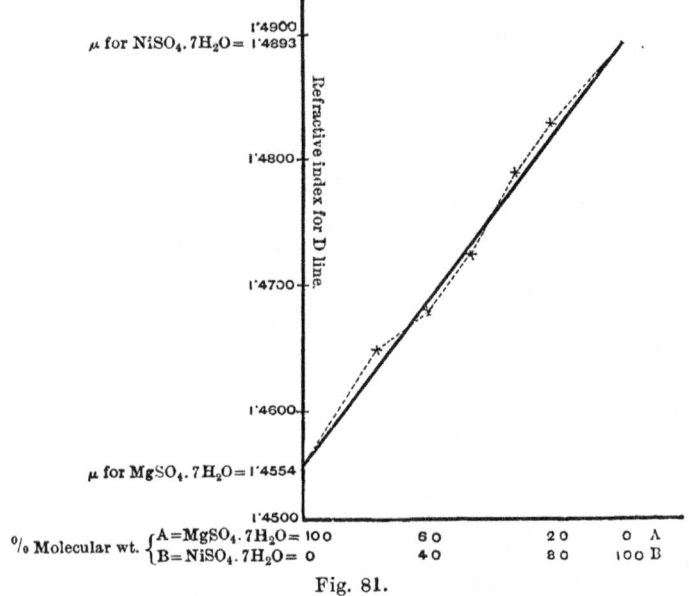

Fig. 81.

Mixture of Magnesium Sulphate and Nickel Sulphate.

Retgers has made extensive investigations of *the relation between the specific gravity and the composition of mixed crystals.*

"On practical grounds the specific gravity is generally
the most suitable property for investigation."

<small>Composition and specific gravity.</small> These researches are marked by extreme care
concerning the chemical and physical purity of the
specimens investigated. Microscopic examination was resorted to

28

F.

in order to ensure that the specimen selected did not enclose mother liquor or air bubbles; that it was an individual, *i.e.* one mixed crystal, and not an aggregation of different possibly dissimilar units; that it was not a twin crystal; and not a case of isomorphous overgrowth. For the accurate measurement of the

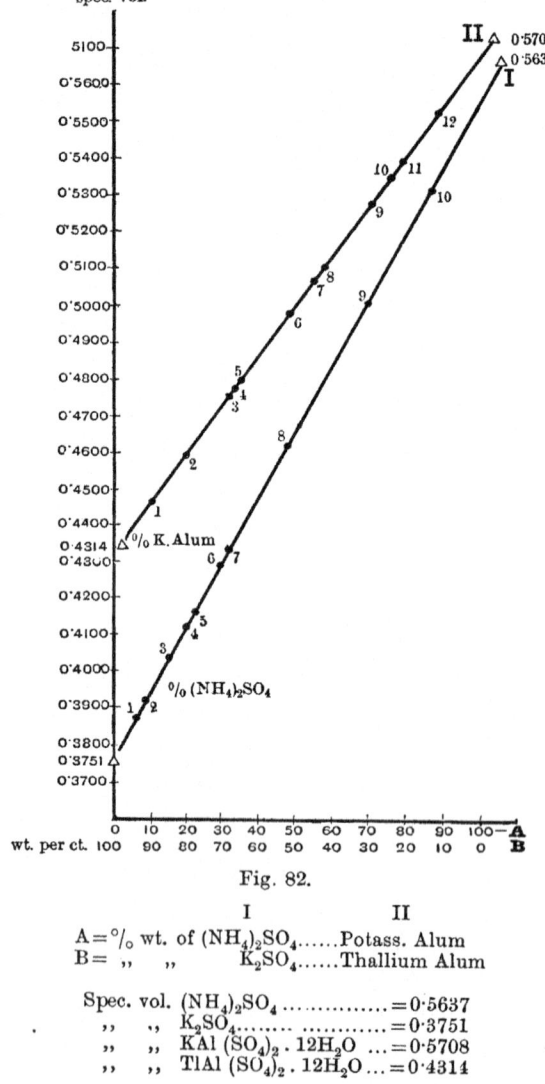

Fig. 82.

	I	II
A =	$\%$ wt. of $(NH_4)_2SO_4$......Potass. Alum	
B =	,, ,, K_2SO_4......Thallium Alum	

Spec. vol. $(NH_4)_2SO_4$ .... = 0·5637
,, ,, K_2SO_4 .... = 0·3751
,, ,, $KAl (SO_4)_2 . 12H_2O$ ... = 0·5708
,, ,, $TlAl (SO_4)_2 . 12H_2O$ ... = 0·4314

specific gravity, the method employed was that of suspension in a heavy liquid (methylene iodide, spec. grav. 3·3, suitably diluted with xylene, spec. grav. ·89). Since only small crystals fulfil the above requirements of physical and chemical purity, the substances investigated had to be so chosen as to be capable of being chemically analysed by accurate and speedily executed methods (NH_4 determined by loss of weight on ignition, MnO_4 estimated volumetrically by Fe″, etc. etc.).

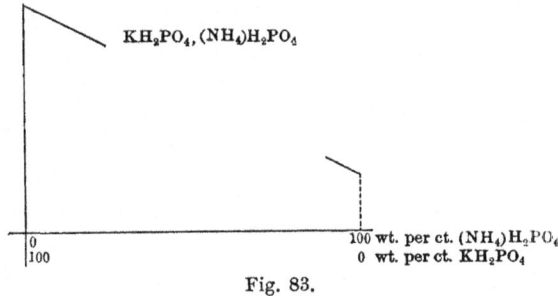

Fig. 83.

He selects for graphic representation the relation between the percentage composition and the specific volume (the

Relation between percentage composition and specific volume represented by a straight line.

reciprocal of the specific gravity, *i.e.* spec. vol. $= \dfrac{1}{\text{spec. grav.}}$), and obtains straight lines with or without breaks (figs. 82, 83), according as a complete series of mixed crystals is formed or not.

The two phosphates would according to Mitscherlich's requirements of complete miscibility not be truly isomorphous, but to Retgers the fact that the two portions of the curve are parts of the same straight line is undoubted proof of the existence

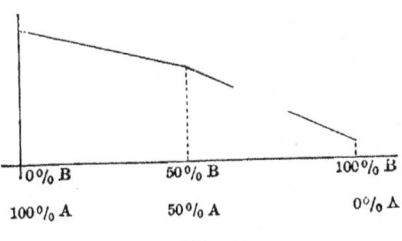

Fig. 84.

of isomorphism between these salts. The difference in this respect between his view and that of Mitscherlich is set forth by diagrammatic representation (fig. 84) of a hypothetical case, the converse of that of fig. 83, in which the miscibility of the two components shows no break, but where the physical properties are not a continuous function of the composition.

The result of these investigations as formulated by Retgers is the proof of the existence of a law according to which, in isomorphous mixtures, there is proportionality between specific gravity (or specific volume) and chemical composition.

Retgers has also examined microscopically *the colour obtained when a coloured and a colourless salt crystallise simultaneously*, and he found that if the salts were isomorphous, *e.g.*

Composition and colour. $KClO_4$ and $KMnO_4$, crystals of all depths of colour were obtained; sometimes even a single crystal was of different colour at its two ends, with all intermediate shades between. But if the two salts were not isomorphous, *e.g.* KCl and $KMnO_4$, the crystals formed showed no transition, but were either colourless or dark red.

Retgers' views on the true nature of isomorphism find a good expression in the generalisations:

Empirical law: physical properties of mixed crystals are continuous functions of their chemical composition.
"All the hitherto better investigated physical properties of isomorphous mixtures are of purely additive nature, continuous functions of the percentage composition." (Brauns, *Chemische Mineralogie.*)

"Two substances are not truly isomorphous except when the physical properties of their mixed crystals are continuous functions of their chemical composition." (Retgers.)

Whilst recognising the enormous advance in precision that we owe to Retgers, it must be admitted that his method is of very limited applicability, and that in the preponderating number of instances in which the physical properties of mixed crystals cannot be measured, a good case may still be made out for the existence of isomorphism, if the substances undoubtedly exhibit similarity of crystalline form, and analogy of chemical composition, and if they form mixed crystals.

In concluding the subject of atomic weight determinations on the basis of isomorphism, a few words must be said concerning the type of cases dealt with, and the accuracy of the results obtained.

3. *Applicability of the method.*

The method differs from that based on Avogadro's hypothesis in not taking its origin from *à priori* considerations concerning the nature of certain conditions of matter; but alike with that based on the law of the constant atomic heat capacity of solids, it has required for its establishment the knowledge of a number of atomic weights obtained by other methods; and in every case dealt with, knowledge of the atomic weight of at least one member of the isomorphous series is essential. The method is therefore only available in conjunction with others. Its applicability is restricted to crystalline *compounds* the crystalline form and chemical composition of which must be investigated. Study of the forms of the elements themselves is of no more avail than is that of their vapour densities in the method which is based on Avogadro's hypothesis (*ante*, pp. 356 *et seq.*).

Types of cases in which atomic weights may be determined from isomorphism.

4. *Accuracy of the results.*

This depends solely on the accuracy of the analysis of the compound containing the element of unknown atomic weight, and on that of the antecedent data, *i.e.* the atomic weights involved in the required calculations. The values so obtained may accordingly directly furnish the standard numbers for the atomic weights investigated, or they may only supply the data necessary for the correct atomistic formulae of the compounds best suited for analysis, that is, give the relation between the equivalent and the atomic weight. Thus Mitscherlich determined the atomic weight of selenium from the recognition of the isomorphism of selenates and sulphates together with the analysis of these salts. He obtained the value 79, but the analyses involved cannot lay claim to sufficient accuracy, and they are rejected by Clarke in his compilation of the data available for calculating the atomic weight of selenium. But Mitscherlich's proof of the isomorphism of selenic and sulphuric acid, together with the determination of the relative amounts of oxygen combined with the *same* amount of selenium in selenious and selenic acid respectively, gave SeO_2 and SeO_3 for the atomistic constitution of these two bodies. Ekman

Accuracy of atomic weight values found by method of isomorphism.

and Pettersson found that in 5 experiments a total of 99·4299 grams of selenious acid yielded on reduction by sulphurous acid 70·7841 grams of selenium.

$$\therefore SeO_2 : Se = 99\cdot4299 : 70\cdot7841$$
$$(Se + 32) : Se = 99\cdot4299 : 70\cdot7841.$$

∴ Se = **79·07**, a number which by chance is practically identical with that of Mitscherlich.

POLYMORPHISM.

Haüy held that every definite crystalline form is characteristic of only *one* substance of definite chemical composition, and conversely, that every substance of definite chemical composition can occur in only *one* definite crystalline form. Under the name of isomorphism are comprised all those cases of substances chemically different but crystallographically identical, or rather very similar, which constitute an exception to the first part of the above generalisation; another set of exceptions, those to the second part of the generalisation, are the phenomena designated by the name of "polymorphism" (from πολύς = many, μορφή = form).

Meaning of term "polymorphism" with examples.

Reference has already been made (*ante*, p. 409) to the establishment of the chemical identity of rhombohedral calcite and rhombic aragonite, two minerals whose chemical composition is CaCO₃.

Mitscherlich in his 1821 memoir describes the occurrence of

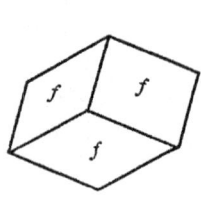

Fig. 85.
Calcite.
$f : f = 101° 9'.$

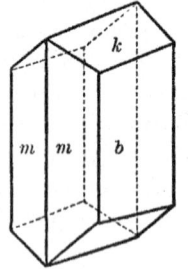

Fig. 86.
Aragonite.
$a : b : c = 0\cdot623 : 1 : 0\cdot721$
$m : m = 63° 48'.$

$NaH_2PO_4 . H_2O$ in two rhombic modifications (*ante*, p. 412), and very shortly after, in 1823, he was able to give an account of his discovery of a similar occurrence in the case of sulphur.

"In my second memoir mention is made of an observation which I published not without great misgiving, but which was so well established by experiment that I saw no reason to account for it by possible

Mitscherlich on rhombic and oblique sulphur.

personal error. This fact, that a substance such as, for example, the biphosphate of soda, assumes two different crystalline forms, had not been observed before in artificial compounds, and Haüy held that it was not supported by the evidence of similar facts from amongst minerals. But nevertheless we know of several cases confirmatory of such an occurrence, only that unfortunately with minerals we are never sure of dealing with compounds free from admixture, and that when a mineral has been found in two crystalline forms, it has always been possible to detect some traces of a foreign substance which might have caused the difference of form[1]. I have worked at this subject, and I consider it now as established that a substance, simple or compound, can assume two different crystalline forms....I will here choose for description sulphur, which being an element, lends itself better to the demonstration of the truth of this observation....I have procured myself crystals of artificial sulphur by different methods, by the evaporation of carbon bisulphide in which a quantity of sulphur had been dissolved, and also by the fusion and subsequent slow cooling of sulphur. The naturally

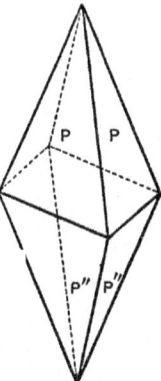 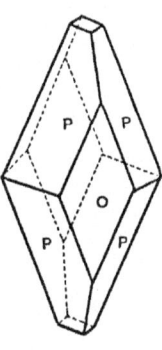

Fig. 87.

Crystals of Rhombic Sulphur.

[1] Hauy and his school attributed the rhombic form assumed by $CaCO_3$ in aragonite to the presence in that mineral of small quantities of $SrCO_3$, which as strontianite crystallises in the rhombic system.

occurring crystals of sulphur and those obtained from the carbide have the same crystalline form with the same modifications; but this form is different from that assumed by the molten sulphur. On fusing native sulphur, crystals are formed which are identical with those derived from the fusion of ordinary sulphur. The primitive form of the crystals of natural sulphur and of those obtained by the evaporation of carbon bisulphide solution is an octahedron[1] with rhombic base....I have shown that the angle between P and P' is 84° 58', and that between P and P'', 143° 17' (fig. 87). The primitive form obtained by the fusion of sulphur is an oblique prism with rhombic base, in which the angle between M' and M'' is 90° 32', and that between P and M, 85° 54' (fig. 88)[2]."

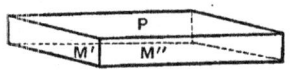 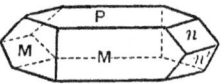

Fig. 88.
Crystals of Oblique Sulphur.

Polymorphism found to be a common occurrence. Mitscherlich's work was quickly followed by the discovery of many new instances, and by the acceptance of many older observations of polymorphism. The recognition of the chemical identity of rutile and anatase, the two tetragonal modifications of TiO_2, dates back to the very beginning of the nineteenth century.

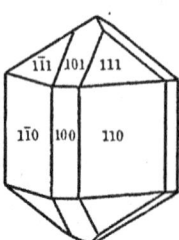

Fig. 89.
Rutile.
$a : c = 1 : 0\cdot6442.$

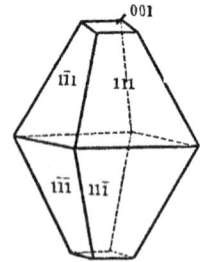

Fig. 90.
Anatase.
$a : c = 1 : 1\cdot7771.$

Mitscherlich in the above quoted memoir refers to the regular hemihedral pyrites and the rhombic marcasite, two varieties of

[1] See footnote 2, p. 411.

The two planes P' are those at the back of the crystal corresponding to P and P.
[2] "Sur les corps qui affectent deux formes cristallines différentes," *Ann. Chim. Phys.* Paris, 24, 1823 (p. 264).

FeS_2. Every year brought additional evidence of the occurrence of almost all chemical substances—products of nature and products of the laboratory alike—in a number of modifications, only one of which may be stable under ordinary conditions, special conditions being required to reveal directly or indirectly the existence of the other unstable modifications.

O. Lehmann[1] has considerably added to our knowledge of the existence, the conditions of formation, and the change into one another of the different modifications of substances.

Lehmann by microchemical method discovers tetramorphism of ammonium nitrate.

His method of investigation has been microchemical. The microscope employed was fitted with arrangements for bringing the slide examined to any desired temperature, and was provided with the necessary contrivances for crystallographic and optical examination. To quote one amongst the many cases investigated by him, we may take that of ammonium nitrate, in which he was able to prove that this substance is *tetramorphous*, that at temperatures below 35° it exists in one rhombic modification, and passes at that temperature with a volume increase of ·033 into another rhombic modification; that at 86° this second rhombic form passes with a volume decrease of ·0143 into a rhombohedral form, which in its turn at 125° changes with expansion into a cubic modification. And he showed further that these four modifications can be changed into one another in either direction by raising or lowering the temperature.

The differences exhibited by polymorphic modifications have been studied and classified, and some attempt has even been made to trace regularities capable of leading to generalisations; but the achievements in this direction are as yet inconsiderable.

R. Brauns in his *Chemische Mineralogie* (pp. 151 *et seq.*) uses the following classification :

1. *Differences in chemical properties.*

Classification of the differences shown by polymorphous substances.
1. Chemical properties.

Calcite is more readily attacked by acids than is aragonite. Diamond is practically not changed by oxidising agents, whilst graphite when treated with a mixture of $KClO_3$ and KNO_3 is oxidised to an insoluble yellow crystalline substance, graphitic acid $C_{24}H_9O_{13}$.

[1] *Molekularphysik*, Leipzig, 1888–9.

2. *Differences in physical properties.*

2. Physical properties.

The specific gravity of diamond is 3·52, that of graphite is 2·25; calcite has a specific gravity of 2·72, aragonite of 2·9; etc.

Diamond is the hardest, graphite one of the softest substances; aragonite is harder than calcite.

The optical properties vary of course with the crystalline system (*ante*, p. 396), but differences are shown in the same system, *e.g.* the double refraction of rutile is positive, that of anatase negative[1].

Rhombic sulphur melts at 114·5°, monoclinic at 120°; yellow phosphorus melts at a lower temperature than the red variety.

3. *Differences in crystalline form.*

3. Crystalline form.

The polymorphous modifications may belong to different crystal systems, as in the case of the rhombohedral (calcite) and rhombic (aragonite) forms of $CaCO_3$, the rhombic and oblique forms of sulphur, etc. Or they may belong to the same system, but have different crystallographic constants; thus in the case of rutile and anatase, two varieties of TiO_2, although the system and the symmetry are the

---

[1] The refraction of light is the most important of the directional properties (*ante*, p. 386) which differentiate the crystallographic systems. Light incident on a transparent crystal belonging to the cubic system (*e.g.* rock-salt or diamond) is transmitted in the same manner as through amorphous, isotropic glass. In the case of calcite, quartz, aragonite, gypsum, etc., the process is different: except along certain definite directions, an incident ray is always resolved into two, and these crystals are said to be *doubly refractive*, or *birefringent*. All crystals except those belonging to the cubic system are doubly refractive. In calcite and quartz and in all the representatives of the tetragonal, hexagonal, and rhombohedral systems there is one direction, that of the principal axis of symmetry, along which light is transmitted in an isotropic medium, *i.e.* is not doubly refracted, and such crystals are said to be *optically uniaxial*; in aragonite, gypsum, and in all the representatives of the rhombic, oblique and anorthic systems there are two such directions, and such crystals are said to be *optically biaxial.*—Moreover comparative examination of the two rays emerging as the result of double refraction shows that in the case of calcite and quartz and other uniaxial crystals, whilst one of the rays called the *ordinary* obeys the ordinary law of refraction, the other called the *extraordinary* is bent according to a quite different law; but that in the case of aragonite and gypsum and all other biaxial crystals, neither ray is transmitted according to the ordinary law of refraction. The double refraction in uniaxial crystals is called *positive*, when as in quartz and rutile, the index of refraction of the extraordinary ray is greater than that of the ordinary; or *negative* when as in calcite and anatase the index of refraction of the ordinary ray is greater than that of the extraordinary. The double refraction of biaxial crystals is also classified into positive and negative, though, with neither ray ordinary, the criterion of distinction is a different one.

same, yet the planes of rutile cannot be referred to the crystallo-graphic axes of anatase, or *vice versâ*.

$$TiO_2 \quad \text{(i)} \quad \text{Rutile, tetragonal,} \quad a:c = 1:0\text{·}6442,$$

$$\text{(ii)} \quad \text{Anatase,} \quad \text{,,} \quad a:c = 1:1\text{·}7771.$$

And similarly for the acid phosphate of soda investigated by Mitscherlich (*ante*, p. 412):

$$NaH_2PO_4 \cdot H_2O \quad \text{(i)} \quad \text{rhombic,} \quad a:b:c = 0\text{·}817:1:0\text{·}50,$$

$$\text{(ii)} \quad \text{,,} \quad a:b:c = 0\text{·}934:1:0\text{·}957.$$

4. *Differences in the conditions of formation.*

Calcite crystallises from cold solutions, aragonite from hot solutions.

4. Conditions of formation.
Monoclinic sulphur crystallises from fusion, rhombic from solution.

It has been stated that $MgSO_4 \cdot 7H_2O$, which separates from slightly supersaturated solutions in the ordinary rhombic form (Epsom salts), crystallises from stronger solutions in an oblique form.

The nature of the solvent, and the presence of foreign substances in the solution, are also of influence on the form assumed by the crystallising substances. A solution of $CaCO_3$ in water which contains carbonic acid and sodium or potassium silicate, deposits calcite exhibiting a great variety of forms; substitution of strontium or lead carbonate for the silicate leads to the formation of aragonite.

5. *Differences in the range of stability, and in the transformation capacity.*

5. Range of stability.
Rhombic sulphur, which melts at 114·5°, is stable at ordinary temperatures, and when heated remains so up to a temperature which accurate measurement by different methods has shown to be 95·5°; at this temperature, termed the *transition point*, the rhombic variety is transformed into the oblique, which is stable between 95·5° and its melting point 120°; but below 95·5° oblique sulphur is unstable, and at ordinary temperatures it changes spontaneously to the rhombic variety. Rhombic aragonite, when heated to a temperature just

below that at which decomposition into lime and carbonic acid begins (about 550°), is transformed into rhombohedral calcite, but the reverse change cannot be brought about. Diamond heated in the electric arc is changed to graphite, but the artificial production of diamonds from graphite has not been realised. The two varieties of sulphur, and the four varieties of ammonium nitrate (*ante*, p. 441) are types of a class of polymorphic substances which have a definite transition point, and for which the change of one variety into the other is reversible. Such substances are termed *enantiotropic* (ἐναντίος = opposite, τρόπος = habit). On the other hand, aragonite and calcite, diamond and graphite, are said to be *monotropic*, because the transformation of the crystalline forms occurs only in one direction, *i.e.* is irreversible[1].

The characteristic differences between the varieties of a few polymorphic substances are summarised in the table on page 445.

Various attempts have been made to account for polymorphism; in accordance with the view held at present of the molecular and atomic structure of matter, it may be due to differences in (i) the complexity of the constituent particles, (ii) the arrangement of the components of each such particle.

Cause of polymorphism.

Hence, using Berzelius' names of isomerism, polymerism, and metamerism (*post*, chap. XVIII), and distinguishing between two orders of magnitudes of the constituent particles, viz. (1) chemical molecules, (2) clusters or aggregates of the chemical molecules, the different types of polymorphism have been classified and explained under the names of:

(1) chemical isomerism, comprising (i) chemical polymerism, (ii) chemical metamerism.

(2) physical isomerism, comprising (i) physical polymerism, (ii) physical metamerism.

The first class with its two divisions is made to include polymorphic substances which exhibit specific differences in chemical properties, and which cannot be made to pass from one variety into another by changes in temperature only.

[1] The consideration of the phenomena of *enantiotropy* and *monotropy* is included in the scope of the "Phase Rule," which within the last years has assumed such enormous importance in the correlation and interpretation of the phenomena of chemical change; see Findlay, *The Phase Rule*, 1904, pp. 31 *et seq.*

Name	System	Crystallographic Constants	Optical Properties	Chemical and Physical Properties	Transition Point and Range of Stability	S.G.	Hardness
Sulphur	Rhombic	$a:b:c$ $= \cdot8130:1:1\cdot9039$	Biaxial	Soluble in carbon bisulphide	95·5°, enantiotropic. Rhombic is stable up to 95·5°; oblique between 95·5° and 120°	2·08	2
	Oblique	$a:b:c$ $= \cdot9958:1:\cdot9998$ $\beta=84°\ 14'$	Biaxial			1·96	
CaCO₃ — Calcite	Rhombohedral		Uniaxial; refractive indices 1·6584, 1·4864.	Soluble in acids	Monotropic At a temperature just below that of decomposition (550°) aragonite changes into calcite	2·72	3
CaCO₃ — Aragonite	Rhombic	$a:b:c$ $= \cdot6224:1:\cdot7206$	Biaxial; refractive indices 1·5301, 1·6816, 1·6859	Less readily attacked by acids		2·9	3½
FeS₂ — Pyrites	Cubic		Opaque; greenish-black streak	Is not ... ed in air at ordinary temperature; has ... pared artificially		5·1	6
FeS₂ — Marcasite	Rhombic	$a:b:c$ $= \cdot7519:1:1\cdot1845$	Opaque; greenish-grey streak	Less ..., as ... in air; has ... been prepared artificially		4·8	6
TiO₂ — Anatase	Tetragonal	$a:c=1:1\cdot7771$	Uniaxial; ... refraction, negative		At a ... not exceeding 860°, anatase ... s separate out; at 1000°, brookite; at ... higher temperatures, rutile	3·9	5½
TiO₂ — Rutile	Tetragonal	$a:c=1:0\cdot644$	Uniaxial; ... refraction, ...			4·3	6¼
TiO₂ — Brookite	Rhombic	$a:b:c$ $= \cdot9444:1:\cdot8416$	Biaxial; double refraction, positive			4	5½
Carbon — Diamond	Cubic		..., transparent; high refractive index, $\mu=2\cdot417$	N ... ... and ... by oxidising agents; ignition ... 760° to 875° ... ignition temperature 575° and above	Mon ... At the temperature of the ... diamond changes into graphite	3·52	10
Carbon — Graphite	Hexagonal? Oblique?		Opaque				

The second class comprises polymorphic substances which do not differ in chemical properties except in the degree to which these may be manifested, and which are monotropic or enantiotropic.

But in the present state of the science, whilst we cannot determine the molecular weight of solids, it is impossible to find conclusive experimental support for the above division, and the almost completely speculative nature of the whole of this classification must not be lost sight of.

ISODIMORPHISM (ISOPOLYMORPHISM).

If a substance A occurs in the two modifications α and β, and if another substance B occurs in the forms α' and β', and if α is isomorphous with α', and if β is isomorphous with β',

Meaning of term isodimorphism, with examples. Direct evidence.

then A and B are said to exhibit *isodimorphism*.

$CoAs_2$ occurs native as cubic smaltite and as rhombic safflorite; $NiAs_2$ occurs native as cubic chloanthite, which is isomorphous with smaltite, and as rhombic rammelsbergite (white nickel), which is isomorphous with safflorite. Hence the diarsenides of nickel and cobalt are isodimorphous.

Native tinstone, cassiterite SnO_2 (p. 402), is isomorphous with rutile, one of the quadratic modifications of TiO_2 (p. 440), and two varieties of artificial SnO_2[1], isomorphous with tetragonal anatase and rhombic brookite respectively, are said to have been prepared; hence the dioxides of tin and titanium are isotrimorphous.

Each of the sulphates $XSO_4 . 7H_2O$ (X = Mg, Zn, Ni, Co, Fe, Mn) is dimorphous, occurring in a rhombic (*ante*, p. 426) and in an oblique modification. The rhombic crystals of all these salts form one isomorphous series, and the oblique crystals form another such series.

But the number of cases in which isodimorphism can be

Indirect evidence for the occurrence of isodimorphism.

established directly by the actual investigation of the forms α, α', β, β', is very small in comparison with that of the cases in which the proof is an indirect one.

---

[1] "SnO_2 is isotrimorphous with TiO_2. SnO_2 crystallises from fusion in borax in the form of rutile, from fusion in phosphoric acid in that of anatase. Heating $SnCl_4$ in steam gives SnO_2 in the form of brookite." (Dammer, *Handbuch der anorganischen Chemie.*)

" If a certain substance A exists in the two modifications a and β, and if a is found to be isomorphous with a second substance B, and if the two substances form mixed crystals of the form β, we infer that the second substance B can exist not only in the form a, but also in the form β. And if this second modification is not met with by itself in well-defined crystals, we should, arguing by analogy from our knowledge of the existence of substances in stable and unstable modifications, consider β a labile modification of B. Thus the first instance of isodimorphism, that of $CaCO_3$ and $PbCO_3$, had been deduced by Johnson from the existence of *plumbocalcite* $(CaPb)CO_3$; pure rhombohedral $PbCO_3$ was not known then, and has not been met with yet." (Arzruni, *Physikalische Chemie der Crystalle*, p. 164.)

$$CaCO_3 \begin{array}{l} \nearrow \text{calcite} \longleftarrow \text{rhombohedral} \longrightarrow \text{plumbocalcite} \searrow \\ \searrow \text{aragonite} \longleftarrow \quad \text{rhombic} \quad \longrightarrow \quad \text{cerussite} \quad \nearrow \end{array} PbCO_3$$

The legitimacy of speculations concerning the members of an isodimorphous series, the existence of which is primarily inferred only indirectly, is illustrated by the following case[1]:

Prognosis of existence of members of series supposed isodimorphous.

The well-known and long established fact that whilst oblique orthoclase $KAlSi_3O_8$ often contains sodium, anorthic albite $NaAlSi_3O_8$ often contains potassium, was in 1874 declared by Groth to be a case of isodimorphism, and the probable existence of the two pure modifications, viz. oblique $NaAlSi_3O_8$ and anorthic $KAlSi_3O_8$, was inferred. This prediction was partially verified two years later by the discovery of microcline, an anorthic form of $KAlSi_3O_8$.

Retgers has contributed more than anyone else to the discovery of such "cryptic" isodimorphism; in the application of his law of

Cryptic isodimorphism discovered by the application of Retgers' law.

direct proportionality between the physical properties and the composition of mixed crystals he found an admirable means for identifying cases of isodimorphism, and for differentiating them from pure

[1] See also Miers, *Mineralogy*, p. 228, or Roscoe and Schorlemmer, *Treatise on Chemistry*, 1894, I., p. 870, for an account of the successive stages in the discovery of all the members of the isodimorphous series As_2O_3 and Sb_2O_3.

Known in the Middle Ages and even in antiquity	a As_2O_3 ...	*arsenic bloom or arsenolite*; cubic, crystallises in octahedra; easily produced artificially from hot aqueous solutions, etc.
	β' Sb_2O_3 ...	*valentinite*; rhombic, $a:b:c=0.3869:1:0.371$; easily produced artificially by sublimation, etc.
Found by Woehler in 1832	β As_2O_3 ...	crystals found in cobalt ore furnaces; rhombic according to Groth, $a:b:c=0.3758:1:0.35$.

Discovered in 1851 ...... a' Sb_2O_3 ... *senarmontite*; cubic, crystallises in octahedra.

isomorphism, or from the formation of double salts. This last phenomenon is characterised, on the chemical side, by the fact that the values for x and y in the formula $xA + yB$ for the composition of the mixed crystals are fixed simple whole numbers; on the physical side by the fact that the properties of the double salt are not the mean of those of its constituents A

and B, and may even be outside these values[1]. The

Difference between double salts and mixed crystals. difference between a double salt and a mixed crystal is the same as that between a chemical combination and a mixture.

Retgers' investigation of the mixed crystals of the vitriols of magnesium and of iron is typical of the method he followed.

"It is known that $MgSO_4 \cdot 7H_2O$ crystallises in the rhombic, and $FeSO_4 \cdot 7H_2O$ in the oblique system ; all the same these salts form mixed crystals, which are rhombic if rich in magnesium, oblique if rich in iron."

Isodimorphism of $MgSO_4 \cdot 7H_2O$ and $FeSO_4 \cdot 7H_2O$. Is this a case of a mixture between oblique and rhombic crystals in which the form is that of the predominant constituent, or is it a case of isodimorphism according to which each of the salts exists in two modifications, one stable, one unstable ?

"The latter view is made very probable by the artificial production of oblique Epsom salts, and by the discovery of the mineral tauriscite, a rhombic modification of iron vitriol. But this does not yet supply a direct proof for the absence of oblique iron vitriol in the rhombic mixed crystals.

I have attempted to solve this problem by an investigation of the specific gravity, and my argument has run thus : If the mixed crystals are aggregates of rhombic Epsom salts and oblique iron vitriol there must be strict proportionality between specific gravity and chemical composition throughout the whole series, because it would simply be a case of direct isomorphism ; but if isodimorphism exists it will most probably show itself by the absence of such proportionality, since the two modifications of the same salt will not have the same specific gravity....If the results are represented graphically by making the percentage weight of $MgSO_4 \cdot 7H_2O$ the abscissae, and the corresponding specific volumes of the mixed crystals the ordinates...it becomes apparent that the observations arrange themselves in two straight lines. Thereby is supplied a most rigorous proof for the occurrence of isodimorphism and for the existence of the labile modifications in the mixed crystals."

[1] Thus:

	Crystalline form	Spec. Grav.
KCl	Colourless transparent cubes	1·989
$CuCl_2 \cdot 2H_2O$	Green rhombic prisms	2·390
$CuCl_2 \cdot 2KCl \cdot 2H_2O$	Blue tetragonal plates	2·41

Inspection of fig. 91 shows that there is a break in the series of the mixed crystals. The values for the specific volumes of the labile modifications of the two salts, *i.e.* ·5914 for $MgSO_4.7H_2O$ and ·5333 for $FeSO_4.7H_2O$, are of course the result not of direct observation but of extrapolation.

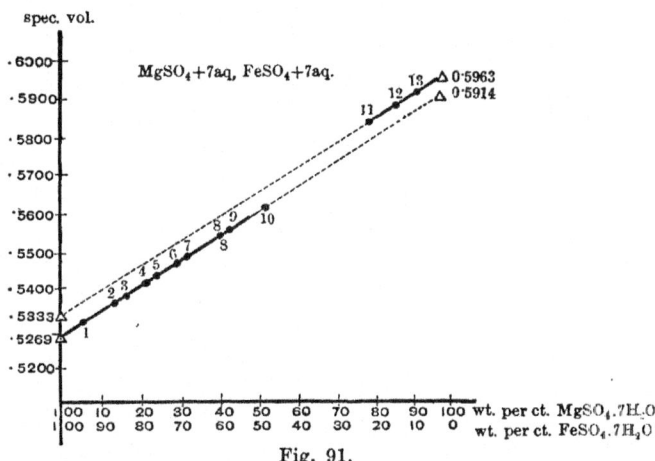

Fig. 91.
Mixture of Magnesium Sulphate and Ferrous Sulphate.

Fig. 92 represents the results of Retgers' investigations of the relations between $NaNO_3$ and $AgNO_3$, which are recognised as another case of isodimorphism. $NaNO_3$ is rhombohedral and has the specific gravity 2·265; $AgNO_3$ is rhombic and has the specific gravity 4·35.

Isodimorphism of $NaNO_3$ and $AgNO_3$.

Mixed crystals were obtained which were rhombohedral; these contained from 0 to 52 per cent. of $AgNO_3$ and gave for the curve representing the relation between specific volume and percentage composition a straight line. If the specific gravity of the $AgNO_3$ present in these rhombohedral crystals is calculated[1] on the supposition of the additive nature of this

---

[1] Specific volume $= \dfrac{1}{\text{specific gravity}} =$ volume occupied by unit weight.

Specific volume of mixed crystals $AB = v$
,, ,, ,, component crystals $A = v_1$,
,, ,, ,, component crystals $B = v_2$,
100 parts by weight of AB contain a of A and $(100 - a)$ of B.

Then on the supposition that for mixtures specific volume is a strictly additive property (*ante*, pp. 431, 435)
$$100\,v = av_1 + (100 - a)\,v_2,$$
and since v, v_1 and a can all be determined directly by experiment, v_2 can be calculated.

property, the value obtained is 4·19, thus pointing to the existence in the mixed crystals of a labile rhombohedral $AgNO_3$. Besides these rhombohedral crystals he obtained also some crystals very rich in $AgNO_3$ (99 to 100 %) and showing the ordinary rhombic form of $AgNO_3$.

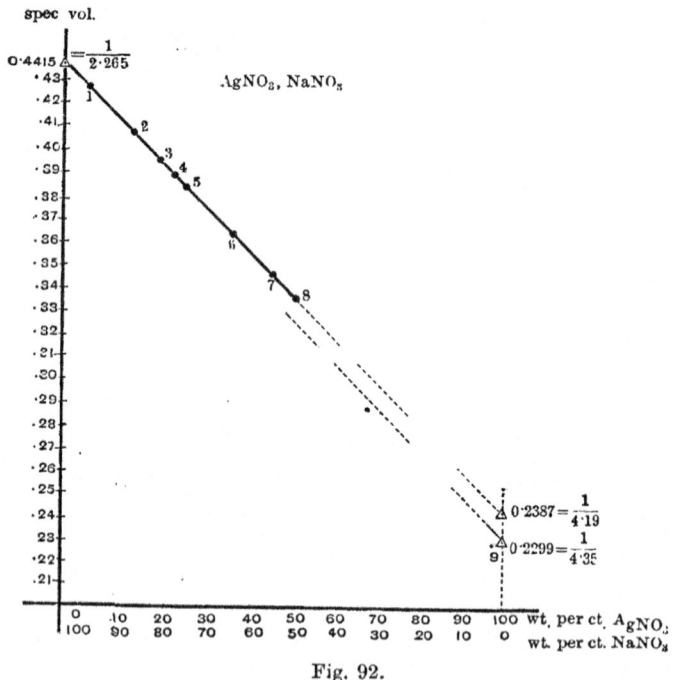

Fig. 92.

Mixture of Sodium Nitrate and Silver Nitrate.

The relations of KNO_3 and $AgNO_3$ represented in fig. 93 are more complicated because of the formation of the double salt $KAgN_2O_6$. Three distinct types of crystals were obtained:

Isodimorphism of KNO_3 and $AgNO_3$ and double salt $KAgN_2O_6$.

(i) Rhombic crystals of the form of KNO_3 $(a : b : c = 0.591 : 1 : 0.701)$ probably containing a small amount of $AgNO_3$.

(ii) Oblique crystals of the double salt $KAgN_2O_6$.

(iii)   Rhombic crystals of the form of $AgNO_3$

$$(a:b:c = 0.943:1:1.3697)^1,$$

probably containing a little KNO_3.

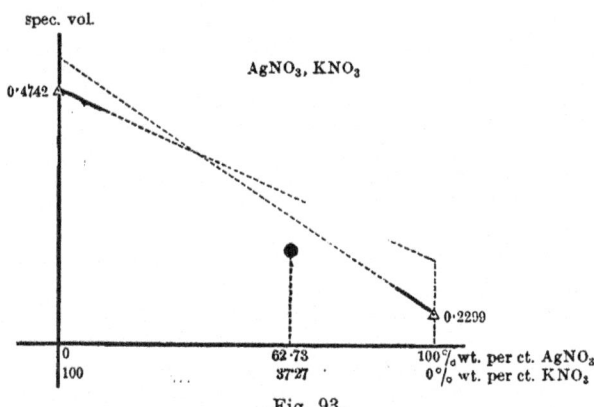

.Fig. 93.

Mixture of Potassium Nitrate and Silver Nitrate.

The above are but two instances of the investigations which comprised the nitrates and the chlorates of all the alkali metals and of silver, and which led Retgers to the expression of the following views:

"We should never forget that the number of the possible modifications of a salt is probably very large....The occurrence of polymorphism is by no

Retgers on the relative importance of isomorphism and isodimorphism.

means restricted to a few substances such as calcite and aragonite, rhombic and oblique sulphur, cases which are referred to in the text-books as exceptional phenomena; on the contrary, it is common, almost without exception, to all solids....And amongst these many forms there would certainly be some which are isomorphous with one another....It is

therefore my opinion that amongst chemical combinations only few directly isomorphous series exist, but that the number of isodimorphous series is very large....The part of direct isomorphism seems to me about played out, at least as far as its utility in chemistry goes; but I hold that isodimorphism, as opposed to direct isomorphism, has a great future before it.

Comparison of the forms of the rhombohedra of $NaNO_3$ and of the rhombic plates of $AgNO_3$ would, supposing the composition of the two

---

[1] Retgers uses these constants in preference to $0.5302:1:0.7263$, for which $c = a$, $a = c$, and $b = 2b$.

substances unknown, never lead to the inference of analogous chemical constitution. But this analogy is made evident by a simple experiment which shows the capacity of the $NaNO_3$ to unite intimately with $AgNO_3$ forming absolutely homogeneous rhombohedra; the same holds for the cubes of $NaClO_3$ and the tetragonal prisms of $AgClO_3$; and finally, if we demonstrate by a simple experiment the formation of mixed crystals of $NaNO_3$ and $NaClO_3$, we shall have proved with sufficient certainty the similarity of composition of these four substances, to all of which belongs the similar formula RQO_3.

We know what great importance Berzelius attached to the original conception of direct isomorphism in its application to the determination of atomic weights, of the valencies of new elements, and of the formulae of new compounds. Now after 70 years of crystallographic investigation we realise how limited the applicability of this principle really is. But when expanded into the conception of isodimorphism, a wide field for fruitful research is opened out....

The quite unexpected discovery of the isodimorphism of the chlorates and the nitrates leads at once to the chemically important conclusion that chloric acid and nitric acid are analogous acids, and that chlorine and nitrogen occur in them as pentavalent elements....If silver happened to be an unknown metal, the formation of mixed crystals with the greater number of sodium salts (the chlorate cubes and the nitrate rhombohedra) would at once supply the proof that we were dealing with a monovalent metal, belonging to the group of the alkalis. The discussion as to whether beryllium is divalent or trivalent had lasted for several years, when Nilson and Pettersson's decisive determination of the vapour density of beryllium chloride settled the matter in favour of the view of its divalency. In spite of the fact that beryllium exhibits in most of its salts differences from the metals of the zinc group, it is probable that simply allowing them to crystallise together would have demonstrated the isodimorphism of the Be and Zn salts and hence also the equal valency of the two metals. Direct isomorphism is in this case of no use whatever, because it is not manifested in any one instance."

In the study of the relations between crystalline form and chemical composition attention is now also being devoted to a set of phenomena which in 1870 Groth named *morphotropy* (from $\mu o \rho \phi \eta$ = form, and $\tau \rho \delta \pi o \varsigma$ = habit).

Meaning of term "Morphotropy," with examples.

The term is used to designate the particular change in crystalline form produced either by the replacement of one element or radicle by another, or by the addition of an element or radicle.

The following may serve as an example of the type of cases dealt with in this branch of the subject:

The substitution of $(OH)'$, or $(NO_2)'$, or $(NH_2)'$, for H in rhombic benzol gives substances which are themselves rhombic, but have different axial ratios; the substitution of Cl' or Br' for hydrogen produces a greater effect, the resulting substitution products crystallising in a system of lower symmetry, the oblique.

C_6H_6 ....rhombic, $a:b:c = 0.891:1:0.799$

$C_6H_4(OH)_2$ ....rhombic, $a:b:c = 0.910:1:0.540$

$C_6H_3(OH)(NO_2)_2$ ......rhombic, $a:b:c = 0.933:1:0.753$

$C_6H_4Cl_2$ ....oblique

$C_6H_2Br(OH)(NO_2)_2$ ...oblique.

"The ultimate object of such investigations may be taken to be the derivation of the crystalline form of a substance from its chemical composition." (Arzruni.)

But this after all is only what has been the ultimate object in any class of investigation concerning the relation between crystalline form and chemical composition.

It must now be considered as established that the most accurate measurements of the chemically pure crystals of any one definite substance would give absolutely constant values characteristic of that substance only; that the corresponding values for a truly isomorphous substance (*i.e.* one characterised by the power of forming with the first mixed crystals, the properties of which are additively those of the components) will be more or less similar, but not identical; and that the differences in the values of the crystallographic constants are functions of the atomic weights of the substituting isomorphous elements. And on the other hand we have come to believe that when the crystalline form of substances of the same percentage composition is different, we are not dealing with identical substances but with isomeric modifications. "No substance possesses more than one crystalline form." (Lehmann.)

General results established for the relation between crystalline form and chemical composition.

Thus it might seem as if the work of a century had only resulted in a return to Haüy's original views. But this is only apparent; the present position marks, not a return to truth after a long lapse into error, but a great advance, far distant though we are from the goal set by Mitscherlich when he wrote in 1830 concerning:

"...the solution of an important problem...how to calculate from the form of two components the form of the resulting compound."

CHAPTER XVI.

MENDELEEFF AND THE PERIODIC LAW[1].

" Les propriétés des corps sont les propriétés des nombres."
DE CHANCOURTOIS, 1863.

GERHARDT'S impassioned advocacy, unsupported though it was by consistent practice on his own part, had paved the way for the adoption of an atomic weight notation in place of the one based on the more vague and shifting conception of the equivalent. Cannizzaro's calm, lucid exposition and perfect formulation of the points at issue could not fail to convert chemists to the advisability of the reform advocated. From 1860 onwards the fundamental constants for chemists became the atomic weights; the nature of these quantities began to be clearly conceived, and definite consistent methods were employed in their determination.

Substitution of atomic for equivalent weights is followed by extended classification of the elements. The legitimacy and the importance of this change were clearly demonstrated when, very shortly after its general adoption, the atomic weights were made the basis of a classification of the elements such as had not been possible before, and which in its results represents one of the greatest triumphs of the science.

Classification, a necessary and most important factor in the development of any science, had in chemistry been employed from early times. The common properties of certain substances now recognised as complex were summarised in the names acids and alkalis; certain of the elements were, in virtue of a number of

---

[1] The history of the "Periodic Law," with very complete references to the literature of the subject, is given in: F. P. Venable, *The Development of the Periodic Law*, 1896; G. Rudorf, *The Periodic Classification and the Problem of Chemical Evolution*, 1900.

common properties, designated as metals, and the alchemists subdivided these into base and noble. A probably unconscious recognition of the fact that in some way the properties of a complex substance depend on the nature of its constituents may have been the cause of the importance always assigned to the classification of these constituents, the elements.

Classification requires the selection of some principle which shall be the means and the measure of division between the objects to be classified. Mendeleeff, to whom chiefly we owe our present system, begins his 1869[1] paper by an historical account of

Mendeleeff on the principles followed in the classification of elements. the classification of elements and a discussion on the nature of the problem involved. He points out that the systematic arrangement of the elements had, in the course of the development of the science, been based on a number of very different principles, *e.g.* the distinction between metal and non-metal, the acidity or basicity of the oxides, electro-chemical potential, relative affinity, valency; but that all these had been found inadequate :

" At the present time there is no undisputed general principle to serve as guide in the estimation of the relative properties of the elements, and as a means for arranging them according to a more or less rigorous system....It is only when dealing with certain groups of elements that we recognise with certainty the existence of an entity, a natural sequence in these similar manifestations of matter. Such groups are : the halogens, the alkaline earths, nitrogen and its congeners....A number of attempts have been made to discover the law which underlies the observed connection between the elements belonging to such groups....But our knowledge of such relations is very incomplete and does not lead to a comprehensive system of the elements; all it does is to justify division into these natural groups....Dumas, Pettenkofer and others have directed attention to the numerical relations between the atomic weights of the elements which constitute a group, but their attempts have not led to a systematic arrangement of all the known elements."

The arrangement of elements in such groups has been the outcome of the study of properties of these elements themselves, and of the properties and composition of their compounds; it represents the accumulated work of many observers. Every text-book of chemistry, recent or antiquated, concise or detailed, sets forth the similarity in properties of chlorine, bromine, and iodine,

---

[1] *The Relations between the Properties of the Elements and their Atomic Weights.*

and of their compounds, in virtue of which these elements are classed together. As far back as 1829 Doebereiner[1] had pointed out that the atomic weight of bromine is very nearly the arithmetical mean of the atomic weights of chlorine and iodine, namely:

Doebereiner's triads.

$$\frac{35\cdot470 + 126\cdot470}{2} = 80\cdot970\,[2].$$

Of this mean he says that though somewhat greater than 78·383, the number actually found by Berzelius, it so closely approximates to it as to justify the hope that repeated accurate determinations of all the atomic weights involved will lead to a disappearance of any difference. He found similar relations for the alkaline earths, the alkalis, and for the group comprising sulphur, selenium, and tellurium, as shown by the equations:

$$\frac{356\cdot019\,(=\mathrm{Ca}) + 956\cdot880\,(=\mathrm{Ba})}{2} = 656\cdot449\,(=\mathrm{Sr}),$$

but experiment gave for strontium[3]   647·285.

$$\frac{195\cdot310\,(=\mathrm{Na}) + 589\cdot916\,(=\mathrm{K})}{2} = 392\cdot613\,(=\mathrm{Li}),$$

but experiment gave for lithium[3]   390·897.

$$\frac{32\cdot239\,(=\mathrm{S}) + 129\cdot243\,(=\mathrm{Te})}{2} = 80\cdot741\,(=\mathrm{Se}),$$

but experiment gave for selenium[2]   79·263.

Pettenkofer[4], Dumas[5], and others have continued the consideration of such *triads*, but cannot be said to have materially advanced the subject from the position in which it was left by Doebereiner. However, when stress had once been laid on the approximate constancy of the differences in the atomic weights of elements forming a group, the ever-dominant desire for simplicity in numerical relations asserted itself. This led to unjustifiable attempts to alter the experimental values in order to make them agree with

[1] " Versuch zu einer Gruppirung der elementaren Stoffe nach ihrer Analogie," *Poggend. Ann.*, Leipzig, 15, 1829 (p. 301).

[2] H = 1.   [3] O = 100.

[4] " Ueber die regelmässigen Abstände der Aequivalentzahlen der sogenannten einfachen Radicale," *Liebig's Ann. Chem.*, Leipzig, 105, 1858 (p. 187).

[5] " Mémoire sur les équivalents des corps simples," Paris, *C.-R. Acad. sci.*, 45, 1857 (p. 709); 46, 1858 (p. 951).

pre-conceived ideas. We know how variable and arbitrary were the criteria (prior to 1860) used in the determination of equivalent and atomic weights in general; but in the special cases of groups of elements there was more uniformity, hence the numbers obtained were comparable, and since the above considerations concerning

Consideration of numerical relations between atomic weights extended from groups to elements in general.

classification applied only to groups of elements, it was possible to bring out within this compass a relation between atomic weight and properties. The extension to the case of elements in general soon followed. This was done in a set of short papers published from 1863[1] onwards by Newlands, the forerunner of Lothar Meyer and Mendeleeff. If measured by the standard of Mendeleeff's treatment of the subject, Newlands' work in its inductive and deductive aspect must be considered as very slight. Mendeleeff from the outset put the principle of classification advocated by him on a basis so firm that practically nothing more remained to be done by others; Lothar Meyer's treatment, though on the same correct lines, failed to convince by reason of its being too concise; and Newlands had only sketched in the merest outline the principle involved. In a paper published in 1865 he expresses himself as follows:

" If the elements are arranged in the order of their equivalents, with a few slight transpositions, as in the accompanying table, it will be observed that elements belonging to the same group usually appear on the same horizontal line.

No.		No.		No.		No.		No.		No.		No.		No.
H	1	F	8	Cl	15	Co Ni	22	Br	29	Pd	36	I	42	Pt Ir	50
Li	2	Na	9	K	16	Cu	23	Rb	30	Ag	37	Cs	44	Tl	53
G[2]	3	Mg	10	Ca	17	Zn	25	Sr	31	Cd	38	Ba V	45	Pb	54
Bo	4	Al	11	Cr	19	Y	24	Ce La	33	U	40	Ta	46	Th	56
C	5	Si	12	Ti	18	In	26	Zr	32	Sn	39	W	47	Hg	52
N	6	P	13	Mn	20	As	27	Di Mo	34	Sb	41	Nb	48	Bi	55
O	7	S	14	Fe	21	Se	28	Ro Ru	35	Te	43	Au	49	Os	51

It will also be seen that the numbers of analogous elements generally differ either by 7 or by some multiple of 7; in other words, members of the same group stand to each other in the same relation as the

Newlands' law of octaves.

extremities of one or more octaves in music. Thus in the nitrogen group, between nitrogen and phosphorus there are

---

[1] *Chem. News*, London, 7—13, 1863—1866; *On the Discovery of the Periodic Law, and on Relations among the Atomic Weights*, London, 1884 (a re-publication of various papers in the *Chemical News*).

[2] G = glucinium = beryllium (Be).

7 elements.; between phosphorus and arsenic 14; between arsenic and antimony 14; and lastly, between antimony and bismuth, 14 also." "The 8th element starting from a given one is a kind of repetition of the first.....This peculiar relationship I propose to provisionally term the 'Law of Octaves.'"

In an earlier table (1864) no mention is yet made of the necessity for "a few slight transpositions." In this same table the difference in the number of elements between phosphorus and arsenic is only 13, an irregularity apparently removed by the subsequent inclusion of indium[1] as number 26, with the atomic weight 72. Further investigations have not justified this course. Indium, because of its relations with aluminium, must be considered as trivalent, and its atomic weight in terms of the equivalent determined by Winkler becomes $3 \times 37.8 = 113.4$, a number confirmed by the value ·057 obtained for its specific heat ($·057 \times 113.4 = 6.46$), and by the vapour densities of its volatile chlorides. In the case of indium, as in that of the places assigned to Y, U (see table), Newlands shows himself to be without that comprehensive and deep insight into the nature of chemical relations which makes Lothar Meyer and Mendeleeff correctly place indium between cadmium and tin, leaving the gaps between zinc and arsenic, since then filled by gallium and germanium.

Newlands, though in his earlier papers still speaking of "equivalent weight," really used atomic weight values throughout and recognised the importance of the difference. He also gave definite proof that his discovery of the law of octaves was not the result of an isolated chance speculation, but of comprehensive investigation guided by a definite purpose. In 1866, at a meeting of the Chemical Society, on the occasion of a discussion of his paper entitled "The Law of Octaves and the Causes of Numerical Relations among the Atomic Weights," a speaker expressed the belief that any arrangement would present occasional coincidences, and enquired of Mr Newlands whether he had ever examined the elements according to their initial letters.

"Mr Newlands said that he had tried several other schemes before arriving at that now proposed. One founded upon the specific gravity of the elements had altogether failed, and no relation could be worked out of the atomic weights under any other system than that of Cannizzaro."

[1] Discovered spectroscopically in 1863 by Reich and Richter in zinc-blende, and subsequently more fully investigated by Winkler.

No. 68 of Ostwald's *Klassiker der exakten Wissenschaften* gives under the name of *Das natürliche System der chemischen Elemente* a collection of the papers of Lothar Meyer and D. Mendeleeff, the two men who are considered the joint discoverers of the periodic law. Practically all the results arrived at by Lothar Meyer are included in Mendeleeff's fuller treatment, but the importance and interest of the subject make it desirable to give quotations from Lothar Meyer's paper, " The Nature of the Chemical Elements as a function of their Atomic Weights "[1] (*Liebig's Ann. Chem.*, Leipzig, 7 Supplement, 1870, pp. 354–364).

" The regular connections which exist between the numerical values of the atomic weights...have been variously represented...; but, since Gmelin's so-

Lothar Meyer on the nature of the chemical elements as a function of their atomic weights.

called equivalents have been replaced by the atomic weights determined according to the laws of Avogadro and of Dulong and Petit, it has become possible to simplify considerably the representation of these relations. In 1864 I was already able to group under one scheme the regularities hitherto found for different families of chemical elements....Mendeleeff has recently shown that we may attain such a classification simply by arranging the atomic weights of the elements, without arbitrary choice, according to their magnitude, in one single series ; by then dividing this series into sections ; and by joining these sections without change of sequence."

" If the series is interrupted at corresponding points, that is, at elements of a similar nature, several short series of analogous construction are formed, which may be so arranged that the elements follow each other in the order of their atomic weights in the horizontal rows, whilst the vertical rows are composed of numbers of the natural families." (*Modern Theories*, p. 118.)

" The following table [reproduced on page 460] is essentially identical with that given by Mendeleeff."

" The table contains, arranged in the order of increasing atomic weights, all the elements (hydrogen excepted) whose atomic weights have so far been determined from the vapour density of their compounds or from their heat capacity, and besides these, Be and In with atomic weights presumptively deduced from their equivalents, making 56 elements in all. The elements not included besides H are Y, Eb (Tb?), Ce, La, Di, Th, U, Jg[2], for which the atomic weight, and in some cases even the equivalent weight, is unknown."

The method followed in the process of arrangement which found its embodiment in the following table, the regularities revealed

---

[1] Some of the quotations are from *Modern Theories of Chemistry*, English translation by Bedson and Williams, 1888, in which Lothar Meyer has incorporated portions of his original paper considerably amplified.

[2] The existence of such an element has not been established.

Lothar Meyer's Table representing the Nature of the Elements as a Function of their Atomic Weights.

I	II	III	IV	V	VI	VII	VIII	IX
	B=11·0	Al=27·3	—	—	—	?In=113·4	—	Tl=202·7
	C=11·97	Si=28	—	—		Sn=117·8	—	Pb=206·4
	N=14·01	P=30·9	Ti=48	As=74·9	Zr= 89·7	Sb=122·1	Ta=182·2	Bi=207·5
	O=15·96	S=31·98	V=51·2	Se=78	Nb= 93·7	Te= 128 ?	W=183·5	—
	F=19·1	Cl=35·38	Cr=52·4	Br=79·75	Mo= 95·6	I=126·5	Os=198·6 ?	—
			Mn=54·8		Ru=103·5		Ir=196·7	
			Fe=55·9		Rh=104·1		Pt=196·7	
			Co=Ni=58·6		Pd=106·2			
Li=7·01	Na=22·99	K=39·04	Cu=63·3	Rb=85·2	Ag=107·66	Cs=132·7	Au=196·2	—
?Be=9·3	Mg=23·9	Ca=39·9	Zn=64·9	Sr=87·0	Cd=111·6	Ba=136·8	Hg=199·8	—

Difference from I to II and from II to III about =16.
Difference from III to V, IV to VI, V to VII fluctuating about 46.
Difference from VI to VIII, from VII to IX =88 to 92.

by this arrangement, and the application of this new principle of classification, may conveniently, though arbitrarily, be considered under four points:

1. *The process of arrangement.*

"Whilst the 9 vertical rows contain the elements in the sequence of their atomic weights, the horizontal rows are made up of elements belonging to natural families."

In order to maintain such an arrangement it had been necessary to alter the natural sequence in certain points:

(i) To leave gaps which in the future might be filled by:

(*a*) elements already discovered, but whose atomic weights were at the time either unknown (*e.g.* those enumerated above) or known incorrectly.

(*b*) elements not yet discovered.

(ii) To transpose some elements on the supposition that either

(*a*) the valency assigned to them was incorrect (*e.g.* Be, *post*, pp. 466, 476).

(*b*) the equivalent weight value was not exact (*e.g.* Te = 128, which is placed before I = 126·5).

Whilst the changes effected under (*a*) may be considerable, the new atomic weight values being $1\frac{1}{2}$, 2, etc. times the original, those under (*b*) can never be very great, not more than an interchange of place with the element just before or just after.

2. *General results of the arrangement.*

The relation between the properties of elements and their atomic weights is recognised as a periodic function; whilst the atomic weight, one of the quantities involved, increases steadily, the properties which constitute the other quantity increase and decrease, showing maxima and minima occurring at fairly regular intervals. The relation between time and the diurnal changes of night and day, or between time and the annual recurrence of the seasons, is the example of periodic variation best known to all.

" If we proceed from the assumption that the atoms are aggregates of one identical kind of matter, and that they differ only in the magnitude of their masses, we can consider the properties of the elements in their dependence on the atomic weight, that is, as functions of the atomic weight. In support of this view, the table shows that the properties of the elements are mostly periodic functions of the atomic weight. The same or similar properties recur when the atomic weight has increased by a certain value, which at first is 16, then about 46, and finally 88 to 92."

3. *Investigation of the change of properties within each period.*

3. Change of properties within each period. " The recognition of the above periodic variation in the properties of the elements is remarkable and attractive, but we are left completely in the dark concerning the variation of the properties within the period, at the end of which recur the characteristics that had been exhibited at its beginning. If, for instance, we start from Li, we find that after an increase of nearly 16 units, its important properties are repeated in Na, and again after another 16 units in K ; but on the way we meet in checkered variety, apparently without any mediation of the changes, first Be, B, C, N, O, F, and then again Mg, Al, Si, P, S, Cl. The saturation capacity of the elements alone rises and falls regularly and equally in the two intervals.

1-valent	2-v.	3-v.	4-v,	3-v.	2-v.	1-v.
Li	Be	B	C	N	O	F
Na	Mg	Al	Si	P	S	Cl

But if we want to represent the nature of the elements in their dependence on the magnitude of the atomic weights, we must trace step by step the change of each property from element to element."

This is done by Lothar Meyer for a number of properties :

(i) Valency. The preceding quotation brings out the periodic nature of this value (see also *post,* chap. XVII).

(i) Valency ;
(ii) Atomic Volume. (ii) Atomic Volume $= \dfrac{\text{atomic weight}}{\text{density}}$, *i.e.* the space occupied by that amount of the element which is taken to represent its atomic weight.

" At present this space, which is known as the atomic volume, cannot be measured absolutely, but a relative measurement may be made by taking such quantities of the different elements as are proportional to their atomic weights, and comparing the space occupied by those quantities. If, as usual, the density of water is taken as unit, and the space occupied by the unit weight of water be the unit of volume, then the values of the atomic volumes are represented by the quotient of the atomic weight by the density of the given element. The atomic weight of lithium = 7·01, the density of the

Lothar Meyer's Curve showing the Relation between Atomic Volumes and Atomic Weights.

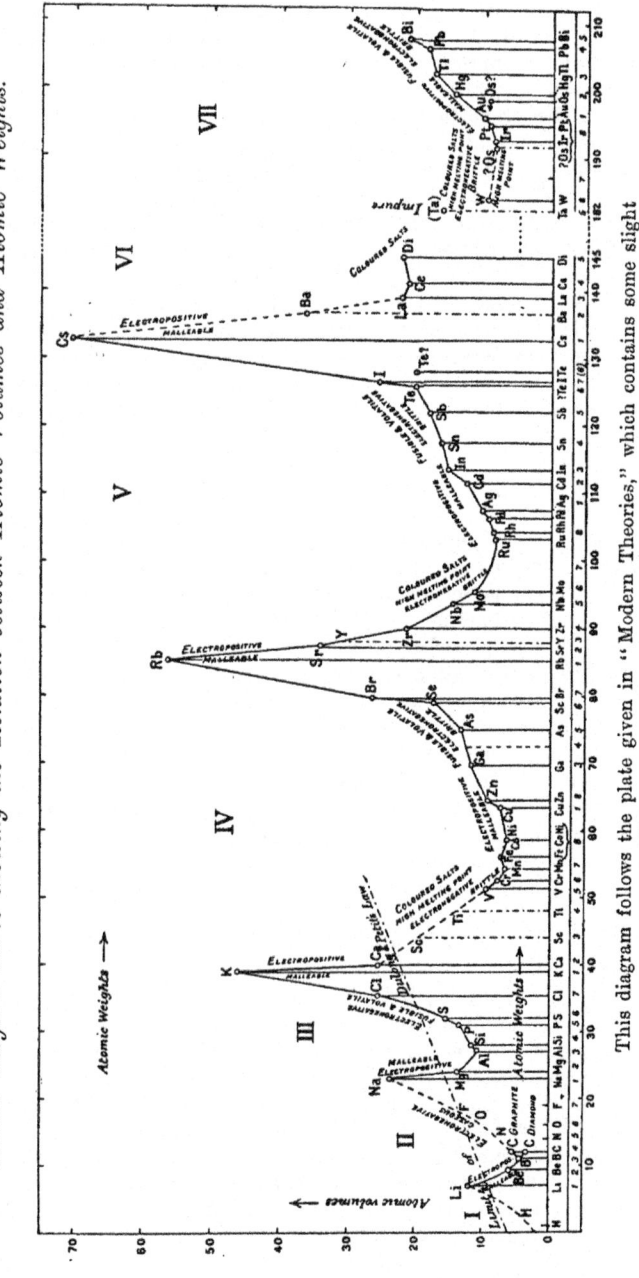

This diagram follows the plate given in "Modern Theories," which contains some slight additions to that appended to the original paper.

metal compared with water is 0·59; the atomic volume is therefore $V = 7·01 \div 0·59 = 11·9$; or expressed in the metric system of weights and measures, 7·01 grammes of lithium occupy a space of 11·9 cubic centimetres." (*Modern Theories*, p. 121.)

"One of the properties which varies fairly regularly with the atomic weight is the space that the elements occupy, *i.e.* the atomic volume. [The table on page 463] gives a graphic representation of the changes of this property dependent on the changes of the atomic weights. The curve has for abscissae lengths proportional to the atomic weights, and for ordinates, lengths proportional to the corresponding atomic volumes of the elements in the solid state....It becomes evident at once that the atomic volume of the elements, just as much as their chemical behaviour, is a periodic function of the magnitude of their atomic weights. As the atomic weight increases and decreases, the curve which represents these changes is divided by 5 maxima into six sections...of which the second and third, fourth and fifth, show strong resemblance, and correspond to almost equal lengths of the axis of the abscissae."

(iii) State of aggregation, ductility, electro-chemical properties.

(iii) Connection between physical properties and position of elements on atomic volume curve.

"The position of an element on this curve is closely connected with its physical and chemical properties, so that similar elements occupy corresponding positions on similar portions of the curve. It is not very remarkable that the maxima of the curve are formed by light metals, and the lowest minima by heavy metals, since it has long been known that the former have very large, and the latter very small atomic volumes. But it is very remarkable that elements with similar atomic volumes exhibit very different properties, depending as to whether their position is on a rising or falling branch of the curve.

Only those elements exhibit the property of malleability which lie in a maximum or minimum of the curve, or immediately follow a maximum or minimum.

Malleability.

The light malleable metals occupy the points of maxima and the contiguous portions of the descending curves (Li, Be; Na, Mg, Al; K, Ca; Rb, Sr; Cs, Ba). The heavy malleable metals are found in the lowest points of sections IV, V, VI, and VII, and in the adjacent sections of the ascending curves (Fe, Co, Ni, Cu, Zn; Rh, Pd, Ag, Cd, In, Sn; Pt, Au, Hg, Tl, Pb)....The brittle heavy metals and semi-metals are found in IV, V, and also in VII...shortly before the lowest points on the descending curves (Ti, V, Cr, Mn; Zr, Nb, Mo, Ru; Ta, W, Os, Ir). Non-metallic and semi-metallic elements are found in each section on the ascending branches of the curve preceding the maximum...(B, C, N, O, Fl; Si, P, S, Cl; As, Se, Br; Sb, Te, I; Bi)." (*Modern Theories*, p. 126.)

"All gaseous elements, and all those elements which fuse easily below a red heat, are found on the ascending portions and at the maxima points of

Volatility.

the atomic volume curve. All infusible and difficultly fusible elements occur at the points of the minima and descending portions of the curve. The fusibility of the elements considered as a function of their atomic weight exhibits a periodicity corresponding to that shown by the atomic volume and the malleability." (*Modern Theories*, p. 128.)

" The electro-chemical behaviour varies regularly; once in each of the sections II and III, and twice in each of IV, V, and VI. In II and III elements on the descending branch of the curve are electro-positive, on the ascending branch they are electro-negative; in IV, V, and VI at the maximum, at the minimum, and directly after these the elements are positive, but shortly before the minimum and maximum they are negative."

" In the descending portion of section II [and III] of the curve, from the maximum to the minimum, are the strongly electro-positive elements Li and Be [and Na, Mg, Al], the hydrated oxides or hydroxyl compounds of which are strong bases. The elements B, C, N, O, and F [and Si, P, S, Cl], which occur on the ascending portion of this section of the curve, from the minimum to the maximum, are mostly electro-negative; their hydrated oxides (and in the case of F its hydrogen compounds) are acids....

In sections IV, V, [and VI] the electro-chemical properties pass through two periods, whilst the atomic volumes pass through one only....On the upper part of the curve, descending from the maximum, the positive

Electro-chemical behaviour.

elements K, Ca, [and Rb, Sr; Cs, Ba] occur....Next follow the more or less negative elements in section IV.,—Ti, Vd, Cr, Mn, and in section V., Zr, Ni, Mo, and Ru [in section VI., Ta, W]. Some of these elements are electro-positive; this is especially true of Cr and Mn, which thus form a link between these and the positive elements situated on the lower parts of the ascending curves, viz. in section IV., Fe, Co, Ni, Cu, Zn; in section V., Rh, Pd, Ag, Cd [in section VI., Ir, Pt, Au; Hg, Tl, Pb]. Then from this point to about the middle of the ascending curve follow strongly negative elements, viz. in section IV., As, Se, and Br; and in section V., after tin, which may be regarded as positive as well as negative, follow Sb, Te, and I [in VI the almost negative element Bi]. And following these, without any intermediate elements, as in sections II and III, is the strongly positive alkali metal caesium, which occupies the maximum." (*Modern Theories*, p. 148.)

Other relations between physical properties and atomic weight are traced by Lothar Meyer in later editions of his " Modern Theories of Chemistry," a book first published in

Other physical properties.

1864, and soon become a classic. Expansion by heat, refraction of light, heat and electrical conductivities, magnetic properties, are all considered in their relations to atomic weight, and the data are given which show these properties to be functions of the atomic weights; the general inference and the final result being:

" Although our knowledge of the interdependence between the physical properties and the atomic weight may still be incomplete, yet sufficient is known to show the necessity for the introduction of a new departure in

F.

30

physical investigation. In physics hitherto...the substance experimented with appeared in the calculations as mass only....It has been shown that the numbers representing the atomic weights are the variables by which the real nature of a substance, and the properties depending upon it, are determined. The atomic weight is therefore the new variable to be introduced into the calculations; the properties of matter or physical phenomena are...to be regarded...as functions of the atomic weights. The mathematical form of this function still remains to be discovered; it will in all probability have to be a very strange one." (*Modern Theories*, p. 152.)

4. *Application.*

4. Application of principle established to: (i) Determination of the relation between equivalent and atomic weight.

"Whilst these and other similar regularities cannot possibly be chance occurrences, we are forced to admit that their empirical investigation alone does not enable us to unravel the mystery of their internal causal connection. But it would seem that at least some beginning has been made towards the investigation of the constitution of the hitherto undecomposed atoms, a guide found for future comparative study of the elements. These scanty data can already be applied in some measure."

(i) To determine which multiple of the equivalent weight shall be taken as the atomic weight in cases where the ordinary methods (the application of Avogadro's law, Dulong and Petit's law) are not available, as was the case then for Be and In.

"According to C. L. Winkler, the equivalent weight of indium is 37·8. Since the density of the metal has been found to be 7·42, the equivalent volume would be 5·1, which cannot be the atomic volume, as this would fall quite outside the curve. If we assume indium to be divalent, In = 75·6, it stands between As and Se, where the ductile electro-positive metal fits as badly as the corresponding atomic volume, 10·2, would fit into the course of the curve. But if we make In = 3 × 37·8 = 113·4, trivalent like aluminium... it falls between Cd and Sn with the atomic volume 15·3, which fits fairly well into the curve."

(ii) To correct and revise the stoichiometric constants, *i.e.* to make more accurate our knowledge of the atomic weight values.

(ii) Revision of stoichiometric constants.

"If the atomic volume corresponding to an atomic weight does not fit in with the regular course of the curve, the existence of some error becomes probable. Thus the atomic weights of tellurium, platinum, iridium, and osmium are probably somewhat too high."

"From the analogy with its associates, tellurium should come before iodine in the series of atomic weights."

Subsequent research has borne out the justice of this inference in the case of platinum, iridium, and osmium,

Atomic weights $(H = 1)$.

	1870	1904
Pt	196·7	193·3
Ir	196·7	191·5
Os	198·6	189·6

whilst in the case of tellurium, we seem to be dealing with a hitherto unexplained exception. Much work has been devoted to the determination of this value, and the results obtained have shown considerable divergence (*ante*, p. 225); but 126·6, the value now accepted as the most probable, still leaves tellurium following instead of preceding iodine, whose atomic weight is 125·9. But having before us an enormous amount of evidence in support of the periodic variation in the relation between the atomic weights of the elements and their properties, we cannot do otherwise than believe that sometime the empirical atomic weight of tellurium will agree with that required by theory, and great will be the interest that will attach to the discovery of the present cause of error.

Perhaps some day a historian will investigate and explain why, having gone so far as he did, having achieved so much in discovering the relationship of the periodic law, in substantiating it by facts, in applying it with a fair measure of success, Lothar Meyer should, in the concluding paragraph of his paper, have characterised his results as uncertain, should have taken so depreciatory a tone.

"It would be premature to make any changes in the accepted values of the atomic weights on grounds so uncertain; moreover we must not for the present attach too much importance to arguments such as those given above, nor should we expect from them a decision as sure as that already obtainable from the determination of the specific heat or the vapour density."

L. Meyer's estimate of the reliability of the new principle of classification.

Of the two papers[1] in which Mendeleeff made public his discovery of the periodic law, the first appeared in 1869; the second,

[1] *The Relations between the Properties of the Elements and their Atomic Weights,* 1869.
The Periodical Regularities of the Chemical Elements, 1871; English translation in *Chemical News,* vols. 40, 41 (1879 and 1880).
Both papers in German translation in Ostwald's *Klassiker,* No. 68.

which is by far the longer and more important, bears the date August, 1871.   The quotation from the 1869 paper already given (p. 455), contains the author's review and criticism of the principles which had, up to that time, been employed in the classification of the elements.   Concerning the considerations which led him to his own choice, he tells us :

"When I undertook to write the text-book, entitled *The Foundations of Chemistry*, I had to decide for some one system, lest, in the classification of the elements, I should have allowed myself to be guided by accidental, and, so to speak, instinctive reasons rather than by an accurate and definite principle."

Mendeleeff on the selection of a principle of classification; superiority of one based on numerical relations.

Mendeleeff then points out that the principles hitherto used in classification had not been of a quantitative nature, and he emphasises the superiority of a system which is based on numerical relations and therefore leaves no scope for arbitrary interpretation.   Consideration of the numerical data available in the case of the elements leads to the rejection as unsuitable of the optical, electrical, and magnetic properties, because these vary with the conditions; and of the vapour density, because this is not known for many elements, and is different for the allotropic modifications.

"...But throughout the changes in the properties of the simple substances[1] one thing remains unaltered...which is the atomic weight of the elements... The atomic weight is, from its very conception, a quantity which has no relation to the temporary condition of a simple substance, but appertains to that material part of it which subsists in the uncombined simple substance as well as in all its compounds.   The atomic weight does not belong to charcoal or diamond, but to carbon.

Gerhardt's and Cannizzaro's method for determining atomic weights is based on principles so correct and unassailable that for the greater number of the elements, especially those whose heat capacity in the solid state is known, there is no uncertainty about their atomic weights, such as did still exist when these quantities were determined according to different, and often quite incompatible principles.

[1] Mendeleeff differentiates between *simple substance* and *element*.   "*A simple substance* is something material, a metal or metalloid possessed of definite physical and chemical properties.   To the conception of simple substance corresponds the molecule consisting of one (Hg, Zn) or several atoms (S_2, S_6, H_2, Cl_2, P_4), which is capable of existing in polymeric modifications and differs from a compound only by the equality of its constituent particles....On the other hand we should designate by the name *elements* the material constituents of the simple and complex substances, to which is due the physical and chemical behaviour of these substances ; the atom therefore goes with the conception of element.   Thus carbon is an element, but charcoal, graphite and diamond are simple substances."

For this reason I have tried to take as basis for my system of classifica_ tion the value of the atomic weight....Beginning with the one of smallest

Its invaria- bility makes the atomic weight a suit- able basis of classification.

atomic weight, I arranged the elements according to the magnitude of their atomic weights, when it became evident that there exists a kind of periodicity in the properties of the simple substances....

Hence in this system of classification the atomic weight of an element determines the place to be assigned to it,...and all the comparative investigations that I have made lead me to the con_ clusion, that the magnitude of its atomic weight determines the character of

Mutual rela- tions between atomic weights and properties of elements named "Peri- odic Law."

an element in the same measure, as the molecular weight determines the properties and many of the reactions of a compound....

I designate by the name of *Periodic Law* the mutual relations between the properties of the elements and their atomic weights, relations which are applicable to all the elements, and which are of the nature of a periodic function."

Of the various tables given by Mendeleeff in illustration of this classification, the one reproduced on page 470 contains the

Mendeleeff's table; series and groups.

arrangement resorted to in its final and most perfect form. The elements appear divided into 12 *series* and 8 *groups*, with suitable places for all the elements

then known, and vacant places for elements yet un- discovered. The series are the horizontal rows, and each comprises a certain number of elements—7 at least, 10 at most—arranged in arithmetical progression according to increasing atomic weight; the groups, which are the vertical rows, contain chemically similar elements.

"At first the properties of the elements change as the atomic weights increase, and afterwards these repeat themselves in a new series, or new period of elements, with the same regularity as in the preceding series."

The following quotations are to show what properties furnished the data for the above generalisation, and to serve as examples of Mendeleeff's method in setting forth his evidence.

1. Proof that the properties of elements are a periodic function of their atomic weights. (i) Valency; variation within series.

(i) *Variation of valency in the series shown by the formulae of* (a) *the hydrides,* (b) *the oxides; and variation of the chemical nature of these substances.*

"Take the case of the light elements whose atomic weight is between 7 and 36, and arrange these in an arithmetical progression according to their atomic weights, thus :

Li $=7$; Be $=9\cdot4$; B $=11$; C $=12$; N $=14$; O $=16$; F $=19$.
Na $=23$; Mg $=24$; Al $=27\cdot3$; Si $=28$; P $=31$; S $=32$; Cl $=35\cdot5$.

Mendeleeff's Table representing the Periodic. Regularities of the Chemical Elements.

Series	Group I R^2O	Group II RO	Group III R^2O^3	Group IV RH^4 RO^2	Group V RH^3 R^2O^5	Group VI RH^2 RO^3	Group VII RH R^2O^7	Group VIII — RO^4
1	H=1							
2	Li=7	Be=9.4	B=11	C=12	N=14	O=16	F=19	
3	Na=23	Mg=24	Al=27·3	Si=28	P=31	S=32	Cl=35·5	
4	K=39	Ca=40	—=44	Ti=48	V=51	Cr=52	Mn=55	Fe=56, Co=59, Ni=59, Cu=63
5	(Cu=63)	Zn=65	—=68	—=72	As=75	Se=78	Br=80	
6	Rb=85	Sr=87	?Yt=88	Zr=90	Nb=94	Mo=96	—=100	Ru=104, Rh=104, Pd=106, Ag=108
7	(Ag=108)	Cd=112	In=113	Sn=118	Sb=122	Te=125	I=127	
8	Cs=133	Ba=137	?Di=138	?Ce=140	—	—	—	—
9	(—)	—	—	—	—	—	—	—
10	—	—	?Er=178	?La=180	Ta=182	W=184	—	Os=195, Ir=197
11	(Au=199)	Hg=200	Tl=204	Pb=207	Bi=208	—	—	Pt=198, Au=199
12	—	—	—	Th=231	—	U=240	—	—

The corresponding members of both series give the same type of combinations, *i.e.* they possess the same valency. The importance of this lies in the fact that the transition of one element to the next is accompanied by a regular change in the type of combination, which change is made evident by comparing the hydrogen and oxygen compounds of these elements. This regularity proves that the above grouping of elements produces a natural series, in which the existence of intermediate members must not be presumed.... Only the last four members can combine with hydrogen, forming, if R represents an element,

$$— \quad — \quad — \quad RH_4, \quad RH_3, \quad RH_2, \quad RH.$$

The greater or lesser stability of these hydrogen combinations, their acid character (*i.e.* the power of exchanging the hydrogen for metals), these and
(*a*) Formulae and chemical nature of hydrides.
other properties change gradually in accordance with the position of the element in the series. Thus HCl is a marked acid of great stability; H_2S is a weak acid easily decomposed by heat; H_3P has lost all acid character and is very easily decomposed, which features are even more strongly marked in H_4Si.

Since all the elements of the second series combine with oxygen, these combinations lend themselves best to tracing the gradual changes in the
(*b*) Formulae and chemical nature of oxides.
properties of the elements which accompany the changes in atomic weight. For the purposes of such a comparison we quote the higher anhydrous oxides which correspond to water, which can combine with water to form hydrates and with one another to form salts....The seven elements of the second series yield the following higher oxides capable of producing salts:

$$Na_2O, \quad Mg_2O_2, \quad Al_2O_3, \quad Si_2O_4, \quad P_2O_5, \quad S_2O_6, \quad Cl_2O_7,$$
$$\text{or } MgO, \qquad \text{or } SiO_2, \qquad \text{or } SO_3.$$

Hence to the 7 members of the series correspond, in definite sequence, 7 generally known types of oxides. Though these latter had been discovered long ago, yet their connection with a fundamental property of the elements had remained unnoticed. The simple process of grouping these 7 types at once shows that in the same order there is a decrease in the basic, and an increase in the acid properties. These elements yield the following normal saline derivatives:

$$NaX, \quad MgX_2, \quad AlX_3, \quad SiX_4, \quad PX_5, \quad SX_6, \quad ClX_7,$$
e.g.
$$Na(NO_3), \quad MgCl_2, \quad Al(NO_3)_3, \quad Si(KO)_4, \quad PO(NaO)_3, \quad SO_2(NaO)_2, \quad ClO_3(KO).$$

At the beginning of the series stand the substances of decided metallic character, at the end the representatives of non-metals; the former possess basic, the latter acidic properties, those intermediate forming the transition. In Li_2O and Na_2O the basic properties appear more strongly than in BeO and MgO; in B_2O_3 and in Al_2O_3 these are only feebly developed, and the compounds begin to show acid properties; CO_2 and SiO_2 possess only acid properties, slightly developed though these are; but the acid character appears more strongly in N_2O_5 and P_2O_5, as well as in SO_3 and Cl_2O_7."

(ii) Physical properties; variation within series. (a) Atomic volume.

(ii) *The changes of physical properties within the series show similar regularity.*

(a) Atomic volume.

	Na	Mg	Al	Si	P	S	Cl
Specific Gravity:	0·97	1·75	2·67	2·49	1·84	2·06	1·33
Atomic Volume:	24	14	10	11	16	16	27

	Na_2O	Mg_2O_2	Al_2O_3	Si_2O_4	P_2O_5	S_2O_6	Cl_2O_7
Specific Gravity:	2·8	3·7	4·0	2·6	2·7	1·9	?
Atomic Volume:	22	22	25	45	55	82	?

(b) Volatility.

(b) Volatility.

"The volatility decreases in the first series from Na to Si, after which it increases; the same may be observed in the case of the oxides, of which MgO, Al_3O_3, SiO_2 are non-volatile."

Having recognised and furnished the proof that the relation between properties and atomic weights is a periodic function, and also that it was not yet possible to discover the accurate expression for this function, Mendeleeff proceeded to investigate the length of each constituent period, *i.e.* the number of elements which go to such a period, and he discovered what Lothar Meyer had failed to notice, namely, a distinct difference between the corresponding members of the odd and the even series.

2. Investigation of the length and character of the periods. (i) Odd and even series.

"The tabular representation given demonstrates the existence and the properties of periods of 7 elements, such as that comprising Li, Be, B, C, N, O, F. We will call this a *short period* or series. Then if H is placed in the first series, Li, etc. come into the second, Na, etc. into the third, etc. But these short series do not accommodate all the elements known to us; and, what is more important still, we find that, excepting the two first series, a distinct difference exists between the corresponding members of odd and even series; the members of one or other of these series showing comparatively greater analogies with one another. An example will fully prove this:

4th series :	K	Ca	—	Ti	V	Cr	Mn
5th series :	Cu	Zn	—	—	As	Se	Br
6th series :	Rb	Sr	—	Zr	Nb	Mo	—
7th series :	Ag	Cd	In	Sn	Sb	Te	I

The members of the fourth and of the sixth series have greater resemblances to one another than the members of the fifth and seventh. Those belonging to the even series are not so markedly semi-metallic as are the elements of the odd series....In their lower oxides the last members of the even series resemble in many respects the first members of the succeeding odd series. Thus Cr and Mn are in their basic oxides similar to the elements Cu and Zn. On the other hand there are marked differences between the last members of the odd series (halogens) and the first members of the next even series (alkalis). Moreover, all those elements which could not be placed within the short periods, fall in accordance with their properties and their atomic weights, between the last members of the even and the first members of the odd series. So between Cr and Mn on the one hand, and Cu and Zn on the other, come Fe, Co, Ni, forming the following transition period

(ii) **Transition periods; group VII**

$$Cr = 52; \quad Mn = 55; \quad Fe = 56; \quad Co = 59 = Ni = 59; \quad Cu = 63; \quad Zn = 65.$$

And just as Fe, Co, Ni follow after the fourth series, so Ru, Rh, Pd follow after the sixth, and Os, Ir, Pt after the tenth.

Two series, one even and one odd, together with an intermediate series of the elements just mentioned, constitute a *long period* of 17 members. Since the intermediate members do not correspond to any of the 7 groups of the short periods, they form an independent group, the eighth. The members of this group

Fe = 56;	Ni = 59;	Co = 59;
Ru = 104;	Rh = 104;	Pd = 106;
Os = 193 (?);	Ir = 195 (?);	Pt = 197;

resemble one another in the same manner as the corresponding members of an even series, as for instance, V, Nb, Ta, or Cr, Mo, W."

Mendeleeff then proceeds to substantiate by data the existence of the similarity enunciated above. The chemical analogies between iron, cobalt, and nickel are, however, facts so well-known that it seems superfluous to follow up this point further, and we can turn at once to the interesting fact which is connected with the recognition of the exceptional character of series 2, the first of the even series.

(iii) **Exceptional character of 2nd series; typical elements.**

" It might seem as if the character of the second series were contrary to the general division into even and odd series, and this because of the following reasons : Certain members of this even series (Li, Be, B, C, N, O, F) have acid properties, give hydrogen and organo-metallic compounds[1], *e.g.* $B(C_2H_5)_3$, $C(C_2H_5)_4$, $N(C_2H_5)_3$, $O(C_2H_5)_2$, $F(C_2H_5)$; some of the elements are gaseous ; in short, they behave like odd series elements. But concerning this second series it should be noted : (1) that it is not followed by an eighth group, as is

---

[1] In a previous passage, not quoted, the formation of such compounds had been recognised as an exclusive characteristic of the odd series.

the case with the other even series; (2) that the atomic weights of its constituent elements differ from the corresponding atomic weights of the following series by about 16, whilst in all the other series this difference is between 24 and 28. Also whilst the difference between the corresponding atomic weights of other successive even series is 46, that between the second and fourth series is only 32 to 36.

Li	Be	B	C	N	O	F;	Na	Mg	Al	Si	P	S	Cl
K	Ca	—	Ti	V	Cr	Mn;	Cu	Zn	—	—	As	Se	Br
—	—	—	—	—	—	—;	—	—	—	—	—	—	—
Diff. 32	31	—	36	37	36	36;	40	41			44	46	45

This not only explains the apparent deviations, but also confirms our fundamental theorem of the connection between the properties of elements and their atomic weights. The difference between the atomic weights being in this case of a different magnitude, the relations between the properties of the elements must also be different......

It is because of this special behaviour that I designate the members of the second series together with H, Na, and Mg, as *typical elements*[1]."

Comparison of Mendeleeff's and Lothar Meyer's treatment of the inductive part of the subject shows that whilst the latter pays greater attention to the consideration of the physical properties, the former devotes himself more to a comparative study of chemical relations; moreover the recognition of the important differences between

3. Mendeleeff's deductive application of his law.

the even and the odd series is due to Mendeleeff alone. But it is especially in the deductive application of the system, that we find the Russian scientist much in advance of the German; the scope of the phenomena encompassed, the definiteness and lucidity of the reasons adduced for the conclusions arrived at, the number and importance of the predictions made together with the marvellous way in which these have been verified, have combined to make this part of Mendeleeff's work one of the greatest scientific achievements of the century, one of the most striking confirmations of the modern method.

"A natural law only acquires scientific importance when it yields practical results, *i.e.* when it leads to logical conclusions which elucidate phenomena hitherto unexplained, when it directs attention to occurrences till then unknown, and especially when it calls forth predictions which may be verified by experiment. In such a case the correctness of the law may be tested and

[1] It has justly been pointed out that this is not a suitable name, giving as it does the impression that in these elements the type or character of the series is especially clearly marked, whilst the very opposite is the case.

its truth made apparent....I will therefore consider in greater detail some of the consequences and applications of the periodic law, selecting the following:

(i)   The classification of the elements.

(ii)   The determination of the atomic weights of elements not yet fully investigated.

(iii) The prediction of the properties of hitherto unknown elements.

(iv)   The correction of atomic weight values.

(v)   The extension of our knowledge of chemical types.

There is considerable overlapping in the nature of the problems thus enumerated; moreover these differ greatly in relative importance and interest. All the same it will be best to follow Mendeleeff's division and sequence, omitting however from consideration the phenomena included under (v) because no special advance has been made in this department through the application of the periodic law.

(i) *The application of the periodic law to the classification of the elements.*

(i) Classification of the elements. "Classification of the elements...serves not only as an aid to memory, for the retention of a .o s arranged and correlated facts, but it is also of scientific importance in so far as it reveals new analogies and thereby points the way to further investigations of the elements."

Having shown the inadequacy of the older methods for this last purpose, Mendeleeff demonstrates theoretically and practically the superiority of the periodic system.

"The connection between the type of the oxide formed and the atomic weight affords to the periodic law invariable numerical values for distribution of the elements, yielding a grouping of elements naturally very similar....In this way the law makes it possible to build up a system of the greatest possible completeness, free from all arbitrariness....The position of an element R in the system is determined by the series and the group to which it belongs, that is by the elements X and Y contiguous to it in the series, as well as by the two elements R' and R'', contiguous to it in the group. It is possible to determine the properties of R from the known properties of X, Y, R', R''. Thus taking in the system the following series

$$(n-2)\text{th series}............ X'R'Y'$$
$$n\text{th series}............ XRY$$
$$(n+2)\text{th series}............X''R''Y''$$

$R'' - R$ is approximately $= R - R' = 45$ (approximately).

We can therefore put down proportions for the determination of the properties of corresponding compounds, and can calculate the intermediate values, owing to the fact that the properties of all elements are really in intimate mutual dependence. This relation of R, on the one hand to X and Y, and on the other hand to R' and R'', is what I term the *atom-analogy* of the element. Thus As and Br on the one hand, and S and Te on the other are *atom-analogues* of Se, whose atomic weight has the intermediate value

$$78 = \frac{(75+80+32+125)}{4}.$$

Accordingly SeH$_2$ stands by its properties midway between AsH$_3$ and BrH, and also midway between SH$_2$ and TeH$_2$, etc....These mutual relations afford the means for the explanation of many otherwise isolated and disputed facts.

Formula of beryllia.

Different views have been held concerning the place to be assigned to beryllium....The magnesia formula has been given to its oxide, but on the other hand glucina[1] has properties similar to those of alumina. The periodic law affords the following proofs in support of the formula BeO: The formula Be$_2$O$_3$ requires that the atomic weight of beryllium should be $\frac{3}{2} \cdot 9{\cdot}4 = 14{\cdot}1$, for which no place could be found in the periodic system, because it would come next to nitrogen, in a position where it should exhibit distinctly acidic properties, forming the higher oxides Be$_2$O$_5$ and Be$_2$O$_3$, which is not the case. If however BeO is taken as the formula of the oxide and 9·4 as the atomic weight of the metal, it finds its place between Li=7 and B=11 which is in accordance with all its properties. For proof of this the following considerations suffice:

(1) Be : Li = B : Be.

In fact the basic properties are much less strongly developed in BeO than in Li$_2$O, and still less so in B$_2$O$_3$; beryllium chloride is more volatile than lithium chloride, and boron chloride is still more so, etc.

(2) Be : Mg = Li : Na = B : Al.

Just as beryllium oxide is a less energetic base than MgO, so also Li$_2$O is weaker than Na$_2$O, and B$_2$O$_3$ is weaker than Al$_2$O$_3$. Glucina dissolves in KOH. Moreover the incomplete isomorphism of the salts of BeO and MgO ...need present no difficulties because the same is found in the case of the salts of Li and Na, of B and Al. The fluoride of beryllium is soluble in water, whilst that of magnesium is not, in complete analogy with the solubility of boron fluoride and the insolubility of aluminium fluoride.

(3) Be : Al = Li : Mg = B : Si.

If in spite of its formula beryllium oxide corresponds in many respects to alumina, this has its counterpart in the similarity of the properties of Li$_2$O and MgO, of B$_2$O$_3$ and SiO$_2$."

---

[1] Glucina (GlO) from γλυκύς, sweet = beryllia (BeO) from the mineral beryl, (Be$_3$ Al$_2$ Si$_6$ O$_{18}$).

The data concerning the equivalent volumes, given by Mende-leeff in support of the above relations, may be tabulated thus :

Equiv. vol. $BeO = \dfrac{25\cdot4}{3\cdot05} = 8\cdot3$, approx. equal to equiv. vol. $\tfrac{1}{3}Al_2O_3 = \tfrac{1}{3} \cdot \dfrac{102\cdot6}{4} = 8\cdot5$

,, ,, $LiCl \ldots\ldots\ldots = 21$,, ,, ,, ,, $\tfrac{1}{2}MgCl_2 \ldots\ldots\ldots\ldots = 22$

,, ,, $\tfrac{2}{3}BCl_3 \ldots\ldots\ldots = 58$,, ,, ,, ,, $\tfrac{1}{2}SiCl_4 \ldots\ldots\ldots\ldots = 56$

,, ,, $\tfrac{1}{3}B_2O_3 \ldots\ldots\ldots = 13$,, ,, ,, ,, $\tfrac{1}{2}SiO_2 \ldots\ldots\ldots\ldots = 13\cdot5$

The case of Cs, of V, of Tl is dealt with in a similar manner to that followed for Be, and the places assigned to these elements in the periodic system are justified by a comparative examination of their properties.

(ii) *The application of the periodic law to the determination of the atomic weights of elements not yet fully investigated.*

(ii) Determination of the relation between equivalent and atomic weight.
"The determination of the atomic weight of an element involves, besides the equivalent weight, the knowledge of some physical property of the simple substance itself (heat capacity) or of its compounds (vapour density, heat capacity), or the existence of isomorphism. Since some of these determinations are accompanied by practical difficulties...the atomic weights of many elements have often been assigned on the ground of very uncertain data....In such cases the periodic law comes to our aid, affording us a new relation between the chemical properties and the atomic weights....If the equivalent E derived from the highest oxide of the element, making this oxide E_2O and the chloride ECl, is multiplied by 1, 2, 3, 4, 5, 6, 7, a number of possible values are obtained for the atomic weight. Of these numbers the one, E_n, which represents the true atomic weight, corresponds to a vacant place in the system, and exhibits in its properties the atom-analogies required by that place. We are led to this inference by the experience that, as far as we know, one element only corresponds to each one place in the periodic system....Take for instance the case of an element whose equivalent weight is equal to 38, whose oxide is not very strongly basic and cannot be further oxidised. The question then arises, what is its atomic weight? or put differently, what is the formula of its oxide? If we assign to the oxide the formula R_2O, then $R = 38$, and the element belongs to the 1st

Criteria for selecting oxide formula.
The atomic weight of indium.
group; but the place is already occupied by $K = 39$, and further, the atom-analogies require for this position a soluble and energetic base. If we make the oxide RO, the atomic weight $= 76$, which again does not fit into the 2nd group, because $Zn = 65$, $Sr = 87$, and all places for elements of low atomic weight are filled in this group. Assuming the oxide formula to be R_2O_3, the atomic weight $R = 114$, and the element goes into the 3rd group, in which there is in fact between $Cd = 112$ and $Sn = 118$ a vacant place for an element with an atomic weight of about 114. Judging by the atom-analogies

with Al_2O_3 and Tl_2O_3 as well as with CdO and SnO_2, the oxide of such an element should possess feebly basic properties. Hence the element should be placed in the 3rd group."

After showing in the same manner that $R = 152 = 4 \times 38$, and $R = 190 = 5 \times 38$ would not fit into the scheme of the periodic law, Mendeleeff supports the correctness of the value 114 by the following considerations:

" The only possible atomic weight is 114, and the formula for the oxide is R_2O_3. Indium is an element answering to this description. Its equivalent weight according to Winkler is 37·8, and hence the atomic weight must be taken to be 113[1], and the composition of the oxide represented by In_2O_3. The atom-analogues in the 3rd group are Al and Tl, and those in the 7th series Cd and Sn....The specific gravity of Cd is less than that of Ag, that of Sb is a trifle less than that of Sn; consequently indium must possess a specific gravity somewhat less than the mean between Cd and Sn, which is really the case. $Cd = 8·6$, $Sn = 7·2$; and the specific gravity of In, which should be less than 7·9, has been found $= 7·42$....Indium and its analogues occur in the odd series, and therefore the higher oxides cannot be strongly basic; the basic character of In_2O_3 must be feebler than that of CdO and Tl_2O_3, and stronger than that of Al_2O_3 and SnO_2. These conclusions are based on the following facts. The oxides of Al and Sn dissolve in alkalis, forming definite combinations, whilst those of Cd and Tl are insoluble. In_2O_3 also dissolves in alkalis, but without forming definite compounds. The oxides of Cd, Sn, Al, and Tl are difficultly fusible powders like In_2O_3. The hydrate of In_2O_3, according to expectation, is colourless and gelatinous.

...In order to test the correctness of this proposed change in the atomic weight and in the oxide formula of indium, I have determined its heat capacity, and have found it $= 0·055$, which is in accordance with the periodic law[2].

Hence without hesitation we can apply the periodic law for correcting the atomic weights of elements which are little known."

The case of uranium and of a number of elements of the rare earths, is dealt with according to the same principles, and oxide formulae are assigned (UO_2, UO_3, $U = 240$; CeO_2, $Ce = 140$; LaO_2, $La = 180$; Di_2O_3, $Di = 138$; Er_2O_3, $Er = 178$) which in most cases though not in all, have been confirmed by subsequent work (UO_2, UO_3, $U = 238·5$; Ce_2O_3, CeO_2, $Ce = 140·25$; La_2O_3, $La = 138·9$;

The atomic weights of the rare earth elements.

---

[1] Up to that time the oxide had been formulated as InO, which made the atomic weight $37·8 \times 2$.

[2] $0·055 \times 114 = 6·27$; Bunsen, using his ice-calorimeter method, had found $0·0577$ for the specific heat, which makes the atomic heat 6·50.

Di_2O_3, $Di = 142 \cdot 3$ [1]; Er_2O_3, Er_2O_5, $Er = 166$). It should be noted that in the case in which Mendeleeff's classification has not been adhered to, he himself had considered the evidence then available as not conclusive, a fact marked by the queries put before the elements Ce, La, Er, Di in the table (p. 470).

(iii) *The application of the periodic law to the determination of the properties of elements as yet undiscovered.*

(iii) Prediction of new elements; eka-boron, eka-aluminium, eka-silicon.

"...As another example of the study of the elements by means of the periodic law, I propose to determine the properties of elements still unknown. Without the periodic law, it is impossible to predict the properties of unknown elements.

... The discovery of new elements has only been a matter of observation and has therefore been due to chance or to the exceptional acuteness of the observer....A new method in this field is provided by the periodic law....All that has been said on this subject goes to prove that the periodic law enables us to draw conclusions concerning the unknown properties of elements whose atom-analogues are known; and the table (p. 470) which represents the periodic relations of the elements shows that several elements required in the series are so far missing. I will therefore describe the properties of some elements which we may expect to discover in course of time. Thus I shall substantiate the periodic law by yet another proof which is manifest and definite, though its confirmation must be reserved for the future....In order not to introduce new names for these unknown elements, I shall designate them by prefixing a Sanscrit number (eka, dwi, tri, etc.) to the name of the nearest lower odd or even atom-analogue in the same group. Thus the unknown elements in the first group will be called eka-caesium, $Ec = 175$, dwi-caesium, $Dc = 220$, etc. If for instance niobium were not known it would be called eka-vanadium.

Then follows the prediction of the elements eka-boron ($Eb = 44$), eka-aluminium ($Ea = 68$), and eka-silicon ($Es = 72$), the scandium ($Sc = 44 \cdot 1$), gallium ($Ga = 70$), and germanium

Comparison between the properties predicted and those found for eka-silicon (germanium).

($Ge = 72$) discovered since, and affording by their properties, physical and chemical, the most brilliant confirmation of the validity of the periodic law classification. The amount of detail given renders it most suitable to select for quotation that part of Mendeleeff's argument which relates to eka-silicon.

---

[1] 142·3 is the value for the atomic weight of the substance considered elementary didymium until it had been shown by Auer von Welsbach in 1885 that this was a mixture of two substances by him termed *praseodymium* ($Pd = 143 \cdot 6$) and neo-dymium ($Nd = 140 \cdot 8$).

EKA-SILICON

(predicted in 1871)

"The two elements missing from the 5th series, and belonging to the 3rd and 4th group respectively, stand in the series between $Zn = 65$ and $As = 75$, and have Al and Si for atom-analogues. The one shall be called eka-aluminium, and the other eka-silicon....The metals should be easily obtained by reduction with carbon or sodium. Their sulphides will be insoluble in water, and EsS_2 will probably be soluble in ammonium sulphide. The atomic weight of eka-silicon will be $Es = 72$. From the atomic volume values $Zn = 9$, $As = 14$, $Si = 18$, we get for the specific gravity of eka-silicon $5 \cdot 5$, and for its atomic volume about 14, numbers which may equally be deduced from a comparison of the volumes of the atom-analogues $Si = 11$, $Sn = 16$, whereby $Es = 13$.

Eka-silicon, with the oxide EsO_2, occupies the place between Ea and As on the one hand, and between Si and Sn on the other. Since all these relations indicate a necessary analogy between Es and Ti, I shall subject Es and Ti to comparative treatment.

It will be possible to prepare eka-silicon by the action of Na on EsO_2 and EsK_2F_6; it will decompose steam with difficulty, and the action on acids will be slight, that on alkalis more pronounced.

The element will be a dirty grey metal, difficultly fusible, and on calcination it will yield a powdery oxide EsO_2, which will be very refractory.

The specific gravity of the oxide will be about $4 \cdot 7$, deduced from the volume, which by analogy with the

GERMANIUM

(discovered in 1886)

This element possesses the following properties, investigated by Winkler :

GeS_2 is completely pp. by H_2S in the presence of mineral acids, and like AsS_3, SnS_2, etc. it is soluble in ammonium sulphide.

Atomic weight of $Ge = 72 \cdot 3$.

Spec. Grav. of $Ge = 5 \cdot 469$.

Ge has been made by the reduction of GeO_2 by carbon, and of K_2GeF_6 by sodium ; it does not decompose water, is not attacked by HCl, but is easily soluble in *aqua regia* ; solution of KOH has no action, but molten KOH oxidises it with incandescence.

The element has metallic lustre, the colour is greyish white ; it does not oxidise in air, but on ignition it forms the oxide GeO_2, a dense white powder which is very refractory.

Spec. Grav. of $GeO_2 = 4 \cdot 703$.

volumes of SiO_2 and SnO_2 will be about 22.

In external appearance and probably also in crystalline form, in properties and reactions, the oxide will resemble TiO_2. For this, as well as for all similar relations, the following proportions hold :

$$Es : Ti = Zn : Ca = As : V$$

and accordingly the basic properties will be less marked in EsO_2 than in TiO_2 and SnO_2, though more so than in SiO_2.

Hence we may expect a hydrate of EsO_2 soluble in acids, though this solution will decompose easily, precipitating a soluble metahydrate.

From its analogy with TiF_4, ZrF_4, and SnF_4, the fluoride of eka-silicon will of course not be gaseous ; ....
on the other hand, eka-silicon chloride, $EsCl_4$, will be a volatile liquid, boiling at 100° (probably somewhat lower) because $SiCl_4$ boils at 57°, and $SnCl_4$ at 115°.....................................

The specific gravity of $EsCl_4$ (at 0°) will be about 1·9, and the volume 113, because the volume of $SiCl_4$ is 112, and that of $SnCl_4$ 115.

A marked difference between Es and Ti will be found in the fact that Es, like Si and Sn, will yield volatile metallo-organic compounds, e.g., $Es(C_2H_5)_4$, which is not the case with Ti, the member of an even series.

Judging from the properties of Sn and Si, $Es(C_2H_5)_4$ will boil at 160°, and its density will be about ·96."

The basic properties of the oxide GeO_2 are very feebly marked, the solubility in acids is slight, though there are indications of the existence of O salts.

Acids do not pp. the hydrate from dilute alkaline solutions ; but from concentrated solutions, acids or CO_2 pp. GeO_2 or a metahydrate.

The fluoride GeF_4 is not gaseous, only volatile.

The chloride $GeCl_4$ is a liquid boiling at 86°.

The specific gravity of $GeCl_4$ at 18° is 1·887.

$Ge(C_2H_5)_4$ is easily obtained.

$Ge(C_2H_5)_4$ boils at 160°, and its density is a little below that of water.

Mendeleeff not only speculates concerning the properties of elements not yet known, but he also considers from the point of view of the periodic law the question of the probable number of such undiscovered elements.

" Is the number of elements limited or unlimited ? In view of the fact

that the system of the known elements is limited and so to speak closed...
that acid properties are gradually effaced as the atomic
weights increase, and that most of the elements of high
atomic weight are heavy metals not readily oxidised, it may
be assumed that the number of elements accessible to us is
very limited."

The total
number of
element is
very limited.

(iv) *The application of the periodic law to the correction of*
atomic weight values.

(iv) Correction
of atomic
weight values.
Only possible
to fix narrow
limits.

"...The periodic law in its present form may be used to
detect such errors in the atomic weights as are comparatively
large....The exact evaluation of the atomic weights requires
an accurate comparative study of the individual properties of
the elements, because these are the cause of the perturbations
in the regular variations of the atomic weights."

Stas's numbers are quoted to show that even where the
differences in the atomic weights of successive members of a group
are very nearly the same, they are not exactly so.

" If $H = 1$ and $O = 15.96$:

$$Li = 7.004$$
$$Na = 22.980$$
difference 15.976.

$$Na = 22.980$$
$$K = 39.040$$
difference 16.060.

$$Cl = 35.368$$
$$Br = 79.750$$
difference 44.382.

$$Br = 79.750$$
$$I = 126.533$$
difference 46.783.

Hence, whilst we must recognise the existence of strict proportionality in
the differences between the atomic weights of corresponding elements of the
system, we also observe individual deviations in these differences. Accordingly
we are led to assume that the elements possess *general* properties, which vary
periodically with the atomic weight, and besides these *individual* properties
which are the consequence of the above-mentioned deviations.

At present we do not really know anything concerning this relation,
except that it is of periodic nature. Hence we have as yet no means for
determining the amount of these deviations and for effecting an accurate
revision of the atomic weights ; all we can do is to fix narrow limits within
which must lie the atomic weight of the element considered."

This principle is then applied to the case of Te, Os, Ir, Pt, and
Au, but there is nothing in Mendeleeff's treatment which calls for
an addition to what has been said in connection with Lothar
Meyer's work on this same point (p. 466).

Lothar Meyer and Mendeleeff between them had supported
the existence of a relationship between the weight of the atom
and its properties by an amount of experimental evidence which

left little to be added by subsequent workers in this field. But from about 1879 onwards—it seems strange that it should have taken nearly ten years for chemists to realise the supreme im_portance of Mendeleeff's work—the literature of chemistry contains an ever-increasing number of contributions on this subject. It is

Increasing tendency to consider all properties in their relation to atomic weight.

shown how properties, not before considered from this point of view, also vary periodically as the atomic weight value of the elements to which they refer; it may almost be said that every new fact concerning the elements and their compounds is critically examined from the point of view of the periodic law; and the number of cases in which the law was applied to the solution of chemical problems increased rapidly.

"It is to Newlands, and especially to Mendeleeff, that we owe a new field of research and a new and powerful method of attacking chemical problems.... The principle proposed...will serve in the future and has done to some extent already, to indicate those directions in which research is most needed and in which there is most promise of interesting results.... It is and will be, in fact, for some time to come, the finger-post of chemical science."

"One of the chief objects of the chemist and physicist of the present day is to refer all the properties of the elements, both chemical and physical, to as few, what we may call *standard properties* as possible, till finally one standard property is obtained to which all the others may be referred in some way or other ; or in other words *we have finally to choose some standard property of which all the others are a function*, so that when we are able to explain this standard property we shall at the same time be able to arrive at the cause of the other properties, and thus be in a position *to predict* the nature and degree of the properties of any given unknown element, or any unknown properties of a known element of which the standard property has been determined numerically.... The tendency at present (and it is no doubt the right tendency) is to take the atomic weight as the ultimate standard, and refer all the other properties to it, for it is capable in most cases of very exact determination, as Stas's classical researches have shown, and for a given element is as far as we know absolutely invariable...."[1]

Carnelley, the author of the quotations just given, has himself done an enormous amount of work on the lines laid down above,

Carnelley's work on the relations between atomic weights and properties of the elements and their compounds.

correlating the available data, many of which had been determined by him for the special purpose. Mendeleeff in a controversial paper concerning priority in the discovery of the periodic law says:

"If anything has been done to further develop the doctrine of the periodic law, it is due to Carnelley, who has shown

[1] Carnelley, *Phil. Mag.*, London, (5), 7, 1879 (p. 305).

31—2

that the melting points and the magnetic properties of substances are a periodic function of their atomic weights."

In papers published by Carnelley in the *Chemical News* and the *Philosophical Magazine* between the years 1879 and 1885, he traces

"the more important relations between the atomic weights or chemical composition, and the various chemical and physical properties of the elements, and more especially of their compounds. In order that this connection might be rendered the more evident, I have limited myself in great part to those properties which are capable of being represented numerically, and consequently to what are generally termed the physical properties."

In 1845 it had been shown by Faraday that all matter was divisible into two classes: (1) paramagnetic substances (+), or those which are attracted by a magnet, and set axially when placed between its poles; (2) diamagnetic substances (−), or those which are repelled by a magnet, and set at right angles to the first position when placed in the magnetic field.

Atomic weight and atomic magnetism.

Carnelley, on examining Faraday's magnetic list, found what seemed a most remarkable confirmation of Mendeleeff's classification of the members of a series into odd and even.

"The following rule *invariably* holds good without a single exception in the case of the 38 elements to which it can be applied. Those elements belonging to the even series are always paramagnetic, whereas those elements belonging to odd series are always diamagnetic."

Extended knowledge of these constants has shown that though there is doubtless a connection between atomic weight and atomic magnetism, this is not of so simple and striking a nature as that enunciated by Carnelley.

"We have data for the classification of 63 elements, of which 37 are diamagnetic and 26 paramagnetic, whilst concerning 7 others we are still uncertain.... In the natural classification according to increasing atomic masses, we find 7 series of consecutively paramagnetic elements, the validity of the inclusion of the first[1] element of each of these series being still doubtful. Alternating with these, and preceding them, we have 7 series of consecutively diamagnetic elements, and it is possible that each of these may

---

[1] These series do not coincide with Mendeleeff's, and hence the "first" elements (Be, Mg, Se, Nb, La, Ta, Th) are not those of group 1 (Li, Na, K, etc.), but belong to other groups of the periodic law classification.

yet be found to have for its last member the doubtful element of the next paramagnetic series....Notwithstanding gaps and uncertainties and exceptions (oxygen, erbium, and ytterbium), we can detect in the magnetic, as in all the other physical and chemical properties, a periodic variation. The series of paramagnetic elements, which on the whole are also very refractory, are found on the descending branches, or in the minima of Lothar Meyer's atomic volume curve [p. 463] for the solid elements, whilst the more easily fusible diamagnetic elements occupy the ascending portions and the maxima of the curve." (H. du Bois, "Propriétés magnétiques de la matière pondérable," *Rapports au Congrès Internationale de Physique*, vol. 2, 1900, p. 460.)

The table, p. 486, gives in the columns marked C the data from which Carnelley generalised, in the columns marked DB the extended and corrected ones used by H. du Bois. The sign (+) signifies that the element to which it is attached is paramagnetic, (−) that it is diamagnetic, and (?) that its magnetic properties in the free state are unknown, or that the validity of the sign before which it is placed, thus: (? −) or (? +), is doubtful.

Mendeleeff had pointed out in his original paper that all those elements which are found in greatest quantity on the earth's surface have small atomic weights, viz. H, C, N, O, Na, Al, Fe, Ca, K, Cl, S, P, Si, Mg, which are all less than 60. Carnelley deals with the same problem from a somewhat different point of view, considering instead of *degree of distribution* the *form of occurrence.*

Atomic weight and form of natural occurrence of the elements.

" Elements belonging to the even series (except C, N, O, and group VIII) never occur in the free state in nature ; whereas elements belonging to odd series generally and sometimes frequently do so occur,"

e.g. Cu, Ag, Au, Hg, As, Sb, Bi, S, Se, Te.

This is of course closely connected with the property of reducibility.

" Elements belonging to the odd series are as a rule easily reducible to the free state, whilst those belonging to the even series are only reducible to the free state with difficulty."

" Elements belonging to odd series usually occur in nature as sulphides or double sulphides ; *i.e.*, in combination with a negative element belonging to an odd series, and only in very few cases as oxides ; whereas elements belonging to even series...usually occur as oxides or double oxides (forming silicates, carbonates, sulphates, aluminates, etc.), *i.e.*, in combination with a negative element belonging to an even series and never (with two exceptions) as sulphides."

The Paramagnetic and the Diamagnetic Elements.

Series	Group I	Group II	Group III	Group IV	Group V	Group VI	Group VII			
2	Li=7 ?	Be=9 ?+	B=11 −	C=12 +	N=14 +	O=16 +	F=19 ?			
3	Na=23 −	Mg=24 ?+	Al=27 ?	Si=28 −	P=31 +	S=32 +	Cl=35·5 −			
	K=39 +	Ca=40 ?	Sc=44 ?	Ti=48 +	V=51 +	Cr=52 +	Mn=55 ++	Fe=56 ++	Ni=59 ++	Co=59 ++
	Cu=64 −	Zn=65 −	Ga=70 ?	Ge=72 −	As=75 −	Se=79 −	Br=80 −			
	Rb=85 ?	Sr=88 ?	Y=89 −	Zr=91 ?	Nb=94 ?−	Mo=96 +		Ru=102 ++	Rh=103 ++	Pd=106 ++
	Ag=108 ?	Cd=112 ?	In=114 ?	Sn=118 −	Sb=120 ?−	Te=127 ?	I=126 −			
	Cs=133 ?	Ba=137 ?	La=138 ?−	Ce=140 ++						
			Yb=173 ?+		Ta=183 ?−	W=184 ?				
	Au=197 −	Hg=200 −	Tl=204 ?+	Pb=207 −	Bi=208 −			Os=191 ++	Ir=193 ++	Pt=195 ++
				Th=232 ?		U=240 +				

The following table contains data illustrative of these relations:

Occurrence of Elements belonging to Even Series:

As oxides or double oxides

Common : Li (lithia, mica, etc.), K (nitre, felspar, etc.), Rb, Cs, Be, Ca, Sr, Ba, B, Sc, Y, La, Yb, C, Ti, Zr, Ce, Th, V, Nb, Di, Ta, O, Cr, Tb, W, Mn.

Frequent : N (in nitre), Mo.

Rare or unknown: None.

The only two of these elements occurring as sulphides are:

Common : Mo.

Very rare: Mn (also O as SO_2 in volcanic gases).

Occurrence of Elements belonging to Odd Series:

As sulphides (selenides, tellurides)

Common : Cu, Ag, Zn, Co, Hg, Ga, In, Tl, Pb, Sb, S, Se, Te.

Frequent : As, Bi, Sn.

Unknown : Au (occurs only[1] in the free state in nature); Na, Mg, Al, Si, P.

As oxides

Common : Na, Mg, Al, Si, P, Sn.

Frequent : Zn, Cu.

Rare : Pb, Sb, Bi, As.

Mendeleeff's original generalisation had taken the form "The properties of the *elements* are a periodic function of their atomic weights," and it is to Carnelley mainly that we owe

Carnelley's extension of the periodic law to the physical properties of compounds.

the extension of the periodic law that the properties of the *compounds* of the elements are a periodic function of the atomic weights of their constituent elements. Among the properties so investigated were the colours of oxides, sulphides, halides, etc., the melting points, boiling points, specific gravities, and heats of formation of the alkyl compounds.

"…However we may arrange the melting points, boiling points, and heats of formation of the normal compounds of the elements, provided only that we arrange them systematically, we always find that certain

Melting points and boiling points of the halogen and alkyl compounds are periodic functions of constituent elements.

definite and regular relations may be traced between them…. …The physical properties of the alkyl compounds of the elements obey exactly the same rules as those of the corresponding halogen compounds."…"If in a series of binary normal compounds, one of the elements[2] be common to all, then the melting points, boiling points, and heats of formation are periodic functions of the atomic weight of the other element."

[1] *Only* is the term used by Carnelley; *chiefly* would be more correct.
[2] The alkyl radicles $(C_nH_{2n+1})'$, *e.g.* CH_3'', C_2H_5', C_3H_7', etc. may for this purpose be treated as elements.

An extremely small portion of Carnelley's extensive experimental data has been compressed into the following table:

	MX / K		MX$_2$ / Ca		MX$_3$ / Sc		MX$_4$ / Ti		MX$_3$ / V		MX$_2$ / Cr		MX / Mn	
	M.P.	B.P.	M.P.	B.P.	M.P.	B.P.	M.P.	B.P.	M.P.	B.P.	M.P.	B.P.	M.P.	B.P.
Chloride	734		719					130						
Bromide	703		676				72	230						
Iodide	634		631				150	360						
Methide														
Ethide														
Propide														

	Cu		Zn		Ga		Ge		As		Se		Br	
	M.P.	B.P.	M.P.	B.P.	M.P.	B.P.	M.P.	B.P.	M.P.	B.P.	M.P.	B.P.	M.P.	B.P.
Chloride	434		262		73	[220]	[86]			132				13
Bromide			394						22	220			−22	58
Iodide	601		446				[350–360]		146	404	[60–70]		36	117
Methide			319				·			373		331		277
Ethide			391							432		381		312
Propide			432											344

	Rb		Sr		Y		Zr		Nb		Mo	
	M.P.	B.P.	M.P.	B.P.	M.P.	B.P.	M.P.	B.P.	M.P.	B.P.	M.P.	B.P.
Chloride	710		825									
Bromide	683		630									
Iodide	642		507									
Methide												
Ethide												
Propide												

	Ag		Cd		In		Sn		Sb		Te		I	
	M.P.	B.P.	M.P.	B.P.	M.P.	B.P.	M.P.	B.P.	M.P.	B.P.	M.P.	B.P.	M.P.	B.P.
Chloride	451		541		[440]			115	72		[209]		25	100
Bromide	427		571				30	201	90		[280]		309	117
Iodide	527		404				156	295	155				110	200
Methide	·							351		354		355		317
Ethide								454		432		373		345
Propide								498						375

3248 cases referring to the halides, of which only 180 or about 5·5 per cent. are exceptions, and 942 data referring to the alkyl compounds, of which only 54 or about 5·7 per cent. are exceptions,

Relations
found between
atomic weights
and melting
points, etc.
supplied the material for the establishment of a number of relations (twelve for the halides and eight for the alkyl compounds) of the type of the following:

"The difference between the melting or boiling points or heats of formation of the bromide and chloride of an element, is less than that between those of its iodide and bromide."

"The differences between the melting points or boiling points or heats of formation diminish as the atomic weight of the positive element increases, except in the case of the melting points of the even members of the second group, for which the opposite relation holds good."

"The boiling point increases, and the specific gravity diminishes, as we pass from the methide to the ethide, and thence to the propide, butide, etc."

"The boiling points and specific gravities of the alkyl compounds of any one group increase as the atomic weight of the positive element increases."

Carnelley's object in these investigations was a twofold one:

(1) To illustrate the truth of the periodic law.

Deductive
application of
periodic law
relations:
1. Carnelley's
calculations of
melting points.

(2) To apply the facts thus obtained deductively.

"It will be readily seen that the relations [established] may be made the basis of a method for calculating (within certain limits) melting and boiling points which have not been *experimentally* determined."

From among the numbers given in illustration of the results so obtained, may be quoted:

Calculated in 1877.	Found.	Authority and date.
Cu_2Br_2 M.P. 774–790	777	Carnelley and Williams 1880
$AsCl_3$ M.P. 244–245	Below 244	Henry 1879
$GaCl_3$ B.P. 496–499	487–492	Boisbaudran 1881.

"There has recently been some dispute as to the atomic weight of Be. From his determination of its specific heat, Emerson Reynolds concludes that its atomic weight is 9·2, and that it is a dyad; whilst Nilson and Pettersson, also Humpidge, from their determination of its specific heat, give to it the atomic weight 13·8, in which case it would be a triad."...."If it be a dyad, the melting points of its halogen compounds must be considerably higher than those of boron, whereas if it be trivalent, these melting points must be comparatively low....The normal chlorides of all triad elements melt below 510 ($BiCl_3$, M.P. = 503, being the highest known)....Now according to calculation, if it has the atomic weight 9·2, the melting point

2. Melting
point of beryl-
lium chloride
agrees with
that calculated
for $BeCl_2$.

of beryllium chloride...ought to be 820–870,...and although this number is somewhat uncertain, yet it is sufficient to determine whether beryllium is really a dyad or a triad.... To throw light on this point, I determined as carefully as possible the melting point of the chloride...of beryllium, and found that the chloride fuses between the melting points of $Ag_4P_2O_7$ (850) and $NaPO_3$ (890), thus agreeing with the number 820–870 calculated for $BeCl_2$ ($Be = 9\cdot2$)."

These results therefore confirm the view that beryllium is a dyad, with atomic weight $9\cdot2$.

Mendeleeff has applied a periodic relation, by him discovered, between the molecular weight of chlorides and the density of their dilute solutions, to the determination of the atomic weight of the metal contained in the chlorides.

"...The density of such solutions of chlorides of metals MCl_n as contain... a large and constant amount of water, regularly increases as the molecular weight of the dissolved salt increases....If the molecule of beryllium chloride be $BeCl_2$, its weight must be 80, and in such a case it must be heavier than the molecule of $KCl(= 74\cdot5)$ and lighter than that of $MgCl_2$ ($=95$). On the contrary, if beryllium chloride is a trichloride $BeCl_3$ ($=120$), its molecule must be heavier than that of $CaCl_2$ ($=111$), and lighter than that of $MnCl_2$ ($=126$). Experiment has shown the correctness of the former formula.

3. Density of aqueous solution of beryllium chloride agrees with that calculated for $BeCl_2$.

Density of $BeCl_2 + 200\ H_2O = 1\cdot0138$
"   " $KCl + 200\ H_2O = 1\cdot0121$
"   " $MgCl_2 + 200\ H_2O = 1\cdot0203$."

(Faraday Lecture, *J. Chem. Soc.*, 1889)

In the above, as well as in similar examples given before (p. 477), the problem for solution was the determination of the valency to be assigned to an element whose equivalent was known with fair accuracy. Comparatively rough values for the physical property measured sufficed, giving, as in the heat-capacity method, a value for the atomic weight which, even if removed by several per cent. from the true value, shows whether the most accurate value that can be deduced from the data available is 1, or 2, or 3, etc. times the equivalent weight.

The following are, on the other hand, examples of how the periodic law has been applied to the determination of *accurate* atomic weight values. In such cases, the chemical properties and analogies of the element considered are so definite that its place in the periodic system, the special group and series to which it belongs, is not a matter of doubt. Then if it so happens that the amount of material available is very small and that the difficulties of separating and purifying the compounds

4. Determination of "accurate" atomic weight values. (i) Atomic weight of germanium.

used for analysis are very great, the value obtained for the equivalent will be uncertain, and recourse may be had to a purely physical method. This is what Lecoq de Boisbaudran did in his determinations of the atomic weights of gallium and of germanium[1].

The law which he applied to the calculation of these values was :

Law applied by Lecoq de Boisbaudran to the calculation of atomic weights from relation of wave-lengths of homologous rays. "In the different natural families the *variation in the increase* in atomic weight is proportional to the *variation in the increase* of the wave-lengths of homologous rays or groups of rays."

The *first* evaluation of the atomic weight of gallium was thus obtained from a comparison of the spectra of Al, Ga, and In on the one hand, and K, Cs, Rb on the other, with the result that :

$$Ga = \textbf{69·86}.$$

Lecoq de Boisbaudran was able to compare this number with those subsequently found by himself by the application of chemical methods, viz. (i) the calcination of the ammonia alum[2], (ii) the calcination of the nitrate obtained from a known weight of the metal[3], processes of which he says :

"The loss of small quantities of matter which occurs in these two operations affects the value of the equivalent weight in opposite directions, and has therefore little influence on the mean value[4]."

The results were :   From (i), Ga = 70·032

From (ii), Ga = 69·698

Mean   Ga = **69·865**.

The spectrum of germanium was investigated by Lecoq de Boisbaudran from the same point of view soon after the discovery of the element, and at a time when a preliminary determination of its atomic weight by Winkler had given 72·75, a number appreciably higher than 72, the value predicted by Mendeleeff.

[1] *Chem. News*, London, 54, 1886 (p. 4).

[2] *Ante*, p. 425.

[3] ·4481 grams of gallium converted into nitrate and ignited gave ·6024 grams of Ga_2O_3.

[4] Calculated by Lecoq de Boisbaudran with the then accepted atomic weight values as antecedent data (H = 1·00, O = 16·00, etc.). Clarke's recalculation in 1896 gave

(i) From the calcination of the alum : if H = 1, Ga = 69·595; if O = 16, Ga = 70·125

(ii) From the calcination of the nitrate: if H = 1, Ga = 69·171; if O = 16, Ga = 69·698

Mean: if H = 1, Ga = **69·383**; if O = 16, Ga = 69·912

Taking it as proved that, from its chemical analogies, germanium should stand in the periodic scheme between silicon and tin, the two groups of atom analogues to be considered are:

	Group III.	Group IV.
Series 3	Aluminium	Silicon
„ 5	Gallium	Germanium
„ 7	Indium	Tin.

Lecoq de Boisbaudran selected for comparison in the spark spectra of these 6 elements the prominent group of rays which consists of a blue and a violet line, and used in his calculations the mean of the wave-lengths of these two rays.

	1st ray	2nd ray	Mean
Si	$\lambda = 412\cdot89$	$\lambda = 389\cdot0$	$401\cdot0$
Ge	$\lambda = 468\cdot04$	$\lambda = 422\cdot6$	$445\cdot3$
Sn	$\lambda = 563\cdot00$	$\lambda = 452\cdot4$	$507\cdot7$.

The whole of the data involved are thus tabulated by him:

Atomic Weights			*Mean of Wave-lengths (of 2 rays)*		
	Differences	Variation		Differences	Variation
Si $= 28$			Si $= 401\cdot0$		
	$90\cdot0$			$44\cdot3$	
Ge $= ?$	between		Ge $= 445\cdot3$		$\dfrac{40\cdot86}{100}$
	Si and Sn			$62\cdot4$	
Sn $= 118\cdot0$			Sn $= 507\cdot7$		
Al $= 27$			Al $= 395\cdot2$		
	$42\cdot4$			$14\cdot9$	
Ga $= 69\cdot9$		$\dfrac{2\cdot8302}{100}$	Ga $= 410\cdot1$		$\dfrac{37\cdot584}{100}$
	$43\cdot6$			$20\cdot5$	
In $= 113\cdot5$			In $= 430\cdot6$		

The details of the method followed in calculating the atomic weight of germanium from these data will show the meaning of the law assumed by Lecoq de Boisbaudran and which has been quoted above.

The wave-lengths of homologous rays for Al, Ga, and In are

$$395\cdot2, \quad 410\cdot1, \quad 430\cdot4;$$

the successive differences are

$$14\cdot9 \text{ and } 20\cdot3;$$

the second difference exceeds the first by

$$5\cdot4\ ;$$

this increase, expressed as a fraction of the first difference, is

$$\frac{5\cdot4}{14\cdot9}\ \text{or}\ 0\cdot3758 \dots\dots\dots\dots\dots(1),$$

which number is what Lecoq de Boisbaudran calls the *variation in the increase*.

Similarly the atomic weights for Al, Ga, and In are

$$27\cdot5,\ 69\cdot9\ \text{and}\ 113\cdot5\ ;$$

the differences are

$$42\cdot4\ \text{and}\ 43\cdot6\ ;$$

and the variation in the increase is

$$\cdot028302\dots\dots\dots\dots\dots\dots(2).$$

The law asserts that the ratio of (1) and (2) will be the same as the corresponding ratio in another family.

But for Si, Ge, Sn we have as homologous wave-lengths

$$401,\ 445\cdot3,\ \text{and}\ 507\cdot7,$$

hence the variation will be

$$\cdot4086 \dots\dots\dots\dots\dots\dots(3).$$

Now if we put X for the unknown atomic weight of germanium, the atomic weights are

$$28,\ X,\ 118\ ;$$

the differences are

$$X-28\ \text{and}\ 118-X,$$

and the variation is

$$\frac{(118-X)-(X-28)}{X-28}\ \dots\dots\dots\dots(4).$$

Hence by the law we have the following proportion to find X,

$$\cdot3758 : \cdot028302 = \cdot4086 : \frac{146-2X}{X-28}$$

which makes X, the atomic weight of germanium, $= \mathbf{72\cdot32}$.

But as Lecoq de Boisbaudran himself pointed out:

"The atomic weights deduced by calculation from the wave-lengths depend of course...on values of wave-lengths and of atomic weights, which may be affected by slight errors, and it is therefore clear that corrections applied to the numbers which have served as basis in the calculations would involve corrections in the values deduced from them."

A comparison of the atomic weight values used by Lecoq de

Boisbaudran with those given by Clarke ten years later reveals differences too considerable to be classed as "slight errors."

		L. de B. 1886	Clarke 1896 $O = 16$
Al	=	27·5	27·11
Ga	=	69·9	69·9
In	=	113·5	113·85
Si	=	28	28·40
Sn	=	118	119·05

The substitution of the modern values in the above calculation gives

$$Ge = \mathbf{73 \cdot 06},$$

showing that even if the law on which the method is based should be exact, its application would not yield reliable results unless the antecedent data were known to a greater degree of accuracy than is the case in the above instance. Lecoq de Boisbaudran at the end of his paper had said:

"It will now be interesting to see whether Winkler will confirm his atomic weight number 72·75, or whether when using very pure material he will obtain a value of about 72·3."

The number that Winkler did get, and which in the absence of any other determination serves as the most *probable value*, is

$$Ge = \mathbf{72 \cdot 44},$$

the following being the experimental investigations involved:

(a) In two[1] experiments, a total of ·5127 grams of germanium chloride was, by indirect titration with standard silver solution, found to contain ·339326 grams of chlorine,

The atomic weight of germanium determined by vapour density and specific heat method.

$$\therefore (X + 35 \cdot 45) : 35 \cdot 45 = \cdot5127 : \cdot339326,$$

and X, the equivalent weight of germanium, $= 18 \cdot 11$.

Atomic weight germanium $= n \times 18 \cdot 11$.

(b) Vapour density (air $= 1$) of the chloride between 301·5° and 739° $= 7 \cdot 44$,

$$\therefore \text{molecular weight} = 7 \cdot 44 \times 28 \cdot 87 = 214 \cdot 7.$$

Since ·5127 grams of the chloride contain ·339326 of chlorine,

$$\therefore \quad 214 \cdot 7 \quad \text{,,} \quad \text{,,} \quad \text{,,} \quad \text{,,} \quad 142 \cdot 1 \text{ of chlorine}$$
and 72·6 of germanium,
$$= 4 \cdot 01 \times 35 \cdot 45 \text{ of chlorine and } 4 \cdot 01 \times 18 \cdot 11 \text{ of germanium,}$$
$$\therefore \quad n = 4,$$

and atomic weight germanium $= 4 \times 18 \cdot 11$

$$= 72 \cdot 44.$$

[1] Winkler made four determinations, but the numbers for the first two given in his papers, and reproduced in references thereto, are obviously disfigured by some misprint.

(b') The value ·0757 found by Nilson and Pettersson for the heat capacity of germanium is in good agreement with this number, giving for the atomic heat ·0757 × 72·44 = 5·47, the chemical analogies between germanium and silicon (p. 480) making a low value probable.

Another example, similar to the above, of the deductive application of the periodic law is found in Runge and Precht's[1]
calculation of the atomic weight of radium from the **(ii) Atomic weight of radium.** data supplied by the atomic weights of similar elements, and the wave-lengths of analogous lines in the emission spectra. The analogy of the line-pairs chosen for comparison was proved by the fact that the resolution in a magnetic field is the same (*post*, p. 607).

"...the strongest lines of the spark spectrum of radium are exactly analogous to the strongest lines of barium, and the corresponding lines of the related elements Mg, Ca, Sr....These lines may be grouped **Radium classed on spectroscopic evidence with Mg, Ca, Sr, Ba.** into three pairs, called...on account of certain analogies with the spectra of the alkalies, the line-pair of the primary series, the line-pair of the first and that of the second secondary series[2]. Measured on the scale of frequency[3], the two lines of each of the three pairs are the same distance apart for any one element.... This distance varies, on the other hand, from one element to another, increasing in a perfectly regular manner with the atomic weight.... This also holds for radium (*infra*), so that radium is to be classed along with Mg, Ca, Sr, and Ba in a group of chemically allied elements,—a conclusion which is supported by the chemical behaviour of radium, in so far as this is known."

	λ	$\dfrac{10^8}{\lambda}$	*Difference*
Primary series	4682·35	21356·8	4858·3
I	3814·59	26215·1	
1st secondary series	4436·45	22540·5	4858·5
II (i)	3649·77	27399·0	
2nd secondary series	5813·9	17200·2	4858·6
II (ii)	4533·33	22058·8	

From one element to another, the distance apart of the lines increases with the atomic weight:

[1] *Phil. Mag.*, London, (6), 5, 1903 (P. 476).
[2] The lines referred to are those marked by I, II (i), II (ii) in the table.
[3] Frequency = number of vibrations in equal times, inversely proportional to the wave-length λ, expressed in the above comparison by $\dfrac{10^8}{\lambda}$.

	Atomic Weight	Distance
Mg	24·36	91·7
Ca	40·1	223
Sr	87·6	801
Ba	137·4	1691
Ra	?	4858·6

" It is suggested to regard the atomic weight as a function of the distance between the lines, and to extrapolate this function for radium....

Law applied by Precht to the calculation of atomic weights from distances between line-pairs.

" In other groups in which line-pairs have been observed, the relation between the width of the line-pairs and the atomic weight is capable of being expressed by the simple formula :

The logarithms of the atomic weights and those of the distances, when plotted as coordinates, lie on a straight line for a chemically related group of elements.

" The following figure illustrates this law :

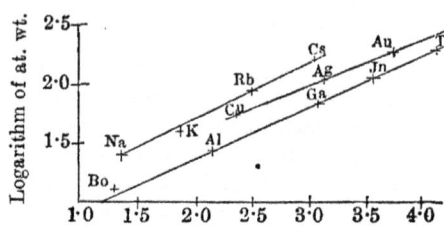

Logarithm of the line-distance
(·01 scale-division = 2·3 per cent. of the value).

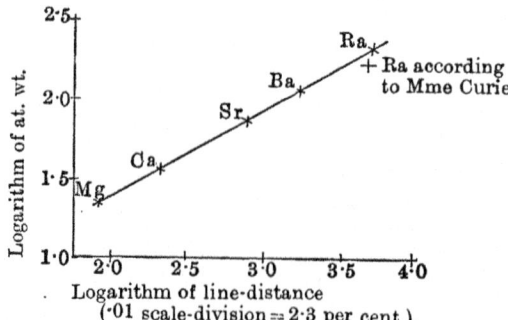

Logarithm of line-distance
(·01 scale-division = 2·3 per cent.).

" ...Extrapolation gives for the atomic weight of radium the value 258,....
The figure shows clearly that the value 225 determined by Madame Curie[1]

---

[1] Her method of determining the atomic weight of radium has been thus described by Madame Curie : " I re-purified the chloride, and obtained a... substance, in the spectrum of which the two strongest barium lines were very faint. Given the sensitiveness of the spectrum reaction of barium, [it has been] estimated

is considerably removed from the straight line....It must be remarked that in view of the close relationship of barium and radium, and the small quantities with which the chemist is forced to work, the complete separation of these two bodies is very difficult, and that if their separation had been incomplete, Madame Curie would have found too small a value for the atomic weight.

Discrepancy between atomic weight of radium so calculated and that obtained by chemical method.

The number 225 is in better correspondence with the periodic system, in so far as it fits the gap between bismuth and thorium in the proper column. According to the value 258, radium would have to be moved two rows further down in the column Mg, Ca, Sr, Ba, and a number of new unoccupied places would be created in the periodic system....On the other hand, Rutherford's remark may be adduced in support of the higher value of the atomic weight. The higher atomic weight is indicative of a more complicated atomic structure, and therefore of an easier splitting up into electrons[1]. The element which gives off electrons most freely should therefore also have the highest atomic weight."

It has been shown how the validity of the periodic law classification was brilliantly established by the discovery of elements whose properties closely agreed with those predicted for them. But many gaps still remain to be filled, and the elements recently discovered have not always been available for this purpose. Though it has been possible to house these fairly satisfactorily within the edifice of the system, it must be recognised

Gaps in the periodic law table. New group made for the inert gases, argon, etc.

that the periodic law had afforded no indication of the existence of the inert gases of the atmosphere[2], or of the radio-active elements,—two classes of substances so intensely interesting because of the fundamental differences between them and all other types of matter.

Helium (H = 4), neon (Ne = 20), argon (A = 39·9), and krypton

that the purified chloride contained only the merest traces of barium, incapable of influencing the atomic weight to an appreciable extent. I made three determinations with this perfectly pure radium chloride. The results were as follows:—

	Anhydrous radium chloride	Silver chloride	M
I	0·09192	0·08890	225·3
II	0·08936	0·08627	225·8
III	0·08839	0·08589	224·0

The mean of these numbers is 225. They were calculated...by considering radium as a bivalent element, the chloride having the formula $RaCl_2$, and taking for silver and chlorine the values $Ag = 107·8$, $Cl = 35·4$."

1 The ultimate constituents of the chemical atoms, which are alike for all substances, and the combination of which, in various large numbers,—between 1000 for hydrogen and ¼ million for radium—form the different kinds of atoms.

2 Dr Johnstone Stoney's diagram, p. 502, excepted.

(Kr = 81·8), which are all monatomic[1] gases characterised by absolute chemical inactivity, have been placed by Sir W. Ramsay in a separate group, designated by the ordinal 0. Whilst the addition of this group has tended to add greatly to the symmetry of the periodic system, providing in these electrically neutral elements, whose valency is apparently zero, a suitable transition from the strongly electro-negative monovalent halogens to the strongly electro-positive monovalent alkalis, the atomic weight value of argon raises the same difficulty as that of tellurium.

[1] Atomicity (*post*, chap. xvii) = number of atoms contained in *one* elementary molecule = $\dfrac{\text{molecular weight}}{\text{atomic weight}}$. The molecular weight is found from the gaseous density, and according to the relations developed on page 349, it is 22·3 × normal density. The atomic weight of these gaseous elements, which are liquefied at very low temperatures only, which form no compounds and therefore have no combining weight, cannot be determined from the vapour density of compounds, nor from the heat capacity of the solid elements, nor from isomorphous substitution. The course followed is the reverse of the usual one : instead of finding the atomicity from the molecular and the atomic weight, the atomic weight is deduced from the molecular weight and the atomicity, which latter value is determined directly by a physical method, namely the measurement of $\dfrac{C_p}{C_v}$, the ratio between the specific heat at constant pressure and the specific heat at constant volume.

In heating a gas, energy is required :

(i) To increase the kinetic energy of the molecules, producing rise of temperature. Call this quantity a. If the explanation of the gaseous laws given by the kinetic theory be correct, then this quantity a is the same for equal volumes of different gases.

(ii) To do external work in overcoming the pressure of the atmosphere when the substance heated is allowed to expand. The coefficient of expansion of all gases being the same, this quantity is a constant for equal volumes, or for equimolecular weights. Let it be called b.

(iii) To produce changes within the molecule, in the relations of the constituent atoms. Call this quantity c.

Hence :

$$\frac{C_p}{C_v} = \frac{a+b+c}{a+c},$$

and the value of this ratio will be the greater, the less the value of c, and will reach a maximum when $c=0$.

From the kinetic theory of gases it has been calculated that when $c=0$, that is, in the case of a monatomic gas, $\dfrac{C_p}{C_v} = \dfrac{5}{3} = 1\cdot66$, and hence that for any gases whose molecules are complex, $\dfrac{C_p}{C_v} < 1\cdot66$, and that it is the smaller the greater the complexity. These theoretical results are borne out by the experimental data for gases whose molecular complexity has been otherwise determined.

Formula :	Hg	H_2	N_2	O_2	CO	CO_2
$\dfrac{C_p}{C_v}$:	1·66	1·41	1·41	1·40	1·41	1·26.

For the gases helium, argon, etc. the ratio of the specific heats has been found = 1·66, and it is therefore concluded that they are monatomic, that the atomic weight is equal to the molecular weight.

According to the principle of arrangement in the order of increasing atomic weights, argon (A = 39·9) comes between potassium (K = 39·15) and calcium (Ca = 40·1), whilst according to properties its place is before and not after potassium. So far it has not been possible to account for this discrepancy by admixture of another inert gas of greater density; fractionation, by diffusion of the gaseous and distillation of the liquefied argon, has not resulted in any separation, and the evidence of the simple elementary character of the inert gas whose atomic weight is 39·9 is of the strongest. For the present therefore argon must be looked upon as an undoubted exception to the principle of the periodic system ; but considering the many and important facts on which the generalisation named " periodic law " is based, and the many ways by which its validity has been confirmed deductively, the natural tendency is to expect from future research a solution of this difficulty, impossible though it is to foretell even the form in which this will come.

Atomic weight of argon forms exception to periodic law.

If we assume that the atomic weight, as in the case of O, N, etc., is identical with the density, there is no place for these elements in the periodic system,...but if we assume that these gases are monatomic, they form a group by themselves. Their atomic weights would then be :

Hydrogen	Helium	Lithium	Beryllium
1	4	7	9
Fluorine	Neon	Sodium	Magnesium
18	20	23	24
Chlorine	Argon	Potassium	Calcium
35·5	40	39	40
Bromine	Krypton	Rubidium	Strontium
80	82	85	87
Iodine	Xenon	Caesium	Barium
127	128	133	137

Since these elements are devoid of all chemical properties, the data which show the periodic relation between atomic weight and properties refer to physical constants only.

" These elements exhibit gradations in properties such as refractive index, atomic volume, melting point and boiling point, which find a fitting place on diagrams showing such periodic relations....Thus the refractive equivalents are found at the lower apices of the descending curves ; the atomic volumes on the ascending branches in appropriate positions ; and the melting and boiling points, like the refractivities, occupy positions at the lower apices.

Although, however, such regularity is to be noticed, similar to that which is found with other elements, we had entertained hopes that the simple nature of the molecules of the inactive gases might have thrown light on the puzzling incongruities of the periodic table....That hope has been disappointed. We have not been able to predict *accurately* any one of the properties of one of these gases from a knowledge of those of the others ; an approximate guess is all that can be made. The conundrum of the periodic table has yet to be solved." (Ramsay and Travers, *Argon and its Companions*, 1900.)

Investigation of properties of inert gases has shown impos- sibility of " accurate pre- diction."

The special feature of the conundrum thus referred to by Prof. Ramsay is how to find the formula for the function which would correlate the numerics of the atomic weight with the properties susceptible of quantitative measurement. Another aspect of it is that of the expression of the atomic weights themselves by means of a general algebraic formula. This problem is an attractive one, and several attempts have been made to solve it, in spite of the fact that the only indications of the direction in which success may be expected are of negative nature. This aspect of the question has been put clearly and forcibly by Mendeleeff himself in the Faraday lecture of 1889 :

Formula to represent the periodic law function.

" The periods do not contain the infinite number of points constituting the curve, but a *finite* number only of such points. An example will better illustrate this view. The atomic weights :

$$Ag = 108$$
$$Cd = 112$$
$$In = 113$$
$$Sn = 118$$
$$Sb = 120$$
$$Te = 125$$
$$I = 127$$

steadily increase, and their increase is accompanied by modification of many properties which constitute the essence of the periodic law. Thus, for example, the densities of the above elements decrease steadily, being respectively 10·5, 8·6, 7·4, 7·2, 6·7, 6·4, 4·9, while their oxides contain an increasing quantity of oxygen :

$$Ag_2O, \; Cd_2O_2, \; In_2O_3, \; Sn_2O_4, \; Sb_2O_5, \; Te_2O_6, \; I_2O_7.$$

But to connect by a curve the summits of the ordinates expressing any of these properties would involve the rejection of Dalton's law of multiple

proportions. Not only are there no intermediate elements between silver,
which gives AgCl, and cadmium, which gives $CdCl_2$, but

Mendeleeff on the special character of the periodic law curve.

according to the very essence of the periodic law, there can
be none; in fact, a uniform curve would be inapplicable in
such a case, as it would lead us to expect elements possessed
of special properties at any point of the curve. The periods
of the elements have thus a character very different from those which are so
simply represented by geometers. They correspond to points, to numbers,
to sudden changes of the masses, and not to a continuous evolution. In
these sudden changes destitute of intermediate steps or positions, in the
absence of elements intermediate between, say, silver and cadmium, or
aluminium and silicon, we must recognise a problem to which no direct
application of the analysis of the infinitely small can be made. Therefore
neither the trigonometrical functions proposed by Ridberg and Flavitsky,
nor the pendulum oscillations suggested by Crookes, nor the cubical curve of
the Rev. Mr Haughton, which have been proposed for expressing the periodic
law, from the nature of the case, can represent the periods of the chemical
elements. If geometrical analysis is to be applied to this subject, it will
require to be modified in a special manner." (*J. Chem. Soc.*, 1889.)

The following may serve as examples of the formulae[1] that
have been proposed to represent the periodic law function:

Dr Mills, who looks upon the elements as the result of the
polymerisation of the primary matter as it cooled, gives

$$y = 15p - 15 \, (\cdot 9375)^x,$$

where $\quad y =$ atomic weight,

$p =$ group number of element considered,

$x =$ an integer different for the different elements.

Carnelley gives

$$A = c \, (m + \sqrt{v}),$$

where $\quad A =$ atomic weight,

$m =$ member of an arithmetical progression depending
on series to which element belongs,

$v =$ number of group of which element is a member,

$c =$ a constant.

Excluding the elements of the first three series, the formula
becomes

$$A = c \, (3 \cdot 5a - 9 + \sqrt{v}),$$

[1] For a complete account of such formulae see: F. P. Venable, *loc. cit.* chaps.
v and vi; G. Rudorf, *loc. cit.*, chap. iii, pp. 60—73, and chap. vi.

where a = number of series to which element belongs,

   c = a constant approximately equal to 6·64, very nearly
   equal to the atomic heat constant.   ⸱

Dr Johnstone Stoney represents the atomic weights, not by lines, but by the volumes of concentric spheres whose radii are the cube roots of the atomic weights. He divides the circles into 16 equal arcs, on which he places chemically similar elements. The connection of all the points so obtained gives a curve (see diagram), which approximates to a logarithmic spiral. In the formula

$$y = k \log (ma),$$
$$\log k = 0\text{·}785, \quad \log a = 1\text{·}986.$$

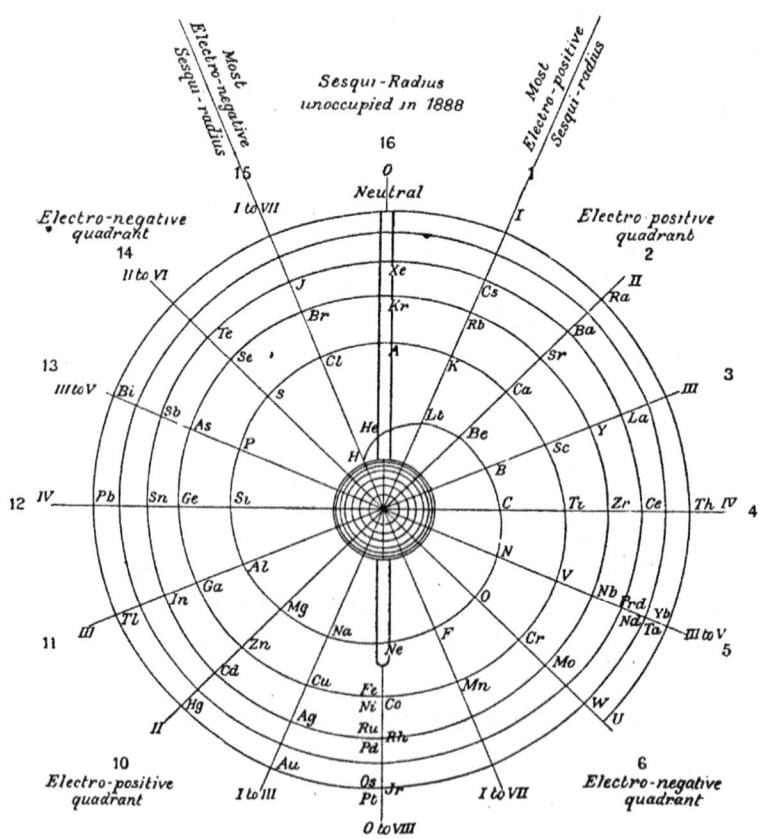

" The curve of the diagram is [a] Logarithmic Spiral.

The cardinal feature of the diagram is that it represents atomic weights by volumes, not by lines.

The *volume* of the globe in the centre represents the atomic weight of hydrogen.

The atomic weights of the other elements are represented by the *volumes* of concentric spheres extending out to the points intended to be indicated by the symbols of the elements.

These points lie along the radii, and so near to their intersections with the logarithmic spiral that [their] deviations are too small to be conveniently represented on the diagram.

The quadrants are alternately electro-positive and electro-negative.

The transition between these states is gradual elsewhere, but *becomes abrupt between sesqui-radius* 15 *and sesqui-radius* 1.

Between these lies the vacant sesqui-radius 16. (By a sesqui-radius is meant a radius along with the inner part of the opposite radius.)

It [has been] proved that the unoccupied sesqui-radius is not arbitrarily introduced into the diagram, but has a real existence in nature.

The natural chemical groupings of the elements come out with conspicuous distinctness, *e.g.* F, Cl, Br, I, on sesqui-radius 15." (G. Johnstone Stoney, 1888.)

"[A] fact that came to light [in 1888] was that in order that it may be possible to represent the atomic weights of the elements by [the above depicted logarithmic spiral], it is essential that we add to the Mendeleeff series, as it was known in 1888, the places indicated on sesqui-radius 16 of the accompanying diagram. The necessity for this addition made it certain that these places have a real existence in nature, although at the time no elements were known that occupied them. The anticipation that this was so has been in the most satisfactory way justified by the discovery in recent years of argon by Lord Rayleigh and Sir William Ramsay, followed by the discovery of four other elementary gases,—helium, neon, krypton, and xenon— by Sir William Ramsay. These new elements occupy the five places which had several years earlier been proved to be a necessary part of the Mendeleeff series. Moreover, there can be little doubt that the very unusual chemical behaviour of these new elements is a consequence of their occupying a position which has sandwiched them between the elements on sesqui-radius 15 (fluorine, chlorine, bromine, and iodine) in which the electro-negative condition rises to its highest intensity, and the elements on sesqui-radius 1 (lithium, sodium, potassium, rubidium, and caesium) which are the most electro-positive known." (G. Johnstone Stoney, *Phil. Mag.* (6), 4, 1902, p. 411.)

Mendeleeff's opinion, expressed in 1889, of the inadequacy of such-like functions, at the present day still represents the consensus on this subject, and it is therefore not necessary to quote for each of them the criticism which has led to its rejection.

The periodic classification has been singularly exempt from

criticism, unfair or even severe. A striking contrast to this

Adverse criti-
cism of peri-
odic law.
Wyruboff.

acceptance—and it is only as such that it attracts notice—has been supplied by Dr Wyruboff[1]. This critic would concede some merit to Mendeleeff's work if it had remained

" a very interesting and highly ingenious table of the analogies and the dissimilarities of the simple bodies—a mere catalogue raisonné of the elements."

Mendeleeff however converted his classification into the " periodic system," nay, into the " periodic law." But a law must be demonstrably free from exceptions, as put by Mendeleeff himself.

"The laws of Nature admit no exception ; therefore the periodic law must be considered as a law of Nature definitely established, which must be accepted or rejected as a whole."

Existence of
exceptions to
principle of
periodic system
which there-
fore should not
be termed
"law."

Dr Wyruboff proceeds to show that when tried by this test, the periodic law is found wanting. His main attack is directed against : 1. The manner in which periodicity of valency is found, 2. The arbitrary selection of atomic weight values.

1. He points out the arbitrariness and the want of a definite guiding principle in the selection, from amongst the various oxides that an element forms, of the one which is taken as typical, and which has the value required by the periodicity in the series from R_2O for group I. to R_2O_7 for group VII. (*ante*, p. 471). This oxide is sometimes the lowest (BaO and not BaO_2), sometimes the highest (Mn_2O_7 in the series MnO, Mn_2O_3, Mn_3O_4, MnO_2, MnO_3); it may be the most stable (BaO more stable than BaO_2), or the least so (Cu_2O less stable than CuO);

2. He makes the accusation that in order to get elements into the group to which they belong by their properties, Mendeleeff does not hesitate to alter the atomic weights at will, irrespective of experimental evidence. The case of La, Ce, and Di is adduced, but tellurium of course offers the best point for attack[2] (*ante*, p. 466).

[1] "The Periodic Classification of the Elements," *Chem. News*, London, 74, 1896 (p. 30).

[2] Argon (atomic weight 39·9) which must be placed so as to precede potassium (atomic weight 39·15) presents a difficulty even greater than does tellurium.

"Professor Brauner, who has made a speciality of the art of causing reluctant elements to enter into the classification of Mendeleeff,...lays before us two alternatives : we must either reject the periodic law, or reject the figure 128, which all authorities have found for the atomic weight of tellurium. But he continues, as the periodic law cannot be rejected, since it is the truth itself, the value 128 must be inaccurate.

He has therefore submitted tellurium to all the tortures which a substance can undergo. He has melted it, sublimed it, oxidised it, hydrogenised it, dissolved it, and precipitated it, and finally arrived at the result which everybody had reached before him, that the atomic weight varies between the wide limits of 125 and 129.

Hence he concludes that we have here a complex body composed of two elements of very different atomic weights. What are these weights, and what are the distinctive properties of tellurium a and tellurium β he does not tell us, for he has not been able to separate them. Still, he takes the number 125 as representing true tellurium, and Prof. Mendeleeff introduces this number in his table, though it has not yet been confirmed by any one[1]."

Dr Wyruboff concludes his attack on the periodic law classification by saying he does not pretend to believe that he will convince anyone. This is a perfectly correct estimate of the situation; chemists, with a degree of unanimity not displayed towards any other of the theses of the science, would give their adhesion to what has been said by Winkler, the discoverer of germanium, Mendeleeff's ekasilicon.

"Investigation of the properties of germanium becomes actually a touchstone for human ingenuity. It would be impossible to imagine a more striking proof of the doctrine of the periodicity of the elements than that afforded by this embodiment of the hitherto hypothetical 'ekasilicon'; this is in truth more than a mere verification of a daring hypothesis, it represents an enormous extension of the chemist's field of view, a mighty stride into the realm of cognition."

Winkler's appreciation of the periodic law.

[1] *Ante*, p. 225.

CHAPTER XVII.

KEKULÉ AND THE DOCTRINE OF VALENCY.

In the chapters dealing with the determination of combining weights and atomic weights an attempt has been made to show that in either case, whether the notation takes for its unit the empirical combining weight or the hypothetical atomic weight, the final criterion for the suitability of the value chosen for representation by the symbol, is the chemical correctness and adequacy of the formulae for the compounds into the composition of which the element enters. And it was said that chemically correct and adequate formulae are such as represent in the simplest possible manner the chemical nature of the compounds, their reactions of formation and decomposition, and their relations to analogous bodies. Berzelius was the first to apply this principle extensively and consistently to the large number of inorganic compounds investigated by him, and in his choice of atomic weights the chief consideration guiding him was how to obtain formulae for the compounds which would fulfil the above requirements. Mention has been made (p. 337) of his indignant repudiation of Dumas' attempt to treble the atomic weight of sulphur, to double that of phosphorus and to halve that of mercury, whereby the formulae H_6S for hydrogen sulphide, H_6P for phosphine, Hg_4O and Hg_2O for the oxides of mercury ceased to show the existing analogies with H_2O, NH_3, Cu_2O and CuO, the corresponding oxygen, nitrogen and copper compounds of great chemical similarity.

Laurent's proportional numbers (*ante*, p. 196) were really in a great number of cases the same as the atomic weights now used and that because his object was the same, namely to represent compounds by formulae chemically correct and adequate.

Symbol weights chosen so as to give suitable formulae for compounds.

The rapid development of organic chemistry, owing to the much greater complexity of the combinations therein dealt with, brought it about that more and more attention was devoted to representing by the manner of writing the formulae, the function of particular atoms or groups of atoms within the molecule, in fact to represent not only what a particular substance was composed of, but also what its chemical metamorphoses were. Of such representations, termed *rational formulae*, Gerhardt says:

.
Rational for-
mulae repre-
sent com-
position and
indicate pos-
sible metamor-
phoses.

"Chemical formulae do not represent, and cannot represent anything beyond relations and analogies; the best are those that make evident the greatest number of such relations and analogies...; it is their object to indicate in the simplest and most exact manner possible the relations between various substances and the chemical changes occurring on their interaction. To represent a substance by a rational formula, is to embody by means of conventional signs a number of reactions in which this substance participates." (*Traité de Chimie*, 4, pp. 563 *et seq.*)

The very popular example of acetic acid, because the simplest, remains the best. The molecular formula is $C_2H_4O_2$ and the following are amongst the most important reactions.

Rationaı ꜰor-
mꭓlaԇ of acetic
acid.

(i) The acid with a metal, or metallic oxide, or carbonate gives a salt formed by the replacement of one-fourth the hydrogen by metal,

$$2C_2H_4O_2 + \quad Fe \quad = (C_2H_3O_2)_2\,Fe \; + H_2$$
$$2C_2H_4O_2 + 2NaOH = 2\,(C_2H_3O_2)\,Na + 2H_2O$$
$$2C_2H_4O_2 + \quad Na_2CO_3 = 2\,(C_2H_3O_2)\,Na + H_2O + CO_2.$$

(ii) The action of phosphorus pentachloride on acetic acid results in the replacement of one atom of oxygen and one atom of hydrogen, *i.e.* of one hydroxyl group, by one atom of chlorine,

$$5C_2H_4O_2 + PCl_5 = 5C_2H_3O \,.\, Cl + H_3PO_4 + H_2O.$$

(iii) Heating of the sodium salt gives sodium carbonate and the volatile substance acetone,

$$2C_2H_3O_2Na = CO_3Na_2 + C_3H_6O.$$

(iv) Heating of a mixture of the sodium salt and of sodium hydrate gives sodium carbonate and the hydrocarbon methane,

$$C_2H_3O_2Na + NaOH = Na_2CO_3 + CH_4.$$

(v)   The products of electrolysis are carbon dioxide, hydrogen and ethane,

$$2C_2H_4O_2 = 2CO_2 + H_2 + C_2H_6.$$

(vi)   The acid is formed by the action of water on methyl-cyanide, the reaction consisting in the substitution of O_2H for N.

$$CH_3CN + 2H_2O = C_2H_4O_2 + NH_3.$$

These reactions find their representation in the rational formulae: (i) $C_2H_3O_2 . H$, (ii) $C_2H_3O . OH$, (iii) and (iv) $CH_4 . CO_2$, (v) $CH_3 : CO_2 . H$, (vi) $CH_3C . O_2H$ or $CH_3C . O . OH$.

"Is a rational formula permanently binding, or in other words, can a substance have one rational formula only? A combination of two or three simple atoms such as HCl or K_2S cannot be decomposed in more ways than one, but if the number of atoms in the molecule is greater, it is evident that the double decompositions which the molecule can undergo can be various. This is specially true of organic substances." (Gerhardt, *ibid.*)

A substance can have a number of rational formulae.

"The rational formulae are decomposition formulae, and in the present state of the science they can be nothing else; they aim at giving a representation of the chemical nature of a substance by a notation which indicates the atomic groups that remain unattacked in a series of reactions (radicles), or by emphasising the components which in certain often recurring metamorphoses play an important part (types). Any formula which represents certain metamorphoses is rational, but amongst different such formulae, that one is most rational which simultaneously represents the greatest number of changes....The formula showing the greatest amount of resolution into its component parts, will most completely represent the nature of the substance to which it belongs." (Kekulé, *Liebig's Ann. Chem.*, 106, 1858, p. 129.)

According to these views any one of the rational formulae given above for acetic acid has its legitimacy, but that which represents it by $CH_3 . C . O . OH$ is the best, because it comprises the greatest number of reactions, in fact all the six there discussed.

Rational formulae represent composition by radicles. Liebig's definition of "radicle."

The rational formulae recognise the presence within the molecule of certain definite *compound radicles*, that is, of groups of atoms which during a whole series of transformations remain together, moving about as a whole from compound to compound. The extensive use of the conception of compound radicles dates from Liebig[1] and Wöhler's[2] classical researches on the radicle of benzoic acid, and to Liebig we owe the definition:

"We...call cyanogen a radicle because it is the never-varying constituent of a series of compounds; because it can be replaced in these by simple bodies; because in its combinations with a simple body, the latter can be separated and replaced by equivalents of other simple bodies.

Of these three chief characteristics of a compound radicle, two at least must be fulfilled, ere it can really be regarded as a compound radicle." (*Liebig's Ann. Chem.*, 25, 1838, p. 1.)

It will serve the present purpose equally well not to follow Liebig's original development of the subject, but to give Gerhardt's exposition of it. The following reactions serve him for examples:

Gerhardt's exposition of the nature of radicles.

Benzoyl chloride and ammonia give benzamide and hydrochloric acid,

$$C_7H_5OCl + NH_3 = C_7H_5O . NH_2 + HCl.$$

Benzoic anhydride and ammonia give benzamide and water,

$$(C_7H_5O)_2O + 2NH_3 = 2C_7H_5O . NH_2 + H_2O.$$

Benzamide and potash give potassium benzoate and ammonia,

$$C_7H_5O . NH_2 + KOH = C_7H_5O . OK + NH_3.$$

" I give the name of *radicle* or *residue* to the components of any substance which can be transported into another substance, or which have been introduced into it by reactions of the above kind. According to this view chloride of benzoyl, anhydrous benzoic acid, and .benzamide contain the radicle C_7H_5O, whilst ammonia, water and potash contain the radicle H. On the other hand, since in the examples quoted the interchange takes place, not only between the benzoyl and the hydrogen, but also between the chlorine and the nitrogen (benzoyl chloride becomes the nitride of benzoyl + hydrogen) as well as between the oxygen and the nitrogen (the oxide of benzoyl becomes nitride of benzoyl + hydrogen, the nitride of benzoyl + hydrogen becomes

[1] Justus Liebig, b. Darmstadt 1803, d. Munich 1873, worked with Gay-Lussac in Paris, was in 1824 made professor at Giessen, and in 1852 accepted a call to Munich. His laboratory in the small university of Giessen, where he founded the first school of chemistry, served for many years as a pattern to the scientific world. His share in the development of organic chemistry warrants the assertion that "the organic chemistry of to-day is grounded mainly upon the pioneering labours of Liebig, and of Liebig and Wöhler together." But this represents only a portion of his work, physiological chemistry, agricultural chemistry, and hygiene having been furthered by him to an equal degree.

[2] F. Wöhler (1800—1882), pupil of Berzelius, co-worker and friend of Liebig, from 1836 professor at Göttingen. His artificial preparation of urea in 1828 is amongst the most striking of his discoveries in organic chemistry; but the greater portion of his work was done in the domain of inorganic chemistry, and comprised investigations of the elements aluminium, boron, and silicon with their compounds.

oxide of benzoyl + potassium), the name of radicles is equally applicable to the chlorine of the benzoyl chloride or of the hydrochloric acid, to the nitrogen of the ammonia and of the benzamide, to the oxygen of the water and of the benzoic acid anhydride, etc.

From this it will be evident that, unlike the majority of chemists, I look upon radicles in a sense of relation, and not in that of substances which are isolated or can be isolated." (Gerhardt, *loc. cit.* p. 568.)

" I hold that radicles are nothing else than the residues which in certain decompositions happen to remain unattacked, and it becomes therefore possible to assume in the same substance a simpler or a more complex radicle, according as to whether a greater or a smaller portion of the atoms is attacked. Thus for sulphuric acid we may have $H_2 . SO_4$ or $(HO)_2 SO_2$ representative of the reactions

The same substance may be represented as composed of different radicles.

$$H_2SO_4 + Zn \doteq ZnSO_4 + H_2,$$

and $$H_2SO_4 + PCl_5 = SO_2 . Cl_2 + POCl_3 + H_2O$$

respectively. Such assumptions of different radicles in compounds though legitimate are one-sided, considering as they do certain decompositions but taking no account at the same time of other equally important metamorphoses." (Kekulé, " Ueber die Theorie der mehratomigen Radikale," *Liebig's Ann. Chem.*, 104, 1857, p. 129.)

These radicles were found to ‑be endowed with a definite and characteristic substituting and combining power. The terms *atomicity, basicity* and *valency* were at that time used as synonymous, and as the measure of the number of hydrogen atoms that could be replaced or held in combination.

" In order to compare the radicles amongst themselves, I propose to refer them all to the radicle of hydrogen, and consequently I name them *monatomic, diatomic, triatomic* according to the quantity of hydrogen which they are capable of replacing in the type H_2O, *i.e.* according as to whether they are equivalent to 1, 2, or 3 atoms of hydrogen, so for instance in alcohol and in ether

Definite substituting and combining value possessed by various radicles.

$$O \begin{matrix} C_2H_5 \\ H \end{matrix} \qquad\qquad O \begin{matrix} C_2H_5 \\ C_2H_5 \end{matrix}$$

C_2H_5, the radicle ethyl, is monatomic because it replaces H (one atom of hydrogen) in the type water.

In anhydrous or hydrated sulphuric acid

$$O . SO_2 \qquad\qquad O_2 \begin{matrix} SO_2 \\ H_2 \end{matrix}$$

SO_2, the radicle sulphuryl, is diatomic because it replaces H_2 (two atoms of hydrogen) in the type water.

In anhydrous or hydrated phosphoric acid

$$\left.\begin{array}{c} PO \\ O_3 \\ PO \end{array}\right. \qquad \left.\begin{array}{c} PO \\ O_3 \\ H_3 \end{array}\right.$$

PO, the radicle phosphoryl is triatomic because it replaces H_3 (three atoms of hydrogen) in the type water.

But since one and the same substance can be represented by two or more different rational formulae according to the special double decomposition which it is intended to indicate, it is evident that a substance can also be formulated in terms of different radicles. So nitric acid can be depicted by the three formulae :

$$\left.\begin{array}{c} NO_2 \\ O \\ H \end{array}\right. \qquad \left.\begin{array}{c} NO \\ O_2 \\ H \end{array}\right. \qquad \left.\begin{array}{c} N \\ O_3 \\ H \end{array}\right.$$

in which the radicles NO_2, NO, and N have different equivalents. NO_2 is the equivalent of H ; NO is the equivalent of H_3, and N is the equivalent of H_5, because these three radicles must be replaced by different quantities of hydrogen in order to form water, thus :

$$\left.\begin{array}{c} H \\ O \\ H \end{array}\right. \qquad \left.\begin{array}{c} H_3 \\ O_2 \\ H \end{array}\right. \qquad \left.\begin{array}{c} H_5 \\ O_3 \\ H \end{array}\right."$$

(Gerhardt, *Traité de Chimie*, 4, p. 600.)

"[Wurtz[1] showed in 1855] that glycerine [$C_3H_8O_3$] may be regarded as the hydrate of the radicle C_3H_5, and its composition represented by the formula

Wurtz on the substituting power of the radicle C_3H_5'''.

$$\left.\begin{array}{c} C_3H_5''' \\ O_3 \\ H_3 \end{array}\right.$$ which is similar to the formula

by which, in accordance with the ideas of Williamson[2], ordinary

$$\left.\begin{array}{c} (PO)''' \\ O_3 \\ H_3 \end{array}\right.$$ phosphoric acid is represented. This radicle C_3H_5''',

[1] C. A. Wurtz (1817—1884), author of the well-known text-book " The Atomic Theory," was a pupil of Liebig and Dumas, and afterwards professor at the École de Médecine and the Sorbonne. He enriched organic chemistry by many discoveries, chief among which are those of the substituted ammonias and the dibasic and tribasic alcohols.

[2] Williamson in 1851 had expressed the view that a large number of compounds may be referred to the type of water, the monobasic acids to one molecule, and the polybasic acids, which are of greater molecular complexity, to a condensed water type. He wrote

$$\left.\begin{array}{c} C_2H_3O \\ H \end{array}\right. O = acetic\ acid\ldots\ldots\ \left.\begin{array}{c} SO_2 \\ H_2 \end{array}\right. O_2 = sulphuric\ acid.$$

" Odling, by an ingenious notation which is still in use, first marked this difference in the capacity for saturation possessed by the acetyl and sulphuryl radicles, by giving to their formulae a different index—

$$\left.\begin{array}{c} (C_2H_3O)' \\ H \end{array}\right\} O \qquad \left.\begin{array}{c} (SO_2)'' \\ H_2 \end{array}\right\} O_2$$

Acetic acid   Sulphuric acid

The idea that the substituting value of sulphuryl is twice that of acetyl is clearly expressed in this notation." (Wurtz, *The Atomic Theory*, p. 198.)

A. W. Williamson (1824—1904) was a pupil of Liebig, and for 38 years professor of chemistry at University College, London. His name is chiefly associated with an epoch-making memoir on the " Theory of Etherification," published in 1850.

which can replace three atoms of hydrogen, is formed by the subtraction of three atoms of hydrogen from the saturated hydrocarbon C_3H_8. The radicle C_3H_7, which can replace one atom of hydrogen, comes from the same hydrocarbon by the loss of a single atom of hydrogen. The subtraction of an atom of hydrogen develops a force in this residue C_3H_7, in virtue of which it is impelled to combine again with this hydrogen atom of which it has been deprived, or with some equivalent to it, and on the other hand, this same force makes it ready to supply the place of an atom of hydrogen wherever it is wanting. Again, the loss of three atoms of hydrogen creates in the residue $C_3H_5 = C_3H_8 - H_3$ a force by which it is ready to replace three atoms of hydrogen. Glycerine is produced in this manner, by the substitution of such a radicle for three atoms of hydrogen in the type of three condensed molecules of water,

$$\left.\begin{matrix} H_3 \\ H_3 \end{matrix}\right\} O_3 \qquad\qquad \left.\begin{matrix} (C_3H_5)''' \\ H_3 \end{matrix}\right\} O_3.$$

There was a gap between the 'monobasic' radicle $(C_3H_7)'$ and the 'tribasic' radicle $(C_3H_5)'''$. The residue C_3H_6 obtained by the subtraction of two atoms of hydrogen from the hydrocarbon C_3H_8 should possess a

Existence of radicle C_3H_6'' inferred from that of C_3H_7' and C_3H_5'''. Consequent discovery of glycols.

substituting or combining value equivalent to these two atoms of hydrogen. This proved to be the case from the study of Dutch liquid[1] and its analogues, which resulted in the discovery of the glycols. This residue or radicle C_3H_6 is propylene, and can replace like its homologue ethylene two atoms of hydrogen in two condensed molecules of water. The bodies possessing this constitution are the glycols

$$\left.\begin{matrix} H_2 \\ H_2 \end{matrix}\right\} O_2 \qquad \underset{\text{Glycol}}{\left.\begin{matrix} (C_2H_4)'' \\ H_2 \end{matrix}\right\} O_2} \qquad \underset{\text{Propylene glycol."}}{\left.\begin{matrix} (C_3H_6)'' \\ H_2 \end{matrix}\right\} O_2}$$

(Wurtz, *The Atomic Theory*, p. 200.)

" The radicles are classified according to their basicity :

$CH_3Cl =$ chloride of a monatomic radicle methyl

$CHCl_3 =$ chloride of a triatomic radicle formyl

$C_2H_5 =$ monobasic ; $C_2H_4 =$ dibasic ; $C_2H_3 =$ tribasic.

Kekulé's classification of elements into mono-, di-, and tribasic.

The molecules of chemical compounds consist of a conjunction of atoms. The number of atoms of another element (or of radicles) that can combine with one atom of a certain element (or radicle) depends on the basicity, or affinity magnitude of these constituents. From this point of view the elements fall into three chief groups :

(i)   Monobasic or monatomic such as H, Cl, Br, K.

(ii) Dibasic or diatomic, such as O, S.

(iii) Tribasic or triatomic such as N, P, As.

---

[1] $C_2H_4 + Cl_2 = C_2H_4Cl_2$; $C_2H_4 + H_2SO_4 = C_2H_5 . H . SO_4$.
Wurtz's researches on the triacid and biacid alcohols are contained in the memoirs : " Theory of Glycerine Compounds," *Ann. Chim. Phys.*, Paris, 43, 1855 (p. 492); " Glycol a Diatomic Alcohol," Paris, *C.-R. Acad. Sci.*, 43, 1856 (p. 199).

The simple compounds representative of this classification are HH, H_2O, H_3N. From these *primary types* the *secondary types* are derived by the replacement of one atom by another atom equivalent to it, *e.g.* HCl, H_2S, H_3P.

Combination between several molecules of the type cannot occur except when the substitution of a polyatomic radicle for two or more atoms of hydrogen supplies a cause for such union. A monatomic radicle can never unite two molecules of the type, but a diatomic radicle can do so.

(i)

$$SO_2 \begin{matrix} Cl \\ Cl \end{matrix} \qquad SO_2 \begin{matrix} OH \\ OH \end{matrix} \qquad CO \begin{matrix} NH_2 \\ NH_2 \end{matrix}$$

[Sulphuryl chloride]   [Sulphuric acid]   [Urea].

Such diatomic radicles may however substitute *two* hydrogen atoms in *one* molecule of the type :

(ii)

$$SO_2 . O \qquad\qquad CO . NH$$

[Sulphur trioxide]   [Cyanic acid]

and similarly

(i)

$$PO \begin{matrix} OH \\ OH \\ OH \end{matrix}$$

[Orthophosphoric acid]

(ii)

$$PO \begin{matrix} OH \\ O \end{matrix}$$

[Metaphosphoric acid]."

(Kekulé, "Ueber die sogenannten gepaarten Verbindungen und die Theorie der mehratomigen Radikale," *Liebig's Ann. Chem.*, 104, 1857, p. 129.)

In the above, Kekulé classifies not only *radicles* but also *elements* according to their substituting power as measured by the number of hydrogen atoms replaced. The first to consider the atoms themselves from such a point of view was Frankland[1]. In the Philosophical Transactions for 1852 appeared the now classical paper on "A New Series of Organic Compounds containing Metals," from which the following passage is a quotation:

"When the formulae of inorganic chemical compounds are considered, even a superficial observer is struck with the general symmetry of their constitution; the compounds of nitrogen, phosphorus, antimony and arsenic especially exhibit the tendency of these elements to form compounds containing 3 or 5 equivalents of other elements, and it is in these proportions

[1] E. Frankland (1825—1899) was a pupil of Liebig and Bunsen, and for many years Professor of Chemistry at the School of Science, South Kensington. His efforts to isolate the compound radicle contained in common alcohol and ether "led to the discovery of the remarkable series of compounds known as organometallic, and to the subsequent recognition of the varying power possessed by the metals and metalloids of uniting with the alcohol radicles, with the halogens and with oxygen."

F.

,that their affinities are best satisfied ; thus in the ternal group we have NO_3, NH_3, NI_3, NS_3, PO_3, PH_3, PCl_3, SbO_3, SbH_3, $SbCl_3$, AsO_3, AsH_3, $AsCl_3$, etc. ; and in the five atom group NO_5, NH_4O, NH_4I, PO_5, PH_4I, etc. Without offering any hypothesis regarding the cause of this symmetrical grouping of atoms, it is sufficiently evident, from the examples just given, that such a tendency or law prevails, and that, no matter what the character of the uniting atoms may be, the combining power of the attracting element, if I may be allowed the term, is always satisfied by the same number of these atoms."

Frankland finds that combining power of an attracting element is always satisfied by the same number of atoms.

He then proceeds to represent the organo-metallic compounds obtained by him by formulae which bring out the analogy with the inorganic types from which they may be supposed to be derived by the substitution of an organic radicle for an oxygen or a sulphur atom. The atomic weights used by Frankland are $O = 8$; $C = 6$; $S = 16$.

Inorganic Types	Organo-metallic Derivatives.
S As S	C_2H_3 As Cacodyl C_2H_3
O As O O	C_2H_3 As C_2H_3 Oxide of Cacodyl O
O O As O O O.	C_2H_3 C_2H_3 As O Cacodylic acid O O
Zn O	Zn C_2H_3 Zinc methylium
O Sb O O	C_4H_5 Sb C_4H_5 Stibethine C_4H_5
O O Sb O O O	C_4H_5 C_4H_5 Sb C_4H_5 Oxide of C_4H_5 Stibethine O	C_4H_5 C_4H_5 Sb C_4H_5 Binoxide of O Stibethine O
Sn O	Sn C_4H_5 Stanethylium
O Sn O	C_4H_5 Sn O Oxide of Stanethylium
I Hg I.	C_2H_3 Hg I Iodide of Hydrargyromethylium

"The oxygen and the methyl in cacodylic oxide and in cacodylic acid supplement each other so that their sum is equal to the number of oxygen atoms in that oxide of arsenic (arsenious acid or arsenic acid) of which the cacodylic oxide and cacodylic acid are the analogous combinations[1]. Similarly we have SbO_5 and $(C_4H_5)_3 SbO_2$ and $(C_4H_5)_4 SbO$; ZnO and C_2H_3. Zn. Frankland finds an explanation for these remarkable facts in the assumption that the saturation capacity of elements forming similar compounds is always satisfied by the same number of atoms independently of their chemical nature. The opinion held is that in the oxides of metals some or all of the oxygen atoms can be replaced by an equal number of atoms of a positive element such as hydrogen, methyl or even by an oxygenous acid radicle." (Kolbe, *Liebig's Ann. Chem.*, 101, 1857, p. 257.)

Important as were Frankland's speculations concerning the substitution value of radicles compared with that of elementary atoms, it is Couper and Kekulé[2], the latter especially, that we must consider to have been the first to propound the views concerning the valency of the elementary atoms which we now hold. Couper had a paper in the *Comptes Rendus* of 1858 entitled : " A new Chemical Theory[3]," Kekulé's paper " On the Chemical Nature of Carbon" came out in the same year in the *Annalen der Chemie und Pharmacie*[4].

Couper and Kekulé consider composition of substances in terms of atoms themselves, and not of groups of atoms.

The views propounded are practically the same, and so is their scope. Both men emphasise the necessity of a more ultimate study of chemical compounds, which should consider the composition of substances not merely in terms of the groups of atoms termed radicles, but in terms of the atoms themselves.

" I go back to the elements themselves, and study their mutual affinities. I believe that this study suffices for the explanation of all chemical com-

[1] Arsenious oxide $= O{\Large\diagdown}_{AsO}^{AsO}$ Cacodylic oxide $= O{\Large\diagdown}_{As\,(CH_3)_2}^{As\,(CH_3)_2}$

Arsenic oxide $= O{\Large\diagdown}_{AsO_2}^{AsO_2}$

Arsenic acid $= OH . AsO_2$ Cacodylic acid $= OH . AsO\,(CH_3)_2$.

[2] August Kekulé (1829—1896), professor in succession at Heidelberg, Ghent, and Bonn. To his merit as one of the founders of structural chemistry must be added his representation of benzene by the hexagon formula, and his active participation in the researches thereby initiated.

[3] "Sur une nouvelle Théorie chimique," Paris, *C.-R. Acad. Sci.*, 46, 1858 (p. 1157); English translation in *Phil. Mag.*, London, 4, 16, 1858 (p. 104).

[4] "Ueber die chemische Natur des Kohlenstoffes," *Liebig's Ann. Chem.*, Leipzig, 106, 1858 (p. 129).

binations." (Couper.) "I consider that it is necessary, and that in the present state of chemical knowledge, it is possible to go back in the explanation of the properties of chemical compounds, to the constituent elements themselves...I do not look upon it as our chief task now, to demonstrate the presence of groups of atoms, which because they possess certain properties may be looked upon as radicles....I hold rather that the consideration may be extended to the constitution of the radicles themselves, that the mutual relations of these radicles should be ascertained, and that their nature as well as that of their combinations should be derived from the nature of the elements." (Kekulé.)

"If we consider the simplest compounds of carbon, CH_4, CH_3Cl, CCl_4, CCl_3H, CO_2, $COCl_2$, CS_2, CNH, we are struck by the fact that that quantity of carbon which chemists have recognised as the smallest

Sum of chemi-cal units com-bined with one atom of carbon is always four. possible entering into the composition of a molecule, *i.e.* as the atom, always combines with four atoms of a monatomic, or two atoms of a diatomic element, or quite generally, that the sum of the chemical units of the elements combined with one atom of carbon is equal to four. This leads to the recognition that carbon is tetratomic. For substances which contain several atoms of carbon we must assume that a portion at least of the atoms is held in combination through the affinity of the carbon, and that the carbon atoms themselves are directly united, whereby of course part of the affinity of one atom is satisfied by an equally large part of the affinity of another atom. ...The simplest and therefore most probable type of such an arrangement of two atoms is, that one affinity unit of the one atom is satisfied by one such affinity unit of the other atom; hence of the 2×4 affinity units of the two carbon atoms, two are used for the holding together of these two atoms themselves, and what is left is six affinity units which can be satisfied by atoms of other elements—that is to say, a group of two atoms of carbon C_2

Some of the affinity units may be used to hold to-gether carbon atoms. will be hexatomic, it will combine with six atoms of a monatomic element, or quite generally with a number of atoms such that the sum of their chemical affinities is equal to six, *e.g.*

General for-mula for the number of chemical units combined with n atoms of carbon.

C_2H_6H	C_2H_5Cl	C_2H_3N
C_2H_4O	C_2H_3OCl	C_2N_2 ... etc.

If more than two carbon atoms unite in the same manner, the basicity of the carbon group will be increased by two units for each one C atom added. The number of H atoms (chemical units) combined with a number n of C atoms united with each other in this manner will be expressed by $n(4-2)+2=2n+2$; for $n=5$, the basicity is 12 (amyl hydride, amyl chloride, amylene chloride $= C_5H_{11}H$, $C_5H_{11}Cl$, $C_5H_{10}Cl_2$, etc.).

So far it has been assumed that all the atoms joined to the carbon are satisfied by the carbon itself, but it is equally legitimate to think that in the case of polyatomic elements (O, N, etc.) only a portion of their affinity, only one of the two units of the oxygen, or only one of the three units of the nitrogen is satisfied by the carbon, that one of the two affinity units of the oxygen, and two of the three affinity units of the nitrogen are left for saturation by other elements, and that these other elements are only in indirect union with the carbons, facts which may be expressed by the typical way of writing the formulae

$$C_2H_5 \diagdown$$
$$O$$
$$H \diagup$$

$$C_2H_5 \diagdown$$
$$H-N$$
$$H \diagup$$

$$C_2H_3O \diagdown$$
$$O$$
$$C_2H_5 \diagup$$

$$C_2H_5 \diagdown$$
$$C_2H_5-N.''$$
$$C_2H_5 \diagup$$

(Kekulé.)

" Carbon exercises its power of combination in two degrees, represented by CO_2 and CO_4[1]....In its elective affinities carbon shows peculiarities....

(i) In order to satisfy its power of combination it unites with equal numbers of equivalents of hydrogen, chlorine, oxygen, sulphur, etc., which can mutually replace each other.

(ii) It enters into combination with itself....This explains the accumulation of carbon atoms in organic compounds... ; the carbon atoms, and not the hydrogen atoms, bind together the elements of organic bodies....If the hydrogen atoms could combine with each other, we ought to have the compounds H_4Cl_4, H_6Cl_6, H_8Cl_8, etc.

I admit that an atom of oxygen in combination may exert a powerful affinity on a second atom of oxygen which itself is combined with another element....The highest power of combination which we know for carbon is four, that of oxygen is two. All the compounds of carbon can be referred to two types represented by the symbols nCM_4 and $nCM_4 - mM_2$, where $m < n$, or rather by $nCM_4 + mCM_2$, where n may become zero. Examples for the first type are supplied by the alcohols, fatty acids, glycols, etc.

Methyl and ethyl alcohols are represented by the formulae

$$C \begin{cases} O...OH \\ \\ H_3 \end{cases}$$

$$C \begin{cases} O.OH \\ H_2 \\ C-H_3 \end{cases}$$

We see that in methyl alcohol the limit of combination for the carbon is four, it being combined with three of hydrogen and one of oxygen... ; this oxygen, of which the power of combination is two, is itself combined with another atom of oxygen united to hydrogen. In ethyl alcohol also, the carbon belongs to the first type, each atom being combined to the second degree, on the one hand with three atoms of hydrogen, on the other hand with two of hydrogen and one of oxygen....In propyl alcohol the power of combination of the middle atom of carbon is reduced to two for hydrogen, since it is combined with each of the other carbon atoms

$$C...O.OH$$
$$|\quad H_2$$
$$C...H_2$$
$$|$$
$$C...H_3.''\quad \text{(Couper, *loc. cit.*)}$$

[1] The atomic weights used by Couper are $C = 12$, $O = 8$.

Gerhardt in 1856 in his *Traité de Chimie*, when discussing the subject of rational formulae, had said:

"It is so prevalent an error to suppose the possibility of representing molecular constitution by means of chemical formulae, or in other words by the actual arrangement of the atoms, that I may find it impossible to persuade certain of my readers of the contrary....*Chemical formulae are not intended to represent the arrangement of the atoms.*"

It is fair to assume from his manner of writing that Gerhardt did not even contemplate a future possibility of such a representation. Kekulé's and Couper's speculations had this very object for their aim. The property by them assigned to each elementary atom, of being able to act and react directly within the molecule with a certain definite number of other atoms, is the basis of a symbolic notation, which, whilst its primary object is the representation of the arrangement of the atoms in the molecule, thereby becomes at the same time the most comprehensive rational formula. All the possible chemical metamorphoses of the molecule are but a consequence of this particular atomic grouping, which itself is a consequence of the characteristic binding capacities of the constituent atoms. If therefore we know the molecular weight of a substance, and if we also know the valency of all the constituent atoms, we can infer what the arrangement of these atoms will be within the molecule, and in those cases when, owing to the presence of several polyvalent atoms, more than one kind of arrangement is possible, what all the different possible arrangements will be. So whilst for the *rational formulae* of a particular molecule there is no theoretical limit to their number other than that of all the various possible permutations of the constituent atoms, the theoretical number of the *structural formulae* is more limited, but these formulae comprise and indicate *all* the functions of possible constituent radicles.

Structural formulae represent actual arrangement of atoms in molecule.

Supposing then that the number of H atoms that can be bound by one carbon atom, one oxygen atom, and one sulphur atom is 4, 2, and 2 respectively, and supposing that we know the molecular formulae for acetic acid and sulphuric acid to be $C_2H_4O_2$ and H_2SO_4, the structural formulae of these substances will be[1]:

---

[1] The other possible structural formulae for molecules $C_2H_4O_2$ are given in the next chapter, p. 566.

Structural
formula of
acetic acid.

$$\begin{array}{ccc} \text{H} & \text{O} & \\ | & || & \\ \text{H--C--C--O--H} & & \end{array} \qquad \begin{array}{c} \text{O--O--H} \\ / \\ \text{S} \\ \backslash \\ \text{O--O--H} \end{array}$$

This acetic acid formula not only comprises all the rational formulae given before (p. 507), but it also indicates that certain molecular splittings like $CH.CO_2'H_3$, and $C_2OH.H_3O$ will not occur. The same holds for the case of the sulphuric acid.

The quotations given so far in this chapter show how various have been the names applied to the combining capacity of the elements and the radicles. Of these different terms *valency* has come to be the only one recognised. *Basicity* and *atomicity* are required for the designation of quite different and quite definite phenomena: basicity for the number of stages in which the replaceable hydrogen of an acid can be substituted by metals, and atomicity for the number of atoms in the molecule of an element. The use of the terms *affinity-units* and *saturation-capacity* cannot be objected to on such grounds, but is not much resorted to.

Name for the measure, in terms of number of hydrogen atoms, of the substituting or combining capacity of atoms or compound radicles.

This being the name, what form should we give to the definition of the property, the nature of which had been recognised and set forth by Frankland, Couper and Kekulé? The following are examples of the form adopted in some important text-books[1]:

I. Definition of valency.

"In the theory of valency it is assumed that each atom possesses a definite limited capacity for combining with other atoms. This capacity is called valency, and the atoms that can combine with one, two, three, four hydrogen atoms (or equivalent atoms or radicles) are said to be univalent, divalent, etc. respectively." (Ostwald, *Outlines of General Chemistry.*)

"The atoms of some elements can only combine with a single atom, but the atoms of some other elements can combine with two, three, four, or more atoms. They have double, treble, etc. the power of the other atom, and are said to be di-, tri-, tetra-, penta- or hexavalent, or they are said to have two, three, or more affinities. Hydrogen again forms the standard of comparison, as it does in the case of the equivalent and the atomic weights." (Lothar Meyer, *Outlines of Theoretical Chemistry.*)

[1] A full account of the present position of the doctrine of valency is given in F. W. Hinrichsen, *Ueber den gegenwärtigen Stand der Valenzlehre* (Ahrensche Sammlung, 7), 1902.

"Atomicity is the capacity of saturation, or the value of substitution possessed by atoms ; and this *valency* is an essentially different thing from the force of combination or the energy which resides in them.   It governs the form of combinations, which varies with each atom.   Thus the hydrogen combinations of chlorine, oxygen, nitrogen and carbon have a different form, and the atoms of carbon are so constituted that they can attract four atoms of hydrogen, whilst nitrogen can only attract three, etc." (Wurtz, *The Atomic Theory.*)

"The extent of the chemical affinity of an elementary atom is denoted by the statement of the number of atoms of a certain element made the basis of comparison, which it can combine with or replace in analogous compounds. If the elements considered *monovaleut* are those of which one atom in its gaseous compounds never combines with or replaces more than one other atom, then the valency of an element is the capacity of one of its atoms to combine with or replace a definite number of monovalent atoms." (Naumann, *Handbuch der Chemie.*)

All that is contained in these variant definitions can be epitomised in the short statement:

The valency of an elementary atom indicates the number of hydrogen atoms or of other atoms equivalent to hydrogen, with which it can unite, or which it can replace in a molecule. Atoms are termed monovalent or monad, divalent or dyad, trivalent or triad, etc. according as to whether they can combine with or replace one, two, three, etc. atoms of hydrogen.

The reason for making hydrogen the basis of comparison is, that in all binary compounds containing this element the number of hydrogen atoms in a molecule is never less than that of the atoms of the second element, *e.g.* HH, HCl, HI, H_2O, H_2S, H_3N, H_3P, H_4C, H_4Si[1], and it is the same in the matter of substitution. In the interchange between hydrogen and other elements the number of the substituting or substituted atoms may be equal to that of the hydrogen atoms or may be less, but it is never greater.

Reason for making the hydrogen atom the standard in valency measurements.

From HCl we can obtain by substitution KCl, $ZnCl_2$, $FeCl_3$, $SiCl_4$, PCl_5.

From H_2SO_4 we can obtain by substitution K_2SO_4, $ZnSO_4$, $Fe(SO_4)_3$, etc.

This choice of hydrogen for the standard is a matter of practical convenience, for thereby the numerical values for valency are

---

[1] Hydrazoic acid, N_3H, is the one inconvenient exception.

never fractional. The smallness of the number of binary hydrides greatly limits the possibility of direct reference to the standard, but recourse can be had to other compounds containing elements equivalent to hydrogen, *e.g.* Cl, Br, I, F. These have been found to possess the same property as hydrogen of never uniting with or replacing a relatively greater number of other atoms, and they replace the standard hydrogen atom by atom, *e.g.*

ClCl	$ZnCl_2$	$FeCl_3$	$SnCl_4$	PF_5	WCl_6
ClBr					
ClI[1]					
IBr					

and HH	H_2O	H_3N	H_4Si
HCl	Cl_2O	Cl_3N	HCl_3Si
HBr			Cl_4Si
HI		I_3N	
HF			F_4Si

The argument given so far is applicable only to the case of elements the molecular formulae of whose simplest compounds are known. Hence the only compounds from which reliable data can be obtained are those whose molecular weights are known. The application of Avogadro's hypothesis enables us to determine this value for substances which are stable in the gaseous condition. Van't Hoff's extension of the gaseous laws to dilute solutions, whilst it has largely added to our general knowledge of molecular weights, has not supplied us with more data available for valency determinations, owing to the fact that the non-volatile simple binary compounds such as metallic halides do not follow the law, their exceptional behaviour being explained by the hypothesis of electrolytic dissociation.

Valency values legitimately deduced only from substances whose molecular formulae are known.

But whilst recognising the inferior legitimacy of inferences drawn from the study of compounds of unknown molecular magnitude, valency values are also deduced from the comparison between non-volatile compounds and chemically analogous volatile ones. The simplest possible formulae of non-volatile chlorides and oxides

Valency values deduced from composition of non-volatile substances.

[1] The substance ICl_3, which is a solid, cannot be volatilised without decomposition into $ICl + Cl_2$.

are thus ranged along with those of the corresponding volatile compounds.

Molecular formula	HCl		Molecular formula	H_2O
Simplest formula	$NaCl$		Simplest formula	Na_2O
„	„	$CaCl_2$	„	„	CaO
Molecular formula	$ZnCl_2$			„	ZnO
„	„	$FeCl_3$		„	Fe_2O_3
„	„	$SnCl_4$		„	SnO_2
	„	SCl_4			SO_2
	„	PF_5		„	P_2O_5
	„	PCl_5			
	„	WCl_6		„	W_2O_3
				„	Mn_2O_7
				„	OsO_4

Having arrived at this conception of valency, what is the work required for its experimental determination? We must ascertain the molecular formulae of compounds containing not more than one atom of the element of unknown valency combined with monovalent atoms only. This involves a knowledge of:

II. The experimental determination of valency.

(i) The quantitative composition of the compound, which is ascertained by a process of analysis suitable to the special case investigated (for chlorides, the determination of the chlorine as silver chloride, etc.).

(ii) The atomic weights of all the elements present, values the determination of which involves extensive work according to the methods dealt with in chapters VIII, XIII, XIV and XV.

(iii) The molecular weight of the compound, derived from its vapour density.

According to the less restricted conception of valency which requires only the formulae of simple combinations such as the halides or oxides, a knowledge of (i) and (ii) without that of (iii) is sufficient.

It should be obvious why the only compounds from which strictly reliable data can be deduced are those that contain only *one* atom of the element considered, united with monovalent

atoms only, or with a single polyvalent atom.. In such cases
only there is the required certainty of direct
action and reaction between the atom investigated
and the number of monovalent atoms taken to be
its valency value, or with the single polyvalent atom
present, *e.g.* $C''O$, $N''O$, $N'''H_3$, $P'''H_3$, etc.

Utilising, as has been done already, the symbolic notation
which represents the valency of an elementary atom by a number
of lines radiating from the symbol, it may be seen from the
following examples how uncertainty arises when different kinds
and different degrees of linking between polyvalent atoms become
possible.

The molecular formula of phosphorus oxychloride is $POCl_3$,
which, on the supposition of the monovalency of chlorine and the
divalency of oxygen, enables us to formulate this compound with
the phosphorus atom as trivalent or as pentavalent.

The molecular composition of nitrous oxide, N_2O, allows the
nitrogen atom to be represented as either monovalent or trivalent,
if oxygen is divalent.

And similarly, taking the case of the use of a non-volatile
compound, such as the chloride or oxide of an alkali metal, for the
determination of the valency of the alkali atom, we have the
simplest formulae KCl and K_2O, which if molecular would for
the chloride prove the monovalency of the potassium atom. But
the true molecular formulae may be K_2Cl_2 or K_3Cl_3 or generally
K_nCl_n, and K_4O_2 or K_6O_3 or ...$K_{2n}O_n$, with valencies according to
the atomic linking represented by:

K = monovalent......$K—Cl$; $K—O—K$

K = divalent... $Cl—K—K—Cl$; $K—K$; $K—K—K—K$

$$K = \text{trivalent} \ldots \ldots Cl—K—K—Cl; \quad O = K—K—K—K = O$$

$$
\begin{array}{ccc}
& \diagdown \diagup & \\
& K & \\
& | & \\
& Cl &
\end{array}
\qquad
\begin{array}{ccc}
& \diagdown \diagup & \\
& K & \\
& | & \\
& K & \\
& \| & \\
& O &
\end{array}
$$

The results of valency measurements lead to an arrangement of the elements in 8 classes, from monovalent to octavalent.

III. Classification of elements according to valency.

These classes are given in the table on page 526 in two divisions, to separate the results deduced from compounds of known molecular magnitude from those based on less certain evidence.

The symbol of the element is followed by the formulae of the compounds from which the valency value is deduced. Not more than one *molecular formula* is given, but in the greater number of cases other volatile compounds of corresponding structure are known, *e.g.* BCl_3, BH_3(?), BF_3, $B(CH_3)_3$, $B(C_2H_5)_3$.

A critical examination of the valency values set out in the table leads to certain generalisations:

Examination of the results of classification according to valency: (1) Maximum value is 6 and 8 respectively.

(1) *The maximum valency deduced from molecular formulae is* 6, but the number of instances in which the higher values are met with is small. We know only one compound, WCl_6, in which a hexavalent atom occurs; the six compounds given in the table as containing pentavalent atoms, together with PCl_5 and NH_4Cl, which can be volatilised without decomposition under very special conditions, comprise all the known representatives of that class. But the number of volatile compounds containing one tetravalent atom united only with monovalent atoms or radicles is considerable; taking silicon alone, we have SiH_4, $SiCl_4$, $SiCl_3H$, $SiBr_4$, $SiBr_3H$, $SiBr_3Cl$, $SiBr_2Cl_2$, $SiBrCl_3$, SiI_4, SiF_4, $Si(CH_3)_4$, $Si(C_2H_5)_4$, $Si(C_2H_5)_3Cl$, $Si(C_2H_5)Cl_3$, $SiH(C_3H_7)_3$.

The study of the oxides gives a range of molecular complexity from R_2O to R_2O_8, in which also the number of representatives of each class gets smaller as the valency displayed increases.

(2) *Valency is a periodic function of the atomic weights.* This had been recognised and emphasised, on the occasion of

the discovery of the periodic law, by both Lothar Meyer and Mendeleeff (pp. 462, 469). Mendeleeff, who elaborated this aspect of the periodic system in greater detail, has pointed out that the valency in the series as exhibited by the hydrides and chlorides increases from 1 to 4 and then decreases from 4 to 1, and it is only the last four members of the series that are capable of forming hydrides. On the other hand, the valency as deduced from the highest oxides or the hydrates shows a continuous increase from 1 to 8.

(2) Valency is a periodic function of the atomic weights.

			RH_4		RH_3	RH_2		RH	
R_2O	$RO\,(R_2O_2)$	R_2O_3	$RO_2\,(R_2O_4)$		R_2O_5	$RO_3\,(R_2O_6)$		R_2O_7	$RO_4\,(R_2O_8)$
LiCl	BeCl$_2$		BCl$_3$	CCl$_4$		NCl$_3$	OCl$_2$		
				CH$_4$		NH$_3$	OH$_2$		FH
Li$_2$O	BeO		B$_2$O$_3$	CO$_2$		N$_2$O$_3$	OO$_2$		
NaCl	MgCl$_2$		AiCi$_3$	SiCi$_4$		PCl$_3$	SCl$_2$		CiCl
				SiH$_4$		PH$_3$	SH$_2$		ClH
Na$_2$O	MgO		Al$_3$O$_3$	SiO$_2$		P$_2$O$_3$	SO$_2$		(Cl$_2$O$_7$).

It is an interesting fact that for all the elements which form hydrides, the sum of the valencies as deduced from these hydrides and from their highest oxides is 8, *e.g.* CH_4 and CO_2, SiH_4 and SiO_2; NH_3 and N_2O_5, PH_3 and P_2O_5; SH_2 and SO_3; ClH and Cl_2O_7.

It has been pointed out before (p. 504) that in order to make evident this periodic change in composition, Mendeleeff had to arbitrarily select from amongst the different oxides of an element one which fits into the scheme, and which however in its chemical relations is not always typical of the element. It is reserved for future investigations to discover a common characteristic of the particular oxides which exhibit this periodic change in valency, a property which we cannot believe to be fortuitous.

(3) The same element may appear in different classes: phosphorus in 3 and 5; nitrogen in 2, 3, and 5; indium in 1, 2, and 3, etc., etc. These facts considered from a purely empirical point of view seem to lead to the necessary inference that valency is a variable property. But theoretical considerations the attractiveness of which is patent, have led to the attempt to look upon valency as an invariable, fundamental property of the atoms, and hence to the necessity of finding some explanation for the large number of cases which constitute apparent exceptions to such a law.

(3) The same element is found in different classes.

Valency of the Elementary Atoms[1].

(The symbol in heavy type denotes the element whose valency is deduced from the formulae in brackets.)

Deduced from: The molecular formulae of *simple volatile* compounds (hydrides, halides, methides and ethides, mon-oxides)			Deduced from: (1) The molecular formulae—indicated by italics—of *more complex volatile* compounds (dioxides and trioxides etc.) (2) The simplest formulae of *typical non-volatile* compounds (oxides and chlorides)	
H (H_2) **K** (KI) **In** $(InCl)$ **Tl** $(TlCl)$	**?F** $(FH$ or $F_2H_2)$ **Br** (BrH) **I** (ICl)	**Cl** (HCl) **Ag** (at $1735°$, $AgCl$) **?Hg** $(HgCl$ or $Hg_2Cl_2)$	Monovalent	**Li** $(LiCl, Li_2O)$ **Cl** (Cl_2O) **Rb** $(RbCl, Rb_2O)$ **Au** $(AuCl)$	**N** (N_2O) **K** (KCl, K_2O) **Ag** (Ag_2O) **Hg** (Hg_2O)	**Na** $(NaCl, Na_2O)$ **Cu** (Cu_2O) **Cs** $(CsCl, Cs_2O)$ **Tl** (Tl_2O)
Be $(BeCl_2)$ **O** (OH_2) **Fe** $(FeCl_2)$ **Cd** $(CdCl_2)$ **Te** (TeH_2)	**C** (CO) **S** (SH_2) **Zn** $(ZnCl_2)$ **In** $(InCl_2)$ **Hg** $(HgCl_2)$	**N** (NO) **Mn** $(MnCl_2)$ **Se** (SeH_2) **Sn** (above $1113°$, $SnCl_2)$ **Pb** $(PbCl_2)$	Divalent	**Be**(O) **Ca** $(CaCl_2, CaO)$ **Fe** (FeO) **Cu** $(CuCl_2, Cu_2Cl_2, CuO)$ **Cd** (CdO) **Ba** $(BaCl_2, BaO)$ **Pb** (PbO)	**F** (F_2H_2) **V** (VCl_2, VO) **Co** $(CoCl_2, CoO)$ **Zn** (ZnO) **In** (InO) **Pt** $(PtCl_2)$	**Mg** $(MgCl_2, MgO)$ **Mn** (MnO) **Ni** $(NiCl_2, NiO)$ **Sr** $(SrCl_2, SrO)$ **Sn** (SnO) **Hg** $(?Hg_2Cl_2, HgO)$
B (BCl_3) **P** (PH_3) **Ga** (above $447°$, $GaCl_3)$ **Sb** $(SbCl_3)$	**N** (NH_3) **Cr** $(CrCl_3)$ **As** (AsH_3) **Bi** $(BiCl_3)$	**Al** $(Al(CH_3)_3)$ **Fe** (above$750°$, $FeCl_3)$ **In** $(InCl_3)$	Trivalent	**B** (B_2O_3) **P** (P_4O_6) **Cr** (Cr_2O_3) **Ga** (Ga_2O_3) **Sb** (Sb_4O_6) **Au** $(AuCl_3)$	**N** (N_2O_3) **Sc** (Sc_2O_3) **Mn** (Mn_2O_3) **As** (As_4O_6) **I** (ICl_3) **Bi** (Bi_2O_3)	**Al** (Al_2O_3) **V** (VCl_3, V_2O_3) **Fe** (Fe_2O_3) **In** (In_2O_3) **Tl** $(TlBr_3, Tl_2O_3)$

			Valency			
C (CH_4) V (VCl_4) Sn ($SnCl_4$) Th ($ThCl_4$)	Si (SiH_4) Ge ($GeCl_4$) Te ($?TeCl_4$) U (UCl_4)	Ti ($TiCl_4$) Zr ($ZrCl_4$) Pb ($Pb(CH_3)_4$)	Tetravalent	C (CO_2) Si (SiO_2) Ti (TiO_2) Ge (GeO_2) Sn (SnO_2) Os ($OsCl_4$, OsO_2) Th (ThO_2)	N (NO_2, N_2O_4) S (SCl_4, SO_2) V (VO_2) Se ($SeCl_4$, SeO_2) Te (TeO_2) Pt ($PtCl_4$) U (UO_2)	O $(O.O_2)$[2] Cl (ClO_2) Mn (MnO_2) Zr (ZrO_2) W (WCl_4, WO_2) Pb (PbO_2)
P (PF_5) Nb ($NbCl_5$)	Mo ($MoCl_5$) Ta ($TaCl_5$)	Sb ($SbCl_5$) W (WCl_5)	Pentavalent	N (NH_4Cl[3], N_2O_5) As ($As(C_6H_5)Cl_4$) Nb (Nb_2O_5)	P (PCl_5[3], P_4O_{10}) Sb (Sb_2O_5) Ta (Ta_2O_5)	V (V_2O_5) I (I_2O_5) Bi ($Bi(C_6H_5)_3Cl_2$)
W (WCl_6)			Hexavalent	S (SO_3) Mo (MoO_3)	Cr (CrO_2Cl_2, CrO_3) W (WO_3)	Mn (MnO_3) U (UO_3)
			Heptavalent	Mn (Mn_2O_7) Cl and I (Cl_2O_7, and I_2O_7, inferred from existence of acids and salts $X'ClO_4$, $X'IO_4$, X'_5IO_6)		
			Octavalent	Ni ($Ni(CO)_4$)	Ru (RuO_4)	Os (OsO_4) Pt (PtH_2Cl_6)

[1] The elements of the rare earths (Ce, La, etc.) and some of group VIII of the periodic law classification are omitted from this table.

[2] Oxygen had until lately been assigned an exceptional position in the group to which it belongs, having been considered as invariably divalent. But it is now recognised that the constitution of a number of organic compounds can be best represented by the assumption of the tetravalency of the oxygen present, an assumption warranted by a number of physico-chemical data. The evidence for the occasional tetravalency of oxygen is contained in Rudorf, *loc. cit.* pp. 80—82; Hinrichsen, *loc. cit.* pp. 69—72.

[3] Special conditions are required to prevent dissociation on volatilisation.

The interest attaching to the discussion, whether valency is constant or variable, has become mainly historical, but as such it

is so great as to justify some account of this phase in the development of the science. Kekulé, one of the founders of the doctrine of the combining capacity of the elementary atoms, made himself the champion of the view of constant valency.

"Atomicity is a fundamental property of the atom, which must be constant and invariable like the weight of the atom itself....To admit that atomicity can vary, is to confound the notion with that of equivalence....We define atomicity as the maximum equivalent or the maximum saturating capacity, and hence because of the existence of the substances NH_4Cl, SCl_4, ICl_3, etc. it has been inferred that the elements N, P, As, Sb are pentatomic; O, S, Se, Te tetratomic; I triatomic : but according to such reasoning, PI_3, TeI_4 would prove that phosphorus is nonatomic, and that the atomicity of tellurium is twelve, which is absurd."

Supporters of a theory of constant valency have always been confronted with the problem of how to explain the considerable

number of instances in which an element does not exhibit the particular combining power which had been chosen by them as the true valency from amongst the various possible values. These exceptions are of two kinds, cases in which (1) a lower, and (2) a higher value is met with.

The valency of carbon is taken as 4. A large number of substances are known of the type of ethylene (C_2H_4), acetylene

(C_2H_2), etc. which on the supposition of the invariable monovalency of hydrogen, cannot be represented as containing tetravalent carbon. Very extensive empirical data showed that in all such cases the number of unsatisfied or inactive affinity units is always 2 or a multiple of 2. This generalisation led to the assumption that the direct action between two carbon atoms may extend over 2 or 3 of the 4 combining units of each, explaining the apparently lower valency of these atoms by so-called *double* or *treble linking,* the graphic representation of which takes the form :

$$\text{Ethylene} = C_2H_4 = \begin{array}{c} H \\ H \end{array}\hspace{-4pt}>\hspace{-2pt}C = C\hspace{-2pt}<\hspace{-4pt}\begin{array}{c} H \\ H \end{array}$$

$$\text{Acetylene} = C_2H_2 = H-C \equiv C-H.$$

The failure of all attempts to prepare a methylene

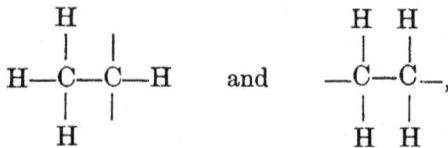

or two substances of the composition C_2H_4, namely

$$
\begin{array}{ccc}
\quad\;\; H & & H \quad H \\
\quad\;\; | & & | \quad\;\; | \\
H-C-C-H & \text{and} & -C-C-, \\
\quad\;\; | \quad\; | & & | \quad\;\; | \\
\quad\;\; H & & H \quad H
\end{array}
$$

is foremost amongst the facts which lend support to this view. But it would seem as if the hypothesis of multiple linking, which for a period has been adequate to the purpose for which it had been devised, might have to be given up before long. The physical and the chemical properties of the supposed doubly or trebly linked carbon atom, as deduced from a study of its compounds, are not such as would be expected theoretically, *e.g.* the atomic volume of the doubly linked atom is greater and not smaller than that of the singly linked, etc.[1] Moreover, the recent announcement of the discovery of the substance $C(C_6H_5)_3$, in which the recognition of the presence of a trivalent carbon atom is inevitable, will, if corroborated, constitute a definite exception to the fundamental generalisation that an *even* number always represents the difference between m, the number of monovalent atoms or radicles actually held in combination by n atoms of carbon, and $2n + 2$, the number of such atoms required by theory on the assumption of the tetravalency of the carbon atoms. But unless it can be shown that the doctrine of multiple linking leads to important deductive inferences which have been verified when put to the test of experiment, even one well-established exception like the above assumes enormous importance. And there is the further fact that some evidence has been adduced in favour of representing certain substances as containing *divalent* carbon. Thus it has been shown from its reaction that the constitutional formula for hydrocyanic acid should be $C = N - H$, rather than $H - C \equiv N$, and that of fulminic acid $C = N - O - H$.

Existence of a substance for which assumption of free carbon valencies is inevitable, and of others for which it is preferable.

[1] For a summary of the evidence against the hypothesis of multiple linking, see Hinrichsen, *loc. cit.*, pp. 34 *et seq.*

Anyhow, in the case of carbon monoxide and of carbon monosulphide, we are confronted with the alternative of looking upon carbon as divalent, or upon oxygen and sulphur as tetravalent, either of which is incompatible with the doctrine of constant valency. But carbon monoxide (or monosulphide), in common with C_2H_4 and C_2H_2, and other substances in which the carbon does not exert its maximum valency, has the property of readily uniting with other substances, elementary or complex, producing addition compounds in which finally each carbon atom holds in combination 4 monovalent atoms or the equivalent of these.

(ii) CO, CS, NO termed unsaturated.

$$CO + O = CO_2; \qquad CO + Cl_2 = COCl_2$$
$$C_2H_4 + Cl_2 = C_2H_4Cl_2; \qquad C_2H_4 + HCl = C_2H_5Cl$$
$$C_2H_2 + Br_2 = C_2H_2Br_2; \quad C_2H_2Br_2 + HBr = C_2H_3Br_3.$$

The name of *unsaturated compounds* has been given to such substances, and it is assumed that, owing to the influence of definite specific conditions, certain of the valencies are prevented from becoming effective, so that of the 4 combining units possessed by the carbon atom in carbonic oxide 2 only are active. It is difficult to see how this can ever have been considered an explanation and consequent proof of the constancy of valency, taking that term in the plain sense of a statement concerning an actual occurrence. Whether further combining power is latent in the carbon atom of the monoxide or not, the fact remains that in this particular substance it is divalent. The same argument applies to the case of nitrogen in nitric oxide, which is undoubtedly divalent, though according to the constant valency doctrine, 3 is the value appertaining to this element.

The valency of phosphorus and nitrogen is taken to be 3, but the formulae of solid phosphoric chloride and of sal-ammoniac are PCl_5 and NH_4Cl, indicating pentavalent action of the phosphorus and nitrogen atoms. These substances, however, when volatilised, break up to a greater or lesser amount into PCl_3 and Cl_2, and into NH_3 and HCl respectively. The instability of these compounds in the gaseous condition led Kekulé to explain these and similar cases in which the molecular complexity is beyond that compatible with the constant value accepted by

2. Valencies apparently higher than true values. Hypothesis of molecular combination.

him, by classing them as a special type of combination, different from the ordinary atomic, and named " molecular."

"We may divide elements into mono-, di-, tri-, and tetratomic. Atoms may combine with others of the same kind or of different kinds. The combinations in which all the elements are united by the affinities of the atoms which saturate each other, may be termed *atomic combinations*; these are the true chemical molecules, and the only ones which can exist in the gaseous state.

Kekulé's differentiation between atomic and molecular combinations.

But besides these atomic compounds we must recognise the existence of a second order of combinations which I will designate as *molecular*. The assumption of the production and the existence of molecular combinations is justified by the following considerations. Attraction should make itself felt even between the atoms which happen to belong to different molecules, and this attraction should produce an approach and a final juxtaposition of the molecules, a phenomenon which always precedes actual chemical decomposition. But when owing to the very nature of the atoms composing these molecules double decomposition becomes impossible, it may happen that the reaction stops at this point, that the two molecules, so to speak, adhere, and form a group endowed with a certain amount of stability, which however is always less than that of atomic combinations. This explains why such molecular compounds do not form vapours, but decompose under the action of heat, regenerating the molecules from which they had been formed." ("Sur l'atomicité des éléments," *C.-R. Acad. Sci.*, 58, 1864, p. 510.)

Kekulé quotes as examples substances which when volatilised break up in the following manner:

Molecular compounds break up under the action of heat into constituent molecules. Examples.

Phosphorus pentachloride breaks up into phosphorus trichloride and chlorine: $PCl_5 = PCl_3 + Cl_2$.

Selenium tetrachloride breaks up into the monochloride and chlorine [probably according to the equation $2SeCl_4 = Se_2Cl_2 + 3Cl_2$].

Iodine trichloride breaks up into iodine monochloride and chlorine: $ICl_3 = ICl + Cl_2$.

Ammonium chloride breaks up into ammonia and hydrochloric acid: $NH_4Cl = NH_3 + HCl$.

Tellurium tetrabromide breaks up into the dibromide and bromine: $TeBr_4 = TeBr_2 + Br_2$.

Tetramethylammonium iodide breaks up into trimethylamine and methyl iodide: $(CH_3)_4NI = (CH_3)_3N + CH_3I$.

Instances are given of other substances which may be considered molecular combinations, such as crystalline salts which under the action of heat break up into an anhydrous salt and water.

How far did this hypothesis of molecular compounds serve the purpose for which it had been devised?

It does not seem as if Kekulé had been led to it inductively from the observation and subsequent classification of some real, important differences between chemical compounds,
but rather that he propounded it with the object of supporting the particular view held by him. concerning the constancy of the binding capacity of the atoms. The natural consequence is to make one's initial attitude towards this hypothesis of molecular compounds. one of suspicion. The ordinary course of scientific procedure in dealing with such a hypothesis is to examine firstly, whether it is inductively adequate, *i.e.* whether it explains all the cases that should be comprised within its scope, and secondly, whether it is deductively true, *i.e.* whether the theoretical influences drawn from it meet with experimental verification.

The use of the conception of molecular combinations in the doctrine of valency.

If a compound is met with in which an element seems to exhibit a higher valency than that assigned to it as constant, the combination must be supposed to be molecular. But as such it must possess the characteristic property of molecular compounds, which is instability, and it must under the action of heat break up into its two component atomic combinations. Of the element phosphorus it is assumed that it has the constant valency 3, that all the combinations of apparently greater valency are molecular and therefore cannot exist in the gaseous state. This is the case with the pentachloride and the pentabromide, but the substance PF_5 discovered and investigated by Prof. T. E. Thorpe has been found to be stable when volatilised. This compound was prepared by the action of phosphorus pentachloride on arsenic trifluoride, and was found to be a gas which obeyed Boyle's law and which could not be liquefied at 7° C. and 12 atmospheres. The analysis (*ante*, p. 351) gave quantities of phosphorus and of fluorine in the ratio of 1 atomic weight of the former to 5 of the latter, and the determinations of the specific gravity of the vapour (H = 1) gave 63·23 as the mean. Hence the molecular weight found is 126·4, whilst that calculated from the formula PF_5 is 125·3.

If molecular, all compounds containing apparently pentavalent phosphorus should be decomposed by heat.

"In phosphorus pentafluoride the phosphorus is undoubtedly penta-valent; this is the first volatile phosphorus compound which contains five

Phosphorus pentafluoride not decomposed when volatilised. monovalent atoms united with one atom of phosphorus. This substance is a gas apparently stable at all temperatures. Wurtz has shown that phosphorus pentachloride can exist as vapour only below a certain temperature; the corresponding bromide at once splits up on heating into bromine and the tribromide; and a pentaiodide of phosphorus does not seem capable of existence at all. Hence it would appear that the chemical valency of an element depends on the weight of the atoms which combine with it." (T. E. Thorpe, *Liebig's Ann. Chem.*, 182, 1876, p. 201.)

The formulation of ammonium chloride as a molecular compound leads to the inference of the possible existence of another

If ammonium chloride is a molecular combination, another substance of formula NH_4Cl should exist. substance of the same formula, NH_4Cl, but of different distribution of the same six atoms within the two constituent atomic combinations, viz. $H_3N.HCl$ and $NH_2Cl.H_2$. This could not be put to the test of experiment for ammonium chloride itself, but has been investigated for substances derived from

ammonium chloride by the substitution of its hydrogen by com-pound radicles C_nH_{2n+1}.

"Meyer attempted to elucidate the constitution of ammonium salts. It was found that the dimethyl diethyl ammonium iodide obtained by the action

Meyer's investigation of its alkyl derivative shows that NH_4Cl is not a molecular compound. of ethyl iodide on dimethylamine is identical with that produced by acting with methyl iodide on diethylamine[1], and no difference can be detected in the character of these salts. As the substances, although identical[2], were obtained by different reactions, it was inferred that they could not be molecular compounds, that is, combinations of a tertiary base with an alkyl haloid, but must contain pentavalent

nitrogen, when by analogy NH_4Cl would be $N\begin{smallmatrix}H\\H\\H\\H\\Cl\end{smallmatrix}$." (T. E. Thorpe, *Victor*

Meyer Memorial Lecture, 1900.)

Thus a deduction from Kekulé's theory of the constantly tri-valent nature of nitrogen has not been verified when put to the

[1] $N(C_2H_5)_2(C_2H_3)_2I$

I $\begin{cases} N(CH_3)_2H + C_2H_5I = N(CH_3)_2C_2H_5 + HI \text{ and} \\ N(OH_3)_2C_2H_5 + C_2H_5I = N(CH_3)_2C_2H_5 . C_2H_5I \end{cases}$

II $\begin{cases} N(C_2H_5)_2H + CH_3I = N(C_2H_5)_2CH_3 + HI \text{ and} \\ N(C_2H_5)_2CH_3 + CH_3I = N(C_2H_5)_2CH_3 . CH_3I \end{cases}$

[2] There are other similar experiments on record, the result of which is not quite as conclusive.

test of experiment, whilst another similar crucial experiment has given a result in support of the variable valency of the phosphorus atom[1].

Preparation and Properties of Substances $PO(C_6H_5)_3$.

Name	Phenoxyl-diphenyl-phosphine	Triphenyl-phosphine-oxide
Formula	$P \diagup \begin{matrix} O \cdot C_6H_5 \\ - C_6H_5 \\ \diagdown C_6H_5 \end{matrix}$	$O = P \diagup \begin{matrix} C_6H_5 \\ - C_6H_5 \\ \diagdown C_6H_5 \end{matrix}$
Preparation	From diphenyl-chloro-phosphine and phenol $(C_6H_5)_2PCl + C_6H_5OH$ $= (C_6H_5)_2 \cdot P \cdot C_6H_5O + HCl$	From triphenyl-bromo-phosphine-bromide and water $(C_6H_5)_3Br_2P + H_2O$ $= (C_6H_5)_3(OH)_2P + 2HBr$ $(C_6H_5)_3(OH)_2P$ heated $= (C_6H_5)_3O \cdot P + H_2O$
Treatment with Br, O, S	Direct combination ensues, the substance behaving like an unsaturated compound	No action occurs, the substance behaving like a saturated compound
State of aggregation	Thick oily liquid, boiling at 62 mm. between 265° and 270°	Solid, melting at 153·5°
Spec. Grav. $_{water=1}$	1·140	1·2124
Spec. Grav. of vapour $_{air=1}$ (i) determined at reduced pressure	(1) 10·07 (2) 9·97	(1) 10·11 (2) 9·788
(ii) calculated for $C_{18}H_{15}PO$	9·68	9·68

If valency is constant, phosphorus oxychloride must be conceived as $Cl - P - O - Cl$; but if valency is variable, there is the

$$Cl - \underset{\underset{Cl}{|}}{P} - O - Cl$$

possibility of the existence of two isomeric compounds in which P is tri- and pentavalent respectively, thus:

$$(1) \ P \diagup \begin{matrix} Cl \\ - Cl \\ \diagdown OCl \end{matrix} \quad \text{and} \quad (2) \ P \diagup \begin{matrix} Cl \\ Cl \\ Cl \\ O \end{matrix}$$

[1] A. Michaelis and W. La Coste, " Ueber die Valenz des Phosphors," *Ber. D. chem. Ges.*, 18, 1885 (p. 2118).

Compound (1) is not likely to exist owing to its probably great instability; but it is a fact that has met with general proof, that the weighting of a molecule, *i.e.* the substitution of C_nH_{2n+1} or other complex groups increases the stability. An investigation of derivatives of $POCl_3$, in which the three chlorine atoms were replaced by phenyl groups, was carried out with the result that the existence of two substances of the same formula $C_{18}H_{15}PO$ was proved, and that the differences between these two were of the kind to be expected, viz. that whilst the one containing trivalent phosphorus easily formed additive compounds, thereby passing into the pentavalent condition, the other one, containing pentavalent phosphorus, behaved as a saturated compound. In the table above are set forth the facts showing the identity of the molecular magnitude of these two substances, the mode of formation which indicates the differences in their composition, and the consequent differences in their properties.

Pentavalency of phosphorus proved by the existence of two substances $PO(C_6H_5)_3$.

Such then are the empirical data concerning valency. They are compatible with the view that valency is simply the observed combining or substituting value exhibited in particular cases, and that as such it is of variable magnitude. But they are equally compatible with the view, that valency as a fundamental invariable property of the atoms, is the power to bind or substitute a certain number of monovalent atoms, and that nothing is really detracted from this power when owing to existing conditions the atoms cannot fully exert it. From this point of view the tetravalency of carbon as shown by the existence of CH_4 and a number of analogous compounds is not contradicted by the existence of CO, nor the pentavalency of phosphorus as deduced from the compounds PCl_5 and PF_5 by the existence of PCl_3 and PH_3, etc. etc.

V. Generalisation from empirical data concerning nature of valency.

What is lacking is some theory, which starting from the assumption of some simple fundamental property of matter, will in terms of this property account for facts such as that the binding capacity of an atom of oxygen is different from that of an atom of carbon, and from that of an atom of nitrogen; that it is the same for oxygen and sulphur, for phosphorus and nitrogen, for carbon and silicon; that it is different for carbon in its

Phenomena requiring explanation by a satisfactory hypothesis as to cause of valency.

two oxides, different for phosphorus in its two chlorides; that the phosphorus atom which can bind five chlorine atoms at one temperature loses this power at a higher temperature; that the phosphorus atom can bind 5 fluorine atoms but only 3 of the equivalent but heavier iodine atoms; that anhydrous white copper sulphate, $CuSO_4$, in which according to the valency values of the constituent atoms, all power of combination should be exhausted, forms with water more complex molecules, producing the crystalline blue $CuSO_4 . 5H_2O$; etc. etc.

" Since the aim of all scientific investigations is to exhibit the most variable phenomena as dependent upon certain active invariable factors taking part in them and in such a manner that each phenomenon appears to be the necessary result of the properties and reciprocal action of these factors; then it is clear, that chemical investigation would be considerably advanced were it possible to prove that the composition of chemical compounds is essentially determined by the valency of the atoms and the external conditions under which these atoms react upon one another. The first necessary step in this direction must consist in the attempt to explain the regularities observed in the composition of chemical compounds, by the assumption of a constant power of saturation or an invariable valency of the atom. The opposite and equally hypothetical assumption that the valency is variable, leads to no advancement. The first step towards progress in the matter would be made, if some hypothesis as to the cause of this variability were proposed....Since...the atoms, so far as our experience goes, have hitherto shown themselves to be throughout constant and invariable in many important respects, it is advisable to regard the atoms for the present, and until the contrary is proved, as invariable magnitudes. It is also advisable to attempt to show that the varying composition of chemical compounds is dependent upon an invariable valency of the atoms and the varying conditions under which it makes itself felt, or by which its action is influenced to a greater or less extent. Should such an attempt be unsuccessful, then in order to found a theory agreeing with facts, the changes which may be supposed to take place in the atoms themselves must be ascertained, and the causes of these investigated." (Lothar Meyer, *Modern Theories.*)

Attempts at hypotheses concerning the cause of valency have not been wanting, but so far everything done in this direction has not gone beyond the stage of the tentative, though there is much to attract in the speculations which would connect the phenomena of valency with the shape and the motion of the atoms. Van't Hoff's name is associated with such a hypothesis:

"The effect which atoms mutually produce at great distances is dependent only on the mass and the distance; the shape of the atoms and their motion have no effect. But in the case of the very small distances within

Van 't Hoff accounts for valency and its change with tempera- ture by shape and motion of atoms.

the molecules, the matter stands differently; the effect of form and of motion becomes the dominant factor, and the simple effect of gravity ceases to be recognised; the result is the manifestation of affinity and valency, *e.g.* the production of chemical effect.

(1) *The influence of Shape.* The phenomena of valency may be referred to the effect of the shape or form of the atoms. A simple consideration will show that any deviation from spherical shape must result in greater manifestations of the attractive force in certain definite directions, the cause being that the atom is so to speak more easy of approach in these places. Each different shape is therefore the cause of the manifestation of a certain number of predominant attraction capacities, that is of valencies. Thus an arbitrarily chosen shape like that represented [in fig. 94] will—if we restrict our consideration to one plane only—exhibit three predominant attraction capacities along OA, OB and OC, and three subordinate ones along Oa, Ob and Oc. An atom of such a shape would therefore comport itself generally as trivalent and occasionally as hexavalent. And it should be noted in connection with this, that just as the number and the kind of the valencies may be deduced from the shape of the atoms, so conversely the

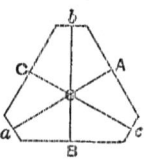

Fig. 94.

shape of the atom might be deduced from an accurate knowledge of its valency. Again, whenever the effect of these attractions at small distances is influenced by the nature of the atoms attracted, we shall find that the number of the effective valencies of the attracting atom will be affected, and a change of valency will be observed on comparing the combinations of that one element with different other elements.

(2) *The influence of Motion.* If an atom moves regularly in all directions about a certain fixed position, a change of external form, and hence also a change of affinity and of valency, is the necessary consequence. If we consider a special case, and if for the sake of sim- plicity we select that of the atom desig- nated above by $AcBaCb$ and considered as before in one plane only, then, when its vibrations have moved that atom from its position of equilibrium by the distance R_1, it will have assumed the external form represented by $A_1c_1B_1a_1C_1b_1$. The result of this change is a twofold one :

(i) All the other atoms will be con- strained to remain at a greater distance from the one considered, and. they will consequently be held by it by a lesser force, *i.e.* the affinity effects of an atom become weaker when it is vibrating. ...

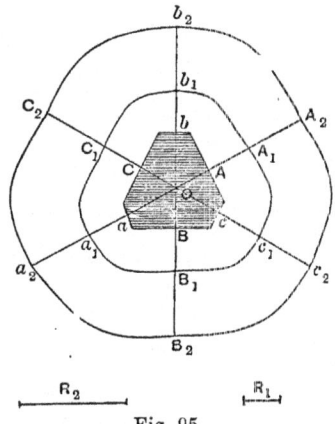

Fig. 95.

(ii) The directions of the predominant attractions are less strongly marked, because it is now a case of the same differences between distances of actually greater value. The three valencies above described as subordinate are thereby made less prominent still, and the atom will behave generally as trivalent. A further increase in the vibration extent to that of the distance R_2, will make the shape of the atom that represented by $A_2c_2B_2a_2C_2b_2$, and the above described effects will become still more strongly marked. The atom gradually loses all chemical power.

If then we admit that the extent of the vibratory motion of the atoms is the result of temperature, the above view leads to the inference, for which we have experimental support, that a rise of temperature diminishes the number of effective valencies, that it weakens the affinity effects, and that it gradually reduces the mutual action of the atoms to that of a simple manifestation of gravitation. And as a matter of fact we know that there are limiting values of temperature, above which chemical action ceases; and on the other hand that as the temperature falls the phenomena get extremely complicated, owing no doubt to the existence and the effect of valencies before overlooked." (*Ansichten über die organische Chemie*, 1881, pp. 3 *et seq.*)

Prof. Armstrong, in his article on Isomerism in Watts' *Dictionary of Chemistry*, says:

"Most discussions on valency are dialectical rather than scientific, in consequence of our powerlessness at present to decide what constitutes a 'valency.' The deduction from Faraday's law of electrolysis to which Helmholtz has drawn the attention of chemists, that definite, as it were 'atomic' charges of electricity are associated with the atoms of matter—that a monad bears a single charge, a dyad two, a triad three—is the only approach yet made to a theory of valency."

Connection between Faraday's law of electrolysis and valency.

The following quotations from the Faraday lecture of 1881 (*J. Chem. Soc.*, 39, 1881, p. 277) contain the pronouncement referred to by Prof. Armstrong, preceded by Helmholtz's exposition of such of Faraday's experimental results as are made the basis of the doctrine propounded.

"When Faraday began to study the phenomena of decomposition by the galvanic current...he put a very simple question,...he asked, what is the quantity of decomposition if the same quantity of electricity is sent through several electrolytic cells? By this investigation he discovered that most important law, generally known under his name, but called by him the law of definite electrolytic action....He compared the amount of decomposition in cells containing different electrolytes, and he found it exactly proportional to the chemical equivalents of the elements, which were either separated or converted into new compounds....According to the...chemical theory of quantivalence, therefore, the same quantity of electricity passing through an electrolyte either sets free, or transfers to other combinations, always the same number of units of affinity at both electrodes ; for instance, instead of $\left.H\atop H\right\}$, either $\left.K\atop K\right\}$, or $\left.Na\atop Na\right\}$, or $Ba\Big\}$, or $Ca\Big\}$, or $Zn\Big\}$, or $Cu\Big\}$ from cupric salts, or $\left.Cu\atop Cu\right\}$ from cuprous salts, etc.

Faraday's discovery of strict proportionality between amount of electrolytic decomposition and the chemical equivalents.

...Products of decomposition cannot appear at the electrodes without the occurrence of motions of the constituent molecules of the electrolyte throughout the whole length of the liquid.... [Faraday] proposed for these atoms or groups of atoms transported by the current through the fluid the Greek word 'ions,' the 'travellers.'...Faraday's law tells us that...'the same definite quantity of either positive or negative electricity moves always with each univalent ion, or with every unit of affinity of a multivalent ion.'...This quantity we may call the electric charge of the atom [affinity unit?].

Hitherto we have spoken only of phenomena. The motion of electricity can be observed and measured. Independently of this, the motion of the chemical constituents can also be measured. Equivalents of chemical elements and equivalent quantities of electricity are numbers which express real relations of material objects and actions. That the equivalent relation of chemical elements depends on the pre-existence of atoms may be hypothetical ; but we have not yet any theory sufficiently developed which can explain all the facts of chemistry as simply and as consistently as the atomic theory developed in modern chemistry.

Now the most startling result of Faraday's law is perhaps this. If we accept the hypothesis that the elementary substances are composed of atoms, we cannot avoid concluding that electricity also, positive as well as negative, is divided into definite elementary portions, which behave like atoms of electricity. The same atom can be charged in different compounds with equivalents of positive or of negative electricity....Faraday often...expressed his conviction that the forces termed chemical affinity and electricity are one and the same....The facts leave no doubt that the very mightiest among the chemical forces are of electric origin. The atoms cling to their electric charges, and opposite electric charges cling to each other....If we conclude

Faraday's law of electrolysis, together with atomic hypothesis, lead to the inference that electricity is divided into definite elementary portions.

from the fact that every unit of affinity is charged with one equivalent either of positive or of negative electricity, they can form compounds, being electrically neutral only if every unit charged positively unites under the influence of a mighty electric attraction with another unit charged negatively....This ought to produce compounds in which every unit of affinity of every atom is connected with one and only one other unit of another atom. This...is the modern chemical theory of quantivalence, comprising all the saturated compounds. The fact that even elementary substances, with few exceptions, have molecules composed of

Every affinity unit is charged with one equivalent of electricity, either positive or negative.

two atoms, makes it probable that even in these cases electric neutralisation is produced by the combination of two atoms, each charged with its full electric equivalent, not by neutralisation of every single unit of affinity. Unsaturated compounds with an even number of unconnected units of affinity offer no objection to such an hypothesis; they may be charged with equal equivalents of opposite electricity."

In 1892 Prof. Armstrong, in the article above quoted from, stated that chemists had not yet considered valency from the point of view suggested by Helmholtz. But since then the development in our knowledge of the connection between electricity and matter has been rapid, and the consequent theoretical speculations and deductive applications are beginning to draw the phenomena of valency within their scope. Not more than a beginning has been made as yet; but whilst formerly it had not even been possible to say what class of conceptions were likely to supply the building material for the adequate hypothesis and theory of valency that is required, it now seems probable that these conceptions will be the indivisible "electron," to which the ions owe their electrical charges and the lines of force connecting the electrons.

The quantity of electricity passed through an electrolyte can be measured accurately, and the same holds for the amount of matter deposited at the electrodes as the result of the decomposition produced; hence the ratio between mass and the charge carried by it is known.

Electrons and their relation to ions.

According to Faraday's law this ratio is the same for chemically equivalent amounts of matter, or is in the simple ratio $1:2:3:4$ for equiatomic amounts of elements. Various physical methods leading to fairly concordant results have supplied us with an approximate value for the absolute weight of the atom (*ante*, p. 346), and therefore we know to a similar degree of approximate accuracy the quantity of electricity associated with the actual atomic masses, and which is the same for an atom of chlorine or bromine or iodine or silver as for an atom of hydrogen, exactly twice as great for an atom of oxygen or ferrous iron, etc., three times as great for an atom of ferric iron, etc.

The extensive researches of Prof. J. J. Thomson[1] and his school have shown that in the conduction of electricity through highly rarefied gases the negatively charged particles shot off from the

[1] *Conduction of Electricity through Gases*, 1903.

cathode, which constitute the so-called cathode rays[1], always carry a charge identical with that conveyed by one single atom of hydrogen or of any other monovalent element when the atom acts as an ion in a solution; but that the mass associated in the cathode ray particles with this constant minimum charge is only about one-thousandth of that of the hydrogen atom, and that it is the same whatever the substance supplying the cathode ray particles may be[2]. These particles, all identical in mass and identical in the charge they carry, have by Prof. Thomson been called *corpuscles*, whilst the more commonly used name of *electrons* has been devised by Dr Johnstone Stoney to designate the natural unit of electricity concentrated on a nucleus, the linear dimensions of which are only about one hundred-thousandth those of an atom.

"Every electric charge is to be thought of as due to the possession of a number of electrons, but a fraction of an electron is at present considered impossible, meaning that no indication of any further subdivision has ever loomed even indistinctly above the horizon of practical or theoretical possibility. The electrification of an atom of matter consists in attaching such a unit to it, or in detaching one from it. An atom of matter [with] an electron in excess [or in defect] is [what is] called an ion." (Sir O. Lodge, *Modern Views on Matter, The Romanes Lecture,* 1903.)

In a letter to *Nature* (vol. 70, 1904, p. 176) Sir Oliver Lodge extends Helmholtz's formulation of the connection between atomic charge and atomic valency, adapting it to the explanation of

---

[1] The terms *radiation* and *ray*, which until quite lately were used exclusively to designate the transmission of light, heat, and electricity radially from their source by the vibration in a surrounding medium, are at present also applied to a new and entirely different class of phenomena, namely, the expulsion of minute particles of matter which travel with enormous velocities along straight lines. The *cathode rays* belong to this latter class. When an electric discharge is passed through a highly exhausted tube, a something starts from the negative electrode which produces various effects on obstacles placed in its path : when its passage is arrested by the glass walls of the vessel, vivid phosphorescence ensues, and the luminous spot produced becomes the source of the phenomenon known as Röntgen radiation ; a thin piece of platinum may be heated to incandescence ; a light windmill is propelled ; an electroscope becomes negatively charged. A solid body placed between the cathode and the walls of the vessel casts a shadow, *i.e.* prevents phosphorescence over a certain area otherwise included in the luminous patch, and the shape and size of this shadow depend only on those of the obstacle, and not on the relative size of the electrode. Finally, the something which produces these different effects can be deflected by a magnet (as shown by the change of position of the phosphorescent patch), and it has the power of penetrating very thin sheets of metal. To satisfactorily explain all these different effects and properties, it has been necessary to assume that minute negatively electrified particles, which travel with enormous velocities along straight lines, are shot out from the cathode in directions at right angles to its surface, and that the mass of these particles is much smaller than that of any known atoms.

[2] This subject will be more fully treated in chap. xix.

residual affinity, that is to that part of the doctrine of valency which recognises a gradation in the manifestations of affinity; which for instance differentiates the binding power between the atoms of copper and sulphur and oxygen in copper sulphate from that which causes further combination with water in crystallised copper sulphate or with more water in copper sulphate solution.

"There appears to be a tendency among chemists to abandon their own doctrine of definite valency, and to recognise an indeterminate and fluctuating number of links connecting atoms with each other.

Electron theory applied to account for existence of different grades of chemical union.

The electron theory of the physicist, which assigns one indivisible unit of charge to a monad, two to a dyad, etc. has therefore encountered some opposition, inasmuch as it seems to tend to harden the old doctrine of 'bonds' whereby atoms were supposed to be linked only in a simple definite and numerical way, no fraction of a bond being contemplated."

The statement of the relation between the electron theory and the doctrine of valency, and the argument given to prove that nothing is involved "inconsistent with the existence of fractions of a bond and any required amount of *residual affinity*," are as follows:

"First, the possession by an atom of a definite charge, numerically specifiable as a simple multiple of an indivisible unit, must be accepted as a physical fact.

Second, this fact corresponds with those other facts which originally led chemists to assert, for instance, that nitrogen was a triad or pentad, carbon a tetrad, etc. a position which it would be absurd to abandon. (Incidentally it may be noted that a monad must be either electro-positive or electro-negative, but that a tetrad need not be either[1], since its pairs of charges may be opposite in sign.)

It has been an occasional habit with physicists when speaking of lines of force to think of a single line of attraction or elastic thread joining each negative electron to its corresponding positive charge; each unit charge, in fact, being regarded as the end of a line of force and nothing else....But...

---

[1] Compare in connection with this the periodic law diagram on p. 502, which gives on sesqui-radius 1 the monovalent electro-positive elements K, Rb, etc.; on sesqui-radius 15 the monovalent electro-negative elements F, Cl, etc.; on sesqui-radius 12 the tetravalent electro-neutral elements C, Si, etc.

"The constant valency of carbon, which in most of its derivatives displays 4 affinities, is especially remarkable....This invariability in the saturation capacity appears most strikingly in the property of the carbon atom to act as a tetrad towards both oxygen and hydrogen, unlike most other elements which, if they can combine with both oxygen and hydrogen at all, do so with a different valency in the two cases. This special property of carbon of standing in the same relation to positive as negative elements, is due to its central position in the periodic system, which itself again is the cause of its electrochemical neutrality" (Hinrichsen, *loc. cit.* p. 100).

there is no evidence that each unit of charge ought to have assigned to it one solitary line of force, it might have a great number; though it is true that on that view it becomes a definite question how many lines a unit charge possesses. On any view electrons are supposed to repel and to be attracted with a force varying as the inverse square of the distance, and this is only consistent with a very large number of lines of force radiating from each and starting out in every direction equally.

When opposite charges have paired off in solitude, every one of these lines start from one and terminate on the other constituent of the pair, and the bundle or field of lines constitutes a full chemical *bond*; but bring other charges or other pairs into the neighbourhood, and a few threads or feelers are at once available for partial adhesion in cross directions also....The charge is indivisible, it is an atomic unit (up to our present knowledge); but the lines of force emanating from it are not indivisible or unified at all. The bulk of them may be occupied with straightforward chemical affinity while a few strands are operating elsewhere; and the subdivision of force may go on to any extent, giving rise to molecular combination and linking molecules into complex aggregates...."

Prof. Percy Frankland sees in Sir Oliver Lodge's suggested electrical interpretation of valency " the possibility of an indefinite number of different grades of chemical union, of which the union by chemical bond, hitherto the only one generally recognised, is to be regarded merely as an extreme case '; and as an example of its application he describes the mechanism of the solution of sodium chloride in water:

"In the case of...sodium chloride, we should in the dry state regard the sodium atom united to the chlorine atom by means of a Faraday tube or

Mechanism of the solution of sodium chloride in water in terms of electron theory of valency.

bundle...the union leading to the great stability of the compound as such. On the addition of water, however, some of the constituent fibres or strands of the bundle become deflected in such a way that the sodium and chlorine atoms become respectively combined with water. With sufficient water present the original union between the sodium

and chlorine atoms will become entirely severed, the Faraday bundle starting with its positive extremity on the sodium atom will terminate at its negative end *by means of a plurality of strands on a number of water molecules*, and similarly the Faraday bundle emanating by its negative extremity from the chlorine atom will terminate at its positive end in a plurality of strands also on a number of water molecules. In such a solution we should thus have independence of the sodium and chlorine atoms[1]....

---

[1] The physical and chemical properties of solutions of substances of the type of sodium chloride all point to the independent presence of two constituent parts of the solute, which are the same as the ions into which it is decomposed on electrolysis. Hence the ionic or dissociation theory of solution assumes the substance dissolved to have been split into two parts.

Moreover the union between sodium and chlorine would be entirely abolished through the complete diversion of the strands of the Faraday bundle formerly uniting them, whilst the union between the oxygen and hydrogen would be but slightly weakened owing to only a small fraction of the total number of strands in the bundles uniting the oxygen and hydrogen in each molecule being diverted by the sodium and the chlorine." (*Nature*, 70, 1904, p. 223.)

Doubt has often been expressed as regards the importance and utility of the study of valency. The unnecessary and unsuccessful attempt made at one time by great masters of the science to class certain valency phenomena as exceptional, and the barrenness of the consequent dialectical discussion concerning constant and variable valency, afford some justification for such a view in the past. It must also be admitted that the study of inorganic chemistry has so far not been advanced to any degree by the doctrine of the different saturation capacity of the elements. But against this must be set the brilliant success achieved in the domain of organic study. Structural chemistry is based on the doctrine of valency; and it seems justified to describe as marvellous the results which have followed from the application of structural chemistry to the explanation of the phenomena of isomerism.

The relative importance of the doctrine of valency in inorganic and organic chemistry.

CHAPTER XVIII.

BERZELIUS AND ISOMERISM.

PROUST'S work had led to the clear recognition that the same constituents may unite in varying ratios, and that the combinations so formed are each endowed with definite characteristic properties, the accompaniment of definite fixed composition. Hence the properties of a substance came to be looked upon as dependent, not only on the qualitative, but also on the quantitative composition of the constituent smallest particles; or, expressed in the terminology which we now use, the properties of the molecules were recognised to depend on the kind and on the relative number of the constituent atoms. But all through the first quarter of the 19th century

Properties of molecules recognised to depend on kind and relative number of constituent atoms, but identity of these supposed to involve identity of properties.

" it had been accepted as an axiom that substances which contain the same elements and the same relative quantities of these must also of necessity have the same chemical properties." (Berzelius, *Jahresbericht*, 1832.)

From 1825 onwards facts began however to accumulate which made such a view untenable. In that year Berzelius, in his "Annual Report on the Progress of Chemistry," says :

" Wöhler has continued his researches on cyanic acid. Silver cyanate[1] when analysed in a variety of ways yielded 77·23 % silver oxide and 22·77 % cyanic acid."

---

[1] Silver cyanate is made by double decomposition between potassium cyanate and a silver salt. Potassium cyanate itself is prepared by the oxidation of potassium cyanide or potassium ferrocyanide.

$$KCN + O = KCNO ; \qquad KCNO + AgNO_3 = AgCNO + KNO_3.$$

In the year following he reports on Liebig's fulminate of silver[1]:

"This substance has been made the subject of a new joint investigation by Gay-Lussac and Liebig, and the result is of the utmost interest. Fulminate of silver dried at 100° loses all its water, and in 3 experiments they found between $16\cdot87$ and $17\cdot38\,\%$ cyanogen, the mean being $17\cdot16\,\%$. The silver oxide was separated by hydrochloric acid and was found to be $77\cdot528\,\%$ of the weight of the salt. These quantities add up to $94\cdot688\,\%$. The deficiency is only $5\cdot312\,\%$, a quantity absolutely equal to that of the oxygen in the silver oxide. Hence the salt was composed of $77\cdot528\,\%$ of silver oxide and $22\cdot472\,\%$ of cyanic acid[2]; but this result is identical with that obtained by Wöhler in his analysis of silver cyanate, in spite of which agreement between the analytical results the two substances have not the same properties. The chief difference is that Wöhler's cyanate of silver when heated by itself does not explode but only burns with moderate intensity, and further that when decomposed by acids it is completely changed into carbonic acid and ammonia..., whilst fulminic acid has explosive properties and gives ammonia and prussic acid when its salts are decomposed by oxyacids. These facts point unquestionably to a difference in composition." (*Jahresbericht*, 1826.)

Cyanic acid and fulminic acid have same composition but different properties.

Berzelius then proceeds to tinker with the analytical results of Gay-Lussac and Liebig, with the object of producing perforce the difference in composition required by the theoretical views of his day. But repetition of the analyses only tended to confirm their accuracy, and Gay-Lussac, after commenting on the identity between his own and Wöhler's numbers, had already said:

"Since these acids are so entirely different in their properties, we are obliged to assume for the purpose of giving an explanation of this difference, that their elements are united in a different manner; but a relation such as this requires to be further investigated." (Footnote to Wöhler's paper, "Recherches analytiques sur l'acide cyanique," *Ann. Chim. Phys.* 27, 1824, p. 200.)

In the *Philosophical Magazine* for 1825 Faraday[3] gave an

---

[1] This is the substance which drew the attention of the boy Liebig to chemical phenomena. "In the market at Darmstadt I watched how a peripatetic dealer in odds and ends made fulminating silver for his pea-crackers....I observed the red vapours which were formed when he dissolved his silver, and that he added to it nitric acid, and then a liquid that smelt of brandy, and with which he cleaned dirty coat collars for the people" (Liebig's Autobiography, *Chem. News*, 63, 1891, p. 265).

[2] Cyanic acid was according to the then general views about the nature of acids (*ante*, p. 172) cyanogen + oxygen.

[3] Michael Faraday (1794—1867), associated from his boyhood with the Royal Institution, preeminent alike as a theoriser, experimenter and lecturer. Chemistry owes him the discovery of the law of electrolytic decomposition, investigations on the liquefaction of gases, and researches on hydrides and chlorides of carbon.

account of his investigation of the liquid found in the cylinders

Faraday's dis-
covery of a gas
the same in
composition
as ethylene,
but different in
spec. grav. and
properties.

containing the gas compressed to 30 atmospheres, which was sent out for illuminating purposes by the " Portable Gas Company." He was able to prove that this liquid was a mixture of different hydro-carbons, which he succeeded in separating from each other by their different volatility. The portion which at atmospheric pressure and at ordinary temperature was gaseous, but at $-18°$ could be condensed to a liquid, was made the subject of a special investigation. He found the specific gravity$_{(H=1)}$ to be between 27 and 28.

"...39 cubic inches introduced into an exhausted glass globe were found to increase its weight 22·4 grains at 60° F., bar. 29·94. Hence 100 cubic inches weigh nearly 57·44 grains [the weight of an equal volume of hydrogen being 2·118 grains]."

He proved that the gas could be absorbed by sulphuric acid, with which it formed a compound, and further that:

" A mixture of 2 volumes of this vapour with 14 volumes of pure oxygen was made and a portion detonated in an eudiometer tube. 8·8 volumes of the mixture diminished by the spark to 5·7 volumes and then by solution of potash to 1·4 volumes, which were oxygen. Hence 7·4 volumes had been consumed consisting of:

Vapour of substance	$=1·1$
Oxygen	$=6·3$
Carbonic acid formed	$=4·3$
Oxygen combining with hydrogen	$=2·0$
Diminution by spark	$=3·1$

This is nearly as if 1 volume of the vapour or gas had required 6 volumes of oxygen, had consumed 4 of them in producing 4 of carbonic acid, and had occupied the other 2 by 4 of hydrogen to form water. Upon which view, 4 volumes or proportionals of hydrogen $=4$, are combined with 4 proportionals of carbon $=24$, to form 1 volume of the vapour, the specific gravity of which would therefore be 28. Now this is but little removed from the actual specific gravity obtained by the preceding experiments; and knowing that this vapour must contain small portions of other substances in solution, there appears no reason to doubt that, if obtained pure, it would be found thus constituted."

After having shown that though in many ways similar to olefiant gas[1], the hydrocarbon investigated was in others essentially different from it, Faraday says:

[1] *Olefiant gas*, spec. grav.$_{(H=1)}=14$; 1 vol. of the vapour requires 3 vols. of oxygen, consumes 2 of them in producing 2 vols. of carbonic acid, and occupies 1 vol. by 2 vols. of hydrogen to form water. Hence 2 vols. of hydrogen $=2$ are combined with 2 proportionals of carbon $=12$, to form 1 vol. of the vapour weighing 14 .....Olefiant gas is absorbed by sulphuric acid, with which it forms a compound, and it unites directly with chlorine.

" Though the elements are in the same proportion as in olefiant gas, they are in a very different state of combination."

He then refers to a previous discovery of Dalton, who in oil gas had found a vapour of greater specific gravity than olefiant gas, requiring a greater amount of oxygen for its combustion, but in other respects very similar to olefiant gas[1], and he quotes from a writer in the *Annals of Philosophy*, who says the substance is

Dalton's discovery of a gas having same composition as ethylene but a greater spec. grav.

"a modification of olefiant gas, constituted of the same elements as that fluid, and in the same proportions, with this only difference that the compound atoms are triple instead of double[2]."

" This I believe is the first time that two gaseous compounds have been supposed to exist, differing from each other in nothing but density...and though the proportion of 3 : 2 is not confirmed, yet the more important part of the statement is, by the existence of the compound described above[3] which though composed of carbon and hydrogen in the same proportions as in olefiant gas is double the density....In reference to the existence of bodies composed of the same elements and in the same proportions, but differing in their qualities, it may be observed that now we are taught to look for them they will probably multiply upon us."

Berzelius in his report published in 1827 (referring to 1825) gives the following summary of Faraday's discovery and of his own attitude towards its theoretical bearings.

Berzelius on Faraday's discovery.

"This memoir contains the interesting discovery that two substances of different properties may have an identical composition both as regards the elements constituting them and their relative quantities, but with the difference that the compound atom of the one contains more atoms of each element than does the compound atom of the other....The two gases are of like constitution, but...a given volume of the one contains twice as many simple atoms as does the other, and this produces a certain dissimilarity in physical and chemical character....Definite knowledge concerning this phenomenon would be of such significance in the doctrine of the

---

[1] " An elastic fluid which agrees with olefiant gas in being condensable by chlorine, but consumes more oxygen and gives more carbonic acid by combustion, and has a higher specific gravity than olefiant gas" (Henry, *Phil. Mag.* 57, 1821, p. 303).

[2] The writer in the *Annals of Philosophy* considers that the gas found by Dalton, the specific gravity of which is 1·395, is a mixture of olefiant gas whose specific gravity is ·972 with another gas of specific gravity 1·486, and that the relation between the composition of these two substances is expressed by the formulae : Olefiant gas $= C_2H_2$; modification of olefiant gas $= C_3H_3$.

[3] The reference is to the substance discovered and just described by Faraday ; this is butylene, the homologue next but one to ethylene :

Berzelius' formulae.   Formulae now used.
CH_2 ....Olefiant gas ....$= $ Ethylene......C_2H_4
                                                  Propylene ...C_3H_6
C_2H_4 ....Faraday's new hydrocarbon $=$ Butylene......C_4H_8

composition of vegetable and animal bodies and would have so important a bearing on organic chemistry, that it must not be accepted as demonstrated until its truth has been subjected to the most severe proof. It is not my intention to dispute the possibility or the actuality of such a fact, but I maintain that before accepting it with confidence, the relation observed by Faraday must be found in a number of other cases."

Towards such a consummation Berzelius made his own contribution. Gay-Lussac had analysed an acid obtained from tartar, and had found that its neutralisation capacity was very nearly the same as that of tartaric acid. This substance which had been named "racemic acid" was investigated by Berzelius with the following results:

Berzelius proves that racemic and tartaric acids, though different in properties, have same composition.

"My analysis shows that [this acid] has the same neutralising power and the same composition as tartaric acid. It differs from tartaric acid by its lesser solubility in water, by the fact that in the crystalline state it contains two atoms of water...and by not giving with potash and soda a double salt crystallising like Rochelle salt[1]....But the greatest difference between these acids is presented by the...lime salt which is so slightly soluble in water that the acid after a time produces considerable cloudiness in a solution of gypsum[2]....These salts are further differentiated from each other by their crystalline form..." (*Jahresbericht*, 1832.)

In the *Jahresberichte* of 1832 and 1833 Berzelius formally adopts the addition to the doctrine of chemical composition necessitated by these discoveries, and suggests the classification and nomenclature of the new phenomena which has become classical:

"In physical chemistry it has long been considered an axiom that substances composed of the same constituents and of the same relative quantities of these must of necessity also have the same chemical properties. Experiments of Faraday seemed to show that an exception to this axiom might occur if two bodies have the same composition, but differ in that, although the relative proportion between their elements is the same, the one contains twice as many simple atoms as does the other. An example of this is afforded by the two hydrocarbons, olefiant gas, CH_2, and that described by Faraday, which is more condensable, has the composition C_2H_4, and therefore a specific gravity twice as great as that of the former. In this case the identity of composition is only apparent, for the

Berzelius recognises existence of substances same in composition, different in properties.
(i) Compound atoms composed of same relative but different absolute number of atoms.

[1] Rochelle salt, the double tartrate of sodium and potassium, $KNaC_4H_4O_6.4H_2O$, also known as Seignette salt, crystallises in beautiful rhombic prisms; it was first prepared in 1672 by a druggist named Seignette, living at Rochelle.
[2] Hence calcium racemate is considerably less soluble than gypsum, $CaSO_4 + 2H_2O$, of which at 15° one part by weight requires 398 of water for solution.

compound atoms are distinctly different, the relative number of the
elementary atoms being equal but the absolute number unequal. More
recent experiments have further shown that the absolute as well as the relative
number of atoms can be equal, and yet that their combination may take
place in a manner so different that the properties of bodies having absolutely
like composition may become unlike. To such a result we have been led only
very gradually. Thus I showed some years ago that two oxides of tin[1] exist,
having the same composition but different properties; soon after, it was discovered that Liebig's fulminic acid and Wöhler's cyanic acid have the same
composition and the same saturation capacity; and finally an analysis made
by me of...racemic acid has proved, in what I would venture to call a decisive
manner, the absolute identity of composition of two substances which have
different properties. Racemic acid has the same composition as tartaric acid;

(ii) Compound atoms composed of same relative and absolute number of atoms, differently united.

it consists of the same elements united in the same atomic
ratio, and it has a neutralisation capacity absolutely the same
as that of tartaric acid. If we wish to embark on hypothetical
speculations concerning such a relation, it would seem as if
the simple atoms of which substances are composed might be
united with each other in different ways....

Mitscherlich's remarkable discovery, that substances composed of different elements but containing these in the same

[1] Berzelius, as always, disregards the water of constitution of acids or basic
hydroxides, and the two substances referred to by him as "oxides" of tin are really
"hydroxides."
" The tin oxide prepared by the action of nitric acid on tin possesses properties
which are not found in the one precipitated from tin chloride by an alkali such as
ammonia. This was the first case of isomerism noted, though of course at that
stage it was natural to account for the difference by a difference in the composition
of the two modifications....From experiments made by me in 1811, I took the oxide
prepared by precipitation from the chloride...to be the sesquioxide, because at that
time it was an axiom in chemistry that identity of composition went with identity
of properties....But on repeating the experiments, I found in both oxides the same
amount of oxygen. This fact, that substances of the same composition could differ
in their properties, had thence to be looked upon as an interesting exception, until
with time, experience led to increase in the number of such cases, and a general
result could be deduced."
Calling the oxide obtained by nitric acid the β modification, and the other one
the α modification, we find :—

α Modification.	β Modification.
Properties much more basic than those of the β variety; the salts dissolve in water easily, and are not decomposed by it.	The salts are difficultly soluble, and are decomposed by water into an insoluble basic salt and free acid.
When moist easily soluble in nitric acid.	Insoluble in nitric acid.
Soluble in dilute sulphuric acid ; the solution does not gelatinise on boiling.	Insoluble in sulphuric acid, even if concentrated.
Easily soluble in hydrochloric acid ; when boiled the solution remains clear.	Unites with hydrochloric acid to a substance insoluble in excess of the acid, but soluble in water ; the aqueous solution gelatinises on boiling.

" Both modifications dissolve in caustic alkalies or alkaline carbonates, and when atomic ratio and arranged in the same manner, crystallise in the same form or are isomorphous, as we now say, has thus received its complement; and this complement consists in the discovery that bodies exist composed of an equal number of atoms of the same elements, arranged, however, in an unlike manner, and possessing therefore different chemical properties and crystalline forms, *i.e.* that they are heteromorphous. If further investigation should confirm this view, an important step would have been taken in the advance- ment of our theoretical knowledge concerning the composition of substances. But since it is requisite that we should be able to express our conceptions by

Name "isomerism" given.

definite and appropriately chosen terms, I have proposed to call substances of the same composition and of different properties 'isomeric,' from the Greek ἰσομερής (composed of equal parts)."

The number of phenomena so designated grew apace; it soon became necessary to subdivide and classify them, and hence to extend the nomenclature.

Differentiation of phenomena at first all de- signated by isomerism : (1) Isomerism. Relative and absolute num- ber of com- ponent atoms the same.

"The ideas which I developed in my last Annual Report on the subject of isomeric substances, that is, of substances which have the same composition but different properties, have not failed to meet with extensive and daily increasing application. Hence in order not to confuse phenomena not absolutely identical, it becomes necessary to define more accurately the meaning of the term *isomeric*. I had pointed out that I had thus designated substances which are com- posed of the same absolute and relative number of the same elements, and which have the same atomic weight[1], as for instance the two oxides of tin, the two phosphoric acids[2], etc. But such

---

reprecipitated by acids, they retain the properties they had before dissolving in the alkalies.

Both modifications can be transformed into one another." (Berzelius, *Lehrbuch*.)

"Two isomeric stannic hydroxides are known, each of which behaves as an acid, yielding a corresponding series of salts. From analogy it would be expected that the acids would correspond to meta- and ortho-silicic acids, and have the composi- tion H_2SnO_3 and H_4SnO_4; [but] both acids exist in all degrees of hydration between the limits required by these two formulae. No satisfactory explanation of the difference in their constitution has...as yet been given. The two acids are distin- guished as stannic and metastannic acids, or as α- and β-stannic acids, and each yields a series of salts from which the original acid may be again obtained by the action of stronger acids." (Roscoe and Schorlemmer, *Treatise on Chemistry*.)

[1] *Atomic weight* as here used corresponds to our *molecular weight*.

[2] "This acid presents the very remarkable phenomenon that under different con- ditions it can assume such different properties that though its composition is not in the least different it can no longer be regarded as the same acid....The relative and absolute weight of the components of the acid remains the same throughout, but presumably the oxygen and phosphorus atoms are differently grouped....We will here call these two modifications α and β phosphoric acids; the properties which distinguish these from one another are:

α Phosphoric acid.   β Phosphoric acid.

Preparation :

(i) Solution of the oxide in cold water.	(i) Aqueous solution of the oxide kept for some days.

cases must not be confounded with others in which the relative number of atoms is the same, but not the absolute number. Thus the relative number of carbon and hydrogen atoms in olefiant gas and "Weinöl" is identically the same (*i.e.*, the number of atoms of hydrogen is twice that of carbon), but one atom of the gaseous substance contains only one atom of carbon and two of hydrogen, $= CH_2$[1], whilst one atom of the oil contains 4 atoms of carbon and 8 atoms of hydrogen, $= C_4H_8$[2]. For the designation of this type of similarity in composition combined with dissimilarity in properties, I would suggest the term *polymeric* (from πολύς, many). But there are yet other relations in which substances appear isomeric in the true sense of the word...without actually being so. Such a case arises when substances consist of two compound atoms of the first order, which are related to one another in different ways and which in consequence can form dissimilar combinations, *e.g.*, S̈nS̈, stannous sulphate[3] and S̈nS̈, stannic sulphite[4], contain the same absolute and relative number of atoms of the same elements and have the same atomic weight, but (in case the latter salt should exist) they could not be considered as one and the same substance. Substances of this kind, with time or change of temperature, suffer a transposition of their components without anything being added or taken away, and a combination of different constitution is formed....In order to precisely differentiate such cases from isomerism, we may use for them the designation of *metameric* substances (from μετά, in the same sense as in *metamorphosis*). Another example of this kind is afforded by the interesting relations between cyanuric acid and hydrated cyanic acid, which can alternately change into one another without anything being taken up or separated out, cyanuric

(marginal notes:)

(2) Poly-merism. Relative number of component atoms the same, absolute number different.

(3) Meta-merism. Same elementary atoms differently distributed among two compound atoms which constitute a more complex compound atom.

(ii) Ignition of the β variety.	(ii) Action of dilute nitric acid on phosphorus.

Precipitate produced by
silver nitrate solution: White. Yellow.

Sodium Salt: The crystalline form and amount of water contained are different.
Does not effloresce in air. Effloresces in air and crumbles to a powder.

Coagulates albumen. Does not coagulate albumen."
(Berzelius, *Lehrbuch*, 1835.)

Graham in 1833 showed that these differences are due to the two substances being acids of different basicity, produced by the combination of the phosphoric oxide with different quantities of water:

$$\alpha \text{ Modification H } PO_3...P_2O_5 + \ H_2O = 2H \ PO_3 \text{ (monobasic)}$$
$$\beta \quad ,, \qquad ,, \quad H_3PO_4...P_2O_5 + 3H_2O = 2H_3PO_4 \text{ (tribasic)}$$

Hence these acids are not really an instance of isomerism.

[1] The formula now accepted is double the above. C_2H_4 = ethylene (olefiant gas), M.P. − 160°, B.P. − 103°.

[2] This substance is a by-product in the preparation of ether from alcohol by sulphuric acid. It is an oil boiling at 280°, and solidifying at − 35°, and is composed of hydrocarbons of the same percentage composition as ethylene, containing two atoms of hydrogen to every one atom of carbon. Berzelius made the molecule 4 times as heavy as that of olefiant gas, but the complexity is doubtless considerably greater.

[3] $SnO.SO_3$. [4] $SnO_2.SO_2$.

acid passing from a compound atom of the first order, or from the oxide of a ternary radicle, into a compound atom of the second order, that is, into cyanic acid chemically united with water. I will therefore call these two substances metameric modifications of one another....On the other hand, I would, at any rate provisionally, designate the cyanuric acid and the white substance into which the aqueous cyanic acid changes, as isomeric oxides of the same radicle. I believe that these distinctions are not without importance in the correct appreciation of the occurrences, and I consider the common derivation of the terms from μέρος a suitable reminder of the generic connection between these special phenomena."

It would seem, therefore, as if Berzelius took as the criteria for the existence of *metamerism* as distinct from *isomerism*:

  (i) the presence of typically different radicles,

  (ii) the transformation of the metameric modifications into one another under the influence of heat or reagents.

But unfortunately Berzelius' terminology cannot be strictly applied in this original sense to the phenomena as they are known at present. Of the examples by him used in illustration of his meaning, the first is of a hypothetical nature, and the second is now known to be a case of polymerism, the molecular formulae for the substances involved being: $HCNO$ (cyanic acid), $H_3C_3N_3O_3$ (cyanuric acid), $H_xC_xN_xO_x$ (cyamelide)[1]. It is no wonder, therefore, that in the history of this doctrine we find the terms isomeric, polymeric, and metameric not always used in a consistent sense. It will be convenient to postpone discussion of the terminology until the theoretical basis of isomerism has been dealt with. The discovery of a further set of phenomena in this branch of the science led Berzelius to the coining of another of our current terms.

Mitscherlich's discovery of the existence of sulphur in two different crystalline varieties has been dealt with in chapter XV

Difficulty about strict application of Berzelius' terminology.

[1] *Cyanic acid*, $H_3C_3N_3O_3$, a tribasic acid which crystallises with $2H_2O$, is obtained by heating urea.

$$3\ {NH_2 \brace NH_2} CO = 3NH_3 + H_3C_3N_3O_3;$$

when subjected to dry distillation cyanuric acid gives:
 Cyanic acid, $HONO$, an unstable liquid, which is a monobasic acid and spontaneously changes to a white porcelain-like solid, the polymer cyamelide.
 Cyamelide $(CONH)_x$ when heated changes back again into cyanic acid.

(p. 439), and data have been given there (p. 445) which show the differences in crystalline form and other physical properties of these two modifications of sulphur and of the two crystalline modifications of carbon. A third variety of sulphur—the plastic form—was also known to exist, and it had been found by Frankenheim that by heating and cooling, these modifications could at definite temperatures be made to change into one another. In the description of these phenomena the term isomerism was used, to which Berzelius objects:

"I feel compelled to call attention to the fact that the word isomerism, which is applied to different substances composed of an equal number of atoms of the same elements, is not compatible with the view as to the cause of the different properties exhibited by the various modifications of sulphur, carbon, silicon[1] etc....Whilst the term still lends itself to the expression of the relation between ethylformate and methylacetate, it is no longer suitable in the case of simple substances which assume different properties, and it might be desirable to substitute for it a better chosen term, *e.g.* allotropy, or allotropic modifications[2]. In accordance with these views, there can be more than one cause for that which we call isomerism, namely :

Allotropy. Occurrence of elements in different modifications.

(i) Allotropy, in which case...the difference between the two sulphides of iron[3] is due to the fact that they contain different modifications of sulphur.

(ii) Differences in the relative position of the atoms in the compound, of which the two kinds of ether (ethylformate and methylacetate) are so striking a proof.

(iii) A combination of (i) and (ii)." (*Jahresbericht*, 1841.)

Berzelius' separation of allotropy from isomerism was based on the ground that it could not be due to the same cause, *i.e.,*

---

[1] "Silicon has in common with carbon and boron the property of passing at a higher temperature into a different allotropic state, when it shrinks, becomes denser and heavier, and darker in colour. The properties are changed to such a degree that I must describe separately the silicon before and after it has been subjected to the influence of a higher temperature :

Si before heating	*Si after heating*
Burns readily in air.	Does not burn in air or oxygen, and is not altered by the blow-pipe flame.
Dissolves in hydrofluoric acid with evolution of hydrogen; is also dissolved on heating with concentrated solution of potash.	Not dissolved, even on boiling, by hydrofluoric acid or potash."

(Berzelius, *Lehrbuch*, 1835.)

[2] Otherwise turned, otherwise formed, from ἄλλος, another, τροπός, habit.

[3] FeS_2 occurs as cubic pyrites and rhombic marcasite (*ante*, p. 445).

a difference in the arrangement of the constituent atoms; but neither in the above nor at any later stage does he express a view as to the theoretical foundation, the cause of the phenomenon. Concerning isomerism, it is evident that he held the view that the cause was difference in the arrangement of the constituent atoms, but this is stated almost incidentally only, and there is no doubt that his treatment of the subject was mainly empirical. The facts were not accepted until much unassailable and corroborative evidence had put beyond doubt the existence of substances which, though identical in composition, are dissimilar in properties. When the number of such known facts had increased, they were classified, and a suitable terminology devised; but it remained for

Theory of isomerism.

a later stage to elaborate the theory of isomerism, which has exerted so profound an influence on the development of the science. It has been shown before (p. 26) how it is one of the requirements of a good hypothesis that it should, without any or with only slight modifications, account for discoveries made after its promulgation within the scope of the phenomena to which it refers. How does the atomic hypothesis acquit itself, from this point of view, towards the occurrences comprised under the name of isomerism?

The atomic and molecular hypotheses were first devised to account for the quantitative laws revealed by the study of chemical combination, and of course, whatever the explanation given, it had to be such as to include in its scope the difference in properties exhibited by different substances. In the simple and adequate explanation offered by these hypotheses, the properties of the molecules (equivalent to the *compound atoms* of Dalton and Berzelius), that is, of the smallest portions which would still exhibit the chemical properties of the whole mass, were determined by the kind and the relative number of their constituent atoms.

But the relative number of the different kinds of atoms constituting the molecule may remain the same, whilst the absolute number varies, producing molecules of different degrees of complexity and of different weight. In the simple case of

Polymerism explained in terms of the atomic hypothesis.

combination between two elements A and B, the composition of such molecules would be represented by

$$(pa + qb), \quad (mpa + mqb), \quad (npa + nqb), \text{ etc.}$$

where a and b are the atomic weights of the elements

A and B, and p, q, m, n, etc. are *simple* whole numbers (*ante*, p. 290). The molecular weights of these substances, whose qualitative and quantitative composition is the same, would be in the ratio of $1 : m : n$; and if they are gaseous or gasifiable, the numerical relation between their gaseous densities would also, in accordance with Avogadro's law, be in the ratio of $1 : m : n$. A substance whose constituent molecules have the composition $(pa + qb)$ should differ in properties from one made up of molecules of composition $m(pa + qb)$, or $n(pa + qb)$; and it may therefore be considered an inevitable deduction from the atomic and molecular hypotheses, that substances of the same percentage composition may differ in properties, provided that they also differ in molecular weight, when the ratio between their molecular weights would be a simple whole number. But the hypotheses do not go beyond making the existence of such substances possible, and perhaps even probable; they do not require their occurrence as an inevitable consequence of the validity of the assumptions made. Chemists for some time after the promulgation of the atomic (and molecular) hypothesis were too fully occupied in supplying for it a firm and extensive inductive basis, to follow up possible deductions from the theory; and it was therefore not until a sufficient number of actual instances had been met with, that they recognised the influence on the properties of a molecule[1] of the *total* number of its constituent atoms. But as soon as the existence of substances the same in percentage composition, but different in properties and molecular weight, had been experimentally established, these phenomena, comprised by Berzelius under the name of "polymerism," could be explained in terms of the atomic and molecular theory without any modification of, or addition to, the original hypotheses.

Molecular weights of polymers show simple numerical relation.

The recognition that the properties of a molecule are determined by the kind of the constituent atoms, and by the relative and absolute number of these, lends itself equally well to the explanation of the phenomena of allotropy. The differences between the various modifications of elements are accounted for by different "atomicity" (*ante*, p. 498). In one case, at any rate, we have direct experimental evidence for the validity of

Allotropy explained by different atomicity of the elementary molecules.

---

[1] Or, in the terminology of those days, of a *compound atom*.

this explanation. The volume changes observed in the partial transformation of oxygen to ozone and in the decomposition of ozonised oxygen indicate that the molecular weight of ozone must be greater than that of oxygen; and Soret's[1] diffusion experiments have shown that it is so in the ratio of 3 : 2. This makes the atomicity of ozone 3, that of oxygen, as determined from its molecular and atomic weight, being 2. The different modifications of sulphur, when volatilised or when dissolved, lose their distinctive character, the allotropy being in this case associated with the solid state only; and hence it is merely by argument from analogy that we might assign to these substances the molecular formulae S_x, S_y, S_z, where x, y, z, are different whole numbers. In the absence of direct experimental evidence, it must remain an open question whether the varying complexity of these molecules is produced by the direct union of different numbers of atoms, or by the association of different numbers of molecules of equal atomicity.

The constituent atoms of allotropic modifications of a substance being all the same, we could not expect the differences in chemical properties to be of the same degree and kind as those exhibited by polymers, where, in the course of the metamorphoses, there may be transference of quite different groups of atoms. With allotropy, the differences are of a more markedly physical nature, though the chemical properties may also show variation, *e.g.* ozone[2] effects oxidations not accomplished by oxygen, the varieties of carbon differ in their behaviour towards oxidising agents (*ante*, p. 441), etc. On the other hand, bromine acting on acetylene, C_2H_2, gives the addition products, $C_2H_2Br_2$ (dibromethylene) and $C_2H_2Br_4$ (tetrabromethane), whilst with benzene, C_6H_6, it gives substitution products and hydrobromic acid:

(marginal note: Chemical differences of allotropic modifications less marked than those of polymers.*)*

$$C_6H_6 + Br_2 = C_6H_5Br + HBr,$$
$$C_6H_6 + 2Br_2 = C_6H_4Br_2 + 2HBr, \text{ etc.}$$

Polymerism, which involves the existence of molecules of increasing complexity, is chiefly found in the case of compounds of carbon, which element by virtue of its high valency and of the

---

[1] Paris, *C.-R. Acad. Sci.*, 61, 1865 (p. 941); 64, 1867 (p. 904).

[2] $2KI + H_2O + O_2$...no effect.
$2KI + H_2O + O_3 = 2KOH + O_2 + I_2$.

power to satisfy some of these combining units by union between carbon atoms themselves, forms an almost endless variety of complex molecules. Many cases of polymerism are known, and some authors have found it desirable to classify them, but the custom has not become general, perhaps owing to the difficulty of consistently applying the principle selected for differentiation. According to the division made, two classes are recognised:

(1) Substances of the type of the cyanic acid, cyanuric acid, and cyamelide dealt with above, which under the action of heat or in the presence of some substance which itself remains unchanged are characterised by being capable of direct transformation into one another. Such another case is that of acetaldehyde C_2H_4O, a liquid which readily passes into paraldehyde, $C_6H_{12}O_3$, another liquid of much higher boiling point, and into metaldehyde $(C_2H_4O)_x$, a white solid, which, like the paraldehyde, by distillation with sulphuric acid passes back again into C_2H_4O. The chemical differences shown by polymers of this type are not of a fundamental nature, and the simple explanation of mere coalescence between several molecules, *i.e.* of direct molecular association, would seem in some cases at least permissible.

Classification of polymers:
(1) Molecular association producing slight chemical differences and enantiotropy.

(2) Different molecular structure causing fundamental differences in properties; direct transformation not possible.

(2) Substances of the type of acetaldehyde C_2H_4O and butyric acid $C_4H_8O_2$, which at any rate by laboratory methods cannot be directly transformed one into another and which differ fundamentally in their properties, as is shown below.

	Acetaldehyde, C_2H_4O	*Butyric Acid*, $C_4H_8O_2$
	Colourless mobile liquid, of aromatic smell, boiling at 21°; neutral	Thick liquid of unpleasant smell, boiling at 163°; acid
Action of soda	Gives insoluble resin	Gives the soluble sodium salt $C_4H_7O_2$. Na
Action of phosphorus chloride	Gives ethylidene chloride, $C_2H_4Cl_2$, one atom of oxygen being replaced by two atoms of carbon	Gives butyryl chloride, C_4H_7O. Cl, one hydroxyl group being replaced by one atom of chlorine
Action of oxidising agents	Readily oxidised to acetic acid by contact with air	Very little effect

To this class would belong the olefenes, *i.e.* the hydrocarbons of the general formula C_nH_{2n} comprising ethylene, C_2H_4, and butylene, C_4H_8, the substances which had furnished one of the earliest instances of polymerism.

Of acetylene, C_2H_2, and benzene, C_6H_6, it would be difficult to say to which of the above two types of polymers they belonged. They certainly exhibit fundamental differences in chemical properties, such as are not shown by the substances comprised under (1). But on the other hand, acetylene can under special conditions be condensed to benzene.

The matter of the classification of polymers is however of no importance in the present state of the theory of the subject. It has been said already how the atomic and molecular hypotheses, whilst supplying an explanation for the occurrence of the phenomenon, would yet be compatible with its non-existence; but further, they do not foretell anything concerning the possibility or impossibility of a certain substance exhibiting polymerism; they do not give any indication of the possible number of polymers in each case, and they do not allow of that comparison between " requirements of the theory " and " experimental results " which, in the study of isomerism, has supplied the most striking proof of the validity and utility of the introduction into the science of these hypothetical magnitudes, the atom and the molecule.

Berzelius, who had at first designated as isomeric, quite generally, substances of the same percentage composition, but of different properties, afterwards separated those named polymeric (percentage composition same, molecular weight different, *e.g.* CH_2 and C_2H_4), and those designated as metameric (percentage composition and molecular weight same, component atom groups different, *e.g.* stannic sulphite and stannous sulphate), thereby restricting the name " isomeric " to substances of the same percentage composition, the same molecular weight, and the same component atom groups in the molecule. The scope of the phenomena to which the term is now applied varies, but it is most usual to comprise under it substances of the same percentage composition and the same molecular weight, that is, to take the course intermediate between the two above quoted denotations.

Berzelian and present day denotation of isomerism.

How is such isomerism to be accounted for? Given two or

more molecules, each composed of the same number of the same kinds of atoms, to what may the differences in their properties be due? Such molecular configurations might differ in two respects,

namely: (i) in the distances between the con-

Two Possible causes of differences in configurations of molecules having same formulae.

stituent atoms, and (ii) in the relative arrangement of these. Either of these differences or both might be the cause of the phenomenon of isomerism, and to decide between these possible explanations requires a comparison in each case between the deductions from the hypothesis and the empirical results. This process quickly leads to the rejection of the first hypothesis.

If variations in the distances between the atoms do occur, and if these exert any influence on the properties of the molecules,

either these changes must be gradual, when they

(1) Differences in distances between the atoms. Deduction from this not supported by facts.

should produce a corresponding gradual change in properties, or they must be sudden, positions of stability occurring at certain definite distances only. This latter case would present a problem which in

the present state of our knowledge,—or rather of our ignorance,—concerning the nature and magnitude of atomic kinetics and atomic attractions, could not be solved theoretically. But considerations quite apart from this show that differences in distance between the atoms cannot afford the simple and adequate explanation sought. If these were the cause of the phenomenon, simple molecules composed of only two atoms should exhibit isomerism, which however is only met with in more complex molecular structures; and moreover, the number of isomeric modifications found increases with the complexity of the molecules. Also, without subsidiary hypotheses, differences in distance cannot be made to account for the specific chemical properties of isomeric modifications, for the different splitting of the molecules in chemical metamorphoses, that is, for the characteristic rational formulae of the isomers. How does the alternative hypothesis acquit itself in all these respects?

To explain isomerism by different arrangement of the parts of the molecule (or compound atom) appeared from the very beginning the most obvious and most simple process. It had suggested itself to Gay-Lussac (*ante*, p. 546); Faraday quoted Gay-Lussac's explanation concerning "bodies composed of the same elements, and in the same proportion, but differing in their qualities," expressing his belief "that now we are taught to look for them they will probably multiply on us." Berzelius, as soon as he had accepted the existence of the phenomenon as definitely established, resorted to the same explanation, which he subsequently refers to and uses whenever dealing with substances exhibiting isomerism; but there is no evidence that he associated with it any definite views concerning the structure of his compound atom and the definite spatial relations of its constituents.

(2) Differences in the arrangement of the atoms; Gay-Lussac initiates this view, afterwards promulgated by Berzelius.

In terms of the radicle theory, the explanation retained the form which had been used by Berzelius in the case of the two sulpho-oxides of tin (*ante*, p. 552) and of metameric substances in general; isomeric modifications were considered as combinations of different radicles, the sum total of whose constituent atoms happened to be the same in kind and number. Inductively this view proved adequate for the explanation of the actually known cases of isomerism, and its connection with the development of rational formulae (*ante*, p. 507) was of course most intimate. Thus, of the various known substances which have the formula $C_3H_6O_2$ one, under the action of potash, gives potassium formate (KO_2CH) and ordinary alcohol (C_2H_6O), and another potassium acetate ($KO_2C_2H_3$) and wood spirit (CH_4O). The rational formulae which summarise these chemical properties are $C_2H_5 . CO_2H$ and $CH_3 . CO_2 . CH_3$ respectively. But on its deductive side, explanation by means of a different distribution of the atoms over the constituent radicles, gave little result. Theoretically there was nothing to indicate what form this distribution would take and could take, that is, nothing to fix the number and kind of the possible isomeric modifications. Kolbe's[1] brilliant prediction of the secondary and tertiary alcohols[2] was of quite exceptional nature. Kolbe looked upon alcohol as derived from

Isomerism and rational formulae.

Kolbe's prognosis of two alcohols isomeric with propyl and butyl alcohol respectively.

---

[1] H. Kolbe (1818—1884), pupil of Wöhler, was for 19 years professor at Leipzig, where he displayed his great powers as a teacher and organiser. His experimental researches were mainly in the province of organic chemistry. He was a prolific writer, and a keen and relentless critic of what he considered erroneous views and unjustified hypotheses, chief amongst which he counted structural chemistry.

[2] *Liebig's Ann. Chem.*, Leipzig, 113, 1860 (p. 293).

F.   36

carbonic acid, C_2O_4,—he used in his formulae the equivalents $C = 6$, $O = 8$,— and gave to it the formula $HO . \begin{Bmatrix} C_2H_3 \\ H_2 \end{Bmatrix} C_2, O.$

"Alcohols are combinations of one atom of water with a derivative of carbonic acid, in which, of three oxygen atoms, one has been replaced by a radicle analogous to hydrogen, the two others by two hydrogen atoms."

$$C_2O_4 \qquad\qquad\qquad HO \begin{Bmatrix} C_2H_3 \\ H_2 \end{Bmatrix} C_2, O$$

Carbonic acid.   Alcohol.

This formula he justifies by the known relations between alcohol, aldehyde, and acetic acid, and by the manner of transformation of these substances into one another.

"If we now consider the formulae by which I express the composition of acetic acid and of the corresponding aldehyde and alcohol, namely :

$$HO . (O_2H_3) [C_2O_2] O \qquad \text{Acetic acid}$$

$$\begin{Bmatrix} C_2H_3 \\ H \end{Bmatrix} [C_2O_2] \qquad \text{Aldehyde}$$

$$HO . \begin{Bmatrix} C_2H_3 \\ H_2 \end{Bmatrix} C_2, O \qquad \text{Alcohol,}$$

it becomes at once evident how it is that in the oxidation...two only of the five hydrogen atoms of the alcohol, and one only of the hydrogen atoms of the aldehyde are substituted. This is so because in the alcohol and the aldehyde it is the *separate* hydrogen atoms which are affected by the oxidation, and which present to the oxygen much more easily accessible points of attack than do the remaining hydrogen atoms, which are more closely linked within the methyl groups."

But this view led to the necessary inference of the existence of two isomeric propyl alcohols, and three isomeric butyl alcohols.

"These views concerning the chemical constituents of the alcohols open up the prospect of the discovery...of a new class of substances, which, closely related in composition to the alcohols, will presumably have certain properties in common with them, but will in some important points behave differently. Let us assume that in the alcohol one or two respectively of the separate hydrogen atoms are substituted for by as many atoms of methyl, ethyl, etc., and the result will be new combinations of the type of alcohols, thus :

$$HO . \begin{Bmatrix} C_2H_3 \\ H_2 \end{Bmatrix} C_2, O \qquad \text{Alcohol}$$

$$HO . \begin{Bmatrix} C_2H_2 \\ C_2H_3 \\ H \end{Bmatrix} C_2, O \qquad \text{Monomethylated alcohol}$$

$$HO . \begin{Bmatrix} C_2H_3 \\ C_2H_3 \\ C_2H_3 \end{Bmatrix} C_2, O \qquad \text{Dimethylated alcohol.}$$

The monomethylated alcohol would only be isomeric, not identical with propyl alcohol :

$$HO . \begin{Bmatrix} C_4H_5 \\ H_2 \end{Bmatrix} C_2, O \qquad \text{Propyl alcohol.}$$

Similarly the dimethylated alcohol contains as great a number of elements as butyl alcohol :

$$HO . \begin{Bmatrix} C_6H_7 \\ H_2 \end{Bmatrix} C_2, O \qquad \text{Butyl alcohol.}$$

Although, so far, none of these compounds of the type of alcohol have been prepared, I am firmly convinced of their existence, and believe that as soon as experimental investigations are undertaken in this direction, they cannot fail to be discovered. Even now several points in their chemical behaviour may be foretold. With the halide acids they will presumably, like the normal alcohols, form halide compounds similar to ethyl chloride ; they will produce sulphur compounds and mercaptanes ; and with sulphuric acid they will form combinations similar to ethyl sulphuric acid. But substances of the type of the twice methylated alcohol cannot by oxidation be transformed into aldehydes and acids, since they do not contain the two *separate* hydrogen atoms which, in the case of the normal alcohols, are affected by oxidation. Likewise the oxyhydrates, which in composition are analogous to mono-methylated alcohol and which retain one separate hydrogen atom, cannot yield acids; but they can undergo the oxidation which transforms the normal alcohols into their aldehydes, though not aldehyde but ketone results as the oxidation product.

$$HO . \begin{Bmatrix} C_2H_3 \\ C_2H_3 \\ H \end{Bmatrix} C_2, O + 2O = \begin{Bmatrix} C_2H_3 \\ \\ C_2H_3 \end{Bmatrix} [C_2O_2]$$

Monomethylated alcohol   Ketone."

Three years later, Friedel[1] found that aqueous acetone, on being heated with sodium amalgam, takes up hydrogen, giving a substance of the same composition as propyl alcohol, and which

---

[1] Paris, *C.-R. Acad. Sci.*, 55, 1862 (p. 53).

behaves like an alcohol, in that it can be etherified[1], and that it yields an iodine compound of its radicle. He left

it an open question whether this substance was identical with propyl alcohol, or only isomeric with it. Kolbe at once identified it as representative of one of the new classes of alcohol by him predicted:

"Concerning the constitution and the chemical nature of this substance I believe it to have the composition and the properties which three years ago I assigned prospectively to the monomethylated ethyl alcohol, which is isomeric with propyl alcohol.

$$\left. \begin{array}{c} C_2H_3 \\ \\ H \end{array} \right\} C_2O_2 + 2H = \left. \begin{array}{c} C_2H_3 \\ \\ H_2 \end{array} \right\} C_2O \cdot HO$$

Aldehyde   Ethyl alcohol.

$$\left. \begin{array}{c} C_2H_3 \\ \\ C_2H_3 \end{array} \right| C_2O_2 + 2H = C_2H_3 \left\} \begin{array}{c} C_2H_3 \\ \\ H \end{array} \right\} C_2O \cdot HO$$

Acetone   Monomethylated alcohol.

To decide whether Friedel's alcohol is really propyl alcohol or, as I suppose it to be, monomethylated ethyl alcohol, a simple experiment should suffice. If the compound is propyl alcohol it should, when treated with ordinary oxidising agents, give propionic acid; but if it should be mono-methylated ethyl alcohol, acetone would be the oxidation product resulting." (Kolbe, *Zs. Chem.*, 5, 1862, p. 687.)

Experiment showed the substance to be the secondary alcohol predicted, and two years later Butlerow's discovery of a tertiary alcohol completed the triumph of Kolbe's hypothesis.

Butlerow, by treating with water the solid resulting from the interaction between acetyl chloride and zinc methide, obtained an alcoholic liquid, which he recognised as tertiary

pseudo-butyl alcohol[2].

"It can hardly be doubted that the pseudo-propyl alcohol obtained by Friedel and my own pseudo-butyl alcohol are methylated methyl alcohols, whilst the normal propyl alcohol is an ethylated methyl alcohol, and the normal butyl alcohol a propylated methyl alcohol." (Butlerow, *Zs. Chem.* 6, 1864, p. 385.)

---

[1] Etherification = replacement of the hydroxyl by an acid radicle.

$$C_2H_5OH + C_2H_3O_2 \cdot H = C_2H_5 \cdot C_2H_3O_2 + H_2O$$
alcohol   acetic acid   ethyl acetic ether

[2] See Note 2 on next page.

The advance from "*rational*" to "*structural*" formulae (*ante*, p. 518) led to a rapid and brilliant development of the theory of isomerism, which in its turn fully justified the validity and the utility of the doctrine of valency on which structural chemistry is based. Given the valency of all the atoms involved, then for any combination of a certain number of certain kinds of atoms, that is, for configurations of known molecular formula, the theory determines with absolute definiteness what different arrangements of the constituent atoms may occur, that is, the number of possible isomers. And further, from the variations in atomic grouping revealed by the theory, we can suitably apportion these structural formulae amongst the known isomers whose chemical behaviour we have studied; we can predict the properties of other isomers not yet known, and from the structural formulae obtain indications of how these may be prepared (*ante*, p. 534). Any text-book of organic chemistry contains on nearly every page examples in proof of these statements. There is no need to devote space here to a subject forming so essential a part of the study of organic chemistry.

Isomerism and structural formulae. Valency of constituent atoms determines number of possible isomers for given molecular formula.

The following simple case, given merely for the sake of symmetry in treatment, is a type of thousands like it:

[2] Kolbe's and Butlerow's formulae :

$$\left(\begin{array}{l} H=1;\ O=8,\ \Theta=O_2=16\ ; \\ C=6,\ \Theta=C_2=12\ ;\ Zn=65. \end{array}\right)$$

Present formulae :

$$\left(\begin{array}{l} H=1 \cdot 008;\ O=16\ ; \\ Zn=65 \cdot 4. \end{array}\right)$$

$HO \cdot C_2H_3 \cdot (C_2O_2)O$ ....Acetic acid ....$CH_3 \cdot COOH$.

$\left\{\begin{array}{l} \Theta H_3 \\ \Theta\Theta Cl \end{array}\right.$ ....Acetyl chloride.............$CH_3 \cdot COCl$.

$\left.\begin{array}{l} \Theta H_3 \\ \Theta H_3 \end{array}\right\} Zn$ ....Zinc methide.................$Zn(CH_3)_2$.

(i) $\left.\begin{array}{l} \Theta H_3 \\ \Theta\Theta Cl \end{array}\right\} + 2 \left.\begin{array}{l} \Theta H_3 \\ \Theta H_3 \end{array}\right\} Zn =$

$= \left.\begin{array}{l} \Theta_4H_9 \\ \Theta\Gamma_3 Zn' \end{array}\right\} \Theta + \left[(\Theta H_3) Zn' \right] Cl$

(i) $CH_3 \cdot CO \cdot Cl + 2Zn(CH_3)_2 =$

$CH_3 \cdot C \left\{\begin{array}{l} O \cdot Zn \cdot CH_3 \\ CH_3 \end{array}\right\} CH_3 + Zn \cdot Cl \cdot CH_3$

zinc methylbutyrate.　　zinc chloric methide.

(ii) $\left.\begin{array}{l} \Theta_4H_9 \\ \Theta H_3 Zn' \end{array}\right\} \Theta + 2H_2\Theta =$

$= \left.\begin{array}{l} \Theta_4H_9 \\ H \end{array}\right\} O + \left.\begin{array}{l} Zn \\ H_2 \end{array}\right\} \Theta_2 + \Theta H_4$.

(ii) $(CH_3)_3C \cdot OZnCH_3 + H_2O$

$= (CH_3)_3COH + ZnO + CH_4$
tertiary butyl　zinc　marsh
alcohol.　oxide.　gas.

Considering only saturated compounds, in which therefore no multiple linking of carbon atoms is assumed, the molecule $C_2H_4O_2$ may, on the assumption of the monovalency of H, the divalency of O, and the tetravalency of C, have the structure represented by the following formulae:

Possible isomers of the molecule $C_2H_4O_2$.

I

$(CH_3.CO_2H)$ ......

$$\begin{array}{cc} H & O \\ | & \| \\ H-C-C-O \\ | & | \\ H & H \end{array}$$

II

$(H.CO_2.CH_3)$ ......

$$\begin{array}{cc} O & H \\ \| & | \\ C-O-C-H \\ | & | \\ H & H \end{array}$$

III

$(CH_2OH.COH)$ ......

$$\begin{array}{cc} H & O \\ | & \| \\ O-C-C-H \\ | & | \\ H & H \end{array}$$

IV

$(CHOH.O.CH_2)$ ......

$$\begin{array}{c} H \\ | \\ H-O-C \\ \quad\quad\; \diagdown O \\ H-C \diagup \\ | \\ H \end{array}$$

V

$(CH_2O)_2$ ....

$$\begin{array}{c} H \\ | \\ O-C-H \\ | \quad | \\ O-C-H \\ | \\ H \end{array}$$

VI

$(CH_2O)_2$ ....

$$\begin{array}{c} H \\ | \\ H-C-O \\ | \quad | \\ O-C-H \\ | \\ H \end{array}$$

Four organic compounds are known which require the formula $C_2H_4O_2$. Of these, one, a liquid boiling at 118·1°, has an acid reaction, and when treated with concentrated potash gives a solid mass of a salt which differs from the original substance in that one of the ·four hydrogen atoms has been replaced by one atom of K.

$$C_2H_4O_2 + KHO = C_2H_3O_2K + H_2O.$$

Another, a liquid boiling at 32·3°, is neutral, and its smell is aromatic, not vinegary like that of the first; when it is warmed with the same potash in a retort, volatile wood spirit, CH_4O, is evolved, and a substance is left behind which from the analysis is potassium formate.

$$C_2H_4O_2 + KHO = CH_4O + CHO_2K.$$

These reactions show that to these substances, in the above order, we must assign the structural formulae I and II respectively.

Another compound which has the formula $C_2H_4O_2$ gives the reactions characteristic of the class of substances named aldehydes, which all contain the group
$$\begin{array}{c} O \\ \| \\ -C-H \end{array}$$
, and hence we must assign to it the formula III.

For molecules of the comparatively simple formula $C_3H_6O_2$ it is easy to ascertain the number of configurations that can occur, but with greater complexity the number **Number of possible isomers calculated.** of possible isomers increases at a rapid rate. The determination of this number is a problem capable of mathematical solution, and the necessary calculations have been made by Cayley in the case of the saturated hydrocarbons of the general formula C_nH_{2n+2}, with results of which some are given in the second column of the following table:

Formula	No. of possible Isomers calculated	No. of Isomers actually known
CH_4	1	1
C_2H_6	1	1
C_3H_8	1	1
C_4H_{10}	2	2
C_5H_{12}	3	3
C_6H_{14}	5	5
C_7H_{16}	9	5
$C_{10}H_{22}$	75	6
$C_{13}H_{28}$	802	1 or 2

It is not essential to our acceptance of the theory of isomerism that all the modifications predicted by it should be actually found. It is sufficient that the number met with should not exceed the theoretical, that in the simpler cases it should be equal to it, and that new discoveries should continually reduce the discrepancy between "number of isomers possible" and "number of isomers known."

Amongst the most interesting chapters in the history of the development of organic chemistry are those which tell how different hypotheses as to the constitution of important substances, such as benzene, anthracene, etc. were tested by the crucial experiment of a comparison between the number of isomeric derivatives actually obtained, and that required by each of the different formulae speculatively assigned. The following is a quotation from Kekulé's classical paper on the benzene formula (*Liebig's Ann. Chem.*, 137, 1866, p. 157):

Number of isomeric derivatives the test of the correctness of a structural formula.

" If we look upon benzene as a close chain consisting of 6 carbon atoms, linked alternately by one and two affinity units, we are at once confronted by a further question: are the 6 hydrogen atoms of benzene equivalent, or do they perhaps owing to their position fulfil different functions? If the 6 hydrogen atoms are absolutely equivalent, then the cause of the difference of all the isomeric modifications of the substitution derivatives can be looked for only in the *relative positions* taken up by the elements or the side chains which replace the hydrogen of the benzene. If however the 6 hydrogen atoms of the benzene are not equivalent, then the isomerism may perhaps be partly accounted for by the difference of the absolute position of the elements or side chains replacing the hydrogen, and the possibility of the existence of a far greater number of isomeric modifications is obvious.

Kekulé's benzene formula.

First hypothesis. The 6 carbon atoms of benzene are linked with one another in an absolutely symmetrical manner, and it may therefore be assumed that they form a perfectly symmetrical ring. In that case, the hydrogen atoms are not only placed perfectly symmetrically in relation to the carbon, but they also occupy perfectly analogous positions in the atomic system [molecule]; they are therefore equivalent. Benzene could accordingly be represented by a hexagon, the six corners of which are formed by hydrogen atoms. It then becomes apparent that the derivatives produced by continued substitution may exist in the following isomeric modifications, *e.g.*, for the bromine substitution products :

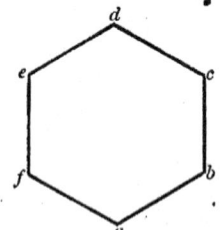

		Modifications	
(i)	Monobrombenzene	1	
(ii)	Dibrombenzene	3	*ab, ac, ad*
(iii)	Tribrombenzene	3	*abc, abd, ace*
(iv)	Tetrabrombenzene	3	*abcd, abce, abde*
(v)	Pentabrombenzene	1	
(vi)	Hexabrombenzene	1	

Second hypothesis. The 6 hydrogen atoms of benzene form three atomic groups, each of which contains two carbon atoms linked by two affinity

.units. The group therefore becomes a triangle, and in addition we may conceive of the constituent carbon atoms as so placed that three hydrogen atoms are within the triangle, and three others on its external sides. Then the 6 hydrogen atoms are alternately of different value, and we could represent benzene by a triangle with three of the six hydrogen atoms at the corners, whereby they are more easily accessible, and the three other hydrogen atoms in the middle of the sides, standing as it were in the interior of the molecule....

This conception opens up the possibility of the existence of a far greater number of isomeric modifications, as may easily be shown by the following example : ·

		Modifications	
(i)	Monobrombenzene	2	*a, b*
(ii)	Dibrombenzene	4	*ab, ac, bd, ad*
(iii)	Tribrombenzene	6	*abc, bcd, abd, abe,*
			ace, bdf
	.etc.		

The experimental investigations required for deciding between these two hypotheses, seemed to Kekulé a task formidable and laborious; but all the same he foresaw the likelihood of its achievement:

"At first it might seem as if a problem such as this could not be solved ; I believe, however, that it is possible to do so by experiment. What is wanted is that the largest possible number of substitution products of benzene should be prepared by the most diverse methods; that they should be most carefully compared with regard to isomerism; that the modifications so found should be counted...."

Kekulé's anticipations have been realised; most extensive investigation of all aromatic compounds has shown that, without exception, mono-, penta-, and hexa-derivatives exist in one modification only, whilst di-, tri-, and tetra-derivatives exist in three modifications, proof of the hypothesis that the structure of benzene is symmetrical.

The generally accepted explanation of isomerism has led to the extension of the original conception of a molecule, and to the recognition that the properties of a molecule depend on the kind, the relative and absolute number, and the arrangement of the constituent atoms.

"A chemical species is a collection of individuals identical in the *nature*, the *proportion*, and the *arrangement* of the elements. All the properties of substances are functions of these three terms, and the object of all our work is to pass, by investigation of the properties, to the knowledge of these three things.

In isomeric bodies, the nature and proportion are the same. The arrangement alone differs. The great interest of isomerism has been to introduce into the science the principle that substances may be, and are, essentially different solely because the arrangement of the atoms is not the same in their chemical molecules.". (Pasteur, 1860.)

There is still a want of agreement in the scope of the phenomena comprised under the name of "isomerism" and in the subdivision of these into classes. It seems desirable to present in tabular form the classifications that have been adopted.

Scope of terms "isomerism" etc.

I

Isomerism

Percentage composition same, properties different

Polymerism	**Metamerism**
Molecular weight *different*	Molecular weight *same*

(Kekulé, *Lehrbuch*, 1866; Fehling, *Neues Handwörterbuch der Chemie*, 1878; Frémy, *Encyclopédie Chimique*, 1882; Ladenburg, *Handwörterbuch der Chemie*, 1887; Lothar Meyer, *Outlines of Theoretical Chemistry*, 1899; Nernst, *Theoretical Chemistry*, 1900.)

II

Polymerism	**Isomerism**
	(in the wider sense)
Percentage composition same, molecular weight *different*, properties different,	Percentage composition same, molecular weight *same*, properties different,
e.g. C_2H_4O (aldehyde) and $C_4H_8O_2$ (butyric acid).	*e.g.* C_3H_8O (propyl alcohol and methyl-ethyl ether).

Metamerism	**Isomerism**
	(in the restricted sense)
The distribution of the atoms amongst the radicles is *different*.	The distribution of the atoms amongst the radicles is the *same*,
(i) The saturation is the *same*, *e.g.* $\dfrac{C_3H_7}{H}\!\!>\!\!O$ and $\dfrac{C_2H_5}{CH_3}\!\!>\!\!O$ (propyl alcohol, B.P. = 97°) (methyl-ethyl ether, B.P. = 10·8°)	*e.g.* $\dfrac{CH_3.CH_2.CH_2}{H}\!\!>\!\!O$ and $\dfrac{(CH_3)_2CH}{H}\!\!>\!\!O$ (normal propyl alcohol, B.P. = 97°) (isopropyl alcohol, B.P. = 83°)
(ii) The saturation is *different*, *e.g.* $CH_3 - CO - CH_3$ (acetone) and $CH_2 = CH - CH_2.OH$ (allyl alcohol).	

(Roscoe and Schorlemmer, *Treatise on Chemistry*, 1894; Bernthsen, *Text-book of Organic Chemistry*, 1894; Richter, *Organic Chemistry*, 1900.)

III

Polymerism	**Metamerism**	**Isomerism**
Percentage composition same, molecular weight *different*.	Percentage composition same, molecular weight *same*, component radicles *different*.	Percentage composition same, molecular weight *same*, component radicles *same*.

(Berzelius, *Jahresbericht*, 1833; Watts' *Dictionary of Chemistry*, 1892.)

Other terms and further subdivisions are met with, such as nucleus isomerism, chain isomerism, isomerism of position, chemical isomerism and physical isomerism, atomic and molecular isomerism, etc., the meaning of which is generally sufficiently indicated by the name, and the extent of application of which is not always fixed.

Until comparatively recently, the text-books relegated to a special class substances said to exhibit *physical isomerism.* The

Physical isomerism; differences are mainly physical.

term was first used by Carius in 1863, and was applied to substances which, whilst almost identical chemically, differed in some physical properties, especially in their action on polarised light[1].

---

[1] The properties of a ray of *ordinary light* are such as to justify the view that the vibrations occur in a plane perpendicular to the direction of propagation, that each vibrating ether particle describes an ellipse about its position of rest as centre, and that the axes of this ellipse are for ever changing in magnitude and direction. In the annexed figure, in which the direction of propagation is perpendicular to the plane of the paper through O, successive positions of an axis of the ellipse are represented by aa', bb', cc', dd', etc. Light called *plane polarised* differs from ordinary light in that each particle moves in a straight line such as aa', and that in all successive vibrations the direction of this straight line remains the same, and such light can be recognised by the difference in properties in the plane of vibration, and in a plane perpendicular to this. The plane passing through OO', the line of propagation, and through aa', the line of vibration, is called the *plane of polarisation.*

The change from ordinary to polarised light can be brought about by (i) reflexion, (ii) passage through certain media, such as Iceland spar (clear rhombohedra of calcite), said to be doubly refracting because a ray of ordinary light is split into two, polarised in planes perpendicular to one another. A Nicol's prism is such a doubly refracting calcite rhombohedron, so modified that of the two rays, one only is transmitted, and one is completely diverted, *i.e.,* the light transmitted is plane polarised, and the plane of polarisation has a definite direction relatively to the prism. In a combination of two such prisms, termed the polariser and analyser respectively, the relative position of these will determine the extent to which plane polarised light emerging from the first is transmitted through the second. When the planes of polarisation are parallel, all the light is transmitted, when at right angles, none is transmitted, and at intermediate positions the effect also is intermediate. When the prisms are so placed that of the light transmitted by the polariser, none emerges from the analyser, the interposition between these of certain solids or liquids or solutions produces light in the field of vision of the analyser, and in order to restore darkness, one or other of the prisms must be turned through an angle either to the right or to the left. The amount of rotation required depends on the length of stratum of matter interposed, on the nature of this matter, and on other factors. If in the figure PP' and AA' are the directions of the planes of polarisation of the polariser and the analyser, and hence also their relative position when no light is transmitted, then if on the interposition of a substance, aa' is the

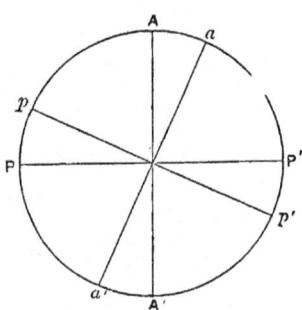

"I have tentatively expressed a view as to the cause of what I call *physical isomerism*. Substances [which exhibit this property] yield, under the same or nearly the same conditions, products which are either identical or physically isomeric. According to our present views, I think it improbable that such substances should have their atoms differently arranged, *i.e.*, that they should be metameric. But it is quite conceivable that

Carius explains physical isomerism by different aggregation of identical molecules.

in the formation of physical isomers, differences of condition may cause the production of substances with the same arrangement of the atoms within the molecule, but with a different aggregation of these molecules; and that thereon depends the difference in their properties....Thus we must consider as certainly only physically isomeric a large number of the substances distinguished by the difference of their action on polarised light, such as the two modifications of amyl alcohol; the tartaric acids, the malic acids, etc." (Carius, *Liebig's Ann. Chem.*, 130, 1864, p. 237 ; 126, 1863, p. 214.)

The name suggested by Carius has passed into the common terminology of the science, but his explanation for the phenomenon designated by it has not proved adequate. In 1873, Wislicenus[1] proposed to substitute the name *geometrical isomerism* for *physical isomerism,* and gave the impetus for the promulgation of a hypothesis which has led to our present theory concerning this type of isomerism.

His researches on the lactic acids revealed the insufficiency of the current theory for the explanation of all the

Wislicenus' researches on the lactic acids.

facts observed. For the understanding of the argument about to be quoted, it will be well to give first the most important of the experimental results to which it refers.

Wislicenus investigated three substances of the molecular formula $C_3H_6O_3$, all of them monobasic hydroxyl acids.

(i) *Ordinary fermentation lactic acid*, optically inactive, oxidises to acetic acid, gives on heating the anhydride lactide, $C_6H_8O_4$, thus:

$$2C_3H_6O_3 = C_6H_8O_4 + 2H_2O.$$

position that must be given to the analyser in order to reestablish complete extinction, it follows that the direction of vibration of the rays which on emergence from the polariser had been along PP', had been rotated by passing through the interposed substance; the direction of their vibration, which must be perpendicular to aa', has become pp'. The phenomenon is called the *rotation of the plane of polarisation*, and the substances that produce it are said to be *optically active*.

[1] Johannes Wislicenus (1835—1902) succeeded Kolbe in 1885 as Professor of Chemistry at Leipzig. Besides his classical-work on aceto-acetic ether, we owe to him the experimental researches and some of the theoretical speculations to which is chiefly due the prominent position now occupied by stereochemistry.

(ii) *Paralactic acid*, rotates the plane of polarisation to the right, is in all chemical properties almost identical with the fermentation acid, except for differences in the solubility of some of the salts (the zinc salt is more soluble, the calcium salt less so).

(iii) *Hydracrylic acid* (ethylene lactic acid), optically inactive, oxidises to carbonic and oxalic acid, when heated splits into water and acrylic acid, thus :
$$C_3H_6O_3 = C_3H_4O_2 + H_2O.$$

These, as well as all the other properties of the three acids, find their expression in the rational formulae assigned to them :

For (i) and (ii)..............$CH_3 . \dot{C}H (OH)(CO_2H)$.

,, (iii)   ....$CH_2(OH) . CH_2(CO_2H)$.

"So far no fact is known which would make it necessary to assign to paralactic acid a structural formula other than that of the fermentation lactic acid. Accordingly we must, at any rate for the time being, assume that paralactic acid is structurally identical with the fermentation lactic acid, *i.e.*, we must represent the sequence in the mutual linkings of the component atoms by the same formula : $CH_3 . CH(OH)(CO . OH)$.

But granting the possibility of the existence of molecules equally composed and structurally identical, though somewhat differing in properties, it is not

Differences between isomeric molecules of same structural formula explained by different spatial arrangement of atoms.

possible to explain such differences otherwise than by the assumption that they are caused merely by a different arrangement in space of the atoms which are united with one another in the same order. It is now a generally accepted view that the chemical properties of a molecule are most decisively influenced by the nature of the component atoms, and the order of their mutual linking. And it seems to me an equally justifiable assumption that, in the case of molecules identical in structure, differences in geometrical sequence,

which in the first instance could produce variations in the magnitude and shape of these molecules, would make themselves noted by variations in the physical properties, and in those properties lying on the boundary line of the physical and chemical relationships, such as solubility, crystalline form, water of crystallisation, etc. Thus we should be led to a conception for which a name has already been introduced into the science, to the strict conception of 'physical isomerism' in contradistinction to the different kinds of 'chemical isomerism.' The name physical isomerism, first used by Carius, was, it is true, applied by him to the designation of all isomerism which is not polymerism or metamerism. Carius thought to explain such isomerism by 'a difference in the aggregation of the molecules,'...but of the cases first comprised under this name, a certain number have since then been explained as isomerism in the restricted sense [*ante*, p. 570]....Hence there is left as the cause of strict physical isomerism only the above-mentioned different arrangement in space of the atoms which are linked to chemically identical structures. But for such cases the term 'geometrical isomerism' would perhaps be more suitable, if, as is usual, the name for the special isomerism is intended to designate, not so much the fact of differences between molecules of quantitatively the same composition, as rather the intrinsic cause of the phenomenon. That this name...is based on a hypothesis is self-evident, and it must be valued accordingly. My conclusion for the present is to declare paralactic acid and the fermentation lactic acid as most probably only geometrically isomeric. Their great similarity, even identity in all chemical properties, the ease of transformation, on heating, of the first into the second, and their differences, particularly in optical behaviour, may all alike be explained on this basis.

Name "geometrical isomerism" substituted for "physical isomerism."

Concerning the special 'how' of this explanation, I am engaged in experimental investigations." (*Liebig's Ann. Chem.*, 167, 1873, p. 343.)

Van 't Hoff, when he supplied the answer to this "how" which is now generally accepted, stated explicitly that it had occurred to him after reading Wislicenus' paper, and that he had since used as a motto the words: "The facts compel us to explain the differences between isomeric molecules possessing the same structural formulae, by a different arrangement of their atoms in space." Before passing on to the exposition of Van 't Hoff's own hypothesis, some account must be given of Pasteur's[1] pioneer work in this field of investigation. Two lectures delivered in 1860 before the Société Chimique de Paris, and published under the title of *Leçons sur la Dissymétrie Moléculaire*[2] contain the result of his now classical research on the tartaric acids. Following up the known fact that

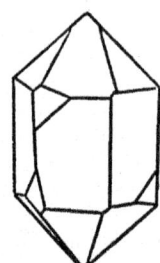
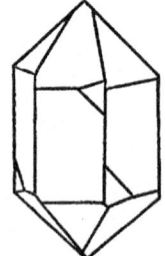

Fig. 96.
Quartz, left-handed.

Fig. 97.
Quartz, right-handed.

"quartz possesses hemihedral faces, and that these faces incline to the right in some specimens, and to the left in others, and that quartz crystals

[1] Louis Pasteur (1822—1895), founder of the microbic theory of disease, owes his position amongst the heroes of chemistry to his researches on optically active crystalline substances and on fermentation, and to the brilliant discoveries in which these resulted.

[2] Reprinted in the English translation *Researches on Molecular Asymmetry*, Alembic Club Reprints, No. 14, from which all the following quotations are taken.

likewise separate themselves into two sets, in relation to their optical properties, one set deviating the plane of polarised light to the right, the other to the left, according to the same laws,»

he studied crystallographically the various tartrates. He found the constant occurrence of hemihedrism, and was led to the

Pasteur estab-lishes relation between hemi-hedrism and optical rota-tion. inference that, as in the case of quartz, there was a relation between the hemihedrism and the optical rotation. An investigation of other organic substances which have the same optical property proved that their crystals also were always hemihedral.

This connection between the hemihedrism of the tartrates and

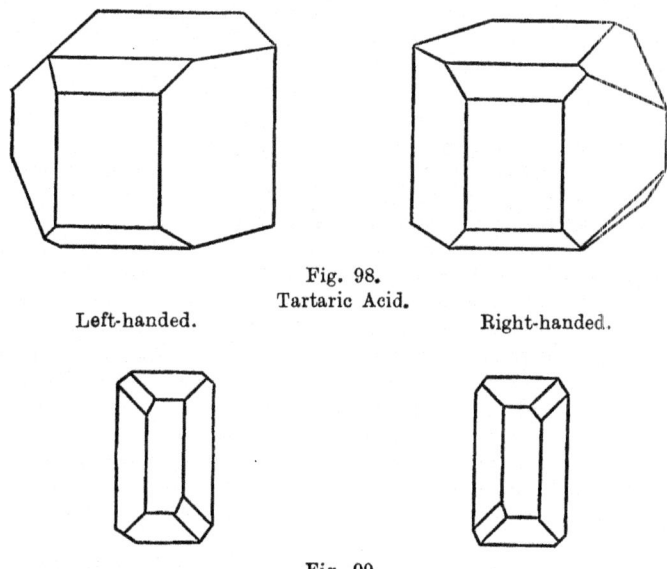

Fig. 98.
Tartaric Acid.

Left-handed.   Right-handed.

Fig. 99.
Ammonium Bimalate.

Left-handed.   .   Right-handed.

their optical activity led Pasteur to attempt an explanation of the anomalous behaviour of ammonium sodium tartrate and ammonium sodium racemate, salts which, though crystallographically and chemically identical, differ in that the solution of

Investigation of the ammo-nium sodium tartrate and racemate. the tartrate is optically active, whilst that of the racemate is not. The presence of hemihedrism in the tartrate and its absence in the racemate should account for this difference. Experiment showed that "as a matter of fact, the tartrate was hemihedral," but the racemate was hemihedral also:

Resolution by crystallisation of ammonium sodium racemate into right-handed and left-handed hemihedral crystals of corresponding optical activity.

" ...Only, the hemihedral faces which in the tartrate were all turned the same way, were in the paratartrate [racemate] inclined sometimes to the right, and sometimes to the left....[He] carefully separated the crystals which were hemihedral to the right from those hemihedral to the left, and examined their solutions separately in the polarising apparatus....The crystals hemihedral to the right deviated the plane of polarisation to the right, and those hemihedral to the left deviated it to the left ; and when [he] took an equal weight of each of the two kinds of crystals, the mixed solution was indifferent towards the light in consequence of the neutralisation of the two equal and opposite individual deviations."

The acids furnished by the two sets of crystals showed the same relations:

" All that can be done with one acid can be repeated with the other under the same conditions, and in each case we get identical, but not superposable products ; products which resemble each other like the right and left hands. The same forms, the same faces, the same angles, hemihedry in both cases. The sole dissimilarity is in the inclination to right or left of the hemihedral facets, and in the sense of the rotatory power."

" On the one hand, the molecular structures of the two tartaric acids are asymmetric, and on the other they are rigorously the same ; with the sole difference of showing asymmetry in opposite senses. Are the atoms of the right acid arranged along the spiral of a right-handed screw, or placed at the corners of an irregular tetrahedron, or disposed according to some particular asymmetric grouping or other ? We cannot answer these questions. But it cannot be a subject of doubt that there exists an arrangement of the atoms in an asymmetric order, having a non-superposable image."

Crystallographic asymmetry and optical activity supposed to be due to asymmetric arrangement of atoms.

In the same year, in 1874, Le Bel in France, and Van 't Hoff[1] in Holland published papers in which they offered an explanation practically the same for cases of isomerism which could not be included under the theories of the time. The views advanced by them proved so apt in the explanation and coordination of facts already known, and so stimulating to further research, that soon there was enough material for separate treatment in book form, and we have had in rapid succession Van 't Hoff's own

Le Bel's and Van 't Hoff's explanation of isomers whose structural formulae are the same.

[1] J. H. Van 't Hoff, b. 1852 at Rotterdam, studied under Kekulé at Bonn and Wurtz in Paris. He held a professorship at the University of Amsterdam from 1878 to 1897, when he was called to a chair created for him in the University of Berlin.

exposition in *La Chimie dans l'Espace* (1875), *Dix Années dans l'Histoire d'une Théorie* (1887), and a number of editions and translations[1], of which each new one reported further veri‐ fications of the theory, extended applications, and reduction in the number of apparent exceptions.

Van 't Hoff tells us that on the whole Le Bel's original paper and his were in accord, but that whilst Pasteur's researches formed Le Bel's starting point, he took for his own " Kekulé's law of the tetravalence of carbon, [to which he] added the hypothesis that the four valencies are directed towards the corners of the tetra‐ hedron in the centre of which is the carbon atom." Van 't Hoff introduced no fundamental change in, or addition to,

Van 't Hoff re‐ tains Kekulé's valency hypo‐ thesis, but considers grouping of atoms in space and not in a plane.

the original valency hypothesis ; a two-dimensional representation of molecular structure could not at any time have been considered as really true to the actual occurrence, but it was legitimate to use it, because of its greater simplicity, as long as it proved adequate to the purpose. And with this recognition and restriction we continue to use plane structural formulae in the majority of cases. Van 't Hoff, following up the suggestion that may be found implied in Pasteur's paper, and that was explicitly stated by Wislicenus, introduces into the science the consideration of the arrangement of atoms in space :

" Stereochemistry [from στερεός, solid], in the restricted sense of the word, comprises chemical phenomena which demand a consideration of the grouping of atoms in space."

Definition of "stereo‐ chemistry."

Carbon compounds only were considered at first, but the scope of the phenomena dealt with has been extended and now includes compounds of trivalent and pentavalent nitrogen, of tin, and of sulphur. But in what follows, the stereoisomerism of carbon alone is dealt with.

"The origin of stereochemical conceptions has been that of all hypotheses : the impossibility of explaining certain phenomena by means of existing theories. It was a question of certain cases of isomerism which the current conceptions on molecular structure could not interpret. As a matter of fact, certain of these, like the tartaric acids, have long been known, but the importance of the phenomenon had been concealed by names such as 'physical isomerism.'

Substances were known, therefore, unmistakably different, which never‐

---

[1] *Die Lagerung der Atome im Raume* (1877); *Chemistry in Space* (1891); *Stéréochimie* (1892); *Die Lagerung der Atome im Raume* (1894); *The Arrangement of Atoms in Space* (1898).

theless were composed of molecules identical according to the existing conception : the same atoms equal in number, linked with one another in the same manner in the two cases. Consequently the difference had to be due, either to the relative position in space of the atoms, or to a difference in the motion of the atoms. I followed the first of these alternatives....

The four groups combined with the carbon atom are Placed at the corners of a tetrahedron. Hence no isomers unless all four groups are different.

In carbon compounds of the type $C(R_1R_2R_3R_4)$, *i.e.* compounds in which four groups of atoms are combined with the carbon, an extra isomerism occurs when these four groups are different, and disappears if but two of them become the same. Assuming a fixed position of the groups round the carbon atom, only a tetrahedral grouping brings us to the same conclusion [as is shown by the figures below],

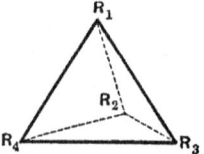 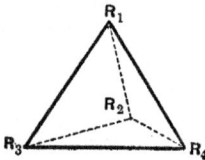

which become identical when R_3 and R_4 become the same, while this leaves the isomerism unaffected if we represent the formulae in one plane [thus]

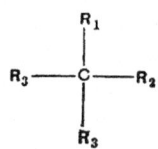   ” (Van't Hoff.)

When $R_1R_2R_3R_4$, the four atoms or groups of atoms, are all different from each other, the carbon atom with which they are united is termed *asymmetric*, and the two possible configurations in space of the combination $CR_1R_2R_3R_4$ give the above figures, which however turned, cannot be made to coincide, being related to one another as an object and its image in a mirror, the sequence of the groups $R_1R_2R_3R_4$ being that of a right-handed and left-handed screw respectively.

Attention must here be drawn to the fact that the "phenomena which it was impossible to explain by means of existing theories" were of two kinds :

Isomerism phenomena for the explanation of which two-dimensional representation inadequate.

(i)   The occurrence of substances unmistakably different, though structurally identical, *i.e.*, the existence of a greater number of isomers than the theoretical. These are referred to in the above quotation.

(ii)   The non-existence of isomers required by the theory.

"That the four valencies of the carbon extend in four symmetric directions in space...was an idea which necessarily occurred to anyone seriously occupied with the problem of the tetravalency of the carbon atom.

For if marsh gas should have the formula

$$C\begin{cases} H \\ H \\ H \\ H \end{cases}$$

we should have to assume the existence of two methyl chlorides,

$$C\begin{cases} H \\ H \\ H \\ Cl \end{cases} \qquad \text{and} \qquad C\begin{cases} H \\ H \\ Cl \\ H \end{cases}$$

but since there exists only one methyl chloride, -bromide, -iodide, etc., such a formula is excluded. If then we express the next probable assumption of a symmetrical extension of the valencies, in the following manner:

$$-\overset{\displaystyle |}{\underset{\displaystyle |}{C}}-$$

we see that this formula cannot be correct either, because though it would require the existence of only one methyl chloride, two chlor-methylenes are indicated, one in which the two chlorine atoms are contiguous, and a second in which they are not,

$$Cl-\overset{\displaystyle Cl}{\underset{\displaystyle H}{C}}-H \qquad \text{and} \qquad H-\overset{\displaystyle Cl}{\underset{\displaystyle Cl}{C}}-H$$

Non-existence of isomers required by theory of arrangement in plane leads to hypothesis same as Van 't Hoff's. But such isomerism does not exist either. This extremely simple consideration, which however was not followed up, has doubtless long ago led many chemists to the assumption that the four hydrogen atoms of marsh gas must be so arranged round the carbon atom as to produce symmetry in space. But of such arrangements the only one possible is the tetrahedral." (V. Meyer, 1890.)

The following classification of the experimental confirmation of Van 't Hoff's fundamental conception closely follows that given in his books above referred to:

.. (1.) *Character of the isomerism due to the asymmetric carbon.*

The near approach to identity in the isomeric forms is such as would be expected when the carbon atom is united to four different groups in different sequence. All the properties, physical and chemical, which depend on molecular dimensions and attractions should be identical: such properties are specific gravity, boiling point, melting point, latent heat, solubility, chemical stability, heat of formation, etc.

Experimental confirmation of Van't Hoff's hypothesis:
(1) Differences in properties of the isomers only slight, and definitely related to supposed differences in structure.

"Comparison of the physical and chemical properties of the corresponding right and left isomers [reveals] the perfect identity of all their properties, excepting always the inversion in their crystalline forms, and the opposite sense of their optical deviations. The physical aspect, lustre of the crystals, solubility, specific weight, simple or double refraction, all these things are not merely alike, similar, nearly allied, but identical in the strictest sense of the word."

The only difference is due to the lack of symmetry, and this is manifested

(i) Physically. Opposite optical activity, the so-called right- and left-handed rotation, is shown by the isomers in the dissolved state—in the state, that is, when the rotation must arise from molecular, not from crystalline structure.

(ii) Crystallographically. Isomers due to asymmetric carbon show an enantiomorphism (ἐναντίος, opposite) corresponding to their molecular structure (figs. 96—99, pp. 574, 575).

(iii) Chemically. The chemical identity emphasised above ceases directly the asymmetric isomers have to do with a substance which is itself asymmetric.

" The identity of properties above described in the case of the two tartaric acids and their similar derivatives, exists whenever these substances are placed in contact with any compound with superposable image, such as potash, soda, ammonia, lime, baryta, aniline, alcohol, ethers,—in a word, with any compounds whatever which are non-asymmetric, non-hemihedral in form, and without action on polarised light.

Chemical relations of asymmetric isomers to non-asymmetric and asymmetric substances respectively.

If, on the contrary, they are submitted to the action of products of the second class with non-superposable image,— asparagine, quinine, strychnine, brucine, albumen, sugar, etc., bodies asymmetric like themselves,—all is changed in an instant. The solubility is no longer the same. If combination takes place, the crystalline form, the specific weight, the quantity of water of crystallisation, the more or less easy destruction by heating, all differ as much as in the case of the most distantly related isomers.

Here, then, the molecular asymmetry of a substance obtrudes itself on chemistry as a powerful modifier of chemical affinities. Towards the two tartaric acids, quinine does not behave like potash, simply because it is asymmetric and potash is not. Molecular asymmetry exhibits itself henceforth as a property capable by itself, in virtue of its being asymmetry, of modifying chemical affinities.

Let us try to illustrate the cause of these identities and differences. Let us imagine a right screw and a left screw separately penetrating two identical blocks of wood with the grain straight. All the mechanical conditions of the two systems will be the same. This will no longer be so from the moment that the same screws are associated with blocks which are themselves twisted in the same sense or in the opposite sense.

Here is a very interesting application of the facts which have just been explained.

Seeing that the right and left tartaric acids formed such dissimilar compounds simply on account of the rotative power of the base, there was ground for hoping that, from this very dissimilarity, chemical forces might result, capable of balancing the mutual affinity of the two acids, and thereby supply a chemical means of separating the two constituents of paratartaric acid....I prepare the paratartrate of cinchonicine[1];...I allow the whole to crystallise, and the first crystallisations consist of perfectly pure left tartrate of cinchonicine. All the right tartrate remains in the mother liquor because it is more soluble. Finally this itself crystallises with an entirely different aspect, since it does not possess the same crystalline form as the left salt. We might almost believe that we were dealing with the crystallisation of two distinct salts of unequal solubility." (Pasteur, *loc. cit.*)

(2) *Agreement between the occurrence of stereoisomerism and the required constitution.* The differentiating properties above enumerated under (i), (ii), and (iii), characteristic of

(2) Optical activity and constitution required to go together.

isomers of this type, always occur together, and hence it is sufficient and simplest to consider optical activity alone, and to trace the connection between its occurrence or non-occurrence, and the presence or absence of carbon atoms which are "asymmetric," that is, linked with four different groups.

_____

[1] Cinchonicine is an alkaloid obtained from cinchona bark; it is isomeric with cinchonine and cinchonidine, in common with which it has the formula

$$C_{19}H_{22}N_2O.$$

(i) All optically active substances have been found to contain one or more asymmetric carbon atoms. In Van 't Hoff's book more than ten pages are filled with the enumeration of cases on which this generalisation is based, and which comprise compounds of most diverse nature, of which a few are given below.

(i) Presence of asymmetric carbon in optically active substances.

Formulae of Optically Active Substances in which the asymmetric carbon atom is always indicated thus: $\mathbf{C}$.

Ethylidene lactic acid...$H—\overset{\displaystyle OH}{\underset{\displaystyle CO_2H}{C}}—OH_3$ Glyceric acid......$H—\overset{\displaystyle OH}{\underset{\displaystyle CO_2H}{C}}—CH_2.OH$

Butyl alcohol ....$H—\overset{\displaystyle OH}{\underset{\displaystyle C_2H_5}{C}}—OH_3$ Amyl chloride ...$H—\overset{\displaystyle Cl}{\underset{\displaystyle C_3H_7}{C}}—CH_3$

Conine ....

Glucose ...$H—\overset{\displaystyle H}{\underset{\displaystyle OH}{C}}-\overset{\displaystyle H}{\underset{\displaystyle OH}{C}}-\overset{\displaystyle H}{\underset{\displaystyle OH}{C}}-\overset{\displaystyle H}{\underset{\displaystyle OH.}{C}}-\overset{\displaystyle H}{\underset{\displaystyle OH}{C}}-\overset{\displaystyle H}{\underset{\displaystyle O}{C}}$

(ii) The supposed activity of bodies containing no asymmetric carbon atom has been disproved.

"In the literature [of chemistry] descriptions could be found of several substances—succinic acid, $CO_2H.OH_2.CH_2.CO_2H$, and styrolene, $C_6H_5.HO=OH_2$, —which do not contain an asymmetric carbon atom, and which, nevertheless, were said to be optically active. Van 't Hoff was so bold as to assert that this could not be correct. He repeated the experiments, and found that as a matter of fact the data were erroneous. Succinic acid and styrolene, when pure, proved under all conditions optically inactive." (V. Meyer, 1890.)

(ii) Supposed activity of substances not containing asymmetric carbon disproved.

"Berthelot found that the substance C_8H_8, the styrol obtained from styrax, was active. For chemists who see no connection between optical activity and constitution there is nothing remarkable in this statement; but from the consideration of formulae in space, I have recently asserted the existence of such a connection, and I have propounded three theorems on this subject. According to my views, the substance $C_6H_5.CH=CH_2$

cannot be active; hence I have investigated it anew....Polymerisation and subsequent fractionation showed that (1) the oil from styrax is active, (2) the activity does not proceed from the styrol present, but from another substance which has a formula not far removed from $C_{10}H_{18}O$." (Van 't Hoff, *Ber. d. Chem. Ges.*, 9, 1876, p. 5.)

(iii)   In derivatives, optical activity and asymmetry persist or disappear together.

(iii) Maintenance in derivatives of connection between asymmetric structure and optical activity.

Le Bel was the first who tried to put the hypothesis of the relation between optical activity and asymmetry to the experimental test of changing the optically active amyl iodide into the symmetrical methyl derivative, and the same method has been successfully applied by others also[1].

"Investigation of the pentane derived from the optically active amyl iodide by the replacement of the I by H, should make it possible to decide whether the destruction of the asymmetry of the tertiary carbon atom also involves the destruction of the rotatory power."

The results set out below are in complete agreement with the requirements of the theory; when the derivative retains the asymmetric carbon atom, it is also optically active (*e.g.* amyl-alcohol and amyl-iodide or amyl-ethide); whilst destruction of the asymmetry yields optically inactive derivatives (*e.g.* amyl-iodide and amyl-methide or amyl-hydride).

$$C_2H_5 \atop H_3C—\overset{|}{\underset{|}{C}}—H \quad +HI \qquad = \qquad C_2H_5 \atop H_3C—\overset{|}{\underset{|}{C}}—H \quad +H_2O$$

<table>
<tr><td>$CH_2 . OH$
Active laevo-rotatory
amyl-alcohol
B. P. = 128°</td><td>OH_2I
Active dextro-rotatory
amyl-iodide
B. P. = 144°</td></tr>
</table>

$$C_2H_5 \atop H_3C—\overset{|}{\underset{|}{C}}—H \quad +ICH_3 \ +2Na = \qquad C_2H_5 \atop H_3C—\overset{|}{\underset{|}{C}}—H \quad +NaI$$

<table>
<tr><td>OH_2I
Active amyl-iodide</td><td>$CH_2 . CH_3$
Inactive methyl-diethyl-
methane
(amyl methide or ethyl butane)
B. P. = 60° (Le Bel)</td></tr>
</table>

[1] Just, *Liebig's Ann. Chem.*, Leipzig, 222, 1883 (p. 146).

$$\underset{\substack{\text{Active amyl-iodide}}}{\overset{\displaystyle C_2H_5}{\underset{\displaystyle OH_2 . I}{H_3C-\overset{\displaystyle |}{\underset{\displaystyle |}{C}}-H}}} \quad +Zn \quad +H_2O = \quad \underset{\substack{\text{Inactive dimethyl-ethyl-}\\\text{methane}\\\text{(amyl hydride or secondary pentane)}\\\text{B. P.}=30^\circ\ (\text{Just})}}{\overset{\displaystyle C_2H_5}{\underset{\displaystyle CH_3}{H_3C-\overset{\displaystyle |}{\underset{\displaystyle |}{C}}-H}}} \quad +Zn \overset{\displaystyle I}{\underset{\displaystyle OH}{}}$$

$$\underset{\substack{\text{Active amyl-iodide}}}{\overset{\displaystyle C_2H_5}{\underset{\displaystyle OH_2 I}{H_3C-\overset{\displaystyle |}{\underset{\displaystyle |}{C}}-H}}} \quad +IC_2H_5+2Na = \quad \underset{\substack{\text{Active dextro-rotatory methyl-}\\\text{ethyl-propyl-methane}\\\text{(amyl ethide or ethyl pentane)}\\\text{B. P.}=91^\circ\ (\text{Just}).}}{\overset{\displaystyle C_2H_5}{\underset{\displaystyle OH_2 . C_2H_5}{H_3C-\overset{\displaystyle |}{\underset{\displaystyle |}{C}}-H}}} \quad +NaI$$

(3) *Cause of the optical inactivity of some asymmetric compounds.*

(i) External compensation.

(3) Explanation of optical inactivity of some asymmetric compounds.

(i) External compensation.

"Substances cannot be optically active unless they contain an asymmetric carbon atom, but the converse does not follow. Substances which do contain an asymmetric carbon atom can occur in an inactive form, and will always do so if they contain a mixture of equal quantities of the dextro- and laevo-rotatory modification. It should therefore not be a matter of surprise if we do not discover activity in a substance containing an asymmetric carbon atom, but we should have to arraign the theory if the case were reversed." (V. Meyer, 1890.)

"Many substances which are apparently not asymmetric, may be like paratartaric acid. [We here deal with] the fact of a double molecular asymmetry concealed by the neutralisation of two asymmetries, the physical and geometrical effects of which rigorously compensate each other." (Pasteur.)

Pasteur's three methods for dividing substances which are inactive owing to external compensation.

Pasteur's separation of inactive racemic acid by mechanical sorting of the crystals of opposite hemihedrism, or by fractional crystallisation of the salt formed with an active base, has been described (pp. 576, 581), and these two methods, together with a third, also due to Pasteur, are still used, and indeed are the only ones known. The third method consisted in the use of organisms such as penicillium, which show a different deportment towards the active isomers.

"The yeast which causes the right salt to ferment leaves the left salt untouched, in spite of the absolute identity in physical and chemical properties of the right and left tartrates of ammonia, so long as they are not subjected to asymmetric action." (Pasteur.)

. .Pasteur's methods have been successfully applied to the division of a considerable number of asymmetric inactive compounds into active components; and it must be considered an important verification of the theory that, in the case of substances which do not contain an asymmetric carbon atom, all such attempts have produced no result.

(ii)   Internal compensation.

When asymmetric carbon occurs in a symmetrical formula, thus,

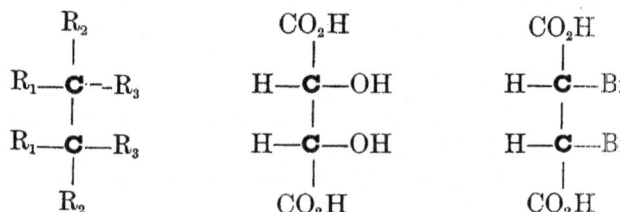

Meso-tartaric acid   Dibromo-succinic acid,

the theory foresees the existence of an indivisible inactive modification.

"There exists an inactive non-asymmetric tartaric acid, quite different from paratartaric acid, which cannot be resolved into a right tartaric acid and a left tartaric acid....A substance...may lose its molecular asymmetry, become untwisted, to use a rough metaphor, and assume in the arrangement of its atoms a disposition with superposable image.   In this way each asymmetric substance offers four varieties; the right body, the left body, the combination of the right and left, and the substance which is neither right nor left, nor formed by the combination of the right and left." (Pasteur.)

(ii) Internal compensation. The four varieties of an asymmetric substance.

For tartaric acid, the formulae representing the structure of these four modifications are the following:

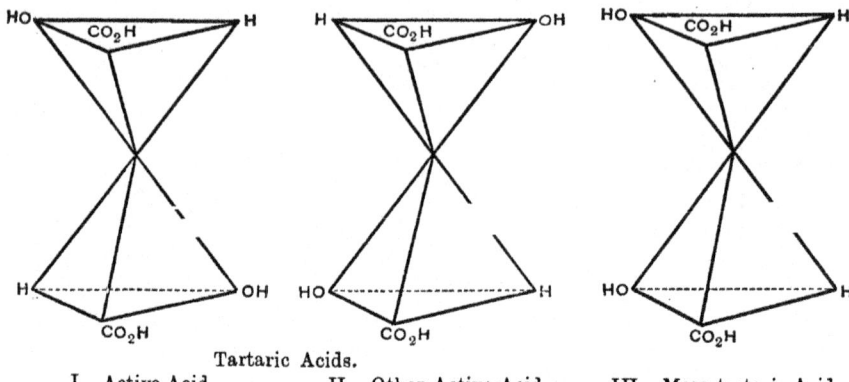

Tartaric Acids.

I. Active Acid.   II. Other Active Acid.   III. Meso-tartaric Acid.
                                                               (inactive, indivisible.)

IV. Racemic Acid.
(inactive, divisible.)

(4) *Comparison between the theoretical and the known number of isomers.*

(4) Theoretical number of isomers and number known.

The number of stereoisomers grows rapidly with the number of asymmetric carbon atoms, and it can be shown that, in accordance with the theory, this number is

4 for 2 asymmetric carbon atoms
8 „ 3   „   „   „
16 „ 4   „
and quite generally 2^n „ n „   „   „

with a possible reduction of this number consequent on symmetry of the formula, thus: for two asymmetric carbon atoms—

No. 1.	No. 2.	No. 3.	No. 4.
R_3	R_3	R_3	R_3
$R_1 \, C \, R_2$	$R_2 \, C \, R_1$	$R_1 \, C \, R_2$	$R_2 \, C \, R_1$
$R_4 \, C \, R_5$	$R_4 \, C \, R_5$	$R_5 \, C \, R_4$	$R_5 \, C \, R_4$
R_6	R_6	R_6	R_6

But in the following, No. 1 is identical with No. 4, and is the inactive indivisible type dealt with under 3 (ii).

No. 1.	No. 2.	No. 3.	No. 4.
R_2	R_2	R_2	R_2
$R_1 \, C \, R_3$	$R_3 \, C \, R_1$	$R_1 \, C \, R_3$	$R_3 \, C \, R_1$
$R_1 \, C \, R_3$	$R_1 \, C \, R_3$	$R_3 \, C \, R_1$	$R_3 \, C \, R_1$
R_2	R_2	R_2	R_2

In the case of some substances with two asymmetric carbon atoms, all the four theoretically possible isomers are known; in the case of more complex structures, such as glucose, which contains four asymmetric carbon atoms, nine of the sixteen possible isomers are known already, and there is every prospect of this number being increased.

No attempt will be made to follow here Van 't Hoff's treatment of the stereoisomerism of unsaturated carbon compounds. The

Stereo-isomerism of unsaturated carbon compounds.

problem is a more complex one, and the isomerism of this class of compounds differs markedly from that of the saturated compounds, as expounded above. J. Wislicenus[1] has made contributions of the first importance to the theory and the study of the unsaturated carbon compounds. He agreed with Van 't Hoff in assuming that when two carbon atoms were linked together, both were capable of rotating in opposite directions about a common axis, but that this possibility ceased when they were doubly or trebly linked. "Wislicenus further called into play the action of certain 'specially directed forces, the affinity-energies,' which determined the relative position of the atoms to one another in the molecule." Among the most important results of the application of his theory may be mentioned the elucidation of the structure of fumaric and maleic acids.

The existence of another distinct class of isomerism has been recognised in recent years, and a large amount of work has been

Tautomerism.

done in the explanation of the phenomena comprised under the name of tautomerism. Little more than mention can be accorded to this subject here. So many different investigators have produced so extensive a literature on this subject, the theoretical speculations are so numerous and have led to such a bewildering multiplicity of terms[2], and the present theory of the subject represents a combination of and a compromise between so many different views, that adherence to the historical method of presentation hitherto followed would involve too lengthy a treatment. Moreover the subject is dealt with

---

[1] "Ueber die raümliche Anordnung der Atome in Organischen Molekulen," *Abh. K. Sächs. Ges. der Wissenschaften*, 14, 1, 1887; English translation in Scientific Memoirs XIII, *The Foundations of Stereochemistry*, New York, 1901.

[2] Tautomerism, pseudomerism, desmotropy, allelotropy, merotropy, tropomerism, dynamical isomerism.

adequately in recent editions of standard text-books of organic chemistry[1].

The occurrence in certain cases of a number of isomers greater than that required by the theory of plane structural formulae, led to the theory of stereoisomerism, which in its funda-

Number of known iso-mers (i) greater, (ii) lesser than the theoreti-cal. Examples of the latter.

mental hypothesis had to consider and account for the typical nature of the differences exhibited by such substances, *i.e.* the near approach to identity in the isomeric forms (*ante*, p. 580). Another class of exceptions to the original doctrine of isomerism is formed by the cases where a less number of isomers is known than that required by theory, and where methods of preparation which, according to the known constitution of the interacting substances and the known usual course of the reaction, should yield the missing isomer, produce not the substance expected, but always another modification.

Thus: the classes of isomeric substances named nitriles and isonitriles, which are derivatives of prussic acid (hydrocyanic acid, HCN), have the general formula RCN, and the differences in the study of their properties[2] find their representation in the structural formulae

$$N \equiv C - R \qquad \text{and} \qquad C \equiv N - R$$
$$\text{Nitrile} \qquad\qquad\qquad \text{Isonitrile.}$$

---

[1] Richter (English translation), 1901, p. 54 ; Berntlisen (English translation), 1894, p. 286.

[2] Acetonitrile or methyl cyanide, C_2H_3N, is a liquid with an agreeable odour, which has a specific gravity of $0{\cdot}789$, boils at $81{\cdot}6°$ C., and solidifies at $-41°$ C. It is prepared by the dehydration of acetamide, effected by distillation with P_2O_5 ; or by the action of potassium cyanide on methyl iodide

$$CH_3 . COO . NH_4 = CH_3 . CN + 2H_2O$$
$$CH_3I + KCN \quad = C_2H_3N + KI ;$$

nascent hydrogen converts it into ethylamine

$$CH_3CN + 4H \quad = CH_3 . CH_2NH_2 ;$$

when treated with aqueous acid or alkali it is saponified according to the equations

$$\begin{cases} CH_3CN + HCl + 2H_2O = CH_3 . CO_2H + NH_4Cl \\ CH_3CN + KOH + H_2O = CH_3 . CO_2K + NH_3. \end{cases}$$

Isoacetonitrile or methyl-isocyanide, C_2H_3N, is a liquid with a disgusting odour, which boils at $59°$ C. It is produced by the action of chloroform and potash on methylamine, or by the action of silver cyanide on methyl iodide

$$CH_3 . NH_2 + CHCl_3 + 3KOH = CH_3 . NC + 3KCl + 3H_2O$$
$$CH_3I + AgCN \quad = CH_3 . NC + AgI ;$$

it is not acted on by alkalis, but is easily saponified by aqueous acids, according to the equation

$$CH_3 . NC + HCl + 2H_2O = CH_3 . NH_2 . HCl + H . COOH$$
$$\text{methylamine} \qquad \text{formic}$$
$$\text{hydrochloride.} \qquad \text{acid.}$$

Accordingly the parent substance should itself exist in two modifications

$$N \equiv C - H \quad \text{and} \quad C \equiv N - H,$$

but only one prussic acid is known.

Again : theory requires the existence of the substance $CH_3 . C_2H_3O$ in the two isomeric modifications

$$\underset{\text{H}_3\text{C} - \text{C} - \text{CH}_3}{\overset{\overset{\text{O}}{\|}}{}} \quad \text{and} \quad \underset{\text{H}_3\text{C} - \text{C} = \text{CH}_2,}{\overset{\overset{\text{OH}}{|}}{}}$$

of which the first containing the $O = C\big\langle$ group, is designated by the name of *keto-form*, whilst the second containing the group $OH - C\big\langle$, is called the *enol-form*[1].

But of these two forms, only the first, which is acetone, is known. The second, which contains the grouping characteristic of an unsaturated alcohol, should be obtained by the usual method of saponification of a monosubstituted unsaturated hydrocarbon, C_nH_{2n} ; but the product obtained is always acetone.

$$\underset{\text{ethyl iodide} +}{\overset{\text{CH}_3}{\underset{\text{CH}_2\text{I}}{|}}} \quad + \quad \underset{\substack{\text{water} \\ \text{(used in excess} \\ \text{at 100°)}}}{\text{H.OH}} \quad = \quad \underset{= \text{ethyl alcohol} + \text{hydriodic acid}}{\overset{\text{CH}_3}{\underset{\text{CH}_2\text{OH}}{|}}} \quad + \quad \text{HI}$$

$$\underset{\text{allyl iodide} +}{\overset{\text{CH}_2}{\underset{\overset{\text{CH}}{\underset{\text{CH}_2\text{I}}{|}}}{\|}}} \quad + \quad \underset{\substack{\text{water} \\ \text{(used in excess} \\ \text{at 100°)}}}{\text{H.OH}} \quad = \quad \underset{= \text{allyl alcohol} + \text{hydriodic acid}}{\overset{\text{CH}_2}{\underset{\overset{\text{CH}}{\underset{\text{CH}_2\text{OH}}{|}}}{\|}}} \quad + \quad \text{HI}$$

$$\underset{\beta\text{-bromopropylene}}{\overset{\text{CH}_2}{\underset{\overset{\text{CBr} + \text{H.OH}}{\underset{\text{CH}_3}{|}}}{\|}}} \quad \underset{\text{should yield}}{} \quad \underset{\beta\text{-allyl alcohol}}{\overset{\text{CH}_2}{\underset{\overset{\text{COH}}{\underset{\text{CH}_3}{|}}}{\|}}} \quad \underset{\substack{\text{but} \\ \text{actually yields}}}{} \quad \underset{\text{acetone}}{\overset{\text{CH}_3}{\underset{\overset{\text{CO}}{\underset{\text{CH}_3}{|}}}{|}}}$$

---

[1] "Enol" from en-, the symbol of double linking, and -ol, the termination for substances containing an alcoholic hydroxyl group.

Another simple case is that of the substance C_2H_4O, which should occur in the two forms

$$
\begin{array}{ccc}
\mathrm{CH_3} & & \mathrm{CH_2} \\
| & & \| \\
\mathrm{H-C=O} & \text{and} & \mathrm{H-C-OH,}
\end{array}
$$

but of which only the first, acetaldehyde, is known.

Instances of this kind, in which in spite of many well-directed experimental researches, isomers required by theory cannot be found, are comprised under the name of tauto-

merism ($\tau a\dot{\upsilon}\tau\dot{a}$, the same), the term being used simply in a descriptive-classificatory sense[1], and without any reference to theoretical explanation of the occurrence.

Obviously such discrepancies between the calculated and the known number of isomers do not necessarily invalidate the theory. The required isomers may really be capable of existence, but the right method of preparing them and the conditions under which they can be isolated are as yet unknown. And there is good evidence for the belief that the non-production of one of two isomeric modifications is often due to its relatively lesser

stability under the conditions of the experiment, to an intramolecular transformation in which, according to the present doctrine of valency, there is a change of linking and a migration of a hydrogen atom; *e.g.* acetone, $CH_3 . CO . CH_3$ is formed in preference to β-allyl alcohol, $CH_3COH . CH_2$.

$$
\begin{array}{cc}
\mathrm{CH_3} & \mathrm{CH_3} \\
| & | \\
\mathrm{C-OH} & \mathrm{C=O} \\
\| \nearrow\!\!\swarrow & | \\
\mathrm{CH_2} & \mathrm{CH_2H.}
\end{array}
$$

Derivatives of the missing isomers are often known in which the hydrogen atom in question is replaced by a compound radicle, and hence presumably the ease of intramolecular migration reduced, *e.g.*

---

[1] W. Wislicenus, *Ueber die Tautomerie* (Ahrensche Sammlung, 2), 1897.

$$
\begin{array}{ccc}
\mathrm{CH_2} & & \mathrm{CH_3} \\
\| & \text{isomeric with} & | \\
\mathrm{H - C - OH} & & \mathrm{H - C = O} \\
\text{Unknown.} & &
\end{array}
$$

$$
\begin{array}{ccc}
\mathrm{CH_2} & & \mathrm{CH_2 \cdot C_2H_5} \\
\| & \text{isomeric with} & | \\
\mathrm{H - C - OC_2H_5} & & \mathrm{H - C = O} \\
\text{Known.} & &
\end{array}
$$

And this assumed power of intramolecular migration of a hydrogen atom and the consequent possible simultaneous presence of the two modifications in relatively varying quantities is supported by the chemical behaviour of a class of substances of which aceto-acetic ether is the oldest and best investigated example. This substance, distinguished by its intense chemical activity, behaves in a certain number of reactions in accordance with the keto-formula

$$
\underset{\mathrm{H_3C - C - C - C - O \cdot C_2H_5}}{\overset{\mathrm{O \quad H_2 \quad O}}{\overset{\| \quad \| \quad \|}{}}}
$$

whilst a number of other facts are more compatible with the isomeric enol-formula

$$
\underset{\mathrm{H_3C - C = C - C - O \cdot C_2H_5.}}{\overset{\mathrm{OH \quad H \quad O}}{\overset{| \quad | \quad \|}{}}}
$$

It would seem therefore as if both kinds of molecules were present, and as if these could so easily and rapidly change into one another that the substance can react according to either of the two structural formulae.

λ   Such a phenomenon is quite compatible with the modern view of chemical equilibrium (*ante*, p. 325), and the phenomena of tautomerism can be looked upon as reversible intramolecular reactions.

"If we consider a mixture of two isomers which...are capable of transformation into one another...and if we assume that the equilibrium between them is established very rapidly, then if we attempted to withdraw one component from such a mixture by means of any chemical method of separation, in consequence of the disturbance of the equilibrium the other component would immediately change into the first, *i.e.* the whole mixture would react as if it were composed of the first component

Tautomerism explained by reversible intramolecular reactions.

only. But if on the other hand we were to use a chemical reagent which acts only on the second component, the opposite would occur, and the whole mixture would behave as if it consisted of the second component alone. Such a mixture would therefore be capable of reacting according to two constitutional formulae, *i.e.* it would...exhibit tautomerism.

According to this...interpretation, prussic acid, for instance, would be a mixture of molecules NCH and CNH, which however at ordinary temperature are so readily transformed into one another that a separation is impossible or at least very difficult; just as at higher temperatures a separation of [other] mixtures would not be possible, because under these conditions the transformation of the two isomers into one another occurs too rapidly. On this hypothesis, lowering the temperature should facilitate the isolation of the two tautomeric forms.

The above view has met with remarkable support in the...separation of the two tautomeric forms of aceto-acetic ether[1].

Of course this does not amount to a final decision as to whether all the facts of tautomerism can be interpreted in this manner, but it may be definitely affirmed that any case of balanced action between two isomers can be made to constitute a case of tautomerism, if by any means (rise of temperature, presence of catalysers) the facility of transformation is sufficiently increased." (W. Nernst, *Theoretische Chemie*, 1898, p. 531.)

---

[1] Schiff (*Ber. D. Chem. Ges.* 31, 1898, p. 205) found that the addition product obtained by the action of benzalaniline, $C_6H_5 . CH = NC_6H_5$, on aceto-acetic ether differed according to the relative quantities of the components, and to the conditions of formation, and that the following three substances could be isolated:

			Aceto-acetic ether.	Benzalaniline.
$C_{19}H_{21}NO_3$	M.P. 95°	obtained by the interaction between	1 mol.	1 mol.
,,	,, 103°	,,	2 mol.	1 mol.
,,	,, 78°.	,,	2 mol.	1 mol.
				+ a few drops piperidine.

"These substances are therefore the three expected isomers....As far as I can judge, that melting at 78° is the keto-form, that melting at 103° the enol-form, and that melting at 95° the keto-enol mixed form. The latter is easily obtained synthetically by recrystallisation of a mixture of the isomers melting at 78° and 103° respectively."

CHAPTER XIX.

THE ULTIMATE CONSTITUTION OF MATTER AND THE GENESIS OF THE ELEMENTS.

" Chemistry forms the connecting link between that kind of knowledge which is founded on quantity, and those kinds of knowledge which rest solely on experience. Now so far as the logic of quantity is applicable, so far are we certain of our conclusions. But when this logic cannot be applied, our conclusions are no longer such as must be, but are only for the most part such as may be."

PROUT, 1834.

SOME mention has been made in chapter XVI of the hitherto unsuccessful attempts to represent mathematically the periodic function which connects the weights of the elementary atoms with their properties. Special interest attaches to the formula suggested by Dr Mills, because it is based on the assumption of polymerisation, on the hypothesis that all elements are condensation products of different complexity of some one common constituent, and because it thereby connects the mathematics of the periodic system with the enquiry into the composition and genesis of the elements, a problem considered from the earliest time of philosophic speculation. The solutions attempted have closely followed the general development from *à priori* speculations to hypotheses which follow on a long train of induction; from something vague, loosely connected with reality at the outset only, to something scientifically definite, made to fit in every detail the qualitative and quantitative occurrences of nature.

The revolt of the human mind against the arbitrary complexity of the creation of some seventy odd distinct elements is not merely a modern phase. It found its earliest expression in the Ionian philosophers' hylozoistic conception of one primordial matter (*ante*, p. 231); Aristotle (p. 253), Bacon (p. 266), Descartes (p. 270), Boyle (pp. 277, 279), widely as they differed in theoretical conception and in actual knowledge of natural phenomena, agreed in

Existence of one kind of primitive matter. Philosophical speculations.

F.   38

assuming the existence of one kind of matter which, in conjunction
with essential properties, or thrown into certain groups and knots,
or differently moved in its parts, or modified by an architectonic
principle, produced the various substances by them considered
elementary, and which by further conjunction gave rise to
the variety of forms that matter can assume. The opposite
view, that of the independent existence *ab initio* of all the
substances considered elementary, propounded in antiquity by
Anaxagoras as well as Leucippus and Democritus (p. 237), was
also held by Dalton (p. 288), who nevertheless indirectly supplied
the impetus for the scientific revival of the belief in the existence
of one kind of primitive matter. The discovery of the laws of
chemical combination and Dalton's atomic hypothesis led to the
determination of those fundamental constants, the atomic weights ;
and it was but in the natural order of scientific development that
these data should forthwith have been made the basis of a
quantitative hypothesis concerning the genesis of the elements.

A paper by W. Prout[1], called "On the Relation between the
Specific Gravities of Bodies in their Gaseous State,
and the Weights of their Atoms[2]," contains tables of
which the following are parts :—

Prout's hypothesis.

Elementary Substances.

Name	Spec. Grav., hydrogen being 1	Wt. of atom, 2 vols. hydrogen being 1
Hydrogen	1	1
Carbon	6	6
Azote	14	14
Phosphorus	14	14
Oxygen	16	8
Sulphur	16	16
Calcium	20	20
Sodium	24	24
Iron	28	28
Zinc	32	32
Chlorine	36	36
Potassium	40	40
Barytium	70	70
Iodine	124	124

[1] William Prout (1785—1850) graduated M.D. at Edinburgh 1811, and afterwards settled in London, where he lectured on chemistry. He is known as the
author of many papers on physiological chemistry, of which he was one of the
pioneers.

[2] *Ann. Phil.* (Thomson), 6, 1815 (p. 321) ; 7, 1816 (p. 111).

Substances stated from analogy, but of which we are yet uncertain.

Name	Spec. Grav., hydrogen being 1	Wt. of atom, 2 vols. hydrogen being 1
Aluminium	8	8
Magnesium	12	12
Chromium	18	18
Nickel	28	28
Cobalt	28	28
Tellurium	32	32
Copper	32	32
Strontium	48	48
Arsenic	48	48
Molybdenum	48	48
Manganese	56	56
Tin	60	60
Bismuth	72	72
Antimony	88	88
Cerium	92	92
Uranium	96	96
Tungsten	96	96
Platinum	96	96
Mercury	100	100
Lead	104	104
Silver	108	108
Rhodium	120	120
Titanium	144	144
Gold	200	200

" On a general review of the tables, we may notice :—

1. That all the elementary numbers, hydrogen being considered as 1, are divisible by 4, except carbon, azote, and barytium, and these are divisible by 2, appearing therefore to indicate that they are modified by a higher number than that of unity or hydrogen. Is the other number 16, or oxygen ? And are all substances compounded of these two elements ?"

" If the views we have ventured to advance be correct, we may almost consider the $\pi\rho\acute{\omega}\tau\eta$ $\H{v}\lambda\eta$ of the ancients to be realised in hydrogen ; an opinion,

All elements compounded of hydrogen.

by the by, not altogether new. If we actually consider this to be the case, and further consider the specific gravities of bodies in their gaseous state to represent the number of volumes condensed into one ; or, in other words, the number of the absolute weight of a single volume of the first matter ($\pi\rho\acute{\omega}\tau\eta$ $\H{v}\lambda\eta$) which they contain, which is extremely probable, multiples in weight must always indicate multiples in volume, and *vice versâ*; and the specific gravities, or. absolute weights of all bodies in a gaseous state, must be multiples of the specific gravity or absolute weight of the first matter (πρώτη ὕλη), because all bodies in a gaseous state which unite with one another unite with reference to their volume."

This conjecture was eagerly taken up by Thomas Thomson. In the book to which reference has been made before[1], his enthusiasm finds expression in the following passage :—

"[Dr Prout's] paper displays a degree of sagacity that has seldom been exceeded in chemical investigations, and shows clearly that the author, if he chose, might rise to the highest eminence as a chemical philosopher. He observed that all [the] atomic weights are multiples of the atomic weight of hydrogen ; indeed, that all of them are multiples of twice hydrogen[2], or 0·25, and most of them of 4 hydrogen, or 0·5.

Th. Thomson's experimental work in support of Prout's hypothesis.

It was this admirable paper that satisfied me that new analytical investigations were still necessary to determine the atomic weights of bodies with perfect accuracy ; and I formed the project of attempting the investigation myself by direct experiment.

My first object was to satisfy myself whether Dr Prout's opinion, that the atomic weights of all bodies are multiples of that of hydrogen, was correct ; because the establishment of its truth would at once give a simplicity to the atomic numbers which had not been suspected, and would place the science of chemistry in a new and much more advantageous situation than it had ever occupied. The very numerous investigations which will be exhibited in the following pages, I flatter myself, fully establish the truth of Dr Prout's sagacious conjecture. For every substance, of which I could procure a sufficient quantity for me to examine it fully, has been found not only a multiple of the atomic weight of hydrogen, but if we except a few compounds into which a single or odd atom of hydrogen enters, they are all multiples of 0·25, or of two atoms of hydrogen."

This revision of atomic weights, undertaken professedly with the object of testing the validity of Prout's hypothesis, is devoid of all value. Thomson, carried away by the attraction of simplicity, lost the impartiality of the judge ; made himself the advocate for one side, the wrong side, that of philosophical speculation as against the experimental facts. Berzelius' own attitude towards Prout's hypothesis and his criticism of Thomson's are in substance and form what might have been expected.

"Prout has tried to show that among the relative weights by means of which we express the atoms of the elements, the weight of the hydrogen atom

---

[1] *An Attempt to Establish the First Principles of Chemistry by Experiment*, 1825.
[2] O = 1.

is a submultiple of the weights of all the rest, so that each of these should
weigh a certain number of times as much as the hydrogen

Berzelius on the accuracy of Thomson's experimental work.

atom. The fact that at present there is not a single reason, either chemical or physical, for believing this to be the case, does not preclude its possibility. O weighs 16, S 32, and P 63 times as much as the hydrogen atom; carbon and nitrogen deviate from the rule.

Thomson at Glasgow has proposed a revision of the atomic weights from this point of view, and the results of his investigations agree to the last decimal place with Prout's hypothesis. But knowing how very widely Thomson's results always differ from the correct ones, whenever he is not able to determine them by previous calculation, it becomes easy to estimate the value of proofs based upon experiments of his." (*Jahresbericht*, 1823.)

"Thomson has published a memoir on the subject of chemical proportions, entitled 'An attempt to establish the First Principles of Chemistry by Experiment.' He has determined the atomic weight of every simple substance and has applied corrections to everything that had been done prior to his work. But his own determinations are all dominated by one special view, namely, that the atomic weights of all substances are exact multiples of that of hydrogen. So what he did was to change all the numbers obtained by his predecessors to the nearest whole multiple of the atomic weight of hydrogen, to calculate the atomic weights of their compounds accordingly, and then to bring together for interaction and precipitation weighed quantities of compounds which were in the ratio required by these corrected atomic weights; he then found that the mutual decomposition was always complete.

This investigation belongs to that very small class from which science can derive no advantage whatever; a good deal of the experimental part, including a number of fundamental experiments, seems to have been worked out at the desk only; and the greatest consideration which contemporaries can show to the author is to treat his book as if it had never appeared." (*Jahresbericht*, 1827, referring to the year 1825.)

In 1827 Berzelius published a table of atomic weights, a very large number of which had been determined by himself, and of which a selection, together with Thomson's corresponding numbers, are given in the table on page 598. The values advocated by Berzelius and generally employed on the Continent could not possibly be looked upon as being without exception whole multiples of the hydrogen unit; but in England Thomson's numbers were current until a special investigation demonstrated their inaccuracy, and thereby proved the incorrectness of Prout's hypothesis, which Thomson considered he had completely established. In 1833 Dr Turner reported to the British Association:

The Atomic Weights
of some of the more common and better investigated elements
as given in the tables of Thomson and Berzelius.

	Thomson (1825) Hydrogen = 1	Berzelius (1827) Hydrogen = 1
Oxygen	8	$8·016 \times 2 = 16·026$
Nitrogen	14	$14·186$
Sulphur	16	$16·119 \times 2 = 32·239$
Chlorine	36	$35·470$
Iodine	124	$123·206$
Carbon	6	$6·125 \times 2 = 12·250$
Arsenic	38	$37·664 \times 2 = 75·329$
Silver	110	$108·305 \times 2 = 216·611$
Mercury	200	$202·863$
Copper	32	$31·707 \times 2 = 63·415$
Lead	104	$103·729 \times 2 = 207·458$
Zinc	34	$32·310 \times 2 = 64·621$
Iron	28	$27·181 \times 2 = 54·363$
Manganese	28	$28·509 \times 2 = 57·019$
Magnesium	12	$12·689 \times 2 = 25·378$
Calcium	20	$20·515 \times 2 = 41·030$
Strontium	44	$43·854 \times 2 = 87·709$
Barium	70	$68·662 \times 2 = 137·325$
Sodium	24	$23·310 \times 2 = 46·620$
Potassium	40	$39·237 \times 2 = 78·515$

Turner on the inaccuracy of Thomson's equivalents, and the consequent want of evidence for Prout's hypothesis.

" He had continued his researches into atomic weights, and had to his own conviction determined the points which had induced him to undertake the enquiry. These were, first to form an opinion of the relative accuracy of the tables of equivalents employed in this country and on the Continent ; and, secondly, to ascertain whether there existed any trustworthy evidence in proof of the hypothesis that the equivalents of bodies are multiples by whole numbers of the equivalent of hydrogen....The general result is, that the atomic weights current in this country are much less exact than those given by Berzelius ;. that though they had been recommended to British chemists as rigidly correct, they were often very inexact....Further, as far as experimental evidence at present goes, the hypothesis above alluded to is unsupported. In some instances the equivalents are so nearly simple multiples of that of hydrogen that they may be taken as such without appreciable error ; but in many other cases the numbers given by experiment cannot be reconciled with the hypothesis."

But there remained the fact, one which at the present day is still inviting explanation, that the atomic weight values which within the limits of experimental error approximate to whole numbers or even actually are whole numbers, represent a far larger fraction of the total than would be the case according to the law of chance only. Calculations have even been made to evaluate the chance of such approximation to whole numbers as is actually found; and quite recently Mr Strutt[1], selecting the eight best known atomic weights and using the values given in Richards' table (*ante*, p. 220)

Evidence for Prout's hypothesis from approximation of atomic weights to whole numbers.

Br = 79·955	N = 14·045
C = 12·001	K = 39·140
Cl = 35·455	Na = 23·050
H = 1·0075	S = 32·065

finds that " if we take the difference between each of these and the nearest whole number, we get as the sum of the differences ·809," and applying a suitable formula to the calculation of the probability of the total deviation having this value, he obtains the result that this is ·001159, or about 1 chance in 1000. An extension of the calculation to 18 other somewhat less certain atomic weight values leads to his final inference:

Strutt's calculation of the Probability of such approximation.

" ...A calculation of the probabilities involved fully confirms the verdict of common sense, that the atomic weights tend to approximate to whole numbers far more closely than can reasonably be accounted for by any accidental coincidence. The chance of any such coincidence being the explanation is not more than 1 in 1000, so that, to use Laplace's mode of expression, we have stronger reasons for believing in the truth of Prout's law than in that of many historical events which are universally accepted as unquestionable."

This " common sense verdict " had been adduced as evidence in the brilliantly conceived vindication of Prout's hypothesis published in 1860 by Marignac[2].

" A glance at [the 9 atomic weight values just published by Stas[3]] is sufficient to show that though they do not absolutely agree with Prout's numbers, they at least approximate to them so closely as to make it

---

[1] *Phil. Mag.*, London (6), 1, 1901 (p. 311).
[2] " Apropos des Recherches de Stas sur les Rapports réciproques des Poids atomiques," Geneva, *Bibl. Univ. Archives*, 9, 1860 (p. 97).
[3] See *post*, p. 602.

impossible to consider this agreement just a mere chance coincidence. Thus for the 9 determinations involved the mean deviation is 0·056, or 1/18 of the equivalent of hydrogen."

Marignac takes as *Prout's number* for chlorine 35·5, having in 1843, as the result of his own determination of the atomic weight of this element (35·456), found it necessary to halve the value of Prout's original common atomic weight unit, a process subsequently repeated by Dumas.

Division of Prout's original atomic weight unit.

" The elements whose equivalents are known with sufficient accuracy divide themselves into 3 distinct sets, in which the equivalents appear as whole multiples of 1, of 0·5, or of 0·25 respectively.

The 1st set may be made to include the following 22 elements :

	Hydrogen = 1	
Carbon = 6	Sodium = 23	Bromine = 80
Oxygen = 8	Iron = 28	Tungsten = 92
Nitrogen = 14	Phosphorus = 31	Mercury = 100
Silicon = 14	Molybdenum = 48	Silver = 108
Sulphur = 16	Cadmium = 56	Antimony = 122
Fluorine = 19	Tin = 59	Iodine = 127
Calcium = 20	Arsenic = 75	Bismuth = 210

The 2nd set comprises 8 elements :

Manganese = 27·5		Barium = 68·5
Cobalt = 29·5	Chlorine = 35·5	Osmium = 99·5
Nickel = 29·5	Tellurium = 64·5	Lead = 103·5

The 3rd set finally is composed of 5 elements :

Aluminium = 13·75		Selenium = 39·75
Copper = 31·75	Zinc = 32·75	Strontium = 43·75."

(Dumas, *Mémoire sur les équivalents des corps simples*, 1859.)

It is interesting to note that the multiplication required to change these *equivalents* into *atomic weights* gives numbers which are all divisible by 0·5.

" M. Dumas, profoundly convinced of the exact validity of Prout's principle, believes that all the atomic weights are multiples of 1·00, or 0·50, or 0·25 that of hydrogen." (Stas.)

The legitimacy of altering the atomic weight of the supposed one common constituent of all the elements has been thus asserted by Marignac :

"...The fundamental principle which has led Prout to the enunciation of his law, that is, the idea of the unity of matter, and all the more or less brilliant conceptions which have been based on this principle, are quite independent of the magnitude of the unit which is to serve as common divisor for the weights of the elementary atoms, and which can therefore be looked upon as expressing the weight of the atoms of primordial matter. If this weight should prove to be that of one atom of hydrogen, or of a half, a quarter atom of hydrogen, or if it should be that of a very much smaller fraction of it, a hundredth or a thousandth for instance, the same degree of probability would attach to it, and the only result would be less simple constitutional relations between the various elements."

The correctness of the idea expressed in the above argument is evident; but it is equally obvious that the smaller the common unit, the less must be the value of the evidence that can be deduced in favour of its existence from the approximation of the atomic weight values to whole numbers.

Marignac's 1860 paper in support of what he termed Prout's "law" had been evoked by the publication in that year of Stas' *Recherches sur les Rapports réciproques des Poids atomiques,* in which nine atomic weight values were given, whose undoubted deviations from whole numbers led the great Belgian chemist to the expression of the following opinion:

"I have arrived at the absolute conviction, the complete certainty, as far as it is possible for a human being to attain to certainty in
Stas' proof of the inaccuracy of Prout's law. such a matter, that the law of Prout, together with M. Dumas' modifications, is nothing but an illusion, a mere speculation definitely contradicted by experience."

The expectation expressed by Stas that if chemists would but sink their prejudices and preconceived notions and be guided by experience alone, they could not fail to share this conviction, was not realised in the case of Marignac. The celebrated Swiss chemist ungrudgingly admitted the unprecedented accuracy of Stas' numbers, pointed out the close agreement with those obtained by himself at an earlier date, and the comparatively much greater differences from those required by Prout's hypothesis.

"......the differences between the results obtained by M. Stas and those required by the law of Prout are certainly very small, but they are considerably greater...than the greatest differences between the results obtained in each set of experiments."

	Stas (1860)	Marignac	Prout
Silver	107·943	107·921	108
Chlorine	35·46	35·456	35·5
Potassium	39·13	39·115	39
Sodium	23·05		23
Ammonium[1]	18·06		18
Nitrogen	14·041	14·02	14
Sulphur	16·037		16
Lead (synthesis of sulphate)	103·453		103·5
Lead (synthesis of nitrate)	103·460		103·5

Marignac expressed his belief that if determinations more accurate than those of Stas should ever be made, they would show no better approximation to whole numbers; but whilst recognising the impossibility of maintaining the absolute validity of Prout's law against such evidence, he proposed to class it as an approximate law, and proceeded to propound certain speculations concerning causes to which the observed deviations from the requirements of an exact law might possibly be due.

Marignac's retention of Prout's hypo-thesis as an approximate law.

"We may believe that the law of Prout (like the laws of Mariotte and Gay-Lussac[2]), though not rigorously confirmed by experiment[3], none the less expresses the relations between the atomic weights of simple substances with an accuracy sufficient for the practical calculations of chemistry; and we may even believe that it perhaps represents the normal relations which would exist between these weights in the absence of perturbing causes, the search for which latter should exercise the sagacity and the imagination of chemists.

In admitting the hypothesis of the unity of matter, might it not be assumed that the unknown cause which in producing certain groupings of the one primordial matter has given birth to our simple chemical atoms, imprinting on each of these aggregations a special character and particular properties, should also have been able to exert some influence on the mode in which these groups of atoms obey the universal law of gravitation, making

Possible causes of approximate nature of Prout's law.

---

[1] This gives for hydrogen $\dfrac{18·06 - 14·04}{4} = 1·005$; but according to the value then generally accepted, if $O = 8·000$, which was Stas' actual practical standard, $H = 1·000$.

"Glancing at the atomic weights of ammonium and of nitrogen, we perceive that they differ by 4·02 instead of 4·00. The undoubted result is either that my syntheses of silver nitrate are inexact...or that the atomic weight of hydrogen itself is incorrect by 5/1000 or 1/200 of its value. The sum of my researches leads me to believe that the error affects the atomic weight of hydrogen rather than that of nitrogen."

[2] Gay-Lussac's law of gaseous expansion, better known in England under the name of Charles' law.

[3] *Ante,* p. 92, *et seq.*

the weight of each one not exactly equal to the sum of the weights of the constituent primordial atoms?"

The following reflection due to another great chemist is quoted from the German translation (1904) of Rudorf, *The Periodic System*.

"It is conceivable that the atoms of all or of many elements should after all consist chiefly of smaller elementary particles of one single kind of primitive matter, perhaps hydrogen; but that their weights do not appear as rational multiples of one another, because in addition to the particles of this primordial matter there may enter into their composition greater or smaller amounts of the matter by us designated as luminiferous ether, which fills all space and which might not be devoid of all weight." (Lothar Meyer, 1896.)

These particular speculations of Marignac and Lothar Meyer concerning the relation between the weight of an elementary atom and that of the sum of the weights of its constituent atoms of primordial matter are hypotheses which in the present state of the science can neither be proved nor refuted; but the same did not

Marignac's suggestion of slight variability of composition disproved by Stas.

apply to Marignac's suggestion of another possible cause for the experimentally established deviations from Prout's law. His doubt concerning the absolute constancy of composition in compounds and Stas' brilliant proof of the law of fixed ratios have been treated fully in chapter V.

Stas showed that "there does not exist a common divisor for the weights of the simple substances which unite to form all the definite combinations," and all the additions made since his time to our knowledge of accurate atomic weight values (*e.g.* $H = 1.0076$) have but gone to confirm this. All the same it may be doubted whether any chemist of the present generation would or could accept in its entirety Stas' final verdict on the results of his experimental work:

Stas' declaration against unity of matter.

"I have come to look upon Prout's hypothesis as a pure illusion, and I consider that all substances reputed undecomposable are distinct beings, with no simple connection of weight between them."

Though forced to recognise that no *simple* connection exists between the weights of the different kinds of atoms, we are not therefore driven to look upon them as *distinct beings*, unless we also retain the original Daltonian conception of their absolute indivisibility.　But for a long time past, even at a period

Philosophical
reasons
against indi-
visibility of
elementary
atoms.

when all experimental support was still lacking, philosophical considerations have been adduced to show
that this view was unnecessarily extreme and rigid,
and that the indivisibility of the entities termed
atoms was probably only relative; that there existed
no *à priori* reason against these units of chemical change being
not really ἄτομοι, but only one definite, though not necessarily the
last stage in the complexity of matter.

"In using the word *atom*, chemists seem to think that they bind themselves to a theory of indivisibility. This is a mistake. The word *atom*
means *that which is not divided*, as easily as it may mean *that which cannot
be divided*, and indeed the former is the preferable meaning. Even when
Lucretius speaks of primordial bodies that cannot be divided, he does not
deny that they have parts, although these...cannot exist by themselves, and
Graham, as well as other atomists, give a similar opinion, that is, that the
original atom may be far down....There comes to us a something indivisible
by us, and it is consistent to call it an atom as it is consistent to call the
smallest particle of alum, with its twenty-four equivalents of water, an atom,
simply because it is the smallest possible portion of alum, which to divide
would be to destroy." (Angus Smith, in *Preface to Physical Researches of
Thomas Graham*, 1876.)

"It is probable that just as those masses which perceptibly occupy space
are composed of molecules, and molecules, or particles of the first order, are
composed of atoms, or particles of the second order, so atoms are composed
of particles of matter of a third and simpler order. This view is supported
by the reflection that if atoms are unchangeable and indivisible, just as many
elementary forms of matter must exist as there are chemical elements. The
existence of some sixty or more entirely different forms of primordial matter
is improbable; the knowledge of certain properties of the atoms, especially
the relations exhibited by the atomic weights of different elements, rendering
this all the more likely." (Lothar Meyer, *Modern Theories*, p. 113.)

Within the last two decades experimental evidence, physical
and chemical, has been accumulating in support of these specula-

Experimental
evidence for
the complexity
of atoms and
the identity
of their con-
stituents.

tions, and the last few years' contributions have
been such as to make the complexity of the atoms
as much of an established fact as that of the
molecular and atomic structure of the masses of
matter that we perceive. Moreover the empirical
results which, taken in their entirety, can almost
be said to have proved this point, have also supplied evidence
which justifies the course hitherto followed by chemists in assigning to the atoms a very special place in the scale of the

complexity of different kinds of matter. It seems that the diversity of matter begins only with the atom: that whilst the component parts of a molecule A are not the same as those of another kind of molecule B, or C, or D, etc., the constituents of an atom M are identical with those of any other different atom P, or Q, or R, etc.; that all atoms are compounded of the same one kind of primordial matter.

Sir W. Crookes, by long continued chemical fractionation[1], separated the rare earth yttria into portions which gave spark spectra identical in every detail, whilst the phosphorescent spectra[2] were different. The explanation of this novel and unique phenomenon involved the assumption of differences in the constituent yttrium atoms, and hence of the complexity of these.

(i) Differences in the phosphorescent spectra of yttria. Crookes' fractionations.

"The old yttrium passed muster as an element. It had a definite atomic weight, it entered into combination with other elements, and could be again separated from them as a whole But now we find that excessive and systematic fractionation has acted the part of a chemical "sorting Demon," distributing the atoms of yttrium into groups, with certainly different phosphorescent spectra, and presumably different atomic weights, though, from the usual chemical point of view, all these groups behave alike. Here, then, is a so-called element whose spectrum does not emanate equally from all its atoms; but some atoms furnish some, other atoms others, of the lines and bands of the compound spectrum of the element. Hence the atoms of this element differ probably in weight, and certainly in the internal motions they undergo." (*Genesis of the Elements*, 1887.)

Faraday in 1862 had investigated whether light produced in a magnetic field is in any way affected by the magnetic force, but had obtained no result. Zeeman in 1897 was able to show that the action of a sufficiently strong magnetic field results in the resolution of bright spectral lines into triplets or quartets or sextets etc. The gradual establishment of this occurrence, the elucidation of its nature and the harmonious interpretation of the apparent

(ii) Magnetic perturbations of the spectral lines. The Zeeman effect.

---

1 "Stated in the briefest way, the operation consists in fixing upon some chemical reaction in which there is the most likelihood of a difference in the behaviour of the elements under treatment, even though the difference be slight, and effecting such treatment incompletely, so that only a certain fraction of the total bases present is separated; the object being to get part of the material in an insoluble, and the remainder in a soluble state."

2 The spectroscopic resolution of the light emitted by substances, when enclosed in highly exhausted tubes through which an electric discharge passes. The phosphorescence produced is supposed to be due to the bombardment of the substance by the particles of gas, which act as the carriers of the electricity.

diversities in its manifestation, constitute a most striking example of the results which the present generation of scientists reaps from being guided in experimental investigations by well-founded theories. This has been forcibly brought out in a Royal Institution Lecture[1] by Prof. Preston.

The chemist owes to the comparative study of the magnetic perturbations of the spectral lines some information concerning the ultimate structure of matter, strong indication of the complexity of the elementary atoms.

" ...if we view the line spectrum of a given substance we find that some of the lines are sharp, while others are nebulous or diffuse, and that some are long, while others are short—in fact, the lines exhibit characteristic differ- ences which lead us to suspect that they are not all produced by the motion of a single unconstrained ion[2]. On closer scrutiny, they are seen to throw themselves into natural groups. For example, in the case of the monad metals sodium, potassium, etc., the spectral lines of each metal form three series of natural pairs, and again, in the case of the dyad group, cadmium, zinc, etc., the spectrum of each shows two series of natural triplets, and so on.

Thus, speaking generally, the lines which form the spectrum of a given substance may be arranged in groups which possess similar characteristics as groups. Calling the lines of these groups A_1, B_1, C_1, ..., A_2, B_2, C_2, ..., A_3, B_3, C_3, ..., we may regard the successive groups as repetitions of the first, so that the A's—that is, A_1, A_2, A_3, etc.—are corresponding lines pro- duced probably by the same ion; while the B's—namely, B_1, B_2, B_3, etc.— correspond to one another and are produced by another ion, and so on. This grouping of the spectral lines has been noted in the case of several substances[3], and it has been a subject of earnest enquiry among spectroscopists for some time past. All such grouping, however, up to the present, has had to depend on the judgment of the observer as to certain similarities in the general character and arrangement of the lines, and similarities which indeed may or may not have any specific relation to the mechanism by which the lines are produced. In fact, such grouping has been effected by guess-work, or by empirical formulae....

...This grouping of the spectral lines...[may be used] to attack the problem of reducing to order the...apparently lawless magnetic effect....The lines in the spectrum of any given substance are not all resolved into triplets by the magnetic field, but some are resolved into triplets, while others become sextets, etc. ; and further, the magnitude of this resolution, that is, the

---

[1] *Nature*, London, 60, 1899 (p. 175).

[2] " The element of matter...is sometimes called an ion, which name is used to imply that [it] carries an electric charge inherently associated with it."

[3] An instance of the application of the comparative study of analogous lines in the emission spectra of elements belonging to the same class has been given in the description of Lecoq de Boisbaudran's determination of the atomic weight of gallium and germanium (*ante*, p. 491, *et seq.*).

interval $\delta\lambda$ between the lateral components, does not appear at first sight to obey any simple law.

According to the prediction of the simple theory, the separation $\delta\lambda$ should be proportional to λ^2, and although this law is not at all obeyed, if we take all the lines of the spectrum as a single group, yet we find that it is obeyed for the different groups if we divide the lines into a series of groups. In other words, the corresponding lines A_1, A_2, A_3, etc., have the same value for the quantity e/m[1], or, as we may say, they are produced by the motion of the same ion. The other corresponding lines B_1, B_2, B_3, etc., have another common value for e/m, and are produced therefore by a different ion.... We are thus led by this magnetic effect to arrange the lines of a given spectrum into natural groups, and from the nature of the effect we are led to suspect that the corresponding lines of these groups are produced by the same ion, and therefore that the atom of any given substance is really a complex consisting of several different ions, each of which gives rise to certain spectral lines, and these ions are associated to form an atom in some peculiar way which stamps the substance with its own peculiar properties.

In order to illustrate the meaning of this, let us consider the spectrum of some such metal as zinc. The bright lines forming the spectrum of this metal arrange themselves to a large extent in sets of three—that is, they group themselves naturally in triplets. Denoting these triplets in ascending order of refrangibility by A_1, B_1, C_1, A_2, B_2, C_2, etc., we find that the lines A_1, A_2, etc., show the same magnetic effect in character and have the same value of e/m, so that they form a series obeying the theoretical law deduced by Lorentz and Larmor. In the same way, the lines B_1, B_2, B_3, etc., form another series, which also obeys the theoretical law, and possess a common value for the quantity e/m, similarly for the lines C_1, C_2, C_3, etc. The value of e/m for the A series differs from that possessed by the B series, or the C series, and this leads us to infer that the atom of zinc is built up of ions which differ from each other in the value of the quantity e/m, that each of these different ions is effective in producing a certain series of lines in the spectrum of the metal. When we examine the spectrum of cadmium, or of magnesium —that is, when we examine the spectra of other metals of the same chemical group—we find that not only are the spectra homologous, not only do the lines group themselves in similar groups, but we find in addition that the corresponding lines of the different spectra are *similarly* affected by the magnetic field[2]. And further, not only is the character of the magnetic effect the same for the corresponding lines of the different metals of the same chemical group, but the actual magnitude of the resolution as measured by the quantity e/m is the same for the corresponding series of lines in the different spectra. This...leads us to believe, or at least to suspect, that the ion which produces the lines A_1, A_2, A_3, etc., in the spectrum of zinc is the same as that which produces the corresponding series A_1, A_2, A_3, etc., in

[1] "The quantity e is the electric charge of the ion, and m is its inertia, and the ratio e/m determines the precessional frequency, or spin, of the ionic orbit round the lines of magnetic force in a given field."

[2] This is the criterion that has been applied by Runge and Precht in their determination of the atomic weight of radium (*ante*, p. 495, *et seq.*).

cadmium, and the same for the corresponding sets in the other metals of this chemical group. In other words, we are led to suspect that, not only is the atom a complex composed of an association of different ions, but that the atoms of those substances which lie in the same chemical group are perhaps built up from the same kind of ions, or at least from ions which possess the same e/m, and that the differences which exist in the materials thus constituted arise more from the manner of association of the ions in the atom than from differences in the fundamental character of the ions which build up the atoms; or it may be, indeed, that all ions are fundamentally the same, and that differences in the value of e/m, or in the character of the vibrations emitted by them, or in the spectral lines produced by them, may really arise from the manner in which they are associated together in building up the atom.

...These observations lend some support to the idea, so long entertained merely as a speculation, that all the various kinds of matter, all the various so-called chemical elements, may be built up in some way of the same fundamental substance." (Preston, *loc. cit.*)

Sir Norman Lockyer, as the result of a most extensive comparative study of the spectra of the terrestrial elements when

(iii) Different spectra yielded by same substance at different temperatures, and spectra of the fixed stars. Lockyer's dissociation hypothesis.

these are produced in the laboratory at successively rising temperatures, and of the spectra given by stars of varying degrees of hotness, has formulated an intensely interesting hypothesis of inorganic evolution[1] (*post*, p. 618). The harmonising of all the most important classes of phenomena involved in the scope of his speculations necessitated the hypothesis of the dissociation at high temperatures of the elementary atoms, hence of course involving the assumption of their complex structure.

"[In the laboratory] changes in spectra [were] observed to accompany changes in the quantity and kind of energy used in the experiments... [Thus] four distinct temperature stages are indicated by the varying spectra of the metals, [which taking] iron as an example are :

1. The flame spectrum, consisting of a few lines and flutings only...

2. The arc spectrum, consisting...of 2000 lines or more.

3. The spark spectrum, differing from the arc spectrum in the enhancement of some of the short lines and the reduced relative brightness of others.

4. A spectrum consisting of a relatively very small number of lines which are intensified in the spark."

[1] *Inorganic Evolution as studied by Spectrum Analysis*, London, 1900.

"At the temperature next lowest [to that of the very hottest stars] we find...metals in the state in which they are observed...when the most powerful jar-spark is employed. At a lower temperature still...the metals exist in the state produced by the electric arc."

"The similar changes in the spectra of certain elements, changes observed in laboratory, sun and stars are simply and sufficiently explained on the hypothesis of dissociation... [The] verdict [of the stars] is that, as in all previous human experience, a higher temperature brings about simplifications." (Lockyer, *loc. cit.*)

Great as is the importance and the interest of the evidence for the complex nature of the atoms supplied by spectroscopy, this is

(iv) The mass of the negatively electrified particles in gases at low pressures. Prof. J. J. Thomson's corpuscles.

far surpassed by that attaching to the proofs derived from the recent extension of our knowledge concerning the connection between electricity and matter. It is mainly to the study of the conduction of electricity through gases by Prof. J. J. Thomson and his school that we owe discoveries which are of intense theoretical interest to the chemist. Even a highly condensed account of the many ingenious investigations which form the links of a strong chain of experimental evidence would occupy more space than is available here; moreover the necessity for a systematic abstract has been removed by the recent publication of a number of books[1] which contain a more or less detailed account of the whole subject. The following gives in the merest outline the argument which leads to the definite inference of the great complexity of the chemical atoms and of the common nature of their constituents.

In the electrolysis of solutions the ions[2], which according to the atomistic view of the structure of matter must be either simple atoms like H, Cl, Na, Zn, or groups of atoms like SO_4, NO_3, HO, carry a charge equal to that of the hydrogen ion (*e.g.* Cl, Na, OH) or twice that charge (*e.g.* Zn, SO_4), and so on. For knowledge concerning the actual mass of these individual ions we have until recently had to rely entirely on calculations involving many assumptions, but which, as is shown by the very fair agreement of the values obtained by different methods, in all probability gave results approximately correct. According to Lord Kelvin's most recent estimate:

---

[1] J. J. Thomson, *The Conduction of Electricity through Gases*, 1903; *Electricity and Matter*, 1904. Rutherford, *Radio-Activity*, 1904. Soddy, *Radio-Activity*, 1904.

[2] *Ante*, p. 538, *et seq.*

N, the number of *molecules* in 1 c.c. of a gas at 0° C.
and 760 mm., which by Avogadro's law is the
same for all gases, is.......................................$= 10^{20}$

Hence

aN, the number of *atoms* in 1 c.c. of a gas of atomicity
a is...$= a \times 10^{20}$

wt. of *one* hydrogen atom ....$= \dfrac{\cdot00009}{2 \times 10^{20}} = 4\cdot5 \times 10^{-25}$ grams

and wt. of *one* atom of an element of atomic weight $A_{H=1} = A \times 4\cdot5 \times 10^{-25}$ grams

Van der Waals' calculation, based on the kinetic theory of gases, gives

N...$= 5\cdot4 \times 10^{19}$

wt. of *one* hydrogen atom.......................................$= 8\cdot3 \times 10^{-25}$ grams[1].

But given N, the value of E, the definite elementary portion of electricity of which the univalent hydrogen atom carries *one unit*, the divalent zinc atom *two units*, etc. (*ante*, p. 539) can be calculated from the relation

$$2NE = \begin{Bmatrix} \text{charge carried by number} \\ \text{of } atoms \text{ present in 1 c.c.} \\ \text{of hydrogen} \end{Bmatrix} = \begin{Bmatrix} \text{current required for the} \\ \text{electrolytic separation} \\ \text{of 1 c.c. of hydrogen.} \end{Bmatrix}$$

The determination of the electrochemical equivalent of hydrogen has given as result that the passage of 1 electromagnetic[2] unit $= 3 \times 10^{10}$ electrostatic[3] units liberates $1\cdot16$ c.c. of hydrogen measured at normal temperature and pressure, and hence, on the basis of Van der Waals' evaluation of N,

$$E = 2\cdot4 \times 10^{-10} \text{ electrostatic units} \atop \text{or} \quad 8 \times 10^{-21} \text{ electromagnetic units} \Big\} \text{(approximately)};$$

and for one atom of hydrogen, in electromagnetic units,

$$\frac{E}{M} = 10^4 \text{ approximately.}$$

The investigation of the properties of a gas which by some means or other—exposure to Röntgen rays, the action of a radio-

---

[1] See also *ante*, p. 346.

[2] An electromagnetic *unit quantity* of electricity is the quantity which a unit of *current* conveys in a unit of time; a unit current being one which, in a wire of unit length bent so as to form an arc of a circle of unit radius, would act upon a unit *magnetic pole* placed at the centre with unit force; and a unit magnetic pole being that which repels a similar pole at unit distance (1 cm.) with unit force (1 dyne).

[3] An electrostatic *unit quantity* is that which exerts the unit of force (1 dyne) on a quantity equal to itself, at a distance of 1 cm. across air.

active substance, etc.—has been endowed with the power of conducting electricity, shows that:

"The constituent to which the conductivity of the gas is due consists of charged particles, the conductivity arising from the motion of these particles in the electric field." (J. J. Thomson, *Electricity and Matter.*)

Since the gas as a whole has no charge, it follows that of these conducting particles or gaseous ions, some must be charged negatively, others positively. It has been shown experimentally that the positive and negative ions are equal in number, from which it follows that the charge carried by a positive ion is equal in value but opposite in sign to that carried by a negative ion. This charge e carried by each particle has been measured by several different methods which have yielded concordant results.

"If at any time there are in the gas n...particles charged positively and n charged negatively, and if each of these carries an electric charge e, we can easily by electrical methods determine ne, the quantity of electricity of one sign present in the gas.... If then we can devise a means of measuring n we shall be able to find e." (*Ibid.*)

The determination of n is based on the discovery that charged particles act as nuclei for the condensation of water in damp and dust-free air in which, but for their presence, no cloud would be formed. The drops cannot be counted directly, but their number, which is equal to n, that of the charged particles, has been estimated indirectly. The first step consists in the calculation of the size of the constituent water-drops from the measured velocity with which the cloud settles, a formula due to Sir George Stokes being used; the second step is the calculation of n, the number of drops, from their size and the indirectly measured quantity of water in the cloud.

Thus, ne and n being known, it was found that

$$e = 3\cdot4 \times 10^{-10} \text{ electrostatic units.}$$

"Experiments were made with air, hydrogen and carbonic acid, and it was found that the ions had the same charge in all these gases." (*Ibid.*)

Moreover, on comparing the values for E and e, namely $2\cdot4 \times 10^{-10}$ and $3\cdot4 \times 10^{-10}$, and remembering the approximate nature of the value for N and hence of that for E, the agreement between the two numbers is sufficient for them to be considered equal, an inference justified by the results of experiments which made direct comparison possible. Hence it may be taken to have been proved that

$$e = E,$$

or that the charge on the gaseous ion is equal to the charge carried by the hydrogen ion in the electrolysis of solutions.

But what is the mass associated with this constant charge in the particles which act as the carriers of electricity in conduction through gases, and how does it compare with that of the carriers which in the case of conduction through solutions are atoms or groups of atoms, and whose absolute masses we know approximately?

Gaseous ions possess the important property of being acted upon by magnetic or electric forces. Thus a particle of mass m carrying a charge e, if subjected to the action of a magnetic force H at right angles to the rectilinear path along which it was moving with velocity v, is deviated into a circular path of curvature ρ. The simultaneous action of an electrostatic force F at right angles to the magnetic force and also at right angles to the direction of motion of the ion will, according to the fall of electric potential in one direction or the other[1], augment or decrease the magnetic deviation produced; and by changing the magnitude of one of them, keeping the other constant, it can be so arranged that the magnetic and the electric force exactly neutralise each other. Then since it can be shown that all the quantities involved are connected by the equations

$$\rho = \frac{mv}{He} \qquad \text{and} \qquad Fe = Hev,$$

it becomes possible by measuring ρ and F and H to calculate v and $\dfrac{e}{m}$.

This method when applied to the determination of e/m for the negatively electrified particles constituting the cathode rays (*ante*, p. 541), as well as for those emitted by metals exposed to ultra-

---

[1] If the electrostatic force is produced by two plates A, B, kept at a constant difference of potential by connecting the one plate to the + terminal of a powerful battery and the other to earth, according as to whether the arrangement is

A ——— to+battery terminal ———   or A ——— to earth ———

B ——— to earth ——— B ——— to+battery terminal ———

violet light or raised to the temperature of incandescence, gave the *constant* value 10^7, e being measured in electromagnetic units. But since in the electrolysis of solutions, in the case of hydrogen, $E/M = 10^4$, and since it has been shown that $E = e$, it follows that:

"the mass of a carrier of the negative charge[1] must be only about one-thousandth part of the mass of the hydrogen atom, which was for a long time regarded as the smallest mass able to have an independent existence." (J. J. Thomson, *Electricity and Matter.*)

Concerning these carriers of the negative charge e in gases at low pressures by him named *corpuscles* (*ante*, p. 541), Prof. Thomson says:

"Whether we produce the corpuscles by cathode rays, by ultra-violet light, or from incandescent metals, and whatever may be the metals or gases present we always get the same kind of corpuscles. Since corpuscles similar in all respects may be obtained from very different agents and materials, and since the mass of the corpuscles is less than that of any known atom, we see that the corpuscle must be a constituent of the atom of many different substances. That in fact the atoms of these substances have something in common.

We are thus confronted with the idea that the atoms of the chemical elements are built up of simpler systems."

Mr Soddy in the first chapter of his recently published book on Radio-activity says: "The chemist's atom is no longer the unit of the subdivision of matter, and the internal struc-

(v) Radio-active changes; atomic disintegration.

ture of the atom is now the object of experimental study." The results obtained within the last few years in the investigation of the phenomena included under the name of *radio-activity* have had a great share in this modification, perhaps more correctly extension, of the chemist's atomistic doctrine of the ultimate constitution of matter. The characteristic property named radio-activity, possessed in the highest degree by radium, and in a lesser degree by uranium and thorium (also polonium and actinium[2]), consists in a continuous and spontaneous radiation (*ante*, p. 541) which is detected and measured by its power of ionising a gas, *i.e.* making it a conductor

[1] The value obtained for e/m for the carriers of the positive charge varies with the nature of the electrodes and with that of the gas through which the discharge occurs; it was found never greater than 10^4, sometimes considerably less, leading to the inference that *atoms* of the different elements available under the different circumstances act as the carriers of the positive charge.

[2] Two other elements discovered by the same method as radium, but hitherto only obtained in extremely minute quantities, and hence little investigated.

of electricity. This radiation has been shown to be complex, made up of three types of rays designated by α, β, γ, respectively, and distinguished by differences in penetrating power and differences in the manner in which they are affected by magnetic and electro-static forces.

The α rays, which account for almost the whole of the ionising effect produced, are the least penetrating; the very slight devia-tion which they experience in a strong magnetic field marks them as being electrically charged particles, the direction of the devia-tion indicating the positive nature of their charge. The measure-ment of the quantity e/m by the method referred to on page 612 gave the value 6×10^3, which on the supposition of the constancy of e for all carriers of electricity, corresponds to a mass somewhat greater than that of the atom of hydrogen for which $E/M = 10^4$, but less than that of helium (He = 4)[1], the element of next higher atomic weight, for which this ratio would be $2 \cdot 5 \times 10^3$. The velocity v was found to have the high value $2 \cdot 5 \times 10^9$, nearly $\frac{1}{10}$th that of light, which is 3×10^{10}.

The β rays are very much more penetrating than the α; they are easily deflected in a magnetic field in the same direction as the cathode rays, and the measurement of e/m gave the value 10^7, thus revealing their corpuscular nature, *i.e.* showing them to be negatively charged particles of a mass about $\frac{1}{1000}$ that of the hydrogen atom. The velocity with which these particles travel is on the average $\frac{2}{3}$ that of light, and for the most penetrating amongst them is nearly equal to it.

The γ rays are the most penetrating of the three kinds, and are not affected by magnetic or electric forces.

The radio-active elements, whose atomic weights are the highest known (Ra = 225, Th = 232, U = 239·5) have therefore been proved to expel with enormous velocities electrically charged particles whose mass in the case of the α rays, though of atomic order, is much smaller than that of the parent atom of which it must be considered a component part liberated by a process of atomic disintegration. The validity of this inference is in complete harmony with another aspect of radio-activity, that of the produc-tion of new and, as far as our identification goes, mostly unstable types of matter.

[1] It has been suggested by Prof. Rutherford that the α particle is an atom of helium.

Taking the special case of radium, it has been found that there is a continuous production, always proportional in amount to the quantity of radium salt present, of a something termed *emanation*; and that there is direct quantitative parallelism between the production of this emanation and the radiation, 25 $°/_o$ of the total α radiation being so accounted for. The emanation, separated in virtue of its volatility from its parent radium salt, has been studied separately and found to behave in every way like a heavy chemi-cally inert gas. The atomic weight of the emanation as deduced from its rate of diffusion is about 160, that is, smaller than that of the radium from which it is produced, another indication of the complexity of the radium atom. The emanation is so small in quantity that it cannot be weighed[1] or recognised by any chemical test, its detection and measurement being by means of radio-activity alone. The activity of the emanation is found to repre-sent 40 $°/_o$ of that of the original salt, but this value decreases steadily at a constant and characteristic rate, whilst the salt, whose activity had been reduced owing to the withdrawal of the emanation, increases in activity, the one gaining as much as the other loses, until when the activity of the emanation has com-pletely decayed, the salt has, by means of the fresh production of an amount of emanation equal to that withdrawn from it, regained its original activity. Neither process can be accelerated nor retarded by any known agency, physical or chemical; they occur at the same rate at the temperature of liquid air as at white heat; the effect of acids, alkalis or oxidising agents is alike *nil*.

The emanation in its turn, concurrently with the emission of α rays, continuously produces a new type of solid matter, also radio-active and only detected in virtue of this property, also of transient existence, as shown by a constant characteristic rate of decay of its activity. This decomposition product of the emana-tion, called *the matter causing the imparted activity*, when not separated from the radium salt to which indirectly through the emanation it owes its existence, supplies 35 $°/_o$ of the total radia-tion of the salt, the loss due to its decay being always made good

· [1] Sir W. Ramsay and Mr Soddy have succeeded in measuring the volume pro-duced from a known quantity of radium in a known time. "The equilibrium quantity of emanation from 60 mg. of radium bromide was found to occupy between ·03 and ·04 cubic mm. at 0° C. and 760 mm. The emanation produced in 5·3 days by 1 gram of radium (element) therefore occupies a volume of about 1·3 cubic mm." (Soddy, *Radio-Activity*.)

by the constant production of equivalent quantities. Presumably this short-lived solid radio-active matter gives as the product of its own disintegration not only α and β particles but also some kind of matter known or unknown, the presence of which has not yet been detected because the quantity produced in the course of ordinary laboratory experiments is too small to be recognised by tests so inferior in sensitiveness to the electrical test[1] for radio-activity as are those available in analytical chemistry. The emanation (at. wt. about 160) also yields, directly or indirectly[2], helium (at. wt. $= 4$), a transformation first suspected on theoretical grounds, which it has been possible to detect experimentally owing to the exceptionally great sensitiveness of the spectroscopic test for helium.

Madame Curie has shown that radio-activity is an atomic property (*ante*, p. 27), and since it has also been established that the production of radiation is concurrent with that of new kinds of matter, it follows that the changing units, which are the atoms of the radio-active elements, must be undergoing chemical change. But with the changing units all chemically the same, change can only consist in: (i) coalescence of some of these units into more complex particles (*e.g.* $3C_2H_2 = C_6H_6$), or (ii) decomposition of each unit into several simpler particles (*e.g.* $I_2 = 2I$; $C_2H_2Br_2 . (CO_2H)_2 = C_2HBr (CO_2H)_2 + HBr$), or (iii) rearrangement of the component parts of each unit (*e.g.* $NH_4CNO = (NH_2)_2CO$). Of these possibilities, the two last involve the complexity of the changing particle. In the first case, that of combination, the amount of change occurring would depend, not only on the number of radio-active atoms present, but also on the chance of the encounter of the number necessary for combination, a chance which depends on the space through which that number is distributed. But experiment has shown that radio-active change is independent of

---

[1] "With suitable precautions an electroscope can...readily measure an ionisation current corresponding to the production of 1 ion per cubic centimetre per second." (Rutherford, *Radio-Activity*.)

[2] "The appearance of helium in a tube containing the radium emanation may indicate either that the helium is one of the final products, which appear at the end of the series of radio-active changes, or that the helium is in reality the expelled α particle. The evidence at present points to the latter being the more probable explanation....The value of e/m determined for the projected α particle points to the conclusion that, if it consists of any known kind of matter, it is either hydrogen or helium." (*Ibid.*)

concentration[1], hence it must belong to the type of reactions known as *monomolecular*, in which each changing unit, whether this be the molecule or as in radio-active change the atom, must either break up into its component parts or undergo internal rearrangement of these components. In the case of radio-active change, the second of these possibilities is excluded because mere internal rearrangement would not account for the simultaneous production from the changing atom of a new type of matter and a material radiation.

"...Radio-active change cannot be of the nature of a combination together, or, as a chemist would say, the polymerisation, of the atoms of the active element, but must be due to their decomposition or disintegration. The term *changing atom* can now be logically replaced by the more definite conception expressed by the use of the term *disintegrating atom*, with considerable advantage to the clearness of the mental picture conveyed.

The term *disintegration* is indeed little more than a convenient and short means of expressing certain experimental facts. It is not until we enquire as to the ultimate cause of radio-activity, and seek a knowledge of the forces at work which bring about the observed disintegration, that we enter a region to which the term *hypothesis* in the ordinary sense of a probable explanation would apply." (Soddy, *loc. cit.*)

Thus investigations of apparently very dissimilar phenomena have led by converging lines of evidence to the recognition that though Prout's hypothesis in its original form, as well as in that given to it by Dumas, has been refuted by Stas' appeal to facts, the fundamental idea underlying it may now be considered as experimentally established. The primary matter, the $\pi\rho\dot{\omega}\tau\eta$ $\ddot{\upsilon}\lambda\eta$, has been shifted down the scale, and hydrogen itself appears as a highly condensed form of matter with each of its atoms containing about 1000 of the truly elemental *corpuscles* (or electrons) of which there is one kind only. We

The genesis of the elements; philosophical speculations.

would seem to have attained to some knowledge concerning the common constituent of all elements, the stones from which these are built. But what about the process by which the different kinds of atoms have been formed?

[1] The effect due to a certain quantity of radium bromide has been found the same when the salt was present in the solid form as when disseminated through a large volume of a solvent; "this result is in general agreement with other observations, for it has not been observed that the decay of activity of any product is influenced by the degree of concentration of that product."

Philosophical speculations concerning the manner in which the homogeneous has changed into the heterogeneous have never been wanting. Some account has been given in chapter IX of cosmogonies conceived by Greek thinkers; "philosophy at the present day demands a genesis of the elements based on evolution."

"The chemical atoms are produced from the true or physical atoms by processes of evolution, under conditions which chemistry has not yet been able to produce." (Herbert Spencer.)

Moreover we have experimental evidence in support of the occurrence of such evolution.

Sir Norman Lockyer, in reading and interpreting the tale by means of which the stars reveal their composition and their temperature, believes that he has found strong evidence for the formation of elements by successive stages in the condensation of hydrogen.

Experimental evidence for inorganic evolution; (i) Successive appearance of elements in cooling stars.

"Do the chemical elements make themselves visible indiscriminately in all the celestial bodies, so that practically from a chemical point of view, the bodies appear to us of similar chemical constitution? This is not so.

Taking the chemical elements as we know them here, we find differences in composition continuously indicated as stars of successively higher temperature are studied....The effect of high temperature in producing simplifications is known to everybody...; the final products of dissociation or breaking up by heat must be the earliest chemical forms. Hence if the various stars...bring before us a progression of new forms in an organised sequence, we must regard the chemical substances which visibly exist in the hottest stars which, so far as we know, bring us in presence of temperatures higher than any we can command in our laboratories, as representing the earliest evolutionary forms.

...The statement can now be firmly made, that in the hottest stars we are brought in presence of a very small number of chemical elements. As we come down from the hottest stars to the cooler ones, the number of spectral lines increases, and with the number of lines the number of chemical elements. ...In the hottest stars of all, we deal with a form of hydrogen which we do not know anything about here (but which we suppose to be due to the presence of a very high temperature), hydrogen as we know it, the cleveite gases[1], and magnesium and calcium in forms which are difficult to get here[2]; we think we get them by using the highest temperatures available in our laboratories. In

---

[1] Cleveite gases, identical with Ramsay's helium, by Lockyer supposed to be a mixture of two substances, helium and asterium.

[2] Presumably dissociated into elements to which Lockyer assigns the atomic weights $\frac{Ca}{4} = 10$ and $\frac{Mg}{2} = 12$.

the stars of the next lower temperature we find the existence of these substances continued in addition to the introduction of oxygen, nitrogen, and carbon. In the next cooler stars we find silicium added ; in the next we note the forms of iron, titanium, copper, and manganese, which we can produce at the very highest temperatures available in our laboratories ; and it is only when we come to stars much cooler that we find the ordinary indications of iron, calcium, and manganese and other metals. All these, therefore, seem to be forms produced by the running down of temperature. As certain new forms are introduced at each stage, so certain old forms disappear." (Lockyer, *Inorganic Evolution*, 1900.)

The transmutation of the elements in the stars, the evidence for which though strong is indirect, takes the course of poly_merisation, of a change from the simpler to the more complex. The production of helium from radium, the first instance on record of directly established transmutation of an element, follows the opposite course, that of the production of a simpler, lighter atom from one more complex and heavier.

(ii) Transmutation of radium into helium.

Empiricism, in the form of the periodic law, points in the same direction, towards the existence of a genetic relation amongst elements.

(iii) The atomic weight relations of the periodic law.

But the atomic weight relations which find their expression in the periodic law, whilst adding precision to the question concerning the genesis of the elements, have also added to the scope of the phenomena to be explained. Why this periodic recurrence of typical properties ? why such changes in the typical properties on their recurrence ? why the limitation to the number of existing elements ?

Sir William Crookes in 1886 propounded a hypothesis to which he was led by the great law of continuity. Like Sir Norman Lockyer, he considers that change of temperature is a dominating factor in the evolution of the elements.

Crookes' pendulum oscillations hypothesis concerning mode of genesis of the elements.

"I venture provisionally to conclude that our so-called elements or simple bodies are, in reality, compound molecules....What existed anterior to our elements, before matter as we now have it, I propose to name *protyle*.

We have now to seek how protyle was converted, not into one kind of matter only, but into many. If we recognise that it contained within itself the potentiality of all atomic weights, how did those potentialities become actual ? We require two very reasonable postulates ; let there be granted an antecedent form of energy having periodic cycles of ebb and swell, rest and activity. Let there also be granted an internal action, akin to cooling, acting slowly in the protyle. The first-born element would, in its simplicity, be most nearly allied to protyle. This is hydrogen, of all known bodies the simplest in structure, and of the lowest atomic weight. For some time hydrogen would be the only existing form of matter (in our sense of the term). Between hydrogen and the next formed element there would be a gap in time, and in the interval the element standing next in order of simplicity would gradually be approaching its birth-point. In this interval we may suppose that the evolutionary process soon to determine the birth of a new element would fix likewise its atomic weight, its affinities, and its chemical position.

In the genesis of the elements, the longer the time taken up in the cooling-down process, during which the hardening of protyle into atoms takes place, the more sharply defined would be the resulting elements; whilst the more rapid and the more irregular the cooling, the more closely the resulting bodies would fade into each other by almost imperceptible degrees. Thus we may conceive that the succession of events which gave rise to such groups as platinum, osmium, and iridium,—palladium, ruthenium, and rhodium,—iron, nickel, and cobalt,—might have produced only one element in each of these three groups if the process had been greatly prolonged. And conversely, had the rate of cooling been much more rapid, elements might have originated still more nearly identical than are nickel and cobalt. Thus may have arisen the closely allied elements of the cerium, yttrium, and similar groups.

Any well-defined element may be likened to a platform of stability, connected by ladders of unstable bodies. In the first coalescence of the primitive stuff there would be formed the smallest atoms; these would then unite, forming larger groups; the gaps between the several stages would gradually be bridged over, and the stable element appropriate to that stage would absorb, so to speak, the unstable rungs of the ladder which led up to it.

The hypothesis just suggested, if taken in conjunction with the diagram [p. 621], enables us to proceed a step or two further along the track of the evolution of the elements. We may trace in the undulating curve the action of two forms of energy, the one acting vertically, and the other vibrating to and fro like a pendulum. Let the vertical line represent temperature gradually sinking through an unknown number of degrees from the dissociation-point of the first-formed element downwards to the dissociation-point of the last member of the scale.

But what form of energy is figured by the oscillating line? We see it swinging to and fro to points equidistant from a neutral centre. We see this divergence from neutrality confer atomicity[1] of one, two, three, or four degrees as the distance from the centre increases to one, two, three, or four divisions. We see the approach to or the retrocession from this same neutral line deciding the electro-negative or electro-positive character of each element; those on the retreating half of the swing being positive, and those on the approaching half negative. In short, we are led to suspect that this oscillating power must be closely connected with the imponderable matter, essence, or source of energy we call electricity.

---

[1] *Atomicity* is here used in the sense of *valency*.

Our pendulum begins to swing from the electro-positive side ; lithium, next to hydrogen in the simplicity of its atomic weight, is now formed, followed by glucinum, boron, and carbon. Each element, at the moment of birth, takes up definite quantities of electricity, and on these quantities atomicity depends. Thus are fixed the types of the monatomic[1], diatomic, triatomic, and tetratomic elements."

The formation of the elements of the first series, from H to Cl, is followed in detail.

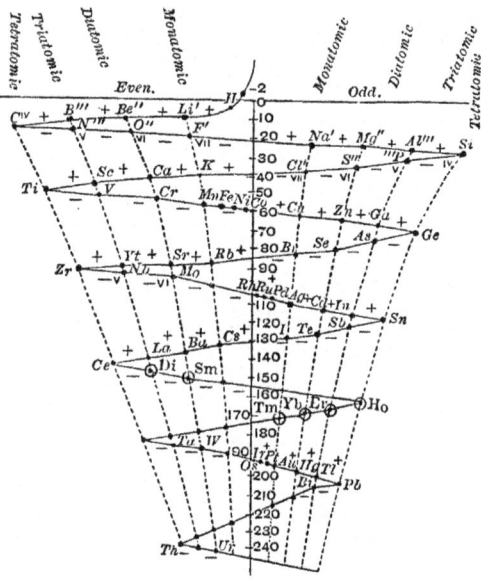

" After the formation of chlorine, the pendulum touches the neutral line, and is in the same position as in the beginning. Had everything remained as at first the next element to appear would again have been lithium, and the original cycle would have been eternally repeated, producing again and again the same fifteen elements. The conditions, however, are no longer the same ; time has elapsed and the form of energy represented by the vertical line has declined ; in other words, the temperature has sunk, and the first element to come into existence when the pendulum starts for its second oscillation is not lithium, but the metal next allied to it in the series, *i.e.* potassium, which may be regarded as the lineal descendant of lithium, with the same hereditary tendencies, but with less molecular mobility and a higher atomic weight.... We have here a phenomenon which reminds us of alternating or cyclical

---

[1] " Monatomic," " diatomic," etc. are here used instead of the now usual terms " monovalent," " divalent," etc.

generation in the organic world, or we may perhaps say of atavism, a recurrence to ancestral types, somewhat modified." ("Genesis of the Elements," *Chem. News*, 55, 1887, p. 83.)

Sir William Crookes, when in 1886 he brought forward the above speculations, had to assume the complexity of the elementary atoms, which he represented himself as "venturing" to do "provisionally." Since then experimental and theoretical knowledge on this point has become much more definite, as is shown by Prof. J. J. Thomson's recent investigations concerning the mode of genesis of the atom from the corpuscle, the structure of the atom and the atomic properties resulting from special atomic structure ("Constitution of the Atom," *Electricity and Matter*, chapter V; "On the Structure of the Atom," *Phil. Mag.* (6), 7, 1904, p. 237). Starting from the hypothesis that the atom is an aggregation of a number of simpler systems, and that these constituent primordial systems are formed by *corpuscles* associated with equal charges of positive electricity, the problem of the aggregation of these rapidly moving units into groups of increasing complexity is investigated in its relation to the gradual decrease in the kinetic energy of the corpuscles. A consideration of corpuscular aggregation with special reference to the genesis of the chemical elements shows this to be compatible with a process of gradual evolution due to the combination of primordial units. It is found that:

Genesis of the elementary atoms by aggregation of corpuscles; structure and consequent properties of these atoms. Prof. J. J. Thomson's investigations.

" The appearance of the more complex systems need not be simultaneous with the disappearance of all the simpler ones...[and] if we regard the systems containing different numbers of units as corresponding to the different chemical elements, then as the universe gets older, elements of higher and higher atomic weight may be expected to appear. Their appearance, however, will not involve the annihilation of the elements of lower atomic weight. The number of atoms of the latter will of course diminish, since the heavier elements are by hypothesis built up of material furnished by the lighter....If, however, there is a continual fall in the [kinetic energy of the corpuscles within the atom]...the lighter elements will disappear in time, and unless there is disintegration of the heavier atoms, the atomic weight of the lightest element surviving will continually increase."

Turning to the problem of the structure of the atoms, that is, of the groups so formed by aggregation, these are conceived as consisting "of a number of corpuscles moving about in a sphere of uniform positive electrification." The mathematical investigation of the manner in which the corpuscles must be arranged in atoms of increasing complexity so as to produce stable groupings is followed by the consideration of the atomic properties which must result from these special arrangements, and it is found that " the properties of the atom will depend upon its atomic weight in a way very analogous to that expressed by the periodic law."

Prof. Thomson shows how amongst the successively more complex configurations there is a recurrence at intervals of the same type of arrangement, giving a number of groupings which may be so related that each is formed by the addition of another ring of corpuscles to the one next simpler, the grouping of 60 corpuscles being that of 40 with an additional ring, that of 40 being itself the 24 arrangement *plus* a ring, etc. Of the resemblances in properties which would naturally be expected to go with these resemblances in structure, the case of the spectra produced by the vibrations of the corpuscles in such related systems is worked out in detail and is found to agree with the similarities actually observed in the spectra of members of the same *group* in the periodic law classification.

Thus the structure of the atom which is conditioned by its complexity, that is, by its weight, accounts for the periodic recurrence of the same properties exhibited by atoms of increasing weight. But the recurrence in arrangement, being only one of similarity and not of identity, also accounts for the fact that the typical properties of the group are exhibited by individual members to different degrees, that is, it explains the variation of properties in the group.

Typical differences in the structure of the atoms are shown to account for differences in their electro-chemical nature, for differences in their valencies, and for the tendency to chemical combination between the different kinds. Moreover, a consideration of atomic structure in the unbroken order of atomic complexity shows not only a sequence which accounts for the variation of these properties in the *series*, but also exhibits occasional sudden changes corresponding to places in the periodic system where the difference in properties of consecutive elements is exceptionally great.

The above is a mere enumeration of the analogies which have been traced by Prof. Thomson between atomic structure and atomic properties, and which supply an explanation for the empirical relations between atomic weight and atomic properties embodied in the periodic law. But an adequate account of all the results arrived at could not be given by means of any short abstract.

In a section of the community usually referred to as the "general public," there seems to be an impression that the recognition of the divisibility of the atom has dealt a death-blow to that atomic theory which was founded by Dalton just a hundred years ago. No misconception could be more complete. Whilst nothing has had to be given up, nothing to be modified, there has been deepening of the foundations, extension of scope, correlation with other sciences. Except that some of the anticipations expressed have since been realised, the situation to-day is exactly what it was in 1867 when Kekulé wrote as follows:

"The question whether atoms exist or not has but little significance from a chemical point of view: its discussion belongs rather to metaphysics. In chemistry we have only to decide whether the assumption of atoms is an hypothesis adapted to the explanation of chemical phenomena. More especially have we to consider the question whether a further development of the atomic hypothesis promises to advance our knowledge of the mechanism of chemical phenomena.

I have no hesitation in saying that, from a philosophical point of view, I do not believe in the actual existence of atoms, taking the word in its literal signification of indivisible particles of matter—I rather expect that we shall some day find for what we now call atoms a mathematico-mechanical explanation, which will render an account of atomic weight, of atomicity, and of numerous other properties of the so-called atoms. As a chemist, however, I regard the assumption of atoms, not only as advisable, but as absolutely necessary in chemistry. I will even go further, and declare my belief that *chemical atoms exist*, provided the term be understood to denote those particles of matter which undergo no further division in chemical metamorphoses. Should the progress of science lead to a theory of the constitution of chemical atoms—important as such a knowledge might be for the general philosophy of matter—it would make but little alteration in chemistry itself. The chemical atoms will always remain the chemical unit; and for the specially chemical considerations we may always start from the constitution of atoms, and avail ourselves of the simplified expression thus obtained, that is to say, of the atomic hypothesis. We may, in fact, adopt the view of Dumas and of Faraday, that *whether matter be atomic or not, thus much is certain, that granting it to be atomic, it would appear as it now does.*"

INDEX.

Academy 247

Accidental errors 83

Accuracy of, atomic weights by Avogadro's law, heat capacity and isomorphism 359, 373, 437; Dalton's atomic weight values 299; experimental results in general 76; law of fixed ratios 145; law of multiple ratios 165; law of equivalent ratios 183; law of combining volumes 312; law of isomorphism 413; law of atomic heat 373; molecular weights from gaseous densities 350; Stas' and Morley's complete syntheses and analyses 101

Accurate atomic weight determination by periodic law 482, 490

Acetaldehyde, Kolbe's formula 562; nonexistence of isomer of 590; polymerism 558

Acetic acid, numerous different formulae 344; Kolbe's formula 562; rational formulae 507; referred to water type 511; structural formula 519; formulae of possible isomers 566

Aceto-acetic ether, keto- and enol-formula 591; separation of tautomeric forms 592

Acetone, production of from β-bromopropylene 589

Acetylene, polymerism with benzene 557; structural formula 528

Acid produced in electrolysis of water 10, 13

Acids, class characteristics 21; neutralisation equivalents 175; oxygen theory of 26

Actinium, a radio-active element 613

Active mass, influence of physical conditions on 120

Additive properties, heat capacity recognised as such 377

Affinity, Bergman's work 112; Berthollet's work 118; Berthollet's inability to measure it 122; classification of substances in order of 110; Geoffroy's tables 111; influence of temperature on 111; mutual, of elementary particles 329; order under different conditions 113; residual, electrical interpretation of 542; Stahl's work 110; units 519; use of term 109

Affinities, single elective, experimental determination 113

Air, absorbed in calcination, nature and specific gravity 51; atmospheric, composition 53; compressibility 92; dephlogisticated 40, 54; eminently pure 54; made the primal element 232; one of the four elements 236, 254, 248; phlogisticated 38; presence required for combustion 35, 38

Albite, member of isodimorphous series 447

Alchemists, symbolic notation 193

Alcohol, Kolbe's formula 562; methyl, ethyl and propyl, Couper's formulae 517; prognosis of secondary and tertiary 28, 562

Alkali, produced in electrolysis of water 10

Alkyl compounds, melting and boiling points 487

Allotropy, explained by differences in atomicity 556; name given by Berzelius 554

Alloys, classed as mixtures 139; instanced in support of variable composition 133

Allyl alcohol, formation of 589

Allyl iodide, saponification of 589

Alums, crystalline form 408; general formula used in atomic weight determinations 408

Aluminium, atomic weight from heat capacity 376; combining weight 204

Amagat, compressibility of air at high pressures 95

F.

CAMBRIDGE: PRINTED BY J. AND C. F. CLAY, AT THE UNIVERSITY PRESS.

CPSIA information can be obtained
at www.ICGtesting.com
Printed in the USA
BVOW09s1458260617
487848BV00011B/248/P